Lecture Notes in Artificial Intelligence   4314

Edited by J. G. Carbonell and J. Siekmann

Subseries of Lecture Notes in Computer Science

W0235428

Christian Freksa   Michael Kohlhase
Kerstin Schill (Eds.)

KI 2006:
Advances in
Artificial Intelligence

29th Annual German Conference on AI, KI 2006
Bremen, Germany, June 14-17, 2006
Proceedings

 Springer

Series Editors

Jaime G. Carbonell, Carnegie Mellon University, Pittsburgh, PA, USA
Jörg Siekmann, University of Saarland, Saarbrücken, Germany

Volume Editors

Christian Freksa
Kerstin Schill
Universität Bremen
FB 3 - Mathematik und Informatik
Cartesium, Enrique-Schmidt-Str. 5, 28359 Bremen, Germany
E-mail: freksa@sfbtr8.uni-bremen.de; kschill@informatik.uni-bremen.de

Michael Kohlhase
International University Bremen
School of Engineering and Science
Campus Ring 1, 28759 Bremen, Germany
E-mail: m.kohlhase@iu-bremen.de

Library of Congress Control Number: 2006940355

CR Subject Classification (1998): I.2

LNCS Sublibrary: SL 7 – Artificial Intelligence

ISSN   0302-9743
ISBN-10   3-540-69911-2 Springer Berlin Heidelberg New York
ISBN-13   978-3-540-69911-8 Springer Berlin Heidelberg New York

Springer is a part of Springer Science+Business Media

springer.com

© Springer-Verlag Berlin Heidelberg 2007
Printed in Germany

Typesetting: Camera-ready by author, data conversion by Scientific Publishing Services, Chennai, India
Printed on acid-free paper   SPIN: 11979364   06/3142   5 4 3 2 1 0

Preface

This volume contains the conference proceedings of the 29th Annual German Conference on Artificial Intelligence (KI 2006) held June 14–19, 2006 at the Convention Center in Bremen, Germany. KI 2006 was organized under the auspices of the AI section of the German Informatics Society (GI), a member society of the European Coordinating Committee of Artificial Intelligence (ECCAI).

This year, we stepped out of our regular pattern of holding the conference at the end of the summer as we decided to accompany the RoboCup 2006 competitions at the Bremen Fair and Convention Center with our scientific AI conference. To avoid an unusually early submission deadline that would have competed with other major AI conferences, we decided to publish post-conference proceedings. We received a large number of excellent contributions: 112 papers from 25 countries were submitted. In a thorough peer-review process, the international Program Committee selected 29 full papers that are published in this volume. The contributions cover a range of topics from ontologies to cognition and emotion, from spatial and spatio-temporal reasoning to machine and robot learning, and from analogies to natural language.

Two speakers were invited for keynote lectures: Ramon López de Mántaras presented his work on applying AI methods to the transformation of musical performances and Ulrich Furbach spoke about applications of automated reasoning. Their written contributions to these topics are also published in this volume.

In addition to the regular conference sessions, six workshop proposals and nine poster contributions were accepted for the conference. The workshop sessions were held on June 14 and on June 19, 2006. The contributions to these sessions were published in separate proceedings volumes.

The month and year of our conference mark the 50th anniversary of the 1956 Dartmouth conference, which is considered the birth of artificial intelligence research. On the occasion of this anniversary, eight special guests — Marvin Minsky, Aaron Sloman, Wolfgang Bibel, Jörg Siekmann, Wolfgang Wahlster, Sebastian Thrun, Hiroshi Ishiguro, and Simon Schmitt — were invited for the public symposium "50 Years AI" on June 17, 2006 after the regular conference sessions. The public symposium was moderated by Wilfried Brauer and presented highlights of AI research from the past 50 years as well as visions for the next 50 years to a scientifically interested public audience. The event was streamed live on the Internet and enabled interested people all over the world to follow the presentations and discussions.

Many individuals and teams contributed to the visions and the realization of this conference. We would like to thank the members of the Program Committee for their careful reviews and their excellent comments to the authors; referee comments are extremely valuable to the authors and constitute an important element to progress in science. We would like to thank the discussants of the golden anniversary celebration for their visions and ideas, in particular Herbert Stoyan for first suggesting the idea. We would also like to thank the members of the various Organizing Committees for their productive and successful contributions, the valuable advice they provided, and the smooth interactions. In particular, we thank our local Chair, Eva Räthe, for her tireless dedication to this project, Marion Stubbemann for her support and assistance, and Sven Bertel for his energy and help with public relations.

The work of the Program Committee and the preparation of the conference proceedings were greatly simplified by Andrei Voronkov's excellent EasyChair system.

September 2006   Christian Freksa
                                                       Michael Kohlhase
                                                       Kerstin Schill

Organization

Conference Organization

Conference Chairs	Christian Freksa, Kerstin Schill
Program Chairs	Michael Kohlhase, Christian Freksa
Honorary Chair	Wilfried Brauer
Workshop Chair	Bernd Krieg-Brückner
Tutorial Chair	John Bateman
Exhibition Chair	Frank Kirchner
Industrial Liasion	Otthein Herzog
Robocup Liaison	Ubbo Visser
Local Chair	Eva Räthe
Press Liaison	Sven Bertel

Program Committee

John Bateman	Barbara Becker
Susanne Biundo	Wolfram Burgard
Stephan Busemann	Ulises Cortés
Rüdiger Dillmann	Klaus Fischer
Ulrich Furbach	Hans Werner Güsgen
Volker Haarslev	Nicola Henze
Joachim Hertzberg	David Israel
Herbert Jäger	Manfred Kerber
Frank Kirchner	Boicho Kokinov
Alexander Koller	Rudolf Kruse
Lars Kulik	Franz Kurfess
Jana Köhler	Longin Jan Latecki
Gérard Ligozat	Ramon López de Mántaras
Rainer Malaka	Katharina Morik
Bernd Neumann	Ian Pratt-Hartmann
Raul Rojas	Thomas Röfer
Alessandro Saffiotti	Ulrike Sattler
Jürgen Sauer	Kerstin Schill
Christoph Schlieder	Tanja Schultz
Laure Vieu	Ipke Wachsmuth
Wolfgang Wahlster	Stefan Wrobel
Jianwei Zhang	Michel de Rougemont
Kai von Luck	

Additional Reviewers

Sven Behnke
Michael Brenner
Jens Claßen
Andreas Eisele
Santiago Franco
Claudia Hess
Kai Hübner
Christoph Kreitz
Franz Kurfess
Tim Laue
Nadine Lessmann
Kai Lingemann
Christian Mandel
Philippe Muller
Andreas Nüchter
Gerhard Paass
Matthew Purver
Bernd Schattenberg
Marc Schröder
Nematollaah Shiri
Spyratos
Stefan Stiene
Rudolph Triebel
Muhammad Umer
Xiaomeng Wang
Michael Wünstel

Stefano Borgo
Stephan Busemann
Christian Döring
Timm Euler
Christian Hahn
Tamas Horvath
Peter Kiefer
Geert-Jan Kruijff
Richard Lassaigne
Esteban Leon
Thorsten Liebig
Cristian Madrigal-Mora
Sebastian Matyas
Jan Murray
Oliver Obst
Ian Pratt-Hartmann
Frank Rügheimer
Martin Scholz
Carl P.L. Schultz
Josenildo Costa da Silva
Klaus Stein
Frieder Stolzenburg
Nicolas Troquard
Anne Vilnat
Michael Wurst
Ingo Zinnikus

Table of Contents

Session 1. Invited Talk

Expressivity-Preserving Tempo Transformation for Music –
A Case-Based Approach 1
    *Ramon López de Mántaras, Maarten Grachten, and
Josep-Lluís Arcos*

Session 2. Cognition and Emotion

MicroPsi: Contributions to a Broad Architecture of Cognition 7
    Joscha Bach, Colin Bauer, and Ronnie Vuine

Affective Cognitive Modeling for Autonomous Agents Based on
Scherer's Emotion Theory... 19
    Christine L. Lisetti and Andreas Marpaung

Session 3A. Semantic Web

OWL and Qualitative Reasoning Models 33
    Jochem Liem and Bert Bredeweg

Techniques for Fast Query Relaxation in Content-Based Recommender
Systems 49
    Dietmar Jannach

Session 3B. Analogy

Solving Proportional Analogies by E–Generalization................. 64
    Stephan Weller and Ute Schmid

Building Robots with Analogy-Based Anticipation 76
    *Georgi Petkov, Tchavdar Naydenov, Maurice Grinberg, and
Boicho Kokinov*

Session 4A. Natural Language

Classification of Skewed and Homogenous Document Corpora with
Class-Based and Corpus-Based Keywords 91
    Arzucan Özgür and Tunga Güngör

Learning an Ensemble of Semantic Parsers for Building Dialog-Based
Natural Language Interfaces   102
     Lappoon R. Tang

Session 4B. Reasoning

Game-Theoretic Agent Programming in Golog Under Partial
Observability   113
     Alberto Finzi and Thomas Lukasiewicz

Finding Models for Blocked 3-SAT Problems in Linear Time by
Systematical Refinement of a Sub-model   128
     Gábor Kusper

Towards the Computation of Stable Probabilistic Model Semantics   143
     Emad Saad

DiaWOz-II – A Tool for Wizard-of-Oz Experiments in Mathematics   159
     Christoph Benzmüller, Helmut Horacek, Ivana Kruijff-Korbayová,
     Henri Lesourd, Marvin Schiller, and Magdalena Wolska

Session 5. Invited Talk

Applications of Automated Reasoning   174
     Ulrich Furbach and Claudia Obermaier

Session 6A. Ontologies

On the Scalability of Description Logic Instance Retrieval   188
     Ralf Möller, Volker Haarslev, and Michael Wessel

Relation Instantiation for Ontology Population Using the Web   202
     Viktor de Boer, Maarten van Someren, and Bob J. Wielinga

Session 6B. Spatio-temporal Reasoning

GeTS – A Specification Language for Geo-Temporal Notions   214
     Hans Jürgen Ohlbach

Active Monte Carlo Recognition   229
     Felix von Hundelshausen and Manuela Veloso

Session 7A. Machine Learning

Cross System Personalization and Collaborative Filtering by Learning
Manifold Alignments   244
     Bhaskar Mehta and Thomas Hofmann

A Partitioning Method for Mixed Feature-Type Symbolic Data Using
a Squared Euclidean Distance 260
  *Renata Maria Cardoso Rodrigues de Souza,
  Francisco de Assis Tenorio de Carvalho, and Daniel F. Pizzato*

Session 7B. Spatial Reasoning

On Generalizing Orientation Information in $\mathcal{OPRA}_m$ 274
  Frank Dylla and Jan Oliver Wallgrün

Towards the Visualisation of Shape Features: The Scope Histogram 289
  Arne Schuldt, Björn Gottfried, and Otthein Herzog

Session 8A. Robot Learning

A Robot Learns to Know People—First Contacts of a Robot 302
  *Hartwig Holzapfel, Thomas Schaaf, Hazım Kemal Ekenel,
  Christoph Schaa, and Alex Waibel*

Recombinant Rule Selection in Evolutionary Algorithm for Fuzzy Path
Planner of Robot Soccer 317
  *Jong-Hwan Park, Daniel Stonier, Jong-Hwan Kim,
  Byung-Ha Ahn, and Moon-Gu Jeon*

Session 8B. Classical AI Problems

A Framework for Quasi-exact Optimization Using Relaxed Best-First
Search 331
  Rüdiger Ebendt and Rolf Drechsler

Gray Box Robustness Testing of Rule Systems 346
  Joachim Baumeister, Jürgen Bregenzer, and Frank Puppe

A Unifying Framework for Hybrid Planning and Scheduling 361
  Bernd Schattenberg and Susanne Biundo

Session 9. Agents

A Hybrid Time Management Approach to Agent-Based Simulation 374
  Dirk Pawlaszczyk and Ingo J. Timm

Adaptive Multi-agent Programming in GTGolog 389
  Alberto Finzi and Thomas Lukasiewicz

Agent Logics as Program Logics: Grounding KARO 404
  Koen V. Hindriks and John-Jules Ch. Meyer

On the Relationship Between Playing Rationally and Knowing How
to Play: A Logical Account 419
    Wojciech Jamroga

Special Event. 50 Years Artificial Intelligence

1956-1966 How Did It All Begin? – Issues Then and Now 437
    Marvin Minsky

Fundamental Questions 439
    Aaron Sloman

Towards the AI Summer 443
    Wolfgang Bibel

History of AI in Germany and The Third Industrial Revolution 445
    Jörg Siekmann

Three Decades of Human Language Technology in Germany 447
    Wolfgang Wahlster

1996-2006 Autonomous Robots 449
    Sebastian Thrun

Projects and Vision in Robotics.................................... 451
    Hiroshi Ishiguro

What Will Happen in Algorithm Country? 455
    Simon Schmitt

Author Index 457

Expressivity-Preserving Tempo Transformation for Music – A Case-Based Approach

Ramon López de Mántaras, Maarten Grachten, and Josep-Lluís Arcos

IIIA, Artificial Intelligence Research Institute
CSIC, Spanish Council for Scientific Research
Campus UAB, 08193 Bellaterra, Catalonia, Spain
Vox: +34-93-5809570; Fax: +34-93-5809661
{mantaras,arcos,maarten}@iiia.csic.es

Abstract. The research described in this paper focuses on global tempo transformations of monophonic audio recordings of saxophone jazz performances. More concretely, we have investigated the problem of how a performance played at a particular tempo can be automatically rendered at another tempo while preserving its expressivity. To do so we have develppoped a case-based reasoning system called *TempoExpress*. The results we have obtained have been extensively compared against a standard technique called uniform time stretching (UTS), and show that our approach is superior to UTS.

1 The Problem of Generating Expressive Music

It has been long established that when humans perform music from score, the result is never a literal, mechanical rendering of the score (the so-called nominal performance). As far as performance deviations are intentional (that is, they originate from cognitive and affective sources as opposed to e.g. motor sources), they are commonly thought of as conveying musical expressivity, which forms an important aspect of music. Two main functions of musical expressivity are generally recognized. Firstly, expressivity is used to clarify the musical structure (in the broad sense of the word: this includes for example metrical structure [Sloboda, 1983], but also the phrasing of a musical piece [Gabrielsson, 1987], and harmonic structure [Palmer, 1996]). Secondly, expressivity is used as a way of communicating, or accentuating affective content [Juslin, 2001; Gabrielsson, 1995].

An important issue when performing music is the effect of tempo on expressivity. It has been argued that temporal aspects of performance scale uniformly when tempo changes [Repp, 1994]. That is, the durations of all performed notes maintain their relative proportions. This hypothesis is called relational invariance (of timing under tempo changes). However, counter-evidence for this hypothesis has been provided [Desain and Honing, 1994; Timmers et al., 2002], and a recent study shows that listeners are able to determine above chance-level whether audio-recordings of jazz and classical performances are uniformly time stretched or original recordings, based solely on expressive aspects of the performances [Honing, 2006]. Our approach also experimentally refutes the relational

C. Freksa, M. Kohlhase, and K. Schill (Eds.): KI 2006, LNAI 4314, pp. 1–6, 2007.
© Springer-Verlag Berlin Heidelberg 2007

invariance hypothesis by comparing the automatic transformations generated by *TempoExpress* against uniform time stretching.

2   TempoExpress

Given a MIDI score of a phrase from a jazz standard, and given a monophonic audio recording of a saxophone performance of that phrase at a particular tempo (the source tempo), and given a number specifying the target tempo, the task of the system is to render the audio recording at the target tempo, adjusting the expressive parameters of the performance to be in accordance with that tempo.

TempoExpress solves tempo transformation problems by case-based reasoning. Problem solving in case-based reasoning is achieved by identifying and retrieving a problem (or a set of problems) most similar to the problem that is to be solved from a case base of previously solved problems (also called cases), and adapting the corresponding solution to construct the solution for the current problem.

To realize a tempo transformation of an audio recording of an input performance, *TempoExpress* needs an XML file containing the melodic description of the recorded audio performance, a MIDI file specifying the score, and the target tempo to which the performance should be transformed (the tempo is specified in the number of beats per minute, or BPM). The result of the tempo transformation is an an XML file containing the modified melodic description, that is used as the basis for synthesis of the transformed performance. For the audio analysis (that generates the XML file containing the melodic description of the input audio performance) and for the audio synthesis, *TempoExpress* relies on an external system for melodic content extraction from audio, developed by Gómez et al. [2003b]. This system performs pitch and onset detection to generate a melodic description of the recorded audio performance, the format of which complies with an extension of the MPEG7 standard for multimedia content description [Gómez et al., 2003a].

We apply the edit-distance [Levenshtein, 1966] in the retrieval step in order to assess the similarity between the cases in the case base (human performed jazz phrases at different tempos) and the input performance whose tempo has to be transformed. To do so, firstly the cases whose performances are all at tempos very different from the source tempo are filtered out. Secondly, the cases with phrases that are melodically similar to the input performance (according to the edit-distance) are retrieved from the case base. The melodic similarity measure we have developed for this is based on abstract representations of the melody [Grachten et al., 2005] and has recently won a contest for symbolic melodic similarity computation (MIREX 2005).

In the reuse step, a solution is generated based on the retrieved cases. In order to increase the utility of the retrieved material, the retrieved phrases are split into smaller segments using a melodic segmentation algorithm [Temperley, 2001]. As a result, it is not necessary for the input phrase and the retrieved phrase to match as a whole. Instead, matching segments can be reused from various retrieved phrases. This leads to the generation of *partial* solutions for the input problem. To obtain the complete solution, we apply *constructive adaptation* [Plaza and Arcos, 2002], a reuse technique that constructs complete solutions by searching the space of partial solutions.

The solution of a tempo-transformation consists in a performance annotation. This performance annotation is a sequence of changes that must be applied to the score in order to render the score expressively. The result of applying these transformations is a sequence of performed notes, the output performance, which can be directly translated to a melodic description at the target tempo, suitable to be used as a directive to synthesize audio for the transformed performance.

To our knowledge, all current performance rendering systems deal with predicting expressive values like timing and dynamics for the notes in the score. Contrastingly, *TempoExpress* not only predicts values for timing and dynamics, but also deals with more extensive forms of musical expressivity, like note insertions, deletions, consolidations, fragmentations, and ornamentations.

3 Results

In this section we describe results of an extensive comparison of *TempoExpress* against *uniform time stretching* (UTS), the standard technique for changing the tempo of audio recordings, in which the temporal aspects (such as note durations and timings) of the recording are scaled by a constant factor proportional to the tempo change.

For a given tempo transformation task, the correct solution is available as a target performance: a performance at the target tempo by a profesional musician, that is known to have appropriate expressive values for that tempo. The results of both tempo transformation approaches are evaluated by comparing them to the target performance. More specifically, let M_H^s be a melodic description of a performance of phrase p by a musician H at the source tempo s, and let M_H^t be a melodic description of a performance of p at the target tempo t by H. Using *TempoExpress* (TE), and UTS, we derive two melodic descriptions for the target tempo from M_H^s, respectively M_{TE}^t, and M_{UTS}^t.

We evaluate both derived descriptions by their similarity to the target description M_H^t. To compute the similarity we use a distance measure that has been modeled after human perceived similarity between musical performances. Ground truth for this was gathered through a web-survey in which human subjects rated the perceived dissimilarity between different performances of the same melodic fragment. The results of the survey were used to optimize the parameters of an edit-distance function for comparing melodic descriptions. The optimized distance function correctly predicts 85% of the survey responses.

In this way, the results of *TempoExpress* and UTS were compared for 6364 tempo-transformation problems, using 64 different melodic segments from 14 different phrases. The results are shown in figure 1. The figure shows the distance of both *TempoExpress* and UTS results to the target performances, as a function of tempo change (measured as the ratio of the target tempo to the source tempo). The lower plots show the significance value for the null hypothesis

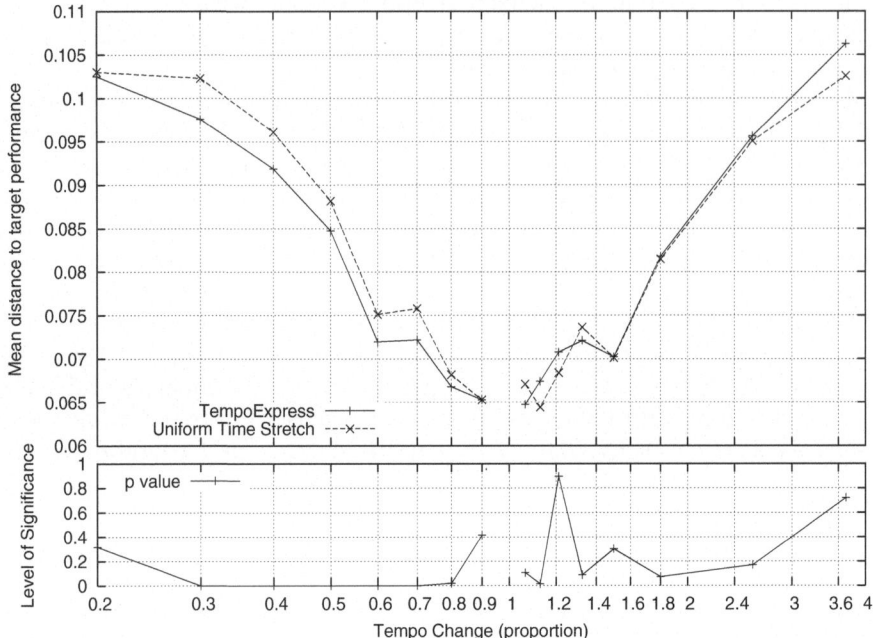

Fig. 1. Performance of *TempoExpress* vs. UTS as a function of the ratio of target tempo to source tempo. The lower plot shows the probability of incorrectly rejecting H_0 for the Wilcoxon signed-rank tests.

that the melodic descriptions generated by *TempoExpress* are not more similar or less similar to the target description than the melodic description generated using UTS (in other words, the hypothesis that *TempoExpress* does not give an improvement over UTS).

Firstly, observe that the plot in Figure 1 shows an increasing distance to the target performance with increasing tempo change (both for slowing down and for speeding up), for both tempo transformation techniques. This is evidence against the hypothesis of relational invariance discussed earlier in this paper. This hypothesis implies that the UTS curve should be horizontal, since under relational variance, tempo transformations are supposed to be achieved through mere uniform time stretching.

Secondly, a remarkable effect can be observed in the behavior of *TempoExpress* with respect to UTS, which is that *TempoExpress* improves the result of tempo transformation specially when slowing performances down. When speeding up, the distance to the target performance stays around the same level as with UTS. In the case of slowing down, the improvement with respect to UTS is mostly significant, as can be observed from the lower part of the plot. Note that the p-values are rather high for tempo change ratios close to 1, meaning that for those tempo changes, the difference between *TempoExpress* and UTS is not statistically significant. This is in accordance with the common sense that

Table 1. Overall comparison between *TempoExpress* and uniform time stretching, for upwards and downwards tempo transformations, respectively

	mean distance to target		Wilcoxon signed-rank test		
	TempoExpress	UTS	p <>	z	*df*
tempo ↑	0.0791	0.0785	0.046	1.992	3181
tempo ↓	0.0760	0.0786	0.000	9.628	3181

slight tempo changes do not require many changes, in other words, relational invariance approximately holds when the amount of tempo change is very small.

Table 1 summarizes the results for both tempo increase and decrease. Columns 2 and 3 show the average distance to the target performance for *TempoExpress* and UTS, averaged over all tempo increase problems, and tempo decrease problems respectively. The other columns show data from the Wilcoxon signed-rank test. The p-values are the probability of incorrectly rejecting H_0 (that there is no difference between the *TempoExpress* and UTS results). This table also shows that for downward tempo transformations, the improvement of *TempoExpress* over UTS is small, but extremely significant ($p < .001$), whereas for upward tempo transformations UTS seems to be better, but the results are slightly less decisive ($p < .05$).

4   Conclusions

In this paper we have summarized our research results on a case-based reasoning approach to global tempo transformations of music performances, focusing on saxophone recordings of jazz themes. We have addressed the problem of how a performance played at a particular tempo can be automatically rendered at another tempo preserving expressivity. Moreover, we have described the results of an extensive experimentation over a case-base of more than six thousand transformation problems. *TempoExpress* clearly performs better than UTS when the target problem is slower than the source tempo. When the target tempo is higher than the source tempo the improvement is less significant. Nevertheless, *TempoExpress* behaves as UTS except in transformations to very fast tempos. This result may be explained by a lack of example cases with fast tempos.

Bibliography

Desain, P. and Honing, H. (1994). Does expressive timing in music performance scale proportionally with tempo? *Psychological Research*, 56:285–292.

Gabrielsson, A. (1987). Once again: The theme from Mozart's piano sonata in A major (K. 331). A comparison of five performances. In Gabrielsson, A., editor, *Action and perception in rhythm and music*, pages 81–103. Royal Swedish Academy of Music, Stockholm.

Gabrielsson, A. (1995). Expressive intention and performance. In Steinberg, R., editor, *Music and the Mind Machine*, pages 35–47. Springer-Verlag, Berlin.

Gómez, E., Gouyon, F., Herrera, P., and Amatriain, X. (2003a). Using and enhancing the current MPEG-7 standard for a music content processing tool. In *Proceedings of Audio Engineering Society, 114th Convention*, Amsterdam, The Netherlands.

Gómez, E., Grachten, M., Amatriain, X., and Arcos, J. L. (2003b). Melodic characterization of monophonic recordings for expressive tempo transformations. In *Proceedings of Stockholm Music Acoustics Conference 2003*.

Grachten, M., Arcos, J. L., and López de Mántaras, R. (2005). Melody retrieval using the Implication/Realization model. MIREX http://www.music-ir.org/evaluation/mirex-results/articles/similarity/grachten.pdf.

Honing, H. (2006). Is expressive timing relational invariant under tempo transformation? *Psychology of Music*. (in press).

Juslin, P. (2001). Communicating emotion in music performance: a review and a theoretical framework. In Juslin, P. and Sloboda, J., editors, *Music and emotion: theory and research*, pages 309–337. Oxford University Press, New York.

Levenshtein, V. I. (1966). Binary codes capable of correcting deletions, insertions and reversals. *Soviet Physics Doklady*, 10:707–710.

Palmer, C. (1996). Anatomy of a performance: Sources of musical expression. *Music Perception*, 13(3):433–453.

Plaza, E. and Arcos, J. L. (2002). Constructive adaptation. In Craw, S. and Preece, A., editors, *Advances in Case-Based Reasoning*, number 2416 in Lecture Notes in Artificial Intelligence, pages 306–320. Springer-Verlag.

Repp, B. H. (1994). Relational invariance of expressive microstructure across global tempo changes in music performance: An exploratory study. *Psychological Research*, 56:285–292.

Sloboda, J. A. (1983). The communication of musical metre in piano performance. *Quarterly Journal of Experimental Psychology*, 35A:377–396.

Temperley, D. (2001). *The Cognition of Basic Musical Structures*. MIT Press, Cambridge, Mass.

Timmers, R.and Ashley, R., Desain, P., Honing, H., and Windsor, L. (2002). Timing of ornaments in the theme of Beethoven's Paisiello Variations: Empirical data and a model. *Music Perception*, 20(1):3–33.

MicroPsi:
Contributions to a Broad Architecture of Cognition

Joscha Bach[1], Colin Bauer[2], and Ronnie Vuine[3]

[1] University of Osnabrück, Institute for Cognitive Science, Osnabrück, Germany
`jbach@uos.de`
[2] Technical University of Berlin, Department for Computer Science, Berlin, Germany
`kolynos@gmail.com`
[3] Humboldt-University of Berlin, Institute for Computer Science, Berlin, Germany
`vuine@informatik.hu-berlin.de`

Abstract. The Psi theory of human action regulation is a candidate for a cognitive architecture that tackles the problem of the interrelation of motivation and emotion with cognitive processes. We have transferred this theory into a cognitive modeling framework, implemented as an AI architecture, called MicroPsi. Here, we describe the main assumptions of the Psi theory and summarize a neural prototyping algorithm that matches perceptual input to hierarchical declarative representations.

1 Introduction

Computational models of cognitive functioning usually emphasize problem solving, not emotion and motivation [1]. Thus they tend to fall short in modeling the interrelations between problem solving and memory functions and the context provided by emotional modulation and motivational priming, and they do not describe the cognitive system as an autonomous agent acting on its environment, but as a module within such an agent – and it is not clear if such a separation is warranted [2]. This has given rise to the suggestion of broader architectures of cognition which tightly integrate motivation and emotion with perceptual and reasoning processes. A very promising approach at such a broad architecture comprises the Psi theory of Dietrich Dörner [3, 4, 5]. Since this theory has not been extensively published in English, we will give a short summary on the following pages.

Psi is routed in a theory of problem solving [6] that makes use of neuro-symbolic models. Representations in the context of Psi are perceptual symbol systems [7], i.e. declarative and procedural descriptions are completely grounded in interaction contexts, which is achieved by using hierarchical spreading activation networks, with the lowest level of the hierarchy addressing sensor and motor systems. Depending on weights and link types, nodes within these hierarchies might carry their semantics individually (localist, symbolic) or as part of a configuration of jointly activated nodes (distributed, sub-symbolic). Cognitive processes are facilitated by control structures that are implemented as procedural representations within the same formalism.

Basic emotions in the Psi theory are understood as modulations of cognition, i.e. they emerge from configurations of various parameters (such as arousal, pleasure/distress signals and resolution level) that determine *how* cognition is carried out,

C. Freksa, M. Kohlhase, and K. Schill (Eds.): KI 2006, LNAI 4314, pp. 7–18, 2007.

and motivation is based on a finite set of competing drives, both physiological and cognitive.

Its focus on emotion, motivation and interaction make Psi very different from contemporary cognitive architectures like ACT-R [8] and Soar [9]. However, the design and integration of the motivational system bears a striking resemblance to the more recent, but independent CLARION architecture [10]. Like CLARION, the Psi theory proposes procedural reinforcement-learning based on pleasure/distress signals originating in the satisfaction and frustration of drives. The suggested cognitive drives however differ somewhat (they are less parsimonious in Psi). On the other hand, representations in Psi differ, because they are not separated into distinct symbolic and sub-symbolic formalisms – they use a single mode of representation for both.

Implementations of the Psi theory by Dörner's group have facilitated the successful evaluation of the emotional model against human emotions in a complex problem solving tasks [11, 12]; Psi is somewhat unique within cognitive architectures in offering such a validated model [13]. However, these implementations are unsuitable for independent experimentation, which we see as a primary requisite to turn the theory into a cognitive architecture, and they do not scale towards the integration of the representational mechanisms proposed by the theory. This demand is addressed by the *MicroPsi model*, by specifying an agent architecture and a scalable implementation framework, albeit not within the context of psychology, but computer science.

2   Assumptions of the Psi Theory

The Psi theory describes cognition in the terms of a homeostatic system: as a structure consisting of relationships and dependencies that is designed to maintain a homeostatic balance in the face of a dynamic environment. It consists of a set of assumptions that could be summarized as follows:

1. Explicit symbolic representations: The Psi theory suggests *hierarchical networks of nodes* as a *universal* mode of representation for declarative, procedural and tacit knowledge: representations in models of the Psi theory (Psi *agents*) are neurosymbolic. These nodes may encode *localist and distributed* representations. The activity of the system is modeled using *modulated* and *directional* spreading of activation within these networks. Plans, episodes, situations and objects are described with a semantic network formalism that relies on a fixed number of pre-defined link types, which especially encode *causal/sequential ordering*, and *partonomic hierarchies* (the theory specifies four basic link-types to denote predecessor und successor, has-part and is-part relations).

There are special nodes (representing neural circuits) that control the spread of activation and the forming of temporary or permanent *associations* and *disassociations*.

2. Memory: The Psi theory posits a world model (*situation image*). The current situation image is extrapolated into a branching *expectation horizon* (consisting of anticipated developments and active plans). Working memory also contains an *inner screen*, a hypothetical world model that is used for comparisons during recognition, and for planning.

The situation image is gradually transferred into an episodic memory (*protocol*). By selective *decay* and *re-inforcement*, portions of this long-term memory provide *automated behavioral routines*, and *elements for plans* (procedural memory). The fundamental atomic element of plans and behavior sequences is a *triplet* of a (partial, hierarchical) situation description, forming a condition, an operator (a hierarchical action description) and an expected outcome of the operation as another situation description.

Object descriptions (mainly declarative) are also part of long-term memory and the product of perceptual processes and affordances. Situations and operators in long-term memory may be associated with *motivational relevance*, which is instrumental in retrieval and reinforcement. Operations on memory content are subject to emotional *modulation*.

3. Perception: Perception is based on conceptual hypotheses, which guide the recognition of objects, situations and episodes. *Hypothesis based perception ('HyPercept')* is understood as a *bottom-up* (data-driven and context-dependent) cueing of hypotheses that is interleaved with a *bottom-down* verification.

The acquisition of schematic hierarchical descriptions and their gradual adaptation and revision can be described as *assimilation* and *accommodation* [14]. Hypothesis based perception is a *universal principle* that applies on visual perception, auditory perception, discourse interpretation and even memory interpretation. Perception is subject to emotional *modulation*.

4. Urges/drives: The activity of the system is directed on the satisfaction of a *finite set* of primary, *pre-defined urges* (drives). All goals of the system are situations that are associated with the satisfaction of an urge, or situations that are instrumental in achieving such a situation (this also includes abstract problem solving, aesthetics, the maintenance of social relationships and altruistic behavior). These urges reflect *demands* of the system: a mismatch between a target value of a demand and the current value results in an *urge signal*, proportional to the deviation, which might give rise to a motive. There are three categories of urges:

- *physiological urges* (such as food, water, maintenance of physical integrity), which are relieved by the *consumption* of matching resources and increased by the metabolic processes (food, water) of the system, or inflicted damage (integrity).
- *social urges (affiliation).* The demand for affiliation is an individual variable and adjusted through early experiences. The urge for affiliation needs to be satisfied in regular intervals by *external legitimity signals* (provided by other agents as a signal of acceptance and/or gratification) or *internal legitimity signals* (created by the fulfillment of social norms). It is increased by social frustration (*anti-legitimity signals*) or *supplicative signals* (demands of other agents for help, which create both a suffering by frustration of the affiliation urge, and a promise of gratification).
- *cognitive urges* (*reduction of uncertainty*, and *competence*). Uncertainty reduction is maintained through exploration and frustrated by mismatches with expectations and/or failures to create anticipations. Competence consists of *task specific competence* (and can be acquired through exploration of a task domain) and *general competence* (which measures the ability to fulfill the demands in general). The urge for competence is frustrated by actual and anticipated failures to reach a goal. The cognitive urges are subject to individual variability and need regular satisfaction.

The model strives for *minimal parsimony* in the specification of urges. For instances, there is no need to specify a specific urge for social power, because this may be reflected by the competence in reaching affiliative goals, while an urge for belongingness partially corresponds to uncertainty reduction in the social domain. The model should only expand the set of basic urges if it can be shown that the existing set is unable to produce the desired variability in behavioral goals. Note that none of the aforementioned urges may be omitted without affecting the behavior.

5. Pleasure and distress: A *change* of a demand of the system is reflected in a *pleasure* or *distress signal*. The strength of this signal is *proportional* to the amount of change in the demand measured over a short interval of time. Pleasure and distress signals are *reinforcement* values for the learning of behavioral procedures and episodic sequences and define *appetitive* and *aversive* goals.

6. Modulation: Cognitive processing in subject to *global modulatory parameters*, which adjust the cognitive resources of the system to the environmental and internal situation. Modulators control behavioral tendencies (action readiness via *general activation* or *arousal*), stability of active behaviors/chosen goals (*selection threshold*), the rate of orientation behavior (*sampling rate* or *securing threshold*) and the width and depth of activation spreading in perceptual processing, memory retrieval and planning (*activation* and *resolution level*). The effect and the range of modulator values are subject to *individual variance*.

7. Emotion: Emotion is not an independent sub-system, a module or a parameter set, but an intrinsic *aspect of cognition*. Emotion is an emergent property of the modulation of perception, behavior and cognitive processing, and it can therefore not be understood outside the context of cognition, that is, to model emotion, we need a cognitive system that can be modulated to adapt its use of processing resources and behavior tendencies. (According to Dörner, this is necessary *and* sufficient.) In the Psi theory, emotions are understood as a configurational setting of the *cognitive modulators* along with the *pleasure/distress dimension* and the assessment of the *cognitive urges*. This perspective addresses *primary emotions*, such as joy, anger, fear, surprise, relief, but not *attitudes* like envy or jealousy, or emotional responses that are the result of modulations which correspond to specific demands of the environment, such as disgust.

The *phenomenological qualities* of emotion are due to the effect of modulatory settings on perception on cognitive functioning (i.e. the perception yields different representations of memory, self and environment depending on the modulation), and to the experience of accompanying physical sensations that result from the effects of the particular modulator settings on the physiology of the system (for instance, by changing the muscular tension, the digestive functions, blood pressure and so on). The *experience of emotion* as such (i.e. as *having an emotion*) requires reflective capabilities. Undergoing a modulation is a necessary, but not a sufficient condition of experiencing it as an emotion.

8. Motivation: Motives are *combinations of urges and a goal* that is represented by a situation that affords the satisfaction of this urge. (Motives are terminologically and conceptually different from urges and emotions. *Hunger*, for instance, is an urge signal, an association of hunger with an opportunity to eat is a motive, and *apprehension* of an expected feast may be an emergent emotion.)

There may be several motivations active at a time, but *only one* is chosen to determine the choice of behaviors of the agent. The choice of the dominant motive depends on the anticipated probability to satisfy the associated urge and the strength of the urge signal. (This means also that the agent may opportunistically satisfy another urge, if it presented with that option.) The *stability of the dominant motive* against other active motivations is regulated using the selection threshold parameter, which depends on the *urgency* of the demand and individual variance.

9. Learning: Perceptual learning comprises the *assimilation/ accommodation* of new/existing schemas by hypothesis based perception. Procedural learning depends on *reinforcing* the associations of actions and preconditions (situations that afford these actions) with *appetitive or aversive* goals, which is triggered by pleasure and distress signals. *Abstractions* may be learned by evaluating and reorganizing episodic and declarative descriptions to generalize and fill in missing interpretations (this facilitates the organization of knowledge according to conceptual frames and scripts). Behavior sequences and object/situation representations are *strengthened by use*. Tacit knowledge (especially sensory-motor capabilities) may be acquired by *neural learning*. Unused associations *decay*, if their strength is below a certain threshold: highly relevant knowledge may not be forgotten, while spurious associations tend to disappear.

10. Problem solving: Problem solving is directed on *finding a path* between a given situation and a goal situation, on completing or *reorganizing mental representations* (for instance, the identification of relationships between situations or of missing features in a situational frame) or it serves an *exploratory* goal. It is organized in stages: If no *immediate response* to a problem is found, the system first attempts to resort to a behavioral routine *(automatism)*, and if this is not successful, it attempts to construct a *plan*. If planning fails, the system resorts to *exploration* (or switches to another motive).

Problem solving is *context dependent* (contextual priming is served by associative pre-activation of mental content) and subject to *modulation*. The strategies that encompass problem solving are *parsimonious*. They can be reflected upon and reorganized according to learning and experience. According to the Psi theory, many advanced problem solving strategies can not be adequately modeled without assuming *linguistic capabilities*.

11. Language: Language has to be explained as syntactically organized symbols that designate conceptual representations and a model of language thus starts with a model of mental representation. Language extends cognition by affording the categorical organization of concepts and by aiding in meta-cognition. (Cognition is not an extension of language.) The understanding of discourse may be modeled along the principles of hypothesis based perception and assimilation/ accommodation of schematic representations.

The Psi theory is largely qualitative, not quantitative, which makes it slightly unusual in contemporary research in cognitive science, but very useful as a frame of thought when addressing cognitive phenomena. After all, most pressing with respect to understanding human intelligent behavior start with "how" and "what" instead of "how much". Yet, to evaluate its proposals, it needs to be implemented as a model, which itself has to include commitments to concrete algorithms, representational formalisms and parameter settings. Dörner's own implementations as partial computer models do not favor such an evaluation, because they do not specify most of these commitments, nor do they separate between theory, architecture and model.

3   MicroPsi

MicroPsi translates the Psi theory into a cognitive architecture that eventually allows the comparison with other approaches. It comprises a development and simulation framework, written in Java, that allows implementing multi-agent systems according to the principles of the Psi theory, and it specifies an agent architecture that is implemented within the framework. MicroPsi is also used as a robot control architecture.

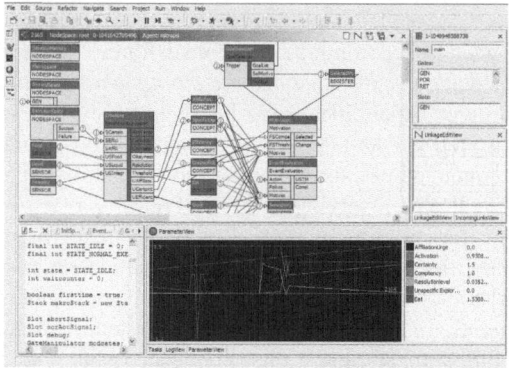

Fig. 1. MicroPsi toolkit

The framework offers an editor for hierarchical spreading activation networks, which is the principal tool for the design of Psi agents, a graphical simulation world that facilitates multi-agent interaction and several customizable environmental designs offering different tasks and tools for the visualization and evaluation of experiments.

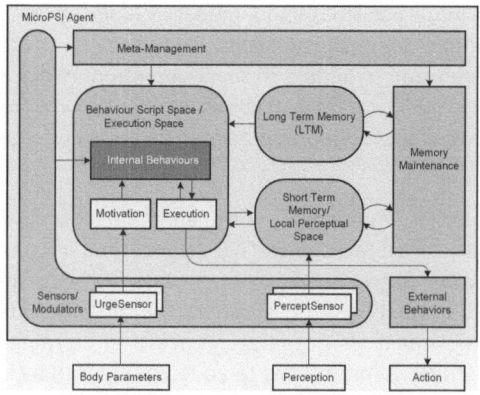

Fig. 2. MicroPsi agent architecture

MicroPsi agents are partial implementations of the Psi theory. They provide a motivational system in which the cognitive processes are embedded. The agent implementations so far address simple hypothesis based perception, means-end analysis, behavior execution, emotional modulation, reinforcement-learning based on satisfaction and frustration of drives, simple neural learning of low-level stimuli and environmental mapping.

3.1 Current Experiments: Neural Prototyping

Our current work deals with the extension of individual components of MicroPsi agents, such as the integration of neural learning of perceptual patterns from camera images with high-level concepts and the acquisition of hierarchical object descriptions. Here, we describe an approach to structure classification using *neural prototyping*.

When recognizing objects, planning and retrieving object hypotheses from long-term memory, MicroPsi agents need to classify hierarchical representations, which is a computationally expensive task, when structure matching is involved. Thus we need a strategy to minimize the structure comparison operations, and we address this need with an algorithm using class prototypes. These prototypes are represented as neural networks, where topology and weights are changed through learning. A first classification is performed by sending activation through these networks and to select only the most active prototypes. These few remaining structures can then be matched using a subgraph matching technique to identify the most similar prototype and with it the most suitable class. The advantages of this approach are that the pruning takes only as many steps as the depth of the largest prototype, and that the expensive structure matching is only used for a very small subset of items in memory.

The first step of the algorithm is to convert the class prototypes (that can be predetermined or acquired through learning) into neural networks. We used the approach developed by Towell and Shavlik [15], who describe the conversion of hierarchical structures into labelled neural networks (KBANN: knowledge-based artificial neural networks) and the properties of those networks. The main idea is to convert the nodes and links into neurons and connections in a network, setting the weights and biases in such a way as to preserve the logic of the structure.

After all prototypes have been converted, the neural classification can be performed on a sample structure. This means that all sensors (the leaf nodes in the hierarchical neural networks) that have the same label as one of the sensors in the example provided get activated and spread their activation through the network. A smooth activation function has to be chosen, otherwise only the examples that are isomorphic to a prototype will be able to activate its root node and thus be classified.

Next, structure matching is performed between the example and the prototypes with the highest activation of their root nodes. This step is necessary, because the neural activation phase is only able to give a rough estimate of similarity. It only takes a very small part of the structures' topologies into account. Since the topology is an important factor for classification, the example is then assigned to the class with the most similar structure. There are many different algorithms for structure matching ([16, 17], an overview in [18]). We decided to use the method described by Schädler and Wysotzki [18, 19], based on a Hopfield-network, where each node corresponds to two nodes whose similarity exceeds a certain threshold, because it can be integrated nicely into the general MicroPsi framework of representation. In our context, the similarity is measured by comparing the sensors of the subtrees rooted at the respective nodes, and the structures of these subtrees. Connections exist between nodes in the Hopfield network if they exist between the nodes in the original structures. In addition, connections with negative weights are added to the network between nodes that correspond to mapping a node in one graph to two nodes in the other. After the construction has been completed, the network is run until it reaches a stable state. The active nodes in this stable state represent a maximum common subgraph of the two original structures.

After the example has been compared to each of the prototypes, there are three possibilities to be considered:

- The example's similarity to one of the prototypes exceeds a predefined threshold. In that case, the example is assigned the class of the prototype and the classification procedure is finished.
- Maximum similarity is high, but not high enough. Here the intuition is that the class of the closest prototype should be correct, but that the prototype is not good enough to capture the individual properties of the example. In this case, the maximum common subgraph computed during structure matching can be used to add the example-specific parts to the prototype. This is achieved by converting the example into a KBANN network, attaching the parts not included in the maximum common subgraph and changing the weights of the network (by backpropagation learning).
- No prototype is sufficiently similar. In that case, no statement can be made about the class of the example, or, in the case of training, a new prototype is added to the example's class.

The algorithm has very interesting features: it does not need a complete set of prototypes in order to work efficiently, because it adds new prototypes or changes existing ones during training. This is required for its application in the MicroPsi architecture, because MicroPsi agents start with very limited or no knowledge about the world and have to build their knowledge base over time with only limited help from the outside. The system works under the supervised as well as unsupervised paradigms. In supervised learning, a training set of labeled examples can be used to generate a good set of prototypes. When dealing with unsupervised learning, the prototypes correspond to a set of clusters that can be built continuously, without having to make assumptions about size or locations of clusters beforehand, as is necessary with other algorithms. The system is very efficient, because the computationally expensive process of pruning the search space is done by neural networks of limited depth that can be run in parallel. Neural learning techniques and the topological modification of these classifying networks enable the system to generalize faster and better than other generalization algorithms that are based on finding the maximum common subgraphs between elements of one class that serve as generalized structures. The fact that the nodes of the prototype networks have semantic meaning can be exploited during training, to achieve faster convergence and less training time.

We ran two different experiments to test the performance of our approach and to compare its performance to other structure classification algorithms. The task of both experiments was to classify visually given objects. To represent them, they were represented as *shock graphs* [20], which are derived by transforming the skeleton of a two-dimensional shape into a hierarchical graph (Fig. 3), where nodes correspond to the vertices and end points of the skeleton and edges to their interrelations. (Using shock-graph was a somewhat arbitrary decision, which aimed at providing a basic abstraction over visual input.) The algorithm first computes the skeleton of the given shape and identifies its shock points. Given the shock points, the so called shock graph grammar can be used to create a hierarchical structure using the shocks as nodes and connecting them according to the grammar rules.

Fig. 3. Skeleton and shock points, sample shock graph taken from [21]

3.1.1 Supervised Learning

We trained the algorithms to classify 8 different shapes from a data-base of visual objects [21]. From each category we chose 5 shapes that were used for training, and the accuracy of each method was measured using 50 objects per class.

The five prototypes of each class, presented one at a time, were used to build the prototype networks. A prototype only became part of the prototype networks, if none of the already present prototypes of the same class got activated during spreading activation. At the end of the training phase, 30 of the 40 prototypes remained. In 10 cases there were other prototypes that already classified the given prototype correctly. Then the 400 examples were presented to the algorithm. For each example, the five prototypes with the highest activation were selected for structural comparison, and the example assigned to the one with the highest similarity.

We have compared our algorithm to several other approaches: to *eigenvalue-based indexing* [22], which represents the structure of the graph as a vector capturing the branching structure and node distribution; similarity is determined by computing the euclidean distance between the eigen-vectors of two graphs; to attributed graph indexing, where the eigenvectors are not derived from the standard adjacency matrix, but from the *attributed graph* – here, the entries in the diagonal are the labels of the nodes, and the other entries the labels of the links connecting the respective nodes; *subgraph prototyping* [19], where – using our own subgraph matching procedure – we chose the prototype with the largest common subgraph. For neural prototyping and the eigenvector-based methods, we use the respective algorithms to retrieve the five most similar graphs for each of the 400 examples and then perform a complete structural comparison. As a base-line, we include linear search, which performs a complete graph-matching with every example (and is prohibitively slow).

Exact runtimes could not be compared because some algorithms used different implementation techniques. Nevertheless, from the experiments we conducted it could be observed that the runtimes of the neural prototyping and eigenvalue-based approaches were comparable, whereas the other algorithms were significantly slower.

The histograms in Figure 4 depict the results for the five algorithms, where the x-axis represents the eight categories and the y-axis shows the percentage of correctly classified examples. The bar graphs represent the percentage of the examples in each class that were correctly classified by the respective algorithm. For example, the neural prototyping algorithm was able to classify 44 of the 50 examples of category 1 correctly, leading to an accuracy of 0.88, as shown in the figure. Since all algorithms were presented with the same test set, their performance can be compared directly by comparing their accuracies across classes.

As expected, linear search gives the best results, but is followed by our algorithm, and then the eigenvector-based approaches. (The results from linear search also show that the examples of the different classes had distinct enough shock graph representations for the structure comparison algorithm to find the correct prototype in most of the cases.) With respect to runtime, neural prototyping is on a par with the eigenvector-based algorithms, because all of them use a pruning strategy to minimize the amount of structural comparisons to be performed. Our algorithm and the subgraph prototyping approach were the only ones that showed generalization effects. In our case, generalization happened during the training phase, where from the total number of 40 prototypes, only 30 generalized prototypes remained and were used for classification.

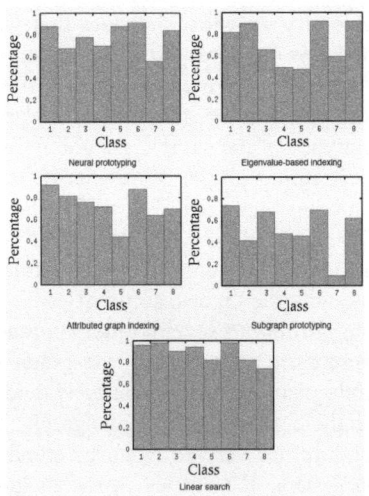

Fig. 4. Results for the different approaches

3.1.2 Unsupervised Learning
25 of the objects of each class that had already been used in the previous experiment were presented to our algorithm one by one without providing class information. The objects in this training pool were used to build a set of prototypes (or cluster points)

for each class. Contrary to the resulting set of prototypes in the previous experiment, the resulting generalized graphs in this case could contain subparts from different, but similar, classes. From a total of 200 prototypes that could be generated, the algorithm constructed 60 partially overlapping prototypes. The other 25 examples from the test pool in the supervised learning setup were used to measure the accuracy of these prototypes. To assess the difficulty of the task, the eigenvalue-based indexing algorithm (algorithm two in the previous experiment) was run on the data. Each graph from the test set was compared to those in the training pool, and the top five matches identified. As in the previous experiment, structure matching was used to select the closest structure out of these candidates. (Linear search in this setting was computationally infeasible.)

Under general circumstances, where no a-priory prototypes are known, the prototype-generating feature of our algorithm should prove advantageous over the eigenvalue-based indexing, where each new object must be compared to all available structures in memory. In our experiment, there were 60 prototypes left at the end of training, as opposed to the originally 200 shapes that were used in the eigenvalue-based approach. In addition, the speed of our algorithm increased over time during training, because as the size of the prototype networks increased, fewer examples needed to modify the topology.

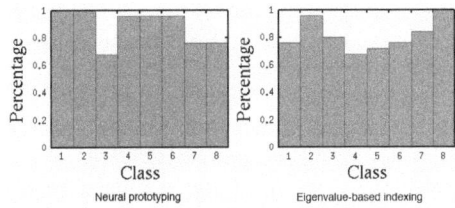

Fig. 5. Results from the unsupervised learning environment

Both algorithms perform well, with some advantage for our algorithm. The inclusion of neural prototyping is a useful extension of MicroPsi, because it enables the agents to explore their world and generate categories without external help. The algorithm generalizes, since only about one fourth of the potential number of prototypes were used to generate prototypes, while maintaining its performance on the test set.

References

1. Detje, F.: Handeln erklären. DUV, Wiesbaden(1999)
2. Clark, A., Grush, R.: Towards a Cognitive Robotics. Adaptive Behavior 7(1) (1999) 5-16
3. Dörner, D.: Eine Systemtheorie der Motivation. Memorandum Lst Psychologie II Universität Bamberg, 2,9 (1994)
4. Dörner, D.: Bauplan für eine Seele. Rowohlt, Reinbeck (1999)
5. Dörner, D., Bartl, C., Detje, F., Gerdes, J., Halcour D., Schaub H., Starker, U.: Die Mechanik des Seelenwagens. Eine neuronale Theorie der Handlungsregulation. Verlag Hans Huber, Bern Göttingen Toronto Seattle (2002)

6. Dörner, D.: Die kognitive Organisation beim Problemlösen. Versuche zu einer kybernetischen Theorie der elementaren Informations-verarbeitungsprozesse beim Denken. Kohlhammer, Bern (1974)
7. Barsalou, L. W.: Perceptual Symbol Systems. In: Behavioral and Brain Sciences, 22,4. (1999) 577-660
8. Anderson, J. R., Lebiere, C.: The atomic components of thought. Erlbaum, Mahwah, NJ (1998)
9. Laird, J. E., Newell, A., Rosenbloom, P. J.: Soar: An architecture for general intelligence. Artificial Intelligence, 33(1) (1987) 1-64
10. Sun, R.: Cognition and Multi-Agent Interaction. Cambridge University Press (2005) 79-103
11. Detje, F.: Comparison of the PSI-theory with human behaviour in a complex task. In: Taatgen, N., Aasman, J. (eds.): Proceedings of the Third International Conference on Cognitive Modelling. Universal Press, Veenendaal, The Netherlands (2000) 86-93
12. Dörner, D.: The Mathematics of Emotion. Proceedings of ICCM-5, International Conference on Cognitive Modeling. Bamberg, Germany (2003)
13. Ritter, F. E., Shadbolt, N, R., Elliman, D., Young, R. M., Gobet, F., Baxter, G. D.: Techniques for Modeling Human Performance in Synthetic Environments: A Supplementary Review. Human Systems Information Analysis Center, State of the Art Report (2003)
14. Piaget, J.: The construction of reality in the child. Basic Books, New York (1954)
15. Towell, G. G., Shavlik, J. W.: Knowledge-based artificial neural networks. In: Artificial Intelligence, 70(1-2). (1994) 119–165
16. Sebastian, T. B., Klein, P. N., Kimia, B. B.: Recognition of shapes by editing shockgraphs. IEEE International Conference on Computer Vision. (2001) 755–762
17. Pelillo, M., Siddiqi, K., Zucker, S. W.: Matching hierarchical structures using association graphs. In: IEEE Transaction on Pattern Analysis and Machine Intelligence, Vol 21. (1999) 1105–1120
18. Schädler, K., Wysotzki, F.: Comparing structures using a hopfield-style neural network. Applied Intelligence, Vol. 11. (1999) 15–30
19. Schädler, K., Wysotzki, F.: A connectionist approach to structural similarity determination as a basis of clustering, classification and feature detection. PKDD. (1997) 254–264
20. Siddiqi, K., Shokoufandeh, A., Dickinson, S., Zucker, S.: Shock graphs and shape matching. IEEE International Journal on Computer Vision. (1998) 222–229
21. Macrini, D.: Indexing and matching for view-based 3-d object recognition using shock graphs. Master's thesis, University of Toronto (2003)
22. Shokoufandeh, A., Dickinson, S.: A unified framework for indexing and matching hierarchical shape structures. 4th International Workshop on Visual Form. (2001) 28–46

Affective Cognitive Modeling for Autonomous Agents Based on Scherer's Emotion Theory

Christine L. Lisetti[1,*] and Andreas Marpaung[2]

[1] Multimedia Communications
Institut Eurecom
Sophia Antipolis, France
lisetti@eurecom.fr
[2] Electrical Engineering & Computer Science
University of Central Florida
Orlando, Florida, USA
marpaung@cs.ucf.edu

Abstract. In this article, we propose the design of sensory motor level as part of a three-layered agent architecture inspired from the Multilevel Process Theory of Emotion (Leventhal 1979, 1980; Leventhal and Scherer, 1987). Our project aims at modeling emotions on an autonomous embodied agent, a more robust robot than our previous prototype. Our robot has been equipped with sonar and vision for obstacle avoidance as well as vision for face recognition, which are used when she roams around the hallway to engage in social interactions with humans. The sensory motor level receives and processes inputs and produces emotion-like states without any further willful planning or learning. We describe: (1) the psychological theory of emotion which inspired our design, (2) our proposed agent architecture, (3) the needed hardware additions that we implemented on the commercialized ActivMedia's robot, (4) the robot's multi-modal interface designed especially to engage humans in natural (and hopefully pleasant) social interaction, and finally (5) our future research efforts.

1 Introduction

Robotic agents have been of great interest for many Artificial Intelligence researchers for several decades. This field has produced many applications in many different fields, i.e., entertainment (Sony Aibo) and Urban Search and Rescue (USAR) (Casper, 2002; Casper and Murphy, 2002) with many different techniques – behavior-based (Brooks, 1989; Arkin, 1998), sensor fusion (Murphy, 1996a, 1996b, 1998, 2000), and vision (Horswill, 1993). As robots begin to enter our everyday life, an important human-robot interaction issue becomes that of social interactions. Because emotions have a crucial evolutionary functional aspect in social intelligence, without which complex intelligent systems with limited resources cannot function efficiently or maintain a satisfactory relationship with their environment, we focus our current contribution to the study of emotional social intelligence for robots. Indeed, the

---

* Part of this work was accomplished while the author was at the University of Central Florida, USA.

C. Freksa, M. Kohlhase, and K. Schill (Eds.): KI 2006, LNAI 4314, pp. 19–32, 2007.

recent emergence of affective computing combined with artificial intelligence has made it possible to design computer systems that have "social expertise" in order to be more autonomous and to naturally bring the human – a principally social animal – into the loop of human-computer interaction.

In this article, *social expertise* is considered in terms of (1) internal motivational goal-based abilities and (2) external communicative behavior. Because of the important functional role that emotions play in human decision-making and in human-human communication, we propose a paradigm for modeling some of the functions of emotions in intelligent autonomous artificial agents to enhance both (a) robot autonomy and (b) human-robot interaction. To this end, we developed an autonomous service robot whose functionality has been designed so that it could socially interact with humans on a daily basis in the context of an office suite environment and studied and evaluated the design *in vivo*. The social robot has been furthermore evaluated from a social informatics approach, using workplace ethnography to guide its design *while* it is being developed (Lisetti et al., 2004)

2 Related Research

There have been several attempts to model emotions in software agents and robots and to use these models to enhance functionality. El-Nasr, (2002) uses a fuzzy logic model for simulating emotional behaviors in an animated environment. Contrary to our approach directed toward robots, her research is directed toward HCI and computer simulation.

Breazeal's work (2000, 2003) also involves robot architectures with a motivational system that associates motivations with both drives and emotions. Emotions are implemented in a framework very similar to that of Velasquez's work but Breazeal's emphasis is on the function of emotions in social exchanges and learning with a human caretaker. Our approach is different from Breazeal's in that it is currently focused on both social exchanges and the use of emotions to control a single agent.

Murphy and Lisetti's approach (2002) uses the multilevel hierarchy of emotions where emotions both modify active behaviors at the sensory-motor level and change the set of active behaviors at the schematic level for a pair of cooperating heterogeneous robots with interdependent tasks.

Our current approach builds on that work, setting the framework for more elaborate emotion representations while starting to implement simple ones and associating these with expressions (facial and spoken) in order to simultaneously evaluate human perceptions of such social robots so as to guide further design decisions.

3 Developing Socially Intelligent Agents

We focus on the study of *social expertise* for artificial agents in terms of:

1. internal motivational goal-based activities, and
2. external communicative behavior

As shown in Figure 1, we are focusing on the Socially Intelligent Agent architecture (within the red circle) of the **Multimodal Affective User Interface (MAUI)** paradigm proposed and developed earlier (Lisetti, 2002; Lisetti and Nasoz, 2004) for the design of affective socially intelligent agents. Our current work within the MAUI framework continues to focus on building user-specific emotional models of the user based on bi-modal bio-sensing of physiological signals associated with emotions – namely heart rate and galvanic skin response (Villon and Lisetti, 2006).

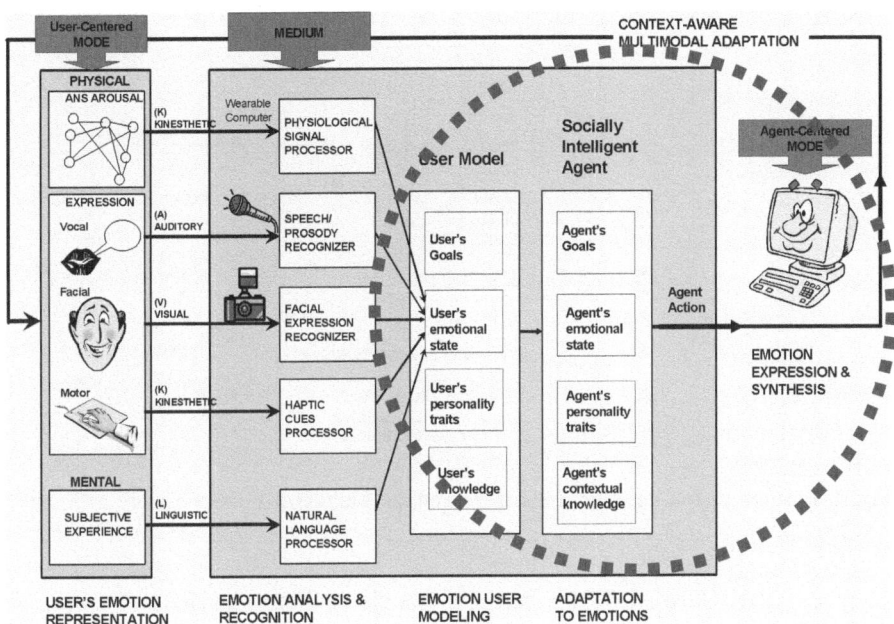

Fig. 1. Overall MAUI Paradigm for Multimodal Affective User Interfaces from (Lisetti and Nasoz, 2004)

We currently propose a psychologically-grounded framework for socially intelligent agents (corresponding to the modules of in the doted circle) based on Scherer's affective-cognitive theory of emotions. This architecture to be used for the development of artificial agents with diverse forms of embodiment such as vocal robots, graphical animated avatars, avatar-based interface on mobile robotic platform, anthropomorphic robotic platforms as shown later.

4 A Three-Layered Emotional State Generator

With recent advances in Psychology, many researchers have proposed theories on the mechanisms of producing emotions in humans. One of the theories of particular interest to us is the *Multilevel Process Theory of Emotion* (Leventhal 1979, 1980, Leventhal and Scherer, 1987), which we chose to inspire the design and the implementation of the Emotion State Generator (ESG) on our commercially available autonomous robot

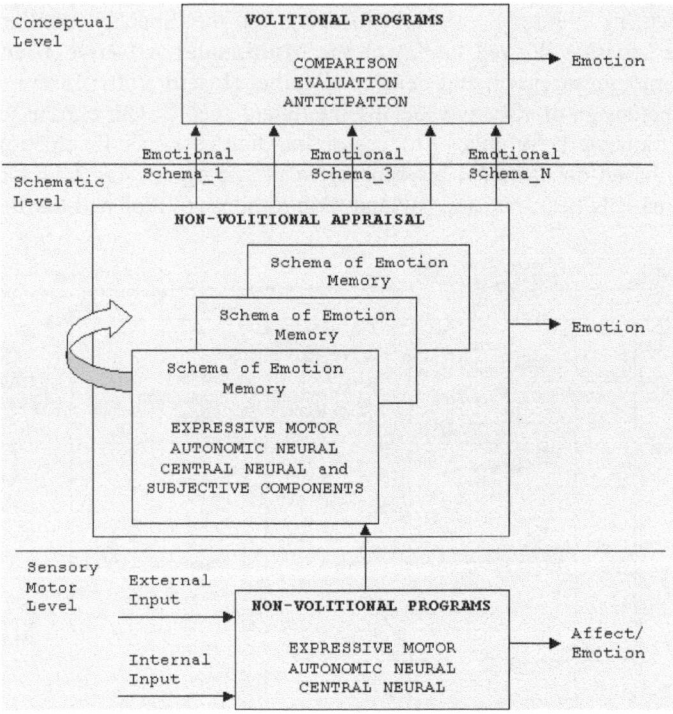

Fig. 2. Emotion State Generator (ESG) based on the Multilevel Process Theory of Emotion (Scherer, 1986)

PeopleBot (ActivMedia, 2002). Figure 1 shows the ESG three-layered architecture we use for generating emotion-like states for our autonomous agents.

Indeed, the Multilevel Process Theory of Emotion postulates that the experience of emotion is a product of an underlying constructive process that is also responsible for overt emotional behavior. It also describes that emotions are constructed from a hierarchical multi-component processing system. In short (Leventhal, 1980):

a. *Sensory motor level* – generates the primary emotion in response to the basic stimulus features in a non-deliberative manner;
b. *Schematic level* – integrates specific situational perceptions with autonomic, subjective, expressive and instrumental responses in a concrete and patterned image-like memory system;
c. *Conceptual level* – corresponds more closely to social labeling processes.

4.1 Sensory Motor Level

The *sensory motor or expressive motor level* is the basic processor of emotional behavior and experience that provides the earliest emotional meaning for certain situations. This level consists of multiple components: (a) a set of innate expressive-motor systems and (b) cerebral activating systems. These components are stimulated automatically by a variety of external stimuli and by internal changes of state that do not require deliberate planning.

Because there is no involvement of the willful planning and learning processes, the lifetime of the emotional reactions caused at this level may be short and will quickly become the focus for the next level, schematic processing. Action in the facial motor mechanism, as part of the expressive motor system, is the source of the basic or primary emotions of happiness, surprise, fear, sadness, anger, disgust, contempt, and interest (Leventhal, 1979). In this project, we are only modeling: happy, surprise, fear, sad and angry.

We briefly describe the schematic and conceptual levels for completeness sake, but we are currently focusing our design on the sensory motor level.

4.2 Schematic Level

The *schematic level* integrates sensory-motor processes with prototypes or schemata of emotional situations in order to create or to structure emotional experiences. But before entering this level, the input needs to be integrated with separate perceptual codes of the visual, auditory, somesthetic (related to the perception of sensory stimuli from the skin), expressive, and autonomic reactions that are reliably associated with emotional experiences.

Schemata - organized representations of other more elementary codes - are built during emotional encounter with the environment and will be conceptualized as memories of emotional-experiences. As shown in Figure 2, humans can activate these schemata by activating any one of its component attributes that is caused by the perception of a stimulus event, by the arousal of expressive behaviors or autonomic nervous system activity, or by the activation of central neural mechanisms that generate subjective feelings. The structure of the schematic memories can be thought of as codes, complex categorical units, a network of memory nodes, or perhaps as memory columns that are conceptualized.

The schematic processing is also automatic and does not require the participation of more abstract processes found at the conceptual level. This schematic level is more complex than the sensory motor level in that it integrates learning processes while building the complexities of schemata. At this level, emotion behavior also has a longer lifetime.

4.3 Conceptual Level

The *conceptual level* can be thought of as the system that can make conscious decisions or choices to some external inputs as well as to internal stimuli (such as stored memories of emotional schemata generated at the schematic level). It is the comparison and abstraction of two or more concrete schemata of emotional memories with certain concepts that will enable the humans to draw conclusions about their feelings to certain events. By comparing and abstracting information from these schemata with conceptual components – verbal and performance component - humans can reason, regulate ongoing sequences of behavior, direct attention and generate specific responses to certain events.

The *verbal components* are not only representing the feelings themselves but they are also communicating the emotional experiences to the subject (who can also choose to talk about his/her subjective experience). On the other hand, the *performance components* are non-verbal codes that represent sequential perceptual and motor responses. The information contained at this level is more abstract than the schematic memories and therefore the representations can be protected from excessive changes when they are exposed to a new experience and can be led to more stable states. Because this level is volitional, components can be more sophisticated through active participation of the agent. When performance codes are present, for example, the volitional system can swiftly generate a sequence of voluntary responses to match spontaneous expressive outputs from the schematic system. This volitional system can anticipate emotional behaviors through self-instruction.

4.4 Stimulus Evaluation Checks (SECs)

In order to produce emotion for each level, many researchers have hypothesized that specific emotions are triggered through a series of stimulus evaluation checks (SECs) (Scherer, 1984; Scherer, 1986; Weiner, Russell, and Lerman, 1979; Smith and Ellsworth, 1985). Inspired by (Lisetti and Nasoz, 2002), we link the SECs system that performs the emotion components' check in the Affective Knowledge of Representation (AKR) that produces a schema of emotion. This schema can be associated with a certain event and emotion and be part of the schema memory for further use. In AKR, each emotion has many components, e.g., valence, intensity, focality, agency, modifiability, action tendency, and causal chains.

Valence: *positive/ negative:* is used to describe the pleasant or unpleasant dimension of an affective state.

Intensity: *very high/ high/ medium/ low/ very low:* varies in terms of degree. The intensity of an affective state is relevant to the importance, relevance and urgency of the message that the state carries.

Focality: *event/ object:* is used to indicate whether the emotions are about something: an event (the trigger to surprise) or an object (the object of jealousy).

Agency: *self/ other:* is used to indicate who was responsible for the emotion, the agent itself *self*, or someone else *other*.

Modifiability: *high/ medium/ low/ none:* is used to refer to duration and time perspective, or to the judgment that a course of events is capable of changing.

Action tendency: identifies the most appropriate (suite of) actions to be taken from that emotional state. For example, happy is associated with generalized readiness, frustration with change current strategy, and discouraged with give up or release expectations.

Causal chain: identifies the causation of a stimulus event associated with the emotion. For example, happy has these causal chains: (1) Something good happened to me, (2) I wanted this, (3) I do not want other things, and (4) because of this, I feel good.

5 Affective-Cognitive Architecture and Embodiment Forms

5.1 Functionalities of Our Robot

Our robot, Petra, has the same tasks as Cherry (Lisetti, et al. 2004) and is designed so that she can socially interact with humans on a daily basis in the office suite environment especially on the second floor of the computer science building at the University of Central Florida. She has a given set of office-tasks to accomplish, from giving tours of our computer science faculty and staff suites to visitors and to engaging them in social interactions. With the sensors that she has (explained below), she is able to roam around the building using her navigational system, recognize someone through her face recognition algorithm, and greet them differently according to their social status (professor, students, staff).

In terms of architectures for autonomous agents and robots, the multi-level theory of emotions currently gets translated into the figure 3 below, of which we have implemented various levels and different types of embodiement forms, as shown in Figures 4 (b-c). We are currently in the process of building a platform independent architecture and an expression control mechanism to adapt to a multitude of robotic and graphical artificial agents such as for example the non-mobile Phillips iCat interactive toy-looking which we are currently working on shown in Figure 4a (Grizard and Lisetti, 2006; Paleari and Lisetti, 2006).

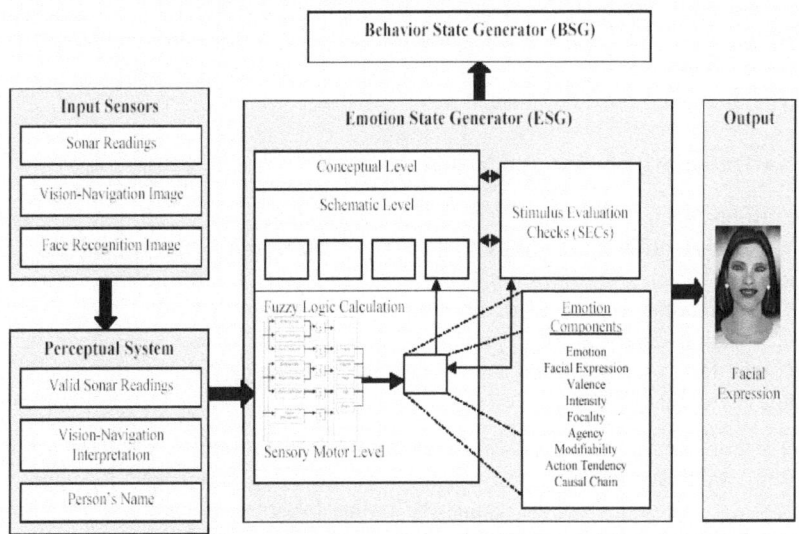

Fig. 3. Affective-Cognitive Three-Layered Architecture

We next describe how our ESG discussed in Section 2 is integrated in the overall affective-cognitive architecture shown in Figure 3 and implemented a mobile ActivMedia PeopleBot (ActivMedia, 2002). We called this robot or project Petra. Currently, Petra has three different sensors - twenty-four sonar, a camera for navigation, and a camera for face recognition to be used during navigation and social interaction. After sensing various stimuli from the real world (e.g., walls, floors, doors, faces), these are sent to the perceptual system. We designed the perceptual system as an inexpensive and simple system so that the information abstracted from the outside world has some interpreted meaning for the robot. For every cycle (in our case, it is 1000 mm travel distance), the sensors send the inputs read to the perceptual

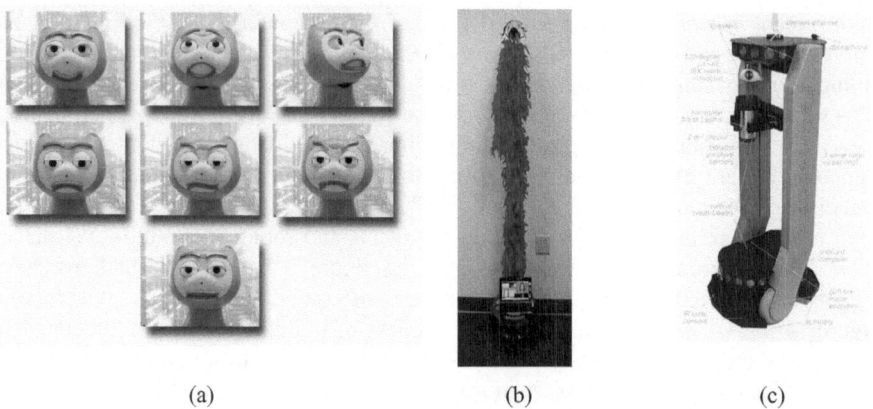

Fig. 4. (a) iCat Platform; (b) Amigobot mobile Platform; (c) Peoplebot mobile Platform

system and these are then processed by the perceptual system as described below. Afterward, the perceptual system sends its outputs (valid sonar readings, vision-navigation interpretation, and person's name) to the sensory motor level, which triggers certain emotion-like states.

5.2 Navigation with Sonar and Vision

Sonar: In our design, the robot performs sonar readings every 200 mm, so for 1000 mm, we get five different readings. Out of these five readings, the system extracts the invalid information out and stores only the good ones for further use in the ESG model. The reading is invalid if the sum of the left-most and the right-most sonar readings are extremely more or extremely less than the distance between the aisle (1,500 mm for our case). And vice versa, the reading is valid if the sum of both readings is around 1,500 mm.

Camera: For every cycle, the camera captures an image and sends it to the vision algorithm. In this algorithm, the image is smoothened and edged by canny edge detector before calculating the vanishing point. In order to calculate the point, in addition to the canny method, we also eliminate the vertical edges and leave the image with the non-vertical ones (edges with some degrees of diagonality). With the edges left, the system can detect the vanishing point by picking up the farthest point in the hall. With this point, represented by the x- and y- coordinate, the system asks the robot to perform course correction, if needed, and uses it as an input for the ESG model. Besides having the capability to center between the aisles of the hallway, the robot is also able to detect some obstacles, i.e, garbage can, boxes, people, etc. When the robot finds the object(s), this detection information is also sent to the ESG model.

5.3 Integration of Face Recognition with Social Status Knowledge

The perceptual system receives input from the eye-level camera only when the robot performs the face recognition algorithm. In our current implementation, this algorithm starts when the robot asks someone to stand next to her and captures an image. Along with the FaceIt technology by Identix (Identix, 2002), our algorithm compares the input with the collection of images in her database of 25 images and when any matching is found, she greets that person. The result, recognized or unrecognized along with the person's name (to be used to greet him/her), is also sent as an input to the ESG model. At this level, the other information of the person whose image was captured and recognized (gender, social status, and social interaction value – the degree of her like/dislike toward that person) is not sent to the sensory motor level, but in the future, this information may be needed for the implementation of the schematic and/or the conceptual level where further learning and information processing will be performed.

6 Sensory Motor Level Design and Implementation

Since the information abstracted from the perceptual system does not go through willful thinking and learning at this level, it may contain some fuzziness to certain degree. Inspired by FLAME (El Nasr, 2002), this level is implemented with the Takagi, Sugeno, and Kang (TSK) fuzzy logic model (Takagi & Sugeno, 1985). Because of its simplicity, it can reduce the number of rules required for this level. Our proposed sensory motor level architecture is shown in Figure 5.

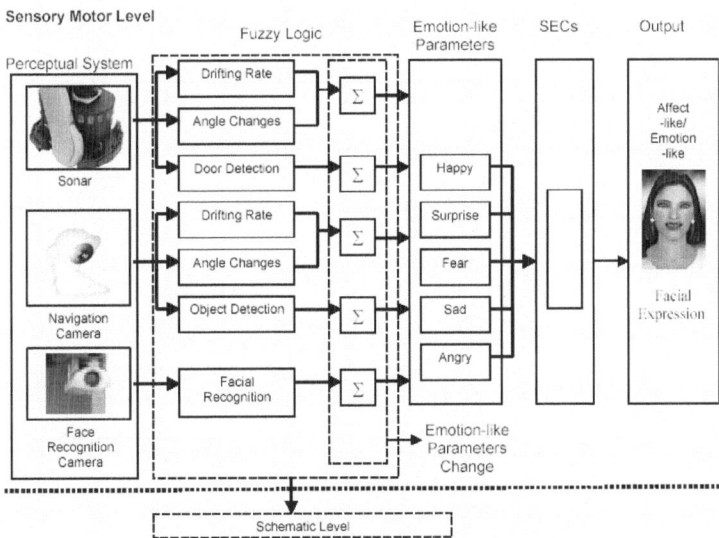

Fig. 5. Sensory Motor Level's sub-Architecture

The information received from the perceptual system is then processed further to determine the drifting rates and angle changes which are represented by five fuzzy values (small, medium-small, medium, medium-large, and large) and the door detection, the object detection, and the face recognition which are represented by boolean values (found and not-found or recognized and not-recognized). Below are the examples of fuzzy representations of the angle changes calculated from the sonar's valid readings (F_{angle_sonar}). Δ is determined by subtracting the current reading from the previous one.

The information (drifting rate, angle changes, door detection, object detection, and face recognition) is then further processed with the TSK model which gives the emotion-like-parameters-change represented by a numerical value which will add/subtract the numerical values of the emotion-like-parameters (happy, surprise, fear, sad and angry) based on the OR-mapping shown on Table 1.

Table 1. Mapping of the emotions' parameter changes

Parameter	Increased if	Decreased if
Happy	- Small to Medium-small value of the processed information from sonar or vision - Open door - Recognize someone	- Medium to Large value of the processed information from sonar or vision - Closed door - Not recognize someone
Surprise[1]	- Large value of the processed information from sonar or vision (on the first detection only)	- The robot is in the happy state
Fear	- Large value of the processed information from sonar or vision (medium repetition)	- The robot is in the happy state
Sad	- Medium to Medium-large value of the processed information from sonar or vision - Closed door - Not recognize someone	- Small to Medium-small value of the processed information from sonar or vision - Open door - Recognize someone
Angry	- Large value of the processed information from sonar or vision (high repetition) - Closed door (repetitively) - Not recognize someone (repetitively)	- Small to Medium-small value of the processed information from sonar or vision - Open door - Recognize someone

---

[1] To show surprise, when the processed information from sonar or vision is large on the first detection, the weight of this emotion is highest among all.

After calculating the emotion-like state, the sensory-motor level performs the Stimulus Evaluation Check (SEC) process to check the emotion appropriate components and create a schema of emotion to be stored in the memory. The checkings are performed by assigning appropriate values to the emotion components (as described in the SEC section above), based on the checks (e.g. *pleasantness*, *importance*, *relevance*, *urgency*). Table 2 shows a schema when an unexpected moving object suddenly appears in the captured navigation-image, i.e, walking students. In this case, surprise will be activated as the final emotion, only for the current cycle.

A sudden appearance of a person in the navigation image is detected as an obstacle that can slow down the navigation process due to the course correction that needs to be performed should the person remain in the navigation image on the next cycle. Thus *intensity* is very high and the *action tendency* is to avoid potential obstacles. Since the face cannot be detected at farther distance, the *valence* is negative. And at current cycle, the *modifiability* is set to its default–medium because the robot has not performed the obstacle avoidance to change the course event.

Table 2. Schematic Representation for Surprise

Components	Values
Emotion	Surprise
Valence	Negative
Intensity	Very High
Focality	Object – walking student
Agency	Other
Modifiability	Medium
Action Tendency	Avoid
Causal Chain	- Something happened now - I did not think before now that this will happen - If I thought about it, I would have said that this will not happen - Because of this, I feel something bad

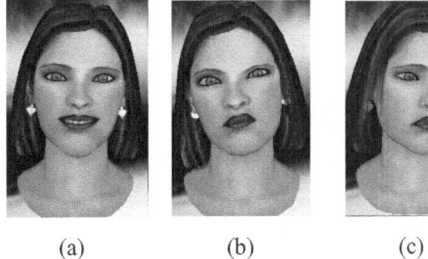

(a)   (b)   (c)

Fig. 6. Facial expressions for some of the modeled emotions a: Happy; b: Angry, c: Sad

After performing the SECs, the robot's facial expression is also adjusted to display her current internal emotion-like state. For every emotion-like that we are modeling, e.g., happy, surprise, fear, sad, and angry, we have designed their facial expressions based on the Facial Action Coding System (FACS) (Ekman and Friesen, 1978) as shown in Figure 6 (a-e).

6.1 Behavior State Generator (BSG)

A behavior is "a mapping of sensory inputs to a pattern of motor actions, which then are used to achieve a task" (Murphy, 2000). After determining the facial expressions, the processed information is sent to BSG. Through these, she can execute different behaviors depending on the input sources (sonar, camera for navigation, and camera for face recognition). Each behavior state is described below:

1. *INIT*: reset the emotion-like, the progress bars, and the starting position.
2. *STAY_CENTER*: center herself between the aisles to avoid the walls.
3. *AVOID_LEFT_WALL*: move right to avoid the left wall. This behavior is triggered when a course correction, calculated by sonar or vision, is needed.
4. *AVOID_RIGHT_WALL*: move left to avoid the right wall. This behavior is also triggered when course correction is needed.
5. *WAIT*: wait for a period of time when the face recognition algorithm cannot recognize anyone or the door is closed (in order to try again to avoid any false positive).

7   Integration on a Robotic Platform with Anthropomorphic Interface

The interface shown in Figure 7 is displayed through the touch screen wirelessly is a modified version of Cherry's (Lisetti et al., 2004). It integrates several components such as the avatar, a point-and-click map, the emotion changing progress bars, several algorithms (navigation system, vision and obstacle avoidance system, and face recognition system), several help menus, i.e., speech text box, search properties, and start-at-room option, and two live-capture frames.

The main improvements on Petra's interface from Cherry's are the progress bars, the two video frames, and navigational and vision algorithms. Through these bars, we

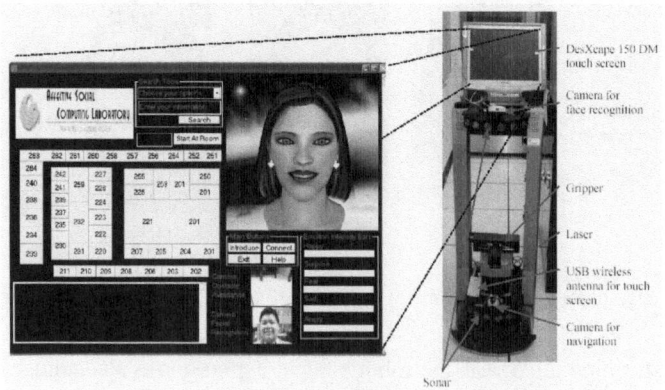

Fig. 7. Petra's Complete Interface and Hardware

are able to show the real-time changes of emotion-like state and which emotion-like state(s) is/ are affected by the stimuli accepted. One of the video streams has the same purpose as Cherry's vision for face recognition, and the other one is used for the vision for navigation system. The other two algorithms (navigation and vision) are designed to have a better and smoother navigational system.

8 Conclusion

The work presented represented a very small milestone toward achieving cognitive-affective architectures for socially intelligent agents. Our intention is to continue to base our work on psychological theories, in particular that of Scherer's because it psychologically links emotion recognition, with emotion generation at the affective-cognitive level and with emotion expression which allows to develop a completely psychologically grounded system for Human-Robot Interaction as depicted in the MAUI (Multimodal Affective User Interface) framework we presented as the basis for our work. Much more remains to be accomplished.

Acknowledgements. The authors would also like to thank the Office of Naval Research for partial funding for this research project. We would also like to thank Eric P. Leger for his help in implementing the navigation and obstacle avoidance algorithms.

References

1. ActivMedia. www.activmedia.com, 2002.
2. Arkin, R. C. *Behavior-Based Robotics*, Cambridge, MA: MIT Press, 1998.
3. Breazeal, C. and Scassellati, B., "Infant-like Social Interactions Between a Robot and a Human Caretaker". *Special issue of Adaptive Behavior on Simulation Models of Social Agents*, guest editor Kerstin Dautenhahn. 2000
4. Breazeal, C. "Emotion and sociable humanoid robots". *International Journal of Human Computer Studies*. Vol. 59. pg. 119 – 155. 2003
5. Brooks, R and Flynn, A. "Fast, Cheap, and Out of Control", AI Memo 1182, MIT AI Laboratory, 1989.
6. Brown, S. Lisetti, C. and Marpaung, A. Cherry, the Little Red Robot with a Mission and a Personality. In *Working Notes of the AAAI Fall Symposium Series on Human-Robot Interaction*, Menlo Park, CA: AAAI Press. Cape Cod, MA, November 2002.
7. Casper, J. "Human-robot interactions during the robot-assisted urban search and rescue response at the World Trade Center." MS Thesis, Computer Science and Engineering, University of South Florida, April 2002.
8. Casper, J. and Murphy, R. "Workflow study on human-robot interaction in USAR." In *Proceedings of ICRA 2002*, pp. 1997 – 2003. 2002.
9. Ekman, P and Friesen, W. "The Facial Action Coding System". Consulting Psychologist Press, San Francisco, CA, 1978.
10. El-Nasr, Magy Seif. Yen, John. Ioerger, Thomas. FLAME - A Fuzzy Logic Adaptive Model of Emotions, *Automous Agents and Multi-agent Systems,* 3, 219-257, 2000.
11. Grizard, A. and Lisetti, C. Generation of Facial Emotional Expressions Based on Psychological Theory. In Notes of the 1st Workshop *Emotion and Computing, 29th Annual German Conference on Artificial Intelligence.* Universität Bremen, Germany, June 2006.

12. Identix Inc. www.identix.com, 2002.
13. Leventhal, H. "A perceptual-motor processing model of emotion." In P. Pilner, K. Blankenstein, & I.M. Spigel (Eds.), *Perception of emotion in self and others. Vol. 5* (pp. 1 – 46). New York: Plenum. 1979
14. Leventhal, H., "Toward a comprehensive theory of emotion." In L. Berkowitz (Ed.), *Advances in experimental social Psychology. Vol. 13* (pp. 139 – 207). New York: Academic Press, 1980
15. Leventhal, H. and Scherer, K. "The relationship of emotion to cognition: A functional approach to a semantic controversy." *Cognition and Emotion.* Vol. 1. No. 1. pp. 3 – 28. 1987.
16. Lisetti, C. Brown, S. Alvarez, K. and Marpaung, A. A Social Informatics Approach to Human-Robot Interaction with an Office Service Robot. *IEEE Transactions on Systems, Man, and Cybernetics - Special Issue on Human Robot Interaction,* Vol. 34(2), May 2004.
17. C. L. Lisetti and F. Nasoz (2002). MAUI: A Multimodal Affective User Interface. In *Proceedings of the ACM Multimedia International Conference 2002,* (Juan les Pins, France, December 2002).
18. Marpaung, A. "Social Robots with Emotion State Generator Enhancing Human-Robot Interaction (HRI)". Master's thesis. University of Central Florida. in progress 2004.
19. Murphy, R. R. Use of Scripts for Coordinating Perception and Action, In *Proceedings of IROS-96 ,* 1996a.
20. Murphy, R. R. Biological and Cognitive Foundations of Intelligent Sensor Fusion, *IEEE Transactions on Systems, Man and Cybernetics,* 26 (1), 42-51, 1996b.
21. Murphy, R. R. Dempster-Shafer Theory for Sensor Fusion in Autonomous Mobile Robots, *IEEE Transactions on Robotics and Automation,* 14 (2), 1998.
22. Murphy, R. R. *Introduction to AI Robotics.* Cambridge, MA: MIT Press, 2000.
23. Murphy, R. R., Lisetti, C. L., Irish, L., Tardif, R. and Gage, A., Emotion-Based Control of Cooperating Heterogeneous Mobile Robots, *IEEE Transactions on Robotics and Automation,* Vol. 18, 2002.
24. Paleari, M. and Lisetti, C. Psychologically Grounded Avatar Expression. In Notes of the 1st Workshop *Emotion and Computing, 29th Annual German Conference on Artificial Intelligence.* Universität Bremen, Germany, June 2006
25. Picard, Rosalind W. Affective Computing, Cambridge, Mass.: MIT Press, 1997.
26. Scherer, K. "Emotion as a multicomponent process: A model and some cross-cultural data." In P. Shaver (Ed.), *Review of personality and social psychology.* Vol. 5. *Emotions, relationships and health* (pp. 37 – 63). Beverly Hills, CA: Sage. 1984
27. Scherer, K. "Vocal affect expression: A review and a model for future research." *Psychological Bulletin.* 99. pp. 143 – 165. 1986
28. Smith, C. A., Ellsworth, P.C. "Patterns of cognitive appraisal in emotion." *Journal of Personality and Social Psychology.* 48. pp. 813 – 838. 1985
29. Villon, O. and Lisetti, C. Toward Building Adaptive User's Psycho-physiological Maps of Emotions using Bio-Sensors. In Notes of the 1st Workshop *Emotion and Computing, 29th Annual German Conference on Artificial Intelligence.* Universität Bremen, Germany, June 2006

OWL and Qualitative Reasoning Models

Jochem Liem and Bert Bredeweg

Human Computer Studies Laboratory, Informatics Institute,
Faculty of Science, Universiteit van Amsterdam, The Netherlands
{jliem,bredeweg}@science.uva.nl

Abstract. The desire to share and reuse knowledge has led to the establishment of the Web Ontology Language (OWL) knowledge representation language. The Naturnet-Redime project needs to share qualitative knowledge models of issues relevant to sustainable development and OWL seems the obvious choice for representing such models to allow search and other activities relevant to sharing knowledge models. However, although the design choices made in OWL are properly documented, their implications for Artificial Intelligence (AI) are part of ongoing research. This paper explores the expressiveness of OWL by formalising the vocabulary and models used in Qualitative Reasoning (QR), and the applicability of OWL reasoners to solve QR problems. A parser has been developed to export (and import) the QR representations to (and from) OWL representations. To create the OWL definitions of the QR vocabulary and models, existing OWL patterns were used as much as possible. However, some new patterns, and pattern modifications, had to be developed in order to represent the QR vocabulary and models using OWL.

1   Introduction

During the development of the Web Ontology Language (OWL), design choices have been made to ensure the language is decidable and not too intricate to implement. Therefore, OWL does not have the expressiveness to formalise things such as default values, arithmetic, string operations, or procedural attachments. Another feature of OWL is that is has an open world assumption. The implications of these design choices on ontology development are still unclear, particularly for advanced applications in Artificial Intelligence (AI). The question is: *"What are the consequences of the OWL design choices on the expressiveness of the language for advanced AI applications?"*

To discover the problems and solutions associated with use of OWL, Garp3 Qualitative Reasoning (QR) models and their vocabulary [3] are formalised. Garp3 unifies three alternative approaches to qualitative reasoning (QPT [5], Envision [4], and QSIM [9]) into a single qualitative reasoning and modelling workbench. There are five reasons this typical AI application is chosen. Firstly, qualitative reasoning predicts the behaviour of systems; a task rather different from the classification task OWL reasoners can solve. Secondly, the knowledge

C. Freksa, M. Kohlhase, and K. Schill (Eds.): KI 2006, LNAI 4314, pp. 33–48, 2007.

representation used in QR models is elaborate and complex. For instance, qualitative models describe both the structure and the behavioural aspects of a system, parts of models may be reused in others, and the requirements for a correct model are restrictive in what kinds of ingredients may be connected. Thirdly, QR models describe their domains in a way which is understandable for non-experts, closely following the naive physics proposal (except for the focus on implementation) [7]. This common-sense view on systems which QR models have, make them interesting for reuse. Fourthly, the OWL community proposes that using an ontology as an information model for design is a typical use case [8]. Since both modelling and design are similar synthesis tasks [13], the formalisation of the QR domain should be a typical application of OWL. This makes the problems discovered during the formalisation of the QR application area of interest to a large group of researchers. Finally, there is a desire within the QR community to share and reuse models through a central online repository (see Figure 1). This goal will be realised within the European NaturNet-Redime project (http://www.naturnet.org/). A requirement to allow users to search for models in which specific concepts or structures are used, is a formalisation of these models in an open semantic format which is processable by query languages. For this purpose, OWL has been chosen since it is the de-facto standard for exchanging ontological models on the web, and has a large user base. Furthermore, well-developed OWL tools are available to facilitate the modelling and model search. Hence, this paper focuses on the question: *"Is OWL expressive enough to formalise the QR vocabulary and models, and which of the QR problems can an OWL reasoner solve?"*

In Figure 1 an overview is given of the desired result of this research. Traditionally, Garp3 can write models to a binary format and can also read them. Functionality has been added to export models to an OWL format and import them again. The models in the OWL format reference to model ingredients defined in the Qualitative Reasoning Vocabulary Ontology. For brevity, in this extra OWL file only the model itself is represented, as the simulations can be

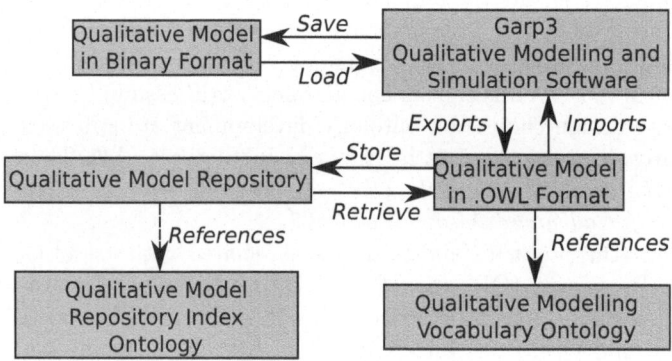

Fig. 1. Garp3's interaction with the binary and OWL files and the qualitative model repository

easily recreated using the software. In the near future an online model repository will be developed in which models can be stored, searched for and retrieved. Models in this repository will reference concepts in an ontology describing the categories within the repository.

The organisation of this paper is as follows. Section 2 explains the QR field and the types of reasoning used. Section 3 describes the implications of the formalisation of general situations using QR ingredients on the use of the OWL reasoner. Section 4 focuses on the reusability of reified relations. Section 5 explains multiple methods of the formalisation of a total order of values. Section 6 describes a pattern to restrict the use of relations for classes with specific conditions. Finally, the results are discussed and conclusions are drawn.

2   Qualitative Reasoning

The aim of qualitative modelling and reasoning [3] is to build models from which the behaviour of systems (in the form of state graphs such as the one in Figure 5a) can be predicted through simulation. Each state describes a specific situation of the system, while each transition represents the changes from one situation to another. QR models require no numerical data. Instead, changeable properties of systems are described as its relevant points and intervals (see Figure 2a and section 5). The size of a population in an environment can be formalised as {zero, positive, max}. This kind of formalisation is particularly advantageous for domains in which it is difficult to obtain numerical data, such as ecology. For experts in these fields, qualitative modelling provides a means to make their knowledge explicit and computer processable. An example of the application of qualitative modelling in ecology is the testing of the succession hypothesis of the Brazilian Cerrado forest [12]. A detailed description of the application of QR in ecology is available in the Ecological Informatics book [2].

An advantage of qualitative modelling is that the causal dependencies between quantities are made explicit (the [I]nfluence and [P]roportionalities in Figure 2b). Next to the ability to predict the behaviour of a system, these causal dependencies make it possible to provide a causal explanation of why a system behaves in a particular way. These features of qualitative simulation provide the opportunity for hypothesis testing and learning.

An important part of a QR model are model fragments, which incorporate model ingredients as either conditions (red) or consequences (blue). Two example model fragments can be seen in Figure 2a and 2b. The first describes a population with a size (which can be read as: if there is a population, it has a size), while the second formalises the consumption process between two populations. In general, the structure of the system is described using conditions and the causal dependencies as consequences, although other model ingredients can also be used as consequences. The model fragment describes causal relations which apply to a general situation within a system. Model fragments are organized in a subtype hierarchy. A child model fragment inherits the model ingredients from their parent and is a specialisation of its parent, because it adds

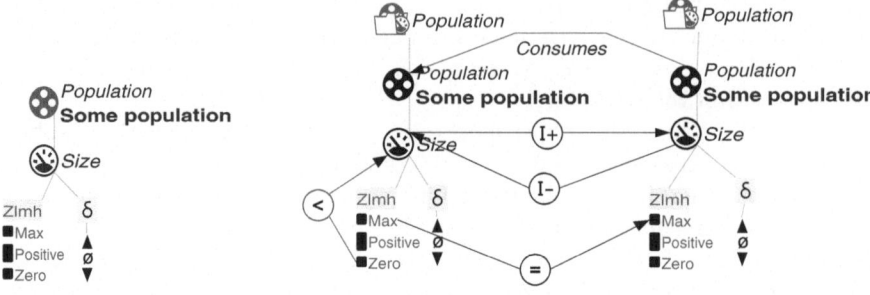

(a) A single population with size zero (point), positive (interval) or maximum (point). The population entity is conditional, while the size quantity is a consequence.

(b) The consumption of a prey by a predator. The population model fragments, the consumes configuration and the size greater than zero inequality are conditional. The causal dependencies and the equality between the max values are consequences.

Fig. 2. The two model fragments in the example model

new model ingredients to the aggregate. It is possible to reuse a model fragment within another model fragment (in Figure 2b the *Population* model fragment is reused twice). Technically speaking, this is similar to the relation between a parent and a child model fragment: the model ingredients of the reused model fragment are incorporated as conditions in the model fragment. This type of reuse allows users to efficiently create qualitative models.

Scenarios are the counterpart of model fragments and describe specific situations of systems (see Figure 3). These aggregates are used to determine the start states of the behavioural graph.

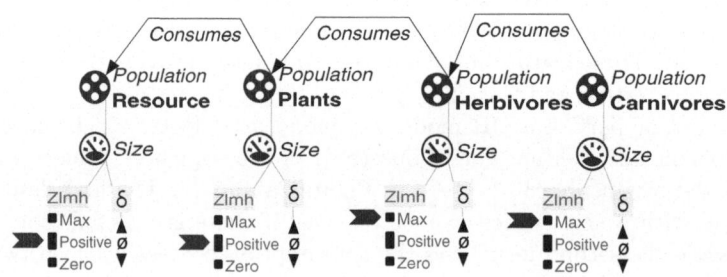

Fig. 3. A scenario plants consuming a resource, herbivores consuming plants, and carnivores consuming herbivores. All the model ingredients are consequences (facts).

The reasoning the QR engine performs can be divided into five parts: classification, inequality testing, consequence merging, influence resolution and prediction.

The first four steps take an incomplete state (which, in the first algorithm iteration, is the scenario) as input and produce a complete state description, i.e. a state containing all consequences of the model fragments applying and with calculated derivatives. The classification task searches for candidate model fragments. Candidate model fragments are the model fragments which structurally match the incomplete state. The behavioural aspects of model fragments (such as known values and inequalities) are ignored when searching for model fragments, as model fragments can contain inequalities as conditions. These conditional inequalities might be true, but have to be derived from the other inequalities which also apply to the state (which is not part of the classification).

The candidate model fragments which result from the previous step might or might not be consistent with the inequalities which are mentioned in their respective model fragments. The reasoner tries to derive the conditional inequalities in the inequality testing step. If the inequalities can be deduced, they are incorporated in the state. If the inequalities are inconsistent with the established inequalities, the candidate model fragment is removed. Model fragments for which both the conditional structure (which can include other model fragments) and the conditional values and inequalities match, become *active*. In the consequence merging step, the consequence model ingredients of active model fragments are added to the state.

The classification, inequality testing and consequence merging steps are repeated with the augmented state until no new applying model fragments can be found. After these steps, candidates are either (1) included in the state because their conditions are true, (2) ignored because their conditions are inconsistent with the state. This results in an augmented state, which incorporates all the consequences of matching model fragments. Simulating the scenario in Figure 3 with the model fragments from Figures 2a and 2b would result in an augmented state as visualised in Figure 4, although the derivatives would still be unknown (the arrows next to the active values indicating the trend).

In the influence resolution step the augmented state is completed by determining the derivatives of the quantities by resolving the influences and proportionalities. Influences are the cause of change within a model, and are therefore said to model processes. Depending on the magnitude value of the source quantity and the type of influence, the derivative of the target quantity either increases or decreases. An influence Q1(I+)Q2 causes the quantity Q2 to increase if Q1 is positive, decrease if it is negative, and remain stable when it is zero (assuming there are no other causal dependencies on Q2). For an influence I- this is just the opposite. Influences are also referred to as direct influences. Proportionalities propagate the effects of a process, (i.e. they set the derivative of the target quantity depending on the derivative of the source quantity). For this reason, they are also referred to as indirect influences. Like influences, proportionalities are either positive or negative. A proportionality Q1(P+)Q2 causes Q2 to increase if Q1 increases, decrease if Q1 decreases, and remain stable if Q1 remains stable. For a proportionality P- the opposite applies. Applying the influence resolution step would result in the completed state description shown in Figure 4.

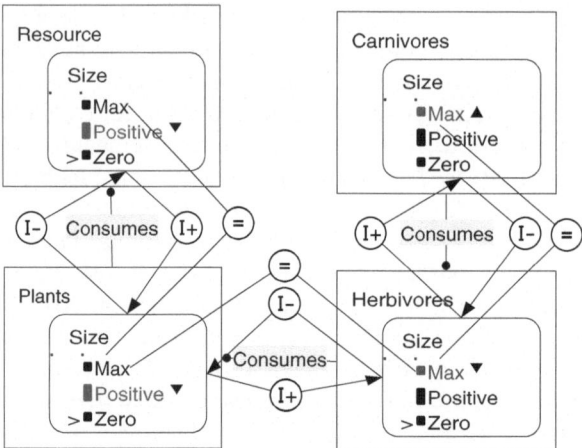

Fig. 4. The completed state description after matching model fragments on the scenario, aggregating the consequences and resolving the causal operators

The prediction algorithm takes the completed state description and identifies the successive states of behaviour and transitions to them. Termination rules are part of the qualitative engine, and indicate under what conditions states change. For example, if the magnitude of a quantity is at a point-value (e.g. the population size is zero), and the derivative of that quantity is positive (e.g. the population size is increasing), then in the next state that quantity has the interval-value directly above its current point-value (e.g. the population size becomes positive). Using this set of rules all the possible terminations of the state are gathered. Not every termination in the set of possible terminations of a state applies. Some terminations have precedence over other terminations. For example, a transition from a point to an interval happens before the transition from a interval to a point. Others occur simultaneously due to correspondences. The final step generates the successive states and transitions using the final set of pruned and merged terminations. For these successive states all the algorithm steps are repeated to generate a complete state graph describing the behaviour of the modelled system.

The state graph resulting from the simulation of the scenario in Figure 3 is shown in Figure 5a. The values of the quantities in each of the states are shown in Figure 5b. The transitions from state 1 to 2 and 3 happen because there is a negative influence (consumption) on the size of the herbivore population which is greater than the positive influence on the herbivores (feeding), since the magnitude of plant population size is smaller than the magnitude of the carnivore population size. This transition has priority over all other transitions because it is a change from a point to an interval, while the other possible transitions are changes from intervals to points. The reason that two states are generated is because the influences on plant population can become either equal, resulting in a stable derivative for the plant population (state 3), or unequal, resulting in the decrease of the plant population size. From state 2, three possible changes may occur. Either only the resource depletes (state 4), both the resource depletes and the plant population becomes zero (state 5), or the resource depletes and the plant and the herbivore populations die out (state 6). From state 3 it is only possible that the resource depletes (state 4), as the derivative of the plant population is stable. As a result of the depletion of the resource, the plant population decreases again, because the positive influence on the plant population from feeding disappears. The other transitions are obvious. From state 4, either both the plant and the herbivore populations die instantly (state 6), or the plant population dies first (state 5), and the herbivore population becomes extinct afterwards (state 6).

(a) The state graph resulting from simulation. (b) The value history corresponding to states in the state graph.

Fig. 5. The state graph and corresponding value history

3   Representing General Situations

The reasonable aim of formalising QR in OWL would be to try to use an OWL classifier to solve the classification task instead of the QR reasoner. Taking that approach, some of the typical inferences made by the QR reasoner such as the inequality testing, consequence merging, influence resolution and prediction tasks, would then still be left to the QR reasoner. To fulfil this 'reasonable' goal the correct formalisation of model fragments is essential. However, as will be pointed out below, this is already rather complex and not adequately solvable with the current version of OWL [1].

The conditional model ingredients in model fragments describe general situations of a system, and scenarios describe specific situations. The classes in OWL describe general concepts, while the instances are specific individuals. OWL reasoners are able to classify instances to classes, therefore the model fragments have to be formalised as classes and the scenarios as instances. As a result the contents of the model fragments has to be formalised using necessary and sufficient conditions. This allows the scenarios to be classified as a certain model fragment. The consequences of model fragments have to be modelled separately, as they cannot be part of the restrictions of the model fragments (they are the consequences of a model fragment firing).

The formalisation of model fragments as classes allows the representation of the subtype hierarchy of model fragments. The subclasses of the model fragments would inherit the restrictions which model the contents of the parents, and add restrictions to describe the new model ingredients of the child model fragment.

There are two problems with the formalisation of the conditions of model fragments as classes. Firstly, it is impractical to specify that a model fragment contains multiple objects of the same type (either as conditions or consequences). It would require two separate relations (one for conditions and one for consequences) for each type of object in combination with a cardinality restriction (to indicate the number of incorporated objects of that type). The unique relations for each object type are required as the cardinality restriction has to apply to precisely one type of object. This problem can be solved by making use of Qualified Cardinality Restrictions (QCR), which indicate a class should have a certain amount of fillers for a specific relation. Although QCR's are not in the first OWL specification, Protégé [6] and RacerPro (Racer Systems GmbH & Co. KG: http://www.racer-systems.com) have already added support for them[1]. Furthermore, QCR's are already mentioned in the drafts of the OWL1.1 specifications[2].

The second problem concerns the impossibility to distinguish between different objects of the same type in restrictions. Consider population x consuming population y, which in turn consumes z (a situation similar to the one described in Figure 3 if it was a model fragment with conditional elements). When formalised as restriction in OWL, this would result in the following definition:

hasCondition exactly 3 Population
hasCondition some (Population and
(consumes some (Population and
(consumes some Population))))

This formalization has multiple interpretations. It could be that population x preys on y, and y preys on x, or that x preys on y, and y preys on z. Furthermore, it becomes hard to formalize the other relations the populations take part in without *naming* the Populations. Without variables to distinguish between two objects of the same type, it is impossible to describe model fragments as classes.

As OWL is not expressive enough to formalize model fragments as classes, there is no choice but to formalize them as instances. This makes it impossible to use an OWL reasoner to classify scenarios on model fragments. On the other hand, using instances does eliminate the requirement of having to separate the conditions and consequences in the formalization of model fragments. Each model fragment can be modelled as an instance. That instance has hasCondition and hasConsequence relations to each of the model ingredients instances it contains. Those ingredient instances in turn have relations which indicate how they are related.

---

[1] http://protege.stanford.edu/mail_archive/msg17798.html
[2] http://owl1_1.cs.manchester.ac.uk/

A problem with the formalization of model fragments as instances is that model fragments can have subclasses and can be reused. Since it is impossible to create instances of instances or subclasses of instances, model fragments cannot be described as instances.

Summarising, model fragments have to be modelled as classes in order to represent the subtype hierarchy of model fragments, and to reuse them in other model fragments. On the other hand, classes are not expressive enough to model the contents of model fragments. Secondly, in order to correctly formalize the contents of model fragments they have to be modelled as instances. However, this would make it impossible to keep track of the reuse of model fragments, and would require a special subclass relation to formalize the model fragment hierarchy.

Both previous results are undesirable. An alternative is to treat model fragment classes as individuals. This is valid, but makes the ontology OWL Full [1]. The model fragment definitions are classes, so a class hierarchy of model fragments can be created. These classes have *hasCondition* and *hasConsequence* relations with instances of the QR ingredients they incorporate. This is ontologically not the most desirable solution, as the conditions in model fragments do not correspond to the restrictions in OWL. However, the solution does correctly represent model fragments and will allow the OWL format to be used for model search and reuse.

4   The Formalisation of Relations

The relations in the QR vocabulary, the *configurations* and *dependencies* (causal, mathematical and correspondence), are not simple binary relations between two objects. For each relation screen information (for visualisation), such as its position, has to be stored. Additionally, the *Calc* relations plus and min potentially link more than two objects, as they model the result of an addition or subtraction of two values, and can have an arbitrary amount of *inequality* relations to compare the result to other values (e.g. it is possible to specify that the sum of the two population sizes in Figure 2b is equal to zero). Since OWL supports only binary relations between objects, the formalisation of n-ary relations and information about relations is an issue. The Semantic Web Best Practices and Deployment Working Group (http://www.w3.org/2001/sw/BestPractices/) describes two variations of a pattern which can be applied to solve this issue [10]. By modelling a relation as a class it is possible to relate multiple objects using only one relation instance. This process is called reification.

However, the existing reification pattern hinders the reuse of reified relations. The calc and inequality relations in the QR vocabulary may only connect specific subsets of elements of quantities (magnitudes, derivatives and points) and other calc relations. For example, magnitudes can only be connected to other magnitudes or points fulfilling certain conditions, while derivatives can only be connected to derivatives or points fulfilling certain other conditions (see Section 6). Thus, the second argument of the relation depends on the first one, but the relations have the same meaning independent of the arguments. It is important that the relations can be reused. Conceptually, there are 3 Calc relations (Plus, Min and their superclass), and 6 inequalities ($<, \leq, =, \geq, >$ and their superclass) in the QR vocabulary. The need to create multiple types of Calc and Inequality relations depending on their arguments creates unwanted complexity in the file format and the QR vocabulary ontology.

If the first variation of the pattern is applied to represent the use of Calc relations by magnitudes, magnitudes are restricted to having an arbitrary amount of Calc relations (*hasCalc only Calc*). Furthermore, the Calc relations have to be related to exactly one magnitude and have an arbitrary amount of inequalities (*hasCalcTarget only Magnitude*; *hasCalcTarget = 1*; *hasInequality only Inequality*). This fixes the target of the Calc relation, making it impossible to reuse the relation for Calc relations between derivatives. At least the Calc relation can be reused by classes with the same target. If the restrictions for the inequalities from Calc relations are ignored, 6 different relations are needed to formalise all the restrictions. For the inequalities 24 reified relations are needed.

The second variation of the pattern only worsens the problem, as the restrictions are all formalised in the Calc relation. These indicate that both its source and its target have to be of a specific type (*hasCalcSource only Magnitude*; *hasCalcSource = 1*; *hasCalcTarget only Magnitude*; *hasCalcTarget = 1*). This prevents reuse of the relation for all other classes. To properly formalise the relation restrictions, 15 reified Calc relations and 36 reified inequality relations are needed.

We developed a new version of the reification pattern to solve the reusability issues of reified relations. Compared to the first version of the reification pattern, the restrictions about the target of the Calc relation and its Inequalities have been moved to the source of the relation (see Figure 6 which uses the Protégé [6] syntax). The formalisation represents that Magnitudes can have an arbitrary amount of hasCalc relations with reified Calc relations, which in turn have exactly one hasCalcTarget relation with another magnitude, and an arbitrary amount of inequality relations with magnitudes. This is achievable as OWL allows the creation of new anonymous class definitions within restrictions (the conjunction in the hasCalc restriction). Using this pattern, it is possible to impose specific usage restrictions for relations in each desired source class. For example, the derivative class can have an arbitrary amount of hasCalc relations with reified Calc relations, which have exactly one other derivative as a target, and an arbitrary amount of inequalities with other derivatives. Note that the real formalisation is more complex, as inequality properties are also reified, and the possible values of the hasCalcTarget and inequality properties are a union of multiple classes.

Another advantage of this new pattern is that it is possible for other users to use the relation defined in the ontology (without having to copy and adapt it), as the usage of the relation is not restricted to specific classes. This is also a disadvantage, as it is possible to abuse the relation with classes for which it does not make sense. A possible work-around is adding a cardinality=0 restrictions

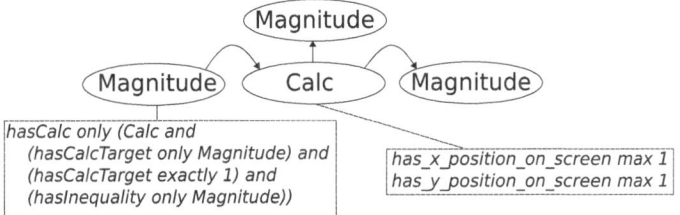

Fig. 6. A reusable reification pattern

to classes for each relation the class cannot be involved with, although this is only an acceptable solution if the number of relations and classes is relatively low.

5 Representing Values

Qualitative values are either points or intervals which are stored in quantity spaces (see Figure 2a). These behavioural ingredients define the possible values of a quantity. A quantity space consists of at least one qualitative value and values adjacent to intervals have to be points and visa versa. The quantity space describes a total order, which means that a magnitude or derivative can only change to a value directly above or below its current value. Qualitative values in a model fragment can participate in (in)equality, correspondence and calc relations. As a result, it is impossible to formalise them as an enumeration of individuals (the 'values as sets of individuals' pattern [11]), as different relations in different contexts would refer to the same value. Therefore, the formalization of qualitative values requires a unique individual for each value instance (i.e. for each quantity space in which the value occurs).

Representing the qualitative values as classes (the 'values as subclasses partitioning a "feature"' pattern [11]) fulfils our requirement of creating a unique value individual for each quantity space instance. The qualitative values can be thought of as a set of subclasses forming a parent class. This class is exactly the superset of all the possible value classes (modelled using *owl:unionOf*). It is necessary to explicitly state that the subclasses are disjoint, as it should be inconsistent to create an individual which is an instance of multiple values.

Representing qualitative values as classes allows individuals to be created for each instance of a specific quantity space. However, the pattern does not model their strict ordering. A possible solution to this problem is to model the values using an RDF collection [10]. Such a collection consists of a number of instances of *rdf:List*. Each of these items is connected to the next in the collection using *rdf:rest*, and points to a qualitative value instance using *rdf:first*. An advantage of this pattern is that OWL editors understand the that the structure is a list, as it is part of the RDF specification. A disadvantage is that using *rdf:List* causes the ontology to become OWL Full. This disadvantage can be remedied by recreating a list structure in OWL [10], but a side effect would be that the editors would not understand that the structure modelled is a list.

The list pattern has two further problems. Firstly, a list has no ontological meaning, as it is a data structure. A list with the cities Amsterdam, Brussels, and Paris has little meaning. They could indicate a travel route, cities with around the same amount of inhabitants, or something else entirely. The relations between the list items is left implicit. Secondly, it becomes impossible to classify a list entry depending on the owner instance of the list, as there is no direct relation between that object and each list item. This would make the "relation restriction through classification" pattern impossible to apply (section 6).

Our solution to formalise quantity spaces makes the ordering of its values explicit using inequalities (see Figure 7). The quantity space is connected to its point and interval instances using *containsQualitativeValue* relations. The ordering is established using reified inequality relation instances originating from the intervals, as inequalities created by the user are required to originate from the points. Each consecutive value in the order must have another type than the previous one. Therefore, the intervals can only have inequalities with points. This solution is a semantic description of values, but OWL editors do not understand it is a list.

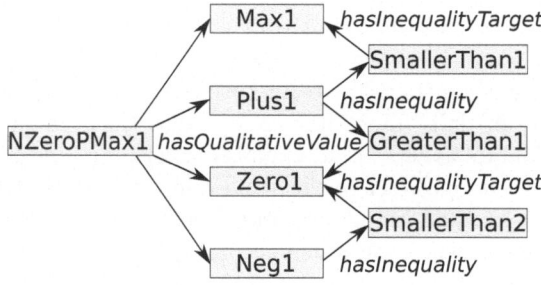

Fig. 7. Representing a quantity space and its values using inequalities

6 Relation Restriction Through Classification

A quantity consists of a magnitude and a derivative, which in turn each have a quantity space (the magnitude is not visualised in Figure 2a). As described above, quantity spaces consist of a set of values in a total order. Roughly speaking, inequalities are ordinal relations which have either magnitude or derivative items (values, calc relations and magnitudes and derivatives themselves) as arguments. Since manually categorising values into either a class for points belonging to magnitudes or a class for points belonging to derivatives would add redundant information to the formalisation, a different solution is desired.

Our inequality formalisation solution introduces two new classes, one for the point-values belonging to magnitudes (PointBelongingToMagnitude), and another for point-values belonging to derivatives (PointBelongingToDerivative).

In these classes necessary (*class* $\implies$ *conditions*), and necessary and sufficient (*class* $\iff$ *conditions*), restrictions are combined. The necessary and sufficient conditions of PointBelongingToMagnitude state that the value belongs to a quantity space which belongs to a magnitude (or a derivative for PointBelongingToDerivative). This allows the OWL reasoner to classify the points as belonging to one of the classes. The necessary conditions specify that all the inequality relations instances which have a PointBelongingToMagnitude as a first argument must have either a PointBelongingToMagnitude or a Calc relation between magnitude items as a second argument.

This pattern is used to restrict the use of inequalities, Calc relations and correspondences. Essentially, a new class is created with necessary and sufficient conditions for one of the special cases, and necessary conditions for the restriction of this special case. This pattern takes away the need to replicate information by using the OWL reasoner.

7   Implementation

We have developed a generic ontology [14] of the QR vocabulary in which all the model ingredients and their usage restrictions are formalised (see Figure 8). Based on this formalisation we have successfully implemented OWL export and import functionality, which has been integrated with the Garp3 qualitative reasoning and modelling tool (http://hcs.science.uva.nl/QRM/) using the SWI-Prolog Semantic Web Library [16]. This functionality is currently used to automatically formalise QR models as domain ontologies, which can be shared using an online model repository. The consistency of these model ontologies was checked using the Triple20 [15] and Protégé [6] ontology editors and the Racer-Pro reasoner. The QR vocabulary ontology and an example QR model in OWL can be found via: http://protege.cim3.net/cgi-bin/wiki.pl?NaturNet_Redime.

8   Conclusions and Discussion

This paper presented the problems encountered during the formalisation of the QR vocabulary and models in OWL. We succeeded in formalising qualitative models in OWL, allowing our formalisation of models to be used as a data format for a central model repository. Due to the limits of the expressiveness of OWL, it was not possible to formalise the model fragments as purely classes with restrictions. Instead, they were formalised as classes with relations to instances (making the ontology OWL Full). This makes it impossible to use the OWL reasoners to classify scenarios on model fragments, since the instances and relations are not necessary and sufficient conditions.

To be able to formalise n-ary relations and represent information about them the reification pattern is used. We have shown that the two existing variations prevent the relations to be reused. Our new pattern solves this reusability issue by moving the relation restrictions to the source class.

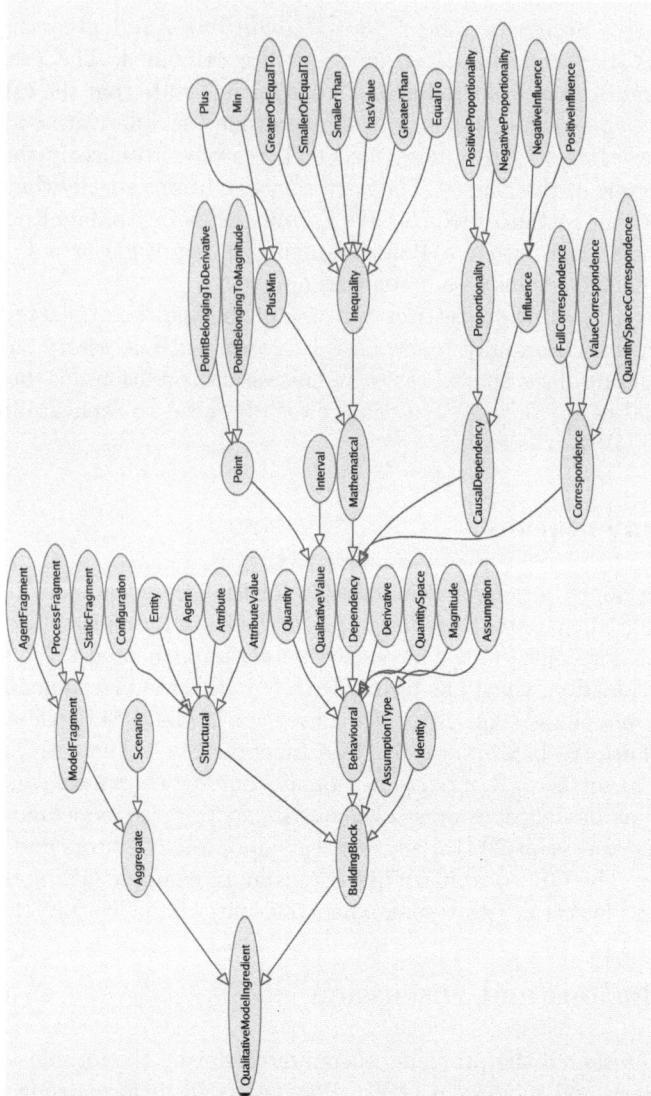

Fig. 8. The taxonomy of qualitative model ingredients

The existing patterns to formalise sequences of values are not enough to represent quantity spaces. The problems with representing them as a enumeration of individuals, a set of classes, and an *rdf:list* of instances were explained. We presented our more semantic representation which solves these problems.

We developed a new pattern to impose relation usage restrictions on classes with certain conditions. New classes are defined with the conditions as necessary and sufficient restrictions. The necessary restrictions are imposed on these classes, and will apply when instances are classified as belonging to the class. This prevents information redundancy, as instances do not have to be explicitly represented as belonging to certain classes.

Future OWL extensions (such as the Semantic Web Rule Language) might make it possible to formalise the model fragments in such a way that it is possible to use an OWL reasoner to solve the QR classification task. For an OWL reasoner to actually replace a QR classification task such as found in Garp3, two additional inferences have to be addressed. Firstly, the reasoner tries to infer if certain conditional inequalities are true in the scenario using inequality reasoning. Secondly, unprovable but possible conditional inequalities and value assignments are assumed by the reasoner. Each mutually exclusive set of these assumptions will result in a new state in the state graph describing the behaviour of the system. Another inference which would be useful to OWL users is adding individuals to a knowledge base using a rule based mechanism. This would allow the replacement of the QR consequence merging task.

The research described in this paper has allowed the implementation of functionality to export qualitative models from the Garp3 application to an OWL file, and import this OWL file again into the workbench. This makes it possible to store these qualitative models in an online repository. This repository should make it possible for the community of practice to (1) share models amongst themselves, (2) search for models in which specific concepts or structures are used, and (3) reuse parts of models, which are all goals of the NaturNet-Redime project. Domain experts can use this repository to search for modelling work related to their own research, for example to compare different formalisations of the same phenomena. Teachers can ask students to download and analyse certain models. Finally, the repository allows modellers to store their results and disseminate them to a larger group.

Acknowledgements

This work is co-funded by the European Commission within the Sixth Framework Programme (2002-2006), project NaturNet-Redime (http://www.naturnet.org), number 004074. We would like to thank Rinke Hoekstra and the reviewers for their valuable comments, and Jan Wielemaker for his extensive programming support.

References

1. S. Bechhofer, F. van Harmelen, J. Hendler, I. Horrocks, D. L. McGuinness, P. F. Patel-Schneider, and L. A. Stein. OWL web ontology language reference. W3C recommendation, February 2004. M. Dean, G. Schreiber (eds.).
2. B. Bredeweg, P. Salles, and M. Neumann. *Ecological Informatics: Scope, Techniques and Applications*, chapter Ecological Applications of Qualitative Reasoning, pages 15–47. Springer, Berlin, 2nd edition, 2006.
3. B. Bredeweg and P. Struss. Current topics in qualitative reasoning (editorial introduction). *AI Magazine*, 24(4):13–16, 2003.

4. J. de Kleer and J. S. Brown. A qualitative physics based on confluences. *Artificial Intelligence*, 24(1-3):7–83, December 1984.

5. K. D. Forbus. Qualitative process theory. *Artificial Intelligence*, 24(1-3):85–168, December 1984.

6. N.F. Noy H. Knublauch, R. Fergerson and M.A. Musen. The protege owl plugin: An open development environment for semantic web applications. In S. A. McIlraith, D. Plexousakis, and F. van Harmelen, editors, *International Semantic Web Conference*, pages 229–243, Hiroshima, Japan, November 2004. Springer.

7. P. J. Hayes. *Formal Theories of the Commonsense World*, volume 1 of *Ablex series in Artificial Intelligence*, chapter The Second Naive Physics Manifesto, pages 1–36. Ablex, Norwood, NJ, June 1985.

8. J. Heflin. OWL web ontology language use cases and requirements. W3C recommendation, February 2004.

9. B. Kuipers. Qualitative reasoning: modeling and simulation with incomplete knowledge. *Automatica*, 25(4):571–585, 1989.

10. N. Noy and A. Rector. Defining n-ary relations on the semantic web. W3C working group note, April 2006. http://www.w3.org/TR/swbp-n-aryRelations/.

11. A. Rector. Representing specified values in OWL: "value partitions" and "value sets". W3C working group note, May 2005. http://www.w3.org/TR/swbp-specified-values/.

12. P. Salles and B. Bredeweg. Qualitative reasoning about population and community ecology. *AI Magazine*, 24(4):77–90, 2003.

13. G. Schreiber, H. Akkermans, A. Anjewierden, R. de Hoog, N. Shadbolt, W. van de Velde, and B. Wielinga. *Knowledge Engineering and Management - The CommonKADS Methodology*. MIT Press, Cambridge, MA, 2000.

14. G. van Heijst, S. Falasconi, A. Abu-Hanna, G. Schreiber, and M. Stefanelli. A case study in ontology library contruction. *Artificial Intelligence in Medicine*, 7(3):227–255, June 1995.

15. J. Wielemaker, G. Schreiber, and B. Wielinga. Using triples for implementation: the Triple20 ontology-manipulation tool. In Y. Gil, E. Motta, V. R. Benjamins, and M. A. Musen, editors, *International Semantic Web Conference*, pages 773–785, Berlin, Germany, November 2005. Springer Verlag. LNCS 3729.

16. J. Wielemaker, G. Schreiber, and B. J. Wielinga. Prolog-based infrastructure for RDF: performance and scalability. In D. Fensel, K. Sycara, and J. Mylopoulos, editors, *The Semantic Web - Proceedings ISWC'03, Sanibel Island, Florida*, pages 644–658, Berlin, Germany, october 2003. Springer Verlag. LNCS 2870.

Techniques for Fast Query Relaxation in Content-Based Recommender Systems

Dietmar Jannach

Institute for Business Informatics & Application Systems
University Klagenfurt
`dietmar.jannach@uni-klu.ac.at`

Abstract. 'Query relaxation' is one of the basic approaches to deal with unfulfillable or conflicting customer requirements in content-based recommender systems: When no product in the catalog exactly matches the customer requirements, the idea is to retrieve those products that fulfill as many of the requirements as possible by removing (relaxing) parts of the original query to the catalog. In general, searching for such an 'maximum succeeding subquery' is a non-trivial task because a) the theoretical search space exponentially grows with the number of the subqueries and b) the allowed response times are strictly limited in interactive recommender applications.

In this paper, we describe new techniques for the fast computation of 'user-optimal' query relaxations: First, we show how the number of required database queries for determining an optimal relaxation can be limited to the number of given subqueries by evaluating the subqueries individually. Next, it is described how the problem of finding relaxations returning 'at-least-n' products can be efficiently solved by analyzing these partial query results in memory. Finally, we show how a general-purpose conflict detection algorithm can be applied for determining 'preferred' conflicts in interactive relaxation scenarios.

The described algorithms have been implemented and evaluated in a knowledge-based recommender framework; the paper comprises a discussion of implementation details, experiences, and experimental results.

1 Introduction

Content-based recommendation approaches employ detailed knowledge about the items in the product catalog. In addition, in particular in interactive and knowledge-based recommender systems, the customer's requirements are in many cases directly or indirectly mapped to product characteristics, which also means that the set of suitable products is determined by dynamically constructing a query to the catalog or database [9]. One of the main problems of such filter-based retrieval methods, however, is that situations can easily arise in which none of the products fulfills all of the customer requirements [2]. 'Query relaxation', i.e., the removal of parts of the query, is beside similarity-based retrieval one of the commonly used approaches to deal with this problem: The main goal of such approaches is to retrieve products that fulfill as many of the customer's

C. Freksa, M. Kohlhase, and K. Schill (Eds.): KI 2006, LNAI 4314, pp. 49–63, 2007.

requirements as possible - a task that can be mapped to the problem of finding a maximum succeeding subquery (XSS) [3] of the original query.

Recently, new algorithms for determining such XSS and Minimally Failing Subqueries (MFS) [3] have been proposed in the context of recommender systems, see e.g., [9,10,12] and it has been shown that query relaxation can be a helpful technique for implementing more intelligent behavior in recommender systems, e.g., for building an interactive relaxation facility or for the generation of explanations for the proposals.

The main problem in finding suitable relaxations, however, lies in the fact that the theoretical search space exponentially increases with the number of atoms (subqueries) of the original query, i.e., if the query can be split into n subqueries there exists 2^n possible combinations.[1] At the same time, the allowed time frame for the system's response is very short in interactive applications, i.e., response times should be below one second: McSherry [9] addresses this problem by precomputing and filtering the set of MFSs by cardinality, which, however, still requires a significant number of database queries. Ricci [12] introduces *Feature Abstraction Hierarchies* for coping with the complexity problem, which results in short response times at the cost of incompleteness. In addition, both approaches primarily rely on the assumption that smaller relaxations (in terms of cardinality) are always preferable, an assumption that might not be true in all application domains.

In this paper, we propose new techniques for determining optimal or preferred relaxations in an efficient way and which help us to overcome the limitations of previous approaches: For the *non-interactive* case (Section 2), in which the system immediately computes a relaxation when the query fails, we propose to evaluate the atoms of the query individually in a preprocessing step and base the subsequent computation of relaxations on these intermediate results. Furthermore, we also show how we can efficiently determine relaxations that lead to *at least n* items, because in some application domains it is desirable that the recommendation comprises more than one single product, such that the end user has a choice of multiple products that he can compare. This computation is again based on the partial query results and is a form of relaxation which is not covered by previous approaches.

For supporting *interactive* relaxation (Section 3), i.e., scenarios in which the end user incrementally states on which requirements s/he is willing to compromise, we show how a recent *conflict-detection* algorithm can be utilized for fast computation of *preferred* conflicts. In contrast to previous approaches, in our approach the conflicts are computed *on demand*, which means that the costly process of finding all conflicts in the requirements in advance (like in [9]) is not required.

In Section 4 finally, we discuss implementation aspects and summarize the experiences gained from several real-world advisory applications in different domains. The paper ends with a short discussion of further and other related work.

---

[1] A detailed complexity analysis for the MFS/XSS problem is given in [3].

2   Non-interactive Relaxation

The basic problem of query relaxation lies both in the size of the search space and in the limited response times: The theoretical search space for the case $n=4$ is illustrated in Figure 1. In fact, in previous approaches [9,10,12] the search for relaxations is based on scanning this lattice; the lattice also served as a basis for the complexity analysis in [3].

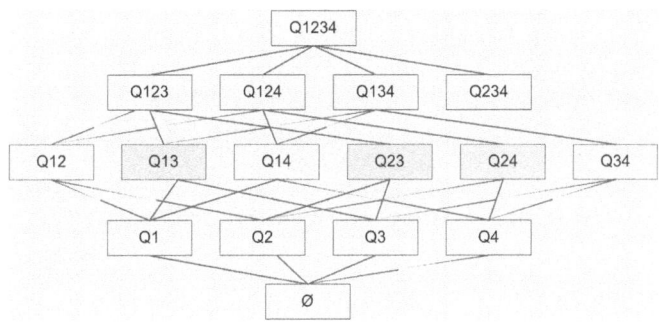

Fig. 1. Lattice of possible subqueries, minmal relaxations are printed in shaded boxes

In contrast to these approaches, we base our first two algorithms on the concept of *product-specific relaxations (PSX)* and the individual evaluation of the atoms of the query: A *PSX* for a product p in the catalog corresponds to the set of atoms of the original query that filters out p from the result set. These sets can be efficiently computed in-memory based on the partial query results, which shall be demonstrated in the following example. Figure 2 depicts a typical product catalog from the domain of digital cameras.

ID	USB	Firewire	Price	Resolution	Make
p1	true	false	400	5 MP	Canon
p2	false	true	500	5 MP	Canon
p3	true	false	200	4 MP	Fuji
p4	false	true	400	5 MP	HP

Fig. 2. Product database of digital cameras

Let the user's query consist of the following requirements (atoms) which results in a theoretical search space of $2^4 = 16$ combinations of atoms.

$Q = \{$ usb = true (Q1), firewire = true (Q2), price < 300 (Q3),
        resolution >= 5 MP (Q4)$\}$

Given the catalog and the user query, we can compute a matrix of zeros and ones (see Figure 3) that shows which atoms of the query filter out which products. Note that for constructing this matrix, we need exactly four database queries; when using bit-set data structures, we only need *nbProducts*nbAtoms* bits for storing this data in memory. In addition, determining the matching set of products for the overall query can be efficiently done by using fast *bitwise-and* operations on the rows of the matrix.

ID	p1	p2	p3	p4
Q1	1	0	1	0
Q2	0	1	0	1
Q3	0	0	1	0
Q4	1	1	0	1

Product-specific relaxation for p1

Fig. 3. Evaluating the subqueries individually

For determining a *PSX* for a given product p_i we can look up in the matrix the fields that have a zero in the corresponding column which directly leads to the set of atoms we would have to relax in order to have p_i in the result set. In the example, the list of $PSXs = \langle \{Q2, Q3\}, \{Q1, Q3\}, \{Q2, Q4\}, \{Q1, Q3\} \rangle$ and the set of minimal relaxations is $\{\{Q1, Q3\}, \{Q2, Q3\}, \{Q2, Q4\}\}$: Considering, for example, $\{\{Q1, Q3\}$ as a relaxation leading to $p1$, we see that when relaxing only $Q1$ or $Q3$ alone, no product will be in the result set (see Figure 1). In general, however, not all PSX's are already *minimal* relaxations: If, for instance, there is a camera $p5$ with neither USB nor Firewire support at a price of 400, the PSX for $p5$ would be $\{Q1, Q2, Q3\}$, which would be a non-minimal relaxation of the problem because it is a superset of another PSX. However, we will subsequently show that the set of *minimal* relaxations always is among the list of product-specific relaxations of the problem and we can determine the minimal relaxations by scanning this list. In addition, we can also easily rank the different relaxations, when we are given a cost function that associates relaxation costs with each part of the query. We base our definitions on the work of [3] and [9], respectively.

Definition 1. *(Query): A query Q is a conjunctive query formula, i.e., $Q \equiv A_1 \wedge ... \wedge A_k$. Each A_i is an atom (condition).*

In the following we denote the number of atoms of the query as $|Q|$ (query length).

Definition 2. *(Subquery): Given a query Q consisting of the atoms $A_1 \wedge ... \wedge A_k$, a query Q' is called a subquery of Q iff $Q' \equiv A_{s_1} \wedge ... \wedge A_{s_j}$, and $\{s_1, ...s_j\} \subset \{1, ..., k\}$*

Lemma 1. *If Q' is a subquery of Q and Q' fails, also the query Q itself must fail.*

We now define the term *relaxation* which is more common in the application domain than *Maximal Succeeding Subqueries* in the sense of [3]. Of course, these things are directly related to each other.[2]

Definition 3. *(Valid relaxation): If Q is a failing query and Q' is a succeeding subquery of Q, the set of atoms of Q which are not part of Q' is called a valid relaxation of Q.*

Definition 4. *(Minimal relaxation): A valid relaxation R of a failing query Q is called minimal, if there exists no other valid relaxation R' of Q which is a subset of R.*

In Figure 1, the minimal relaxations for the example problem are depicted in shaded boxes.

Definition 5. *(Maximal succeeding subquery - XSS): Given a failing query Q, a Maximal Succeeding Subquery XSS for Q is a non-failing subquery of Q and there exists no other query Q' which is also a non-failing subquery of Q for which holds that XSS is a subquery of Q'.*

Lemma 2. *Given a maximal succeeding subquery XSS for Q, the set of atoms of Q which are not in XSS represent a minimal relaxation R for Q.*

Lemma 3. *If the product catalog P is not empty, a relaxation R for Q will always exist.*

A product-specific relaxation can be defined as follows:

Definition 6. *(Product-specific relaxation - PSX): Let Q be a query consisting of the atoms A_1, ..., A_k, P the product catalog, and p_i an element of P. $PSX(Q, p_i)$ is defined to be a function that returns the set of atoms A_i from $A_1, ..., A_k$ that are not satisfied by product p_i.*

Lemma 4. *The set of atoms returned by $PSX(Q, p_i)$ is also a valid relaxation for Q.*

We have to show that all minimal relaxations are contained in the PSXs of the products.

Proposition 1. *Given a failing query Q and a product database P containing n products, at most n minimal relaxations can exist and all minimal relaxations are among the PSXs for Q and P.*

---

[2] The term 'query relaxation' is also used in the context of XML databases, where the goal is to find *approximate* answers to user queries. These approaches, however, have little relation with our work and mainly aim at relaxing structural constraints in general XML-specific query languages [1,7].

ALGORITHM: *MinRelax*
In: A query Q, a product catalog P
Out: Set of minimal relaxations $minRS$ for Q

MinRS $= \emptyset$
forall $p_i \in P$ **do**
    PSX = Compute the product-specific relaxation $PSX(Q, p_i)$
    % Check relaxations that were already found
    SUB $= \{r \in MinRS \mid r \subset PSX\}$
    if $SUB \neq \emptyset$
        % Current relaxation is superset of existing
        continue with next p_i
    endif
    SUPER $= \{r \in MinRS \mid PSX \subset r\}$
    if $SUPER \neq \emptyset$
        % Remove supersets
        $MinRS = MinRS \setminus SUPER$
    endif
    % Store the new relaxation
    $MinRS = MinRS \cup \{PSX\}$
endfor
return MinRS

Fig. 4. Algorithm for determining all minimal relaxations

Proof. For each product $p_i \in P$ there exists exactly one subset PSX of atoms of Q which p_i does not fulfill and which have to be definitely relaxed altogether in order to have p_i in the result set. Given n products in P, there exist exactly n such PSXs. Thus, any valid relaxation of Q has to contain all the elements of at least one of these PSXs for obtaining one of the products of P in the result set. Consequently, any relaxation which is not in the set of all PSXs of Q has to be a superset of one of the PSXs and is consequently no longer a minimal relaxation. This finally means that any minimal relaxation must be contained in the PSXs of all products and not more then $|PSX| = n$ such minimal relaxations can exist.

Computing the optimal relaxation. The computation of the optimal relaxation (in terms of relaxation costs) can be done by a simple scan of the *PSXs* of the relaxation problem: Given an arbitrary cost function that takes for instance the cardinality of the relaxation and/or individual costs for the individual atoms into account, we only have to determine the *PSX* that minimizes that function. The only assumption for that is that the costs for a superset of a given *PSX* must not be lower than the costs of that *PSX* itself.[3]

---

[3] Within the ADVISOR SUITE system (see later), the 'costs' for relaxing a certain subquery are defined a-priori by a domain expert. Other forms of acquiring the cost function, e.g., by analyzing the user behavior, are also possible.

Computing all minimal relaxations. The set of all minimal relaxations (comparable to the *Recovery Set* from [10]) can be computed with the help of Algorithm *MinRelax* by removing supersets from the set of all PSXs.

Proposition 2. *Algorithm MinRelax is sound and complete, i.e., it returns exactly all minimal relaxations for a failing query Q.*

Proof. The algorithm iteratively processes the product-specific relaxations PSXs for all products $p_i \in P$. From Lemma 4 we know that all these PSXs are already valid relaxations. Minimality of the relaxations returned by *MinRelax* is guaranteed by the algorithm, because a) supersets of already discovered PSXs are ignored during result construction and b) already discovered PSXs that are supersets of the current PSX are removed from the result set. As such, there cannot exist two relaxations $R1$ and $R2$ in the result set for which $R1$ is a subset of $R2$ or vice versa. In addition, we know from Proposition 1 that all minimal relaxations are contained in the PSXs of the products of P. Since *MinRelax* always processes all of these elements, it is guaranteed that none of the minimal relaxations is missed by the algorithm.

Finding relaxations with at least n products. In practical applications, it is sometimes not desirable just to present one single product to the user but rather have at least a few products in the proposal that could for instance be used for comparison purposes. Note that the relaxations computed with the algorithms described above only guarantee that at least one product will be in the result set. In the following paragraphs we thus show how we can compute such relaxations that have *at least* n products based on our in-memory data structures without the need for further database queries.

We use the example shown in Figure 5 (with seven products and four subqueries) for illustrating a corresponding algorithm for determining such relaxations. In this example, the list of product-specific relaxations is as follows:
$PSXs = \langle \{f4\}, \{f1, f4\}, \{f3\}, \{f2, f3\}, \{f1\}, \{f4\}, \{f1, f3\} \rangle$

ID	p1	p2	p3	p4	p5	p6	p7
f1	1	0	1	1	0	1	0
f2	1	1	1	0	1	1	1
f3	1	1	0	0	1	1	0
f4	0	0	1	1	1	0	1

Fig. 5. Problem setting for n products

Algorithm *NRelax* (Figures 5 and 6) computes relaxations that lead to at least n products based on the list of *PSXs*, meaning that no further database queries are required. Note that *NRelax* starts with the full list of the original *PSXs*, because using only the minimal relaxations would be not sufficient for our purposes and the optimal relaxation for 'at least n' products could be lost.

The algorithm works by incrementally exploring combinations of the individual *PSXs*: The algorithm starts with the initial list of *PSXs* and systematically constructs the possible combinations in order of their cardinality. When two *PSXs* are to be combined, the union of the involved atoms is generated, the number of products for the relaxation is determined, and the corresponding cost function is calculated. When a combination is found that leads to enough products, we remember the cost value and subsequently prune the search space by removing combinations that cannot lead to a better result anymore.

The following example shall illustrate the details of the algorithm. We use a simple cost function, i.e., relaxing the filter condition f_n shall lead to costs of n.

In the first step, we remove the duplicates from the list of *PSXs* and annotate each element with the corresponding costs and the number of products that will result in that relaxation and sort the elements according to the costs. Determining the number of products for a certain relaxation can be done by checking for subset relations in *PSXs*, e.g., $\{f1, f3\}$, will have costs of "4" and will result to 3 products, since $\{f1\}$ and $\{f3\}$ are also elements of the list of *PSXs*. The collapsed list of *PSXs* named *CPSX* for our example therefore is the following, where $\{f3\}(3/1)$ denotes that the relaxation *f3* has costs of 'three' and results in one product.

$$CPSX = \langle\ \{f1\}\ (1/1),\ \{f3\}\ (3/1),\ \{f4\}\ (4/2),\ \{f1,f3\}\ (4/3),\ \{f1,\ f4\}$$
$$(5/4),\ \{f2,\ f3\}\ (5/2)\ \rangle$$

Starting with this initial list, we now compute the combinations of the elements of *CPSX* and use the data structure *RNode* for storing costs and number of products associated with an element in *CPSX*, i.e.,

```
struct RNode:
    atoms: List of atoms of PSX
    cost: costs of node
    nbProducts: number of products
    closed: flag, if node was closed
endstruct
```

and use an 'agenda' (list) of such nodes to remember the combinations that still have to be explored.

Figure 6 illustrates how the combinations are generated and how the search space can be pruned. The different aspects of the algorithm are marked with numbers in the Figure: At (1), a new node $\{f1,f3\}$ is constructed from $\{f1\}$ and $\{f3\}$ respectively. At (2), the node $\{f1,f3\}$ from the current agenda (at the first level in Figure 6) can be closed as it will be further explored in the next round of expansion. In Figure 5 this fact is indicated with an 'x'; At (3), the successor of $\{f1\}$ and $\{f1,f3\}$ would be $\{f1,f3\}$. However, we already found that node in (1) and can ignore it for further exploration; in Figure 5, nodes that are pruned from the search space in that way are marked with a rhombus. At (4) we have found a relaxation that comprises all possible atoms, which means that we do not explore that node any further.

Note that the number of nodes on the top level (and more importantly, overall) is limited to the number of possible relaxations for the given query, i.e. if

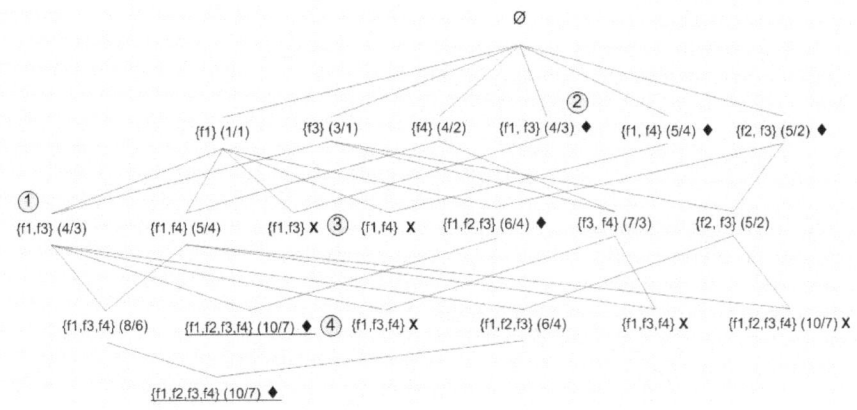

Fig. 6. Searching for at least n products

$|query| = n$, there can only be $2^n - 1$ nodes in the worst case, independent of the number of products in the catalog. Still, if there are already $2^n - 1$ different nodes on the first level, no further expansion will be required since all possible combinations are already contained in this first level.

All the computations can be done on the basis of the pre-evaluated partial results and do not require any further database queries. Also, compared with an approach that works by constructing all 2^n combinations of all possible atoms of the original query, we can also restrict the search space based on the already existing partial results and can leave out those that will definitely lead to more products: In our example, {f2} is not a product-specific relaxation and we therefore will never consider useless combinations like {f1,f2}, {f2,f4} and so forth, since we already now that no additional product will be in the result set when adding {f2} alone. Still, the completeness of the algorithm is still guaranteed by the systematic construction of all possible combinations of the *PSXs*.

Finally, in practical and more complex examples, the described cost-based tree pruning techniques will significantly reduce the number of nodes to be explored, which is not shown in the example i.e., only small fractions of the theoretical search space will be explored. If we, for instance, search for a relaxation with at least 3 products for the given example, we will find {f1,f3} to be the best relaxation in the original agenda and will not have to add a second-level element to the agenda due to cost-optimality of the node.

3   Interactive Relaxation

The alternative approach to immediately computing a relaxation is to let the user decide interactively on which requirements (s)he is willing to compromise. [9] proposes a corresponding algorithm, which is based on the concept of Minimally Failing Subqueries:

ALGORITHM: *NRelax*
In: A set of product-specific relaxations *PSXs*, threshold *n*
Out: An optimal relaxation for *n* products

CPSX = Collapse and sort *PSXs* as sequence of *RNodes*
bestNode = new *RNode* with infinite costs.
return *NRelaxInt(CPSX,n, bestNode)*

---

function NRelaxInt(agenda,n, bestNode)
In: *agenda:* Sequence of RNodes to explore, *n*: threshold,
     bestNode: currently best node
Out: An optimal relaxation for *n* products

---

if $|agenda| = 0$ **return** bestNode
% Check for new optimum
newBest = node from *agenda* with lowest costs for which *node.costs*
     are lower than *bestNode.costs* and nbProducts $> n$
if *newBest* $\neq$ *null* **then** *bestNode* = *newBest* **endif**
% Set up new agenda
newAgenda = $<>$
% Combine elements of given agenda
for i=0 **to** $|agenda| - 1$
     for j=i+1 **to** $|agenda|$
     n1 = agenda[i]
     n2 = agenda[j]
     % ignore closed nodes
     if n1.closed **or** n2.closed **continue** with next *j* **endif**
     newNode = combine atoms, costs, products of *n1* and *n2*
     if not exists $n \in newAgenda$ where *n.atoms* = *newNode.atoms*
      Close nodes *n in* agenda, for which *n.atoms=newNode.atoms*
      if newNode.costs $<$ bestNode.costs **and**
         newNode does not contain all possible atoms
          add *newNode* to *newAgenda* **endif**
     endif
     endfor
endfor
NRelaxInt(newAgenda,n, bestNode)
return *bestNode.atoms*

Fig. 7. Algorithm for finding relaxation with *n* products

Definition 7. *(Minimal Failing Subquery - MFS): A failing subquery Q^* of a given query Q is a minimally failing subquery of Q if no proper subquery of Q^* is a failing query.*

As an alternative to McSherry's approach which relies on the possibly costly computation of *all* MFSs before starting the interactive relaxation process, we propose to apply Junker's recent QUICKXPLAIN [6] algorithm for computing MFSs *on demand*: The overall scenario is that when we have the situation of unfulfillable user requirements, we aim at finding a preferable and minimal conflict in these requirements and let the user decide how to proceed. *Preferred* means that we shall try to identify those conflicts (among possibly many conflicts) that contain requirements for which we assume that a typical user might be willing to compromise. In the digital camera domain we could for instance assume or learn that experts in digital photography searching for cameras supporting 'firewire' connectivity are rather willing to compromise on the price than on the technical requirement.

In general, the required priority values for each requirement can either be annotated in advance or they can be learned from the interaction history of different users over time. The implementation and evaluation of such a module that dynamically adapts priorities over time is part of our ongoing work. Originally, QUICKXPLAIN was developed for finding conflicts (corresponding to MFSs) in Constraint Satisfaction problems, but its general, non-intrusive nature allows us to adapt it for our purposes (Figure 8). QUICKXPLAIN is based on a *divide-and-conquer* strategy: In the decomposition phase it partitions the problem into subproblems of smaller size (thus pruning irrelevant parts of the problem) and subsequently tries to re-add individual elements and checks for consistency while at the same time taking preferences into account. Depending on the number of atoms in the query n, the size of the preferred conflict k, and the splitting point (e.g., $n/2$), QUICKXPLAIN in the best case only needs $log(n/k) + 2k$ consistency checks (database queries in our case) and $2k * log(n/k) + 2k$ in the worst case [6]. When we consider our initial example from Section 2, we see that there are three minimal conflicts in the requirements, i.e., $\{usb = true, firewire = true\}$, $\{firewire = true, price \leq 300\}$, $\{price \leq 300, resolution \geq 5MP\}$. Let us assume that the partial order $\prec$ (in the sense of [6]) among the attributes in the requirements is "*price $\prec$ firewire $\prec$ usb $\prec$ resolution*". Given these priorities, our adapted QUICKXPLAIN (Figure 8) will immediately split the atoms (denoted as P,F,U and R for short) of the query into $\{P,F\}$ and $\{U,R\}$. In the first recursive call, the algorithm will detect that $\{P,F\}$ contains conflicting requirements and thus proceeds by further analyzing this subset alone, which means that half of the atoms in that example do not have be taken into consideration in subsequent steps. Next, QUICKXPLAIN proceeds with the next sets of atoms $\{P\}$ and $\{F\}$ in our example can immediately determine that both $\{P\}$ and $\{F\}$ have to be in the minimal conflict, since the number of the remaining atoms $|A| = 1$ in both cases. Thus, the algorithm returns the preferred conflict $\{P,F\}$, because given the priorities in the example, it is preferable for the user to give up the price or the firewire requirement than to give up the requirement on the desired resolution.

A general algorithm that shows how QUICKXPLAIN can be integrated into an interactive relaxation procedure is sketched in Figure 9. Please note that with the help of conflicts computed with QUICKXPLAIN we can also compute the set of minimal or optimal relaxations (or XSSs) based on Reiter's [11] Hitting-Set Algorithm (see also [5]): This can be seen as an alternative to our approach based on product-specific relaxations from the previous section. The run-time performance of such an adapted Hitting-Set algorithm has been evaluated in [5] for different problem instances: The results showed that even if we do not rely

---

ALGORITHM: *mfsQX*
In: A failing query Q
Out: A preferred conflict of Q

A = *sorted list of atoms of Q*
return $mfsQI(\emptyset, A)$

---

function *mfsQI* (BG, A)
In: BG: List of atoms in background
    A: List of atoms of failing query
Out: A preferred conflict of Q

    % Construct and check the current set of atoms
    $query = \bigwedge_{b \in BG}(b)$
    if *query* is not successful
        return $\emptyset$
    endif
    if $|A| = 1$
        return A
    endif
    % Split remaining atoms into two parts
    $C1 = \{a_i \in A | i < (|A|/2)\}$
    $C2 = A \setminus C1$
    % Evaluate branches
    $\Delta_1 = mfsQI(BG \cup C1, C2)$
    $\Delta_2 = mfsQI(BG \cup \Delta_1, C1)$
    return $\Delta_1 \cup \Delta_2$

Fig. 8. Using QuickXPlain for computing preferred MFS

on the pre-computation of the *PSXs* no more than one second is required for finding the optimal relaxation also for the hard instances.

4   Implementation and Evaluation

All of the described algorithms and techniques have been implemented within the Java-based ADVISOR SUITE system [4], a knowledge-based framework for the rapid development of interactive advisory systems. Within that system, product retrieval is initially[4] based on if-then-style *filtering rules*, like *'If the user is interested in high-connectivity, propose products that support firewire'* or *'Only propose products that are cheaper than the customer-specified limit'*. Note that the consequent of the rules can contain arbitrary boolean formulae, i.e., also disjunctions. The rules are maintained by the domain expert or knowledge engineer with the help of graphical tools and can also be annotated with corresponding relaxation costs and explanatory texts, both for the case that the rule could be applied and for the case of relaxation (see Figure 10).

---

[4] After the initial determination of suitable products, these products are sorted based on a utility-based approach.

ALGORITHM: *InteractiveRelax*

In: Sorted list of atoms A of failing query Q

$query = \bigwedge_{a \in A}(a)$
if *query* is not successful
 % Compute a minimal preferred conflict
 $conflict = mfsQX(\emptyset, A)$
 $remaining = conflict$
 % Set up the choice points
 do $|conflict|$ **times**
  $choice = $ *Ask user to select an option from*
       remaining or 'backtrack'
  if $choice = $ *'backtrack'* **return**
  $remaining = remaining \setminus \{choice\}\}$
  % Remove the choice and try again
  $interactiveRelax(A \setminus \{choice\})$
 end do
else
 Minimize the relaxation and compute results
 Report success and show proposal to user.
 $response = $ Ask user if happy with result
 if $response = $ *'yes'*
  exit function
 % backtrack to last choice point
 else return

Fig. 9. Basic algorithm for interactive relaxation

We tested our algorithms with several real-world knowledge-bases from different domains and with different complexities; an average case would be a setting where we have 10-15 atoms (filter rules) in the query to be relaxed and a few hundred products in the catalog. All of the relaxation problems (see [5] for more details on running times) could be easily solved within the targeted time frame of one second, most of them much faster: Remember that for computing the user-optimal relaxation we only need $|query|$ database queries and such a query typically requires 5-10 msecs. The in-memory search process for the optimum can be done in a few milliseconds. Even more, we can also exploit 'cross-session' re-use and caching of partial query results, e.g., the set of products that fulfil *usb* = *true* remains static as long as no new products appear in the catalog. If no variables are used in the filter rules (like a customer-specified value), relaxations can be computed even without further database accesses.[5]

In our system, the relaxations are also used to 'explain' the proposal, i.e., we use the explanatory text (fragments) for assembling a human-readable explanation and, furthermore, let the user interactively state his actual preferences on the compromises by giving him the possibility to enforce the application/relaxation

---

[5] To the author's knowledge, no 'benchmark' problems are yet available for comparing these running times.

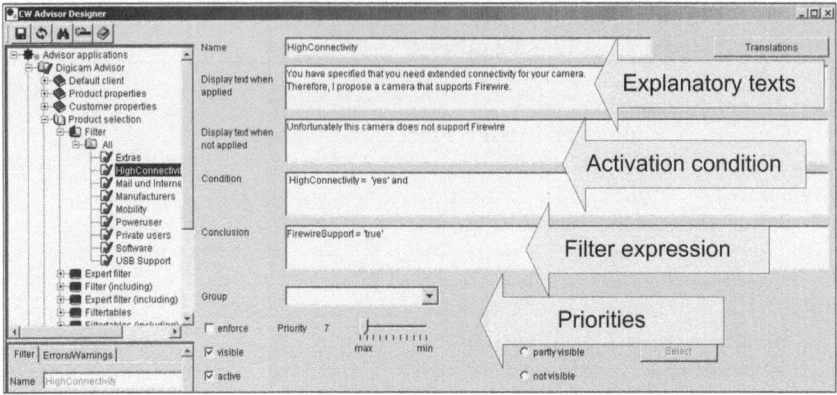

Fig. 10. Graphical editing tool for filter rules

of a rule. An evaluation of a real-world application [4] also indicates that the explanations provided by the system are a well-appreciated feature of advisory applications. Finally, the choice whether an incremental procedure is appropriate has to be decided based on the individual application, e.g., asking additional questions after a longer preference elicitation dialog may for instance be problematic.

5   Conclusion and Future Work

We have presented new techniques and algorithms for fast query relaxation for content-based recommender systems that particularly aim at minimizing the number of required database queries and take a-priory or learned preferences into account. Based on the initial work and formalisms from [9,12] and [3] we have shown how we can a) compute *preferred* conflicts for an interactive relaxation procedure with the help of a recent, general-purpose conflict detection algorithm and b) how the individual evaluation and caching of partial queries allows us to compute optimal relaxations in-memory at the cost of only slightly increased memory requirements. Overall, our approach also continues recent research from the field of Case Based Reasoning (CBR) recommender systems aiming at overcoming the typical shortcomings of such systems [8,13] like for instance the lack of adequate explantation mechanisms.

In our future work, we aim at going one step further than viewing relaxation only as a problem of 'removing' parts of the query: We consider the current approach of fully giving up individual requirements only as a first step in intelligent, content-based recommender systems. In future systems, however, we will therefore aim at building systems that are also capable of coming up with a personalized proposal of how to 'soften' the requirements, e.g., by proposing to increase the price limit by a certain amount.

References

1. S. Amer-Yahia, L. V. S. Lakshmanan and S. Pandit. FleXPath: flexible structure and full-text querying for XML, Proceedings ACM SIGMOD International Conference on Management of Data, Paris, 2004, pp. 83-94.

2. D. Bridge. Product recommendation systems: A new direction. In R. Weber and C. Wangenheim, eds., Workshop Programme at 4th Intl. Conference on Case-Based Reasoning, 2001, pp. 79-86.

3. P. Godfrey. Minimization in Cooperative Response to Failing Database Queries, International Journal of Cooperative Information Systems Vol. 6(2), 1997, pp. 95-149.

4. D. Jannach. ADVISOR SUITE - A knowledge-based sales advisory system. In: Proceedings of ECAI/PAIS 2004, Valencia, pp. 720-724.

5. D. Jannach, J. Liegl. Conflict-directed relaxation of constraints in content-based recommender systems, Proc. 19th International Conference on Industrial, Engineering & Other Applications of Applied Intelligent Systems (IEA/AIE'06), Annecy, France, 2006 (forthcoming).

6. U. Junker. QUICKXPLAIN: Preferred Explanations and Relaxations for Over-Constrained Problems. Proceedings AAAI'2004, San Jose, 2004, pp. 167-172.

7. D. Lee. Query Relaxation for XML Model, Ph.D Dissertation, University of California, Los Angeles, June 2002.

8. D. McSherry. Explanation of Retrieval Mismatches in Recommender System Dialogues, ICCBR Workshop on Mixed-Initiative Case-Based Reasoning, Trondheim, 2003, pp. 191-199.

9. D. McSherry. Incremental Relaxation of Unsuccessful Queries, Proc. of the European Conference on Case-based Reasoning, In: P. Funk and P.A. Gonzalez Calero (Eds.) LNAI 3155, Springer, 2004, pp. 331-345.

10. D. McSherry. Maximally Successful Relaxations of Unsuccessful Queries. Proceedings of the 15th Conference on Artificial Intelligence and Cognitive Science, Castlebar, Ireland, 2004, pp. 127-136.

11. R. Reiter. A theory of diagnosis from first principles, Artificial Intelligence, 32(1), 1987, pp. 57-95.

12. F. Ricci, N. Mirzadeh and M. Bansal. Supporting User Query Relaxation in a Recommender System, Proceedings of the 5th International Conference in E-Commerce and Web-Technologies EC-Web, Zaragoza, Spain, 2004.

13. B. Smyth, L. McGinty, J. Reilly, K. McCarthy, Compound Critiques for Conversational Recommender Systems, IEEE/WIC/ACM International Conference on Web Intelligence(WI'04), Maebashi, China, pp. 145-151.

Solving Proportional Analogies by *E*–Generalization

Stephan Weller and Ute Schmid

Department of Information Systems and Applied Computer Science,
Otto-Friedrich-University, Bamberg
{Stephan.Weller,Ute.Schmid}@wiai.uni-bamberg.de

Abstract. We present an approach for solving proportional analogies of the form $A : B :: C : D$ where a plausible outcome for D is computed. The core of the approach is *E*–Generalization. The generalization method is based on the extraction of the greatest common structure of the terms A, B and C and yields a mapping to compute every possible value for D with respect to some equational theory. This approach to analogical reasoning is formally sound and powerful and at the same time models crucial aspects of human reasoning, that is the guidance of mapping by shared roles and the use of re-representations based on a background theory. The focus of the paper is on the presentation of the approach. It is illustrated by an application for the letter string domain.

1 Introduction

An often quoted observation by the psychologist William James more than a 100 years ago is that "a native talent for perceiving analogy is ... the leading fact in genius of every order" (see [1]). Accordingly, the process of analogy making is studied extensively in cognitive psychology as well as in artificial intelligence [2]. The most fundamental kind of analogies are so called proportional analogies of the form A : B :: C : D. They are studied in verbal settings (*Lungs are to humans as gills are to* [fish]), with geometric figures [3,4], and in the letter string domain [5] (*abc : abd :: kji :* [kjj]).

The core processes of all computational approaches to proportional analogy are (1) construction of a structured representation of the given patterns, usually in the form of terms, (2) identification/calculation of the relation between terms A and B, (3) mapping of terms A and C, and (4) application of the relation found between A and B to term C using substitutions based on the mapping of A and C. Some approaches, namely PAN [4] and Copycat [5], additionally present a mechanism for re-representation of terms, addressing the fact, that the outcome of mapping is dependent on the perceived structure of the given terms. For example, the string *abc* can be perceived as an arbitrary sequence of letters, or as an ascending sequence of three letters.

Our approach differs from the approaches named above in two respects: First, mapping is determined by the common *gestalt*, that is the structural commu- nalities of the base (A) and target (C) terms. Second, arbitrary background

C. Freksa, M. Kohlhase, and K. Schill (Eds.): KI 2006, LNAI 4314, pp. 64–75, 2007.
© Springer-Verlag Berlin Heidelberg 2007

knowledge can be considered when comparing the structures of terms. The first characteristic is covered by the method of syntactic anti-unification. The second characteristic is covered by an anti-unification modulo equational theory or E–Generalization.

In the following section we will introduce syntactic anti-unification and E–Generalization and discuss their computational advantages as well as relations to human analogy making. Afterwards we will introduce the letter string domain as example domain. We present our algorithmic approach and illustrate it for letter string analogies. We will conclude with an evaluation and further work to be done.

2 Syntactic Anti-unification and E–Generalization

In this section we will introduce the notion of E–Generalization, as used in [6] and [7]. To facilitate the understanding of E–Generalization, we will first introduce syntactic anti-unification, which is a proper subset of E–Generalization.

2.1 Syntactic Anti-unification

Unification is a well known and widely used technique, the probably most prominent application being the programming language *Prolog*. It computes the *most general unifier (MGU)* of two or more terms, i.e. the most general term, such that both terms can be reduced to the *MGU* by applying a substitution. Anti-unification is the dual concept to unification. Instead of computing the most general unifier, it computes the most specific generalization. It can be defined as follows:[1]

Definition 1 (Anti-instance). *Let u and $t_i, i = 1, \ldots, n$ be terms and σ_i substitutions for each term, such that $t_i = u\sigma_i \forall i = 1, \ldots, n$. Then u is called an anti-instance of the terms $t_1, \ldots, t_n$.*

u is called the most specific anti-instance of $t_1, \ldots, t_n$, if for each term u' which is an anti-instance of $t_1, \ldots, t_n$ there exists a substitution θ, such that $u = u'\theta$.

In contrast to unification, anti-unification is always possible and there is always a single most specific solution (up to variable renaming).

Algorithms for computing the anti-unification of n terms effectively were introduced in [8] and [9] independently.

Anti-unification can be used to establish a relation between two terms in the following way: If we anti-unify two terms, and those terms share some common structure, the result will be a non-trivial term containing some variables. Let us for example consider figure 1 [2], an example of anti-unifying two terms, $5 \cdot 3 + 7$ and $8 \cdot 3 + 9$. The two terms are anti-unified resulting in $x \cdot 3 + y$. The result of the

---

[1] As commonly used $t\sigma$ denoted the application of the substitution σ to the term t.

[2] Standard arithmetic rules are assumed, i.e. $5 \cdot 3 + 7$ is to be read as $+(\cdot(5, 3), 7)$.

anti-unification conserves as much of the structure of the terms as possible. The generalized structure reflects the roles the objects are playing in the respective expressions. For example, the variable x describes the role that 5 plays in the first term and 8 in the second, namely, that of the first factor in the term. The second term can be obtained from the first one by applying the substitution τ_1 inversely and then applying τ_2.

This is a contrast to a direct mapping approach used in many models for analogies. A direct mapping approach aims at computing the one term directly out of the other, without using intermediate results. This sometimes requires stochastic elements, as used in Copycat ([5]) or the use of heuristics, such as the systematicity principle in SME ([10]). Those may be powerful in solving the analogy, however a stochastical approach is psychologically hardly plausible.

Anti-unification allows for a mapping by using the abstract description of the "roles" some subterms fulfil. In one word, it allows for analogy via abstraction, which has some psychological motivation ([11]) and also yields a formally sound approach.

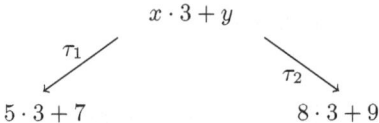

Fig. 1. Simple example for syntactic anti-unification

2.2   *E*–Generalization

Sometimes, for constructing a suitable abstraction, it might be necessary to include knowledge about the domain. E. g., for $t_2' = 9 + (3 \cdot 8)$ and t_1 as above, only the overly general anti-instance $u + v$ can be obtained. But we know, that addition is commutative and therefore, we can rewrite t_2' into its original form t_2. Knowledge about the equality of terms can be represented by an equational theory. The laws for addition constitute such a theory:

$$x + y =_E y + x$$
$$x + (y + z) =_E (x + y) + z.$$

If we use knowledge in form of equational background theories for rewriting terms before syntactical AU is performed, we speak of E–generalization. Models for solving proportional analogies, such as Copycat [5] or PAN [4], allow for an arbitrary sequence of rewritings of the initial representation (re-representation) to find a suitable solution. In contrast, E–generalization allows us to perform abstraction while modeling equivalent representations by appropriate equations between terms. All equivalent representations are considered *simultaneously* in the abstraction process. Therefore, abstraction becomes insensitive to representation changes.

The basic idea is to anti-unify *regular tree grammars* instead of terms. Regular tree grammars are a language class developed in 1968 (cf. [12] and [13]). This language class is located in the Chomsky-Hierarchy between regular and context-free languages (for a very comprehensive introduction to regular tree grammars see [14]).

Regular tree grammars allow for the representation of equivalence classes of terms, it would for example be possible to represent the terms $3 \cdot 8 + 9, 8 \cdot 3 + 9, 9 + 3 \cdot 8$, and $9 + 8 \cdot 3$ by one regular tree grammar, assuming the given background knowledge.

The construction of regular tree grammars from a background theory (for example a canonical equational theory) can in some cases be done automatically (cf. [15] and [16] for criteria when this is possible).

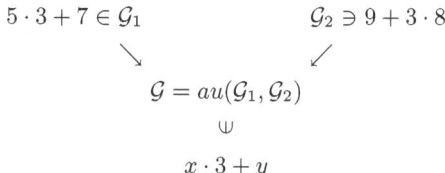

Fig. 2. Simple example for E–Generalization

Assuming regular tree grammars for our terms, we can now anti-unify these regular tree grammars by an algorithm originally developed in [6] and refined in [7]. This process is depicted in figure 2. Unfortunately, this algorithm needs exponential time [3] in general, but [7] shows that in some cases an efficient computation is nevertheless possible.

It should be noted that the result of this E–Generalization process is not a term, but a regular tree grammar of terms. But this is only a natural consequence of representing equivalence classes of terms as regular tree grammars. The result has to be an equivalence class of terms itself. In the next section we will see, that this will make it possible to compute *all* solutions of a proportional analogy in one step.

For an in-depth description of the algorithm mentioned (and also its implementation and application so proportional analogies) see [17].

E–Generalization may be used as a model for analogies even more than syntactic anti-unification. The features described in the last subsection are fulfilled by E–Generalization as well and additionally, we are not limited to one representation. E–Generalization may account for a change rerepresentation of the terms, which is necessary in many cases (see for example [18] for a justification of this claim). We can therefore hope to find a method of solving proportional analogies by E–Generalization.

---

[3] Exponential in the size of the grammars used.

3   Letter String Analogies

We will illustrate our approach for the letter string domain which has been widely investigated in cognitive science as well as in artificial intelligence [19,20,21,5,1]. This domain has several characteristics which makes it interesting: First, it is very simple and any number of analogies can be constructed. Second, for many examples, there are different plausible solutions. Which solution is generated (by a program or a human subject) is dependent on how the perceived structure of the other strings. Third, in principle any type of proportional analogy problem which can be represented in form of terms can be mapped to the letter string domain.

For example, geometrical patterns can be described by a system of terms, thus matching it to the letter-string-domain. This possibility was described already in 1971 in [22] and is known as *Structural Information Theory (SIT)*.

It was first introduced as a coding system for linear one-dimensional patterns. Leeuwenberg represents perceptual structures by three operators named *iteration, alteration,* and *symmetry*. Iteration is supposed to reflect some kind of repetitive process (e.g. $Iter(xy, 3) := xyxyxy$). Symmetry should represent the reversed repetition of a term t after a second term s ($Sym(xyz, ()) := xyzzyx$). Finally, alteration describes the interleaving of a term into a list of terms, such as $Alt(a, (x, y, z)) := axayaz$.

On those operators, Leeuwenberg introduces the notion of *information load,* which is supposed to describe the complexity of an operator. Leeuwenberg claims that the descriptions using the minimal information load correspond to perceptual gestalts (for an introduction to gestalt theory, see [23]). His claim is therefore, that the gestalt principle can be explained by even simpler principles, such as his information load.

A more algebraic version of Structural Information Theory can be found in [21]. Here, even some computational modelling of proportional analogies is done.

4   Solving Proportional Analogies by E–Generalization

4.1   Illustration of the Approach

In the following we will show how to apply the method of E–Generalization to solve a proportional analogy of the form $A : B :: C : D$ (read: A is to B as C to D), where D is to be computed.

As an example, let us assume we want to solve the proportional string analogy $abc : abd :: ihg : D$. where D is unknown. Using a representation language similar to *SIT* ([21]) we could represent the term abc as $Iter(a, succ, 3)$, meaning that abc is established by iterating the successor operation three times on the constant a. Another possible representation would be of the form $a \cdot succ(a) \cdot succ(succ(a))$, where $\cdot$ means concatenation and $succ$ is the successor relation.

Our first aim is now to compute the common structure of the terms A and C, or, in our example the terms abc and ghi. At this point it should be noted, that this common structure could be extracted straightforward by syntactic anti-unification, *if we had knowledge about the structure of the terms* abc *and* ghi. Let us for a moment assume, that we know that the structure of our both terms is $Iter(a, succ, 3)$ and $Iter(i, pred, 3)$ respectively [4]. In this case, an application of syntactic anti-unification would yield the common structure $Iter(x, y, 3)$ and the two substitutions $\tau_1 = \{x \leftarrow a, y \leftarrow succ\}, \tau_2 = \{x \leftarrow i, y \leftarrow pred\}$. Let us further assume the structure $Iter(a, succ, 2) \cdot succ(succ(succ(a)))$ for the B-term abd. Given this, we could apply τ_1 inversely to the B-term, yielding a new term Q of the form $Iter(x, y, 2) \cdot y(y(y(x)))$. Applying τ_2 to this term would yield the result $Iter(i, pred, 2) \cdot pred(pred(pred(i)))$ which describes the term ihf, which is one possible solution of the analogy.

Seeing this, one possibility to solve a proportional string analogy would be to compute some representation of the participating terms and using syntactic anti-unification (and inverse and normal substitution application) to compute a result. The decision on some representation will thus determine, which result we will obtain.

But instead of choosing one particular representation at the start, we can take the process one step further and use E–Generalization instead of syntactic anti-unification. The complete process is shown in figure 3.

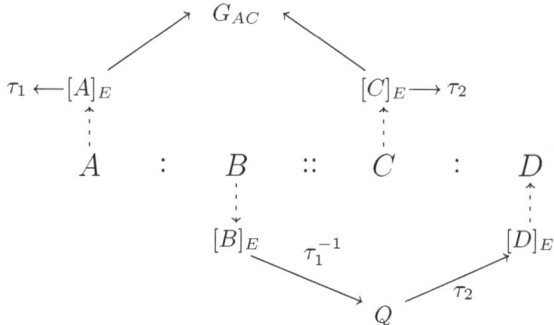

Fig. 3. Solving a proportional string analogy

To every ground term A, B, and C a regular tree grammar representing the equivalence class of all representations of the term is built up. Those regular tree grammars are denoted by $[A]_E, [B]_E$, and $[C]_E$. The process of building the regular tree grammars is denoted by the dotted arrows. Next, the E–Generalization algorithm from [7] is used on the two regular tree grammars $[A]_E$ and $[C]_E$, thus extracing their common structure as a regular tree grammar G_{AC}. This grammar is not needed in the further process, it is a byproduct of the algorithm representing a form of abstraction from the ground terms. What *is* needed in the next step, are the substitutions τ_1 and τ_2 also produced by the E–Generalization step.

---

[4] Actually, the example is not completely formally correct, as it intermixes first and second order terms for *succ* and *pred*, which is of course not valid, but simplifies the example a lot.

The substitution τ_1 is inversely applied to $[B]_E$ [5] , resulting in a regular tree grammar describing the common structure between B and D. It is denoted by Q in the figure. Again, this grammar forms an abstraction from the actual terms that can be seen as a byproduct of the analogical process.

Finally, the substitution τ_2 is applied to Q, leading to a final grammar $[D]_E$ with the following properties:

- It shares the structure of B, as it is derived by application of (inverse) substitutions to $[B]_E$.
- The constants occuring in the term A are replaced by those in C, as the two (inverse) substitutions τ_1^{-1} and τ_2 are applied.
- It can thus be described as the result of "doing the same thing" to B as it was done to A to get C.
- And therefore, it can be seen as a valid solution of the proportional analogy $A : B :: C : D$.

As mentioned before, the result of this process is not a single term, but a regular tree grammar and as such a whole set of terms. However, it describes *every* result that can be obtained by replacing constants in any representation of B by their counterparts with respect to every possible representation of A and C.

Naturally the question arises, how to extract a term from this tree grammar (when a single term is desired rather than a set of terms). This process is an enumeration of the regular tree grammar, which is in general not possible completely, as infinitely many terms are described by the grammar (imagine for example a grammar for arithmetical operations allowing for the addition of 0 - this alone leads to result grammars describing infinitely many terms in nearly every case). Nevertheless, it may be desirable to enumerate the first n terms according to some ordering relation. The enumeration is not a problem, if the ordering relation is defined. But finding a suitable ordering relation is not a simple task. However, using simple relations, like ordering by number of occuring constants, "depth" of the term etc. are possible. Those might even represent some kind of simplicity used by humans to decide which answer to choose in solving a proportional analogy.

Which ordering relation would correspond to the preference humans use is an empirical question and probably requires deep insight in the cognitive processes used in analogical thinking.

4.2   Algorithmic Realization of E-Generalization

The algorithm used to implement E–Generalization was originally developed in [7]. Computing the E–Generalization of two terms is split up in three parts: Computing *universal substitutions*, *lifting* a grammar and finally intersection of regular tree grammars.

---

[5] Note that (inverse) application of substitutions is well-defined on regular tree grammars. It is a special case of an inverse tree homomorphism. See [14] for details.

Regular tree grammars are defined as a quadruple $\mathcal{G} = (\Sigma, \mathcal{N}, S, \mathcal{R})$. Σ is a signature, i.e. a set of function symbols f, where each f has a fixed arity. If the arity of one f is 0, it is called a constant. $\mathcal{N}$ is a finite set of nonterminals, $S \in \mathcal{N}$ is a starting symbol. Finally, $\mathcal{R}$ is a finite set of rules of the following form:

$$\mathcal{N} ::= f_1(N_{11}, \ldots, N_{1n_1}) | \ldots | f_m(N_{m1}, \ldots, N_{mn_m})$$

Let us first assume, that the substitutions are already known. This subproblem is also known as *constrained E–Generalization*. In this case, only two steps are needed: *Lifting* the grammars and intersecting them. Lifting the grammar is a process to incorporate the knowledge about variables that could be used by the substitution. In other words, from a grammar G we want to derive a grammar G^σ, such that $G^\sigma \sigma = G$, where G^σ containts all "suitable" variables from the substitution σ. To this end, we define the following algorithm:

Algorithm 1 (Lifting). *For a regular tree grammar* $\mathcal{G} = (\Sigma, \mathcal{N}, S, \mathcal{R}$ *and a substitution* σ *define a new grammar* $G^\sigma = (\Sigma \cup \text{dom } \sigma, \{N^\sigma | N \in \mathcal{N}\}, S^\sigma, \mathcal{R}^\sigma)$,[6] *where* N^σ *is a new nonterminal, one distinct nonterminal is introduced for each old nonterminal. The same is done to S and the rules $\mathcal{R}^\sigma$ are derived from $\mathcal{R}$ as follows:*

For every rule

$$N ::= \Big|_{i=1}^{m} f_i(N_{i1}, \ldots, N_{in_i})$$

from $\mathcal{R}$ we introduce a new rule

$$N^\sigma ::= \Big|_{i=1}^{m} f_i(N_{i1}, \ldots, N_{in_i}) | \Big|_{x \in \text{ dom } \sigma, x\sigma \in \mathcal{L}_{\mathcal{G}}(N)} x$$

Where $\mathcal{L}_{\mathcal{G}}(N)$ describes all terms in the grammar $\mathcal{G}$, that can be reached when using N as a starting symbol.

To compute the intersection of the regular tree grammars, a standard algorithm from [14] is used. It is similar to intersection algorithms for regular languages.

Lifting two grammars as defined above and intersecting them afterwards will now result in the E–Generalization of those two grammars. However, to allow for *unconstrained E–Generalization*, we still need some means to extract the substitions required to apply constrained E–Generalization.

It is possible, to find two *universal substitutions* τ_1 and τ_2, which are universal in the following sense: For any two substitutions σ_1, σ_2 we can find a substitution σ, such that $t\sigma_i \in \mathcal{L}(N) \implies t\sigma\tau_i \in \mathcal{L}(N)$ for $i = 1, 2, N \in \mathcal{N}$ and any t from the union of both domains.

The existence of such substitutions is proven in [7]. They are constructed in the following two steps:

---

[6] dom σ denotes the domain of σ, i.e. the set of all terms occuring on the left-hand side of a substitution.

Algorithm 2 (Universal substitutions)

1. *Construct* $\mathcal{N}_{\max}$:

   (a) *Set* $N = \emptyset$ *and* $\mathcal{N}_{\max} = \emptyset$
   (b) *For each Nonterminal* $n \in \mathcal{N}$, *compute* $(\bigcap_{x \in N} \mathcal{L}(x)) \cap \mathcal{L}(n)$ *and if the result is not empty, add* n *to* N.
   (c) *Add* N *to* $\mathcal{N}_{\max}$, *remove all elements in* N *from* $\mathcal{N}$, *and if* $\mathcal{N} \neq \emptyset$, *set* $N = \emptyset$ *and continue with step* (b).

2. *For a nonterminal* N, *define* $t(N)$ *as an arbitrary term from* $\mathcal{L}(n)$. *For each pair* $(N_1, N_2) \in \mathcal{N}_{\max} \times \mathcal{N}_{\max}$ *introduce a new variable* $v(N_1, N_2)$ *and define* $\tau_i = \{v(N_1, N_2) \leftarrow t(N_i)\}$ *for* $i = 1, 2$.

Using those substitutions and applying the constrained E–Generalization will yield the unconstrained E–Generalization. This completes the algorithm:

Algorithm 3 (Unconstrained E-Generalization). *Let* $\mathcal{N}_1, \mathcal{N}_2$ *be two regular tree grammar. Their unconstrained E-Generalization is computed with the following steps:*

1. *Compute two universal substitutions* τ_1, τ_2 *for the regular tree grammars* $\mathcal{N}_1$ *and* $\mathcal{N}_2$ *respectively by algorithm 2.*
2. *Compute the "lifted" grammars* $\mathcal{N}_1^{\tau_1}$ *and* $\mathcal{N}_2^{\tau_2}$ *by algorithm 1.*
3. *Compute the intersection* $\mathcal{N} := \mathcal{N}_1^{\tau} \cap \mathcal{N}_2^{\tau}$ *by a standard algorithm, for example from [14].*

4.3   Using E-Generalization to Solve Proportional Analogies

We have now all ingredients to describe our overall approach algorithmically:

Algorithm 4 (Solving proportional analogies). *Let a proportional analogy of the form* $A : B :: C :?$ *be given by regular tree grammars* $[A]_E$ *for* A, $[B]_E$ *for* B *and* $[C]_E$ *for* C. *Compute a solution* D *by the following steps:*

1. *Compute universal substitutions* τ_1, τ_2 *for* $[A]_E$ *and* $[C]_E$ *respectively, using algorithm 2.*
2. *Lift the grammar* $[B]_E$ *with respect to the substitution* τ_1, *using algorithm 1 to get* $Q := [B]_E^{\tau_1}$.
3. *Apply the substitution* τ_2 *to* Q *to get the final result* $[D]_E := Q\tau_1$.

4.4   Implementation

As mentioned above, the E–Generalization algorithm is exponential in the size of the grammars used in the general case. A proof-of-concept implementation of the algorithm was done, which can be used for small examples. Grammars describing terms like abd have typically a size of 30 to 40 rules, one for each letter of the alphabet and several more for the operators (see [17] for example grammars).

The implementation was done in Moscow-ML, an implementation of Standard ML, which is a stricly functional language. This language was selected due to it's support for pattern matching, which enables it to interact with trees and terms very straightforwardly.

Sample grammars were generated for very small mathematical problems and proportional string analogies.

The program was run with prototypical examples, such as $abc : abd :: ghi :?$. Performance time is quite reasonable, typically about 30 sec. The size of the returned grammar was in between 40 and 50 rules. Nearly all of the computing time was spent in the process of generating the universal substitutions. This coincides with the theoretical results ([7]), as this is the step that requires exponential computation effort. More statistics on the program were not done, due to its prototypical nature.

An algorithm to enumerate the resulting grammar is not yet available. As the grammar represents *all* solutions with respect to a given background knowledge, such an enumeration would reflect the *preference* human subjects would have when choosing a solution. The question of enumerating the grammar is therefore more a psychological than an algorithmical one. It is possible to sort the terms in the grammar by complexity (i.e. the depth of the terms). However, wether such an enumeration would correspond to human preference is still an open question.

A more extensive description of the implementation can be found in [17].

For an application to a task which requires more than very small grammars, some restriction to the algorithm is inevitable. Such restrictions would of course depend on the application.

5   Conclusion and Further Work

We have introduced the idea of anti-unification and its extension with background knowledge to E–Generalization. Then we have shown how this method can be applied to solve proportional string analogies in a generic way, that is, without incorporating domain knowledge into the algorithm. We demonstrated the approach for the letter string domain, which can be seen as a representative domain for all other domains accessible to term representation. That is, our approach is applicable to all kinds of proportional analogy problems. The only restriction is given by the fact that the background knowledge has to be represented as a canonical equational theory, which is not always possible (for criteria cf. [6]).

Our approach can be applied to more complex analogical reasoning tasks as well. For example, it can be applied to solve predictive analogies in the domain of naive physics which are addressed in the cognitive model SME [10] as we demonstrated in [11]. Furthermore, an extension to second-order generalization can be applied to the domain of program construction by analogy [24,25].

Solving proportional analogies is only one domain where E–Generalization can be applied. There are for example applications in the field of lemma generation (cf. [6]) or in the completion of number series, as they are used often in intelligence tests. The latter were also worked on by [5] (cf. chapter 2.2), also making use of analogies. The application of E–Generalization to this problem has been done in [15].

An important property of this method is the calculation of the common structure of the terms as a byproduct of the analogy solving. In contrast to most other models of analogies, this method does account for the emergence of abstract knowledge without any extra computation. The creation of the abstract knowledge about the common structure is not gained by an extra step, but rather as an intrinsic property of the process.

At least in this aspect this is similar to the way humans solve analogies. One cannot "suppress" the abstraction from the concrete terms. To learn something about the common structure of the terms is inherent in the process of solving the analogy.

To investigate further in human solving of proportional analogies, empirical research is necessary. Our next step will therefore be to conduct an empirical study. The aim of this study will be to check wether the results chosen by humans correspond to a certain ordering relation of the terms in the computed grammar or wether terms occur not covered by the grammar at all (this can of course not be ruled out, as human decisions might not be explicable by background knowledge but rather based on intuition or other non-rational processes).

Acknowledgements

We would like to thank Jochen Burghardt for his support of our work.

References

1. Mitchell, M.: Analogy-Making as Perception: A Computer Model. MIT Press, Cambridge, MA (1993)
2. French, R.: The computational modeling of analogy-making. Trends in Cognitive Sciences **6** (2002) 200–205
3. Evans, T.G.: A Program for the Solution of a Class of Gemetric-Analogy Intelligence-Test Questions. In Minsky, M., ed.: Semantic Information Processing. MIT Press (1968) 271–353
4. O'Hara, S.: A model of the redescription process in the context of geometric proportional analogy problems. In: Int. Workshop on Analogical and Inductive Inference (AII '92). Volume 642., Springer (1992) 268–293
5. Hofstadter, D., the Fluid Analogies Research Gr.: Fluid Concepts and Creative Analogies. BasicBooks (1995)
6. Heinz, B.: Anti-Unifikation modulo Gleichungstheorie und deren Anwendung zur Lemmagenerierung. Technical report, GMD - Forschungszentrum Informationstechnick GmbH (1996)
7. Burghardt, J.: E-generalization using grammars. Artificial Intelligence Journal **165** (2005) 1–35
8. Plotkin, G.: A note on inductive generalization. In: Machine Intelligence. Volume 5. Edinburgh University Press (1970) 153–163

9. Reynolds, J.: Transformational Systems and the Algebraic Structure of Atomic Formulas. In: Machine Intelligence. Volume 5. Edinburgh University Press (1970)
10. Falkenheimer, B., Forbus, K.D., Gentner, D.: The structure-mapping engine: Algorithm and examples. Artificial Intelligence **41** (1989) 1–63
11. Schmid, U., Gust, H., Kühnberger, K.U., Burghardt, J.: An algebraic framework for solving proportional and predictive analogies. In Schmalhofer, F., Young, R., Katz, G., eds.: Proceedings of the First European Conference on Cognitive Science (EuroCogSci03), Mahwah, NJ, Lawrence Erlbaum (2003) 295–300
12. Brainerd, W.: The minimalization of tree automata. Information and Control **13** (1968) 484–491
13. Thatcher, J., Wright, J.: Generalized finite automata theory with an application to a decision problem of second–order logic. Mathematical Systems Theory **2** (1968)
14. Comon, H., Dauchet, M., Gilleron, R., Jacquemard, F., Lugiez, D., Tison, S., Tommasi, M.: Tree automata techniques and applications. Available on: `http://www.grappa.univ-lille3.fr/tata` (1997) release October, 1rst 2002.
15. v. Thaden, M., Weller, S.: Lösen von Intelligenztestaufgaben mit E-Generalisierung (Solving intelligence tasks by E-Generalization). In: Tagungsband der Informatiktage 2003, Gesellschaft für Informatik e.V. (2003) 84–87
16. Emmelmann, H.: Code Selection by Regularly Controlled Term Rewriting. In: Proc. of Int. Workshop on Code Generation. (1991)
17. Weller, S.: Solving Proportional Analogies by Application of Anti-Unification modulo Equational Theory. Available on `http://www-lehre.inf.uos.de/~stweller/ba/` (2005) Bachelor's Thesis, unpublished.
18. Yan, J., Gentner, D.: A theory of rerepresentation in analogical matching. In: Proc. of the 25th Annual Conference of the Cognitive Science Society, Mahwah, NJ, Erlbaum (2003)
19. Burns, B.: Meta-analogical transfer: Transfer between episodes of analogical reasoning. Journal of Experimental Psychology: Learning, Memory, and Cognition **22** (1996) 1032–1048
20. Cornuejlos, A.: Analogy as minimization of description length. In Nakhaeizadeh, N., Taylor, C., eds.: Machine Learning and Statistics. The Interface. Wiley, New York (1997) 321–335
21. Dastani, M., Indurkhya, B., Scha, R.: An Algebraic Approach to Modeling Analogical Projection in Pattern Perception. In: Proceedings of Mind II. (1997)
22. Leeuwenberg, E.: A perceptual coding language for visual and auditory patterns. American Journal of Psychology **84** (1971) 307–349
23. Goldstein, E.B.: Sensation and Perception. Wadsworth Publishing Co., Belmont, California (1980)
24. Hasker, R.W.: The Replay of Program Derivations. PhD thesis, Univ. of Illinois at Urbana-Champaign (1995)
25. Schmid, U., Sinha, U., Wysotzki, F.: Program reuse and abstraction by antiunification. In: Professionelles Wissensmanagement – Erfahrungen und Visionen, Shaker (2001) 183–185 Long Version: `http://ki.cs.tu-berlin.de/~schmid/pub-ps/pba-wm01-3a.ps`

Building Robots with Analogy-Based Anticipation*

Georgi Petkov, Tchavdar Naydenov, Maurice Grinberg, and Boicho Kokinov

Central and East European Center for Cognitive Science
New Bulgarian University
21 Montevideo Str., Sofia 1618, Bulgaria
gpetkov@cogs.nbu.bg,
tnaydenov@gmail.com,
{mgrinberg,bkokinov}@nbu.bg

Abstract. A new approach to building robots with anticipatory behavior is presented. This approach is based on analogy with a single episode from the past experience of the robot. The AMBR model of analogy-making is used as a basis, but it is extended with new agent-types and new mechanisms that allow anticipation related to analogical transfer. The role of selective attention on retrieval of memory episodes is tested in a series of simulations and demonstrates the context sensitivity of the AMBR model. The results of the simulations clearly demonstrated that endowing robots with analogy-based anticipatory behavior is promising and deserves further investigation.

1 Introduction

Our everyday behavior is based on explicit or implicit employment of predictive models. If we are looking for an object it may simply happen that we see it by chance and go to take it (reactive behavior) or we can imagine where it could possibly be and go to that place to check whether it is there (anticipatory behavior).

IT systems will be closely related to everyday environments in the near future and they will have to perform complicated tasks related to these environments, including searching for lost objects. In order for them to be successful and autonomous, to deal with novel and dynamic environments, to be pro-active and trustworthy in supporting people in their activities, such robots and devices need to have sophisticated cognitive capabilities based on *anticipation*.

Some models, related to this goal, use connectionist networks that generalize from past experience and predict the future on the basis of these generalizations. For example, ALVINN [16] not only reactively responds to the environment but also predicts what will be seen in the next step. The Anticipatory Learning Classifier Systems (ALCs) ([17], [4], [5]) form a class of models that combine reinforcement learning with online generalization and thus propose diverse anticipatory mechanisms. The DYNA-PI systems [18] use reinforcement learning mechanisms to plan on the basis of a model of the world. Planning iteratively generates "chains of predictions" starting from the current state and using the model of the environment.

---

* This work has been supported by the MIND RACES project funded by the 6th FP of the EC (IST Contract 511931).

C. Freksa, M. Kohlhase, and K. Schill (Eds.): KI 2006, LNAI 4314, pp. 76–90, 2007.

Recently these models implement this planner with connectionist networks ([1], [2]). Finally, the architecture of Rodney Brooks [3] proposes a set of layered modules that interact dynamically with each other and with the environment. Thus, "behavior-based robots" based on the concept of the "action circuit" are proposed. However, the architecture is mainly reactive and is not able to anticipate future events.

This paper proposes an alternative approach towards anticipation based on reasoning by analogy. Thus, anticipations are formed not only by capturing the regularities in the world, but by using a single past episode. Sometimes the analogy between the current situation and an episode in memory can be very superficial. For example, the new and the old episodes can share the same set of objects but they can be placed differently, an analogy of this type would be: last time the bone was behind the red block, so the robot may expect the bone to be behind the red object again. However, sometimes the analogy could be much deeper and thus generating non trivial predictions. For example, suppose that in the past episode the bone was behind an object with unique color (the only green object among many red ones). If in the target situation all the objects are having the same color, but there is an object with unique form (the only cube among many cylinders), then the robot can predict by analogy that the bone is now behind this unique object.

Many share our assumption that analogy-making is not only a specific and rare human capability but is fundamental for human cognition [7]. A number of cognitive models exist modeling various parts of the analogy-making process. SMT [6] assumes that only the relations are important and thus the attributes are completely ignored. The analogy is based on mapping identical relations from the target to the base situation. On the contrary, ACME [8], LISA [9], and AMBR [10] allow for mapping of relations with different names. LISA differs from AMBR in the sense that the former is not capable of mapping too complex structures. ACME constructs all possible correspondences and thus the model can not be scaled up. None of the models presented so far has been used for prediction and anticipation. We decided to extend the AMBR model in order to build anticipatory capabilities in robots based on the DUAL cognitive architecture ([11] [15]).

In this paper we present a first exploratory step in this direction based on a simulation of a robot in the Webots environment. We use AMBR as a reasoning core of the system and connect it to the simulated robot and physical environment.

As it became clear from the above discussion, our long-term project is to design robots that are able to demonstrate anticipatory behavior based on analogy. However, in order to explore in detail how analogy can be used in anticipation we started with a relatively simple example based on a simulated house-like environment (see **Fig.** 1a) and a simulated AIBO robot. The simplicity of the task makes it possible to investigate the role of the various mechanisms involved in analogy as assumed by the AMBR model.

The house-like environment consists of several rooms and doors between some of them. In the rooms there are various objects like cubes, balls, cylinders, etc. The goal of the AIBO robot is to find a bone (or bones) hidden behind an object (Figure 1a). In a more complicated task there could be many robots: some of the robots should find and collect some 'treasures', whereas other robots play the role of 'guards' that try to keep the treasures and hide them dynamically or block the way of the treasure-hunters. (**Fig.** 1b).

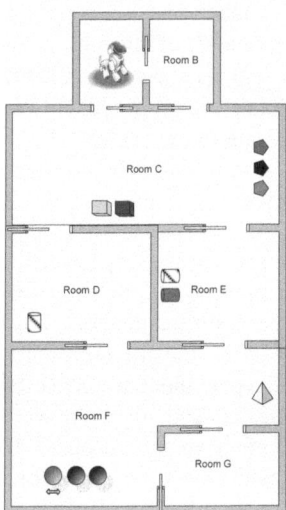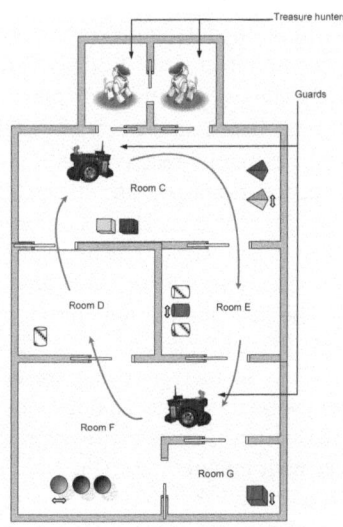

Fig. 1. The micro-domain of a house-like environment. a) The goal of the robot is to find the bone. b) There are 'treasure-hunters' and 'guards' with different goals and strategies.

The strategy of exhaustive search would be very inefficient in real time and sometimes simply impossible especially when the environment is dynamic and changes over time as in the case of Figure 1b. Moreover, we believe that in structured environments like the one shown in Figure 1, anticipation based on analogy will be the most efficient approach, because it will take full advantage of the structure of the environment memorized in previous episodes. A simulated robot and its simulated environment – a room in which there are three cubes and possibly a bone hidden behind one of them – is shown in Figure 2.

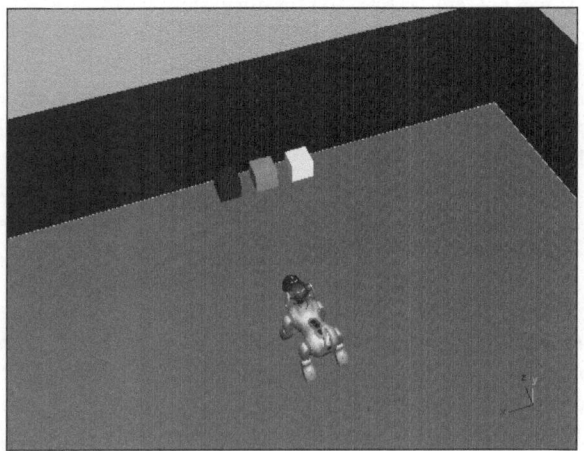

Fig. 2. Robot-environment interaction (real or simulated)

2 The DUAL Architecture

2.1 Basic Properties

DUAL is a cognitive architecture, launched by Kokinov in 1994 ([11], [15]). It consists of memory structures and processing mechanisms, organized around the following principles:

- *Hybridity* – DUAL combines the symbolic with connectionist approaches, by integrating them at the micro-level.
- *Emergent computations* – the global behavior of the architecture emerges from local interactions among a huge number of small entities, called DUAL-agents. There is no central executor that monitors the whole system.
- *Dynamics and context–sensitivity* – The behavior of DUAL changes continuously in response to the influence of the dynamic changes in context. There is no clear-cut boundary between the task and its context. Instead, the context is assumed to be the state of the system in any certain moment, i.e. the pattern of activation over the set of DUAL-agents. This pattern is assumed to reflect the relevance of the various pieces of knowledge in the current context. Some DUAL-agents might be relevant because the corresponding elements are currently perceived from the surrounding environment, others – because they reflect the current goals of the system, and finally, some agents might be relevant because they were recently used and thus they have some residual activation.

2.2 DUAL–Agents

The basic structural and functional element in DUAL is the DUAL–agent. It is hybrid in two ways – it has both connectionists and symbolic aspects, and it serves both as a representational and a functional unit.

Connectionist's Aspect. From the connectionist's perspective, each agent is a node in a localist neural network. It continuously receives activation, updates its current activation level, and spreads it through associative links to other agents. An important feature of DUAL is that it distinguishes the semantic meaning of the agents from their relevance, considering them as independent. The *activation level* is a numeric value that codes the *relevance* of each agent. The pattern of activation does not represent any concept or scheme, but just the current context.

Symbolic Aspect. From the symbolic point of view, DUAL–agents are organized in a semantic network. Each agent 'stands' for something – an object, property, relation, etc. It has its own *micro-frame*. The micro-frames have *slots*, which in turn may have *facets*. There are two kinds of slots – general ones (*G-slots*), and frame-specific ones (*S-slots*).

G-slots have labels, the meaning of which is invariant across the agents, for example *:type*, *:subc*, etc. S-slots also have labels, but their labels are arbitrary, i.e., *:slot1* in one agent may mean something very different from *:slot1* in another agent. S-slots (and only S-slots) have facets, i.e., slots within slots. The same set of labels applies to both G-slots and facets.

Symbolic Processing. DUAL – agents interact with each other. These interactions are relatively simple – they always involve two agents – one of them *sends* some information, and the other one *reads* it.

Each DUAL – agent has a *symbolic processor*. It can receive or construct symbolic structures, transform them, store them in its own *local memory*, and send them to neighboring agents. A typical symbolical transaction involves receiving a symbolic structure, comparing it with the other symbolic structures in the local memory, storing it, transforming it via specific to the agent's type routines, and sending its modification. Each one of these steps is discrete. DUAL – agents manipulate symbols sequentially, one after another, with a frequency that reflects the relevance of the respective agents.

Relationships between Connectionist and Symbolic Processing. All aspects of the agents are merged in a single whole and each one influences the others.

Only a small number of agents whose activation exceeds a certain threshold form the *Working Memory* (WM). The agents that are outside of the WM cannot perform any symbolic operations – they are assumed invisible. For the agents that are involved in the WM, the activation level determines the *speed* of the symbolic processing. Each elementary symbolic operation (namely *read, send, modify*, etc.) has a 'price' that is paid with activation. Whenever an agent wants to perform such an operation, it begins to accumulate activation in order to pay the required price. Only after it is ready, it can perform the operation. Therefore, the most active agents work rapidly, the less active ones – slowly, and the inactive ones do not work at all. In this manner, the relevance influences the symbolic part of the architecture.

There is also an opposite dependence. The symbolic operations cause new agents to be born, and new connections to be established. These operations change the overall pattern of activation and thus, the symbolic operations influence the pattern of relevance too.

Types of agents. Each DUAL-based model consists of nothing but of DUAL-agents with various types and with various properties. For the purpose of the AMBR-based robots with anticipatory capabilities are used the following types of agents:

Concept–Agents (for short *concepts*) represent classes of homogeneous entities and are organized in a semantic network. They are interconnected via vertical links, for pointing respectively to their super-classes and some of their sub-classes. They are interconnected also horizontally, pointing to some associations and prototypical relations. Note that concepts can stand for objects as well as for relations and abstract terms.

Instance-Agents (*instances*) represent individual entities. Each instance has a G-slot that points to its respective category concept. There are also links in the opposite direction that connect the concepts to some of their instances. Some instances are permanent – they are part of LTM and represent concrete memory traces. Other instances are temporary – they are constructed on the spot because of certain inferences.

Hypothesis-Agents represent possible correspondences between entities. They are temporary agents; they do not participate in LTM; and if they lose their relevance, they disappear. Hypotheses are organized in a constraint satisfaction network. Each hypothesis has its 'life cycle' – it can transform itself from *embryo* to *mature hypothesis*, and then to *winner*. These sub-types reflect the degree to which the hypotheses are novel and attractive.

Anticipation-Agents represent possible entities and relations, predicted by the system. They are also temporary agents. The predictions can be confirmed or rejected by the perceptual systems. In the first case the anticipation agents would mutate into permanent ones, whereas in the second case they would disappear.

Cause-agents represent a special kind of relations. They are equipped with a special procedure that allows them to make decisions whether particular movements to be performed. For example, if the system anticipates that the bone is behind the blue cylinder, then the knowledge transferred from the past situation that movement to the blue cylinder would cause finding the bone would be represented by a cause relation. In turn, the concrete action (in the example – movement to the blue cylinder) would be actually performed.

Action-Agents code procedural knowledge about particular movements. Note that if an action-agent becomes relevant this does not imply that the particular movement would be necessarily performed. The decision should be taken by a cause-agent.

2.3 The Coalitions of Agents

DUAL – agents are very simple and some of the more important properties of the architecture could be observed if looked at from a distant perspective.

DUAL – agents form *coalitions*, i.e., sets of agents, together with the pattern of interactions among them. Coalitions represent more complex entities like propositions or situations. However, the coalitions are not part of the strict computational description of the architecture. Instead, they enhance the conceptual understanding only.

Three important properties of the architecture become visible only at the level of coalitions. The coalitions are *decentralized*, *emergent*, and *dynamic*. None of these properties is presented in the individual agents. The coalitions vary in the intensity of the interactions among their members in comparison to the intensity of interactions with the outside agents. 'Tight' and 'loose' coalitions could be distinguished in respect to this ratio and there is a whole continuum between these two extremes.

Coalitions do not have clear-cut boundaries. Instead, the same agent can participate in two or more coalitions and to a different extent. In the course of time, the coalitions can become 'tighter' or 'looser', can break up, and new coalitions can emerge.

3 The AMBR Model

3.1 Main Ideas

AMBR is a model for analogy making based on DUAL. It treats analogy making as an emergent result of the common work of several overlapping sub-processes – perception, retrieval, mapping, transfer, evaluation, and learning. However, AMBR is a long-term project and unfortunately, at the current stage only the processes of retrieval, mapping, and transfer are modeled and integrated.

AMBR is capable of capturing some similarities between local structures of agents and mapping them, creating *hypotheses for correspondences*. It is a pressure for these mappings to grow, involving other agents. In this way, a Constraint Satisfaction Network is formed. Just as in the process of crystallization, the system strives to a stable equilibrium, changing quantitatively itself. Because the structure–based mappings emerge locally and grow, often some inconsistent hypotheses meet and compete with each other and sometimes blending between episodes occurs.

3.2 Mechanisms Used in AMBR

Spreading Activation. The sources of activation are two special nodes – INPUT and GOAL. Their activation level is always equal to the maximal value. The node INPUT represents the perception of the system, whereas the node GOAL – the current tasks of the system. More sophisticated perceptual and goal-analyzing modules are under construction now. They should replace the INPUT and GOAL nodes. AMBR's work begins when some agents that represent the environment, are attached to INPUT, and some other agents, which represent the task – to GOAL. Various context or priming objects can be attached or removed from INPUT at any moment in time. The activation then spreads through the Long-Term Memory (LTM) network.

The spreading activation mechanism defines the working memory as the set of all agents, which activation level exceeds a certain threshold; determines the speed of the symbolic processes performed by each individual agent; and underlies the relaxation of the constraint satisfaction network.

Marker Passing. Generally speaking, the marker-passing mechanism serves to find out whether a path between two agents is present or not. All symbolic interactions between agents, i.e. exchanging messages, are local and involve only two neighbors. The marker passing mechanism is not an exception, but it allows information about agents to be carried through longer distance as an emergent result.

AMBR marks the instance-agents, which enter in the WM; they in turn mark their respective concepts; then the markers spread to their neighbors that are up in the class hierarchy, and so on.

The main purpose of the mechanism is to justify some semantic similarities between two agents. Whenever two markers cross somewhere, AMBR creates a hypothesis about a correspondence between the two marker origins. The justification for this hypothesis is the fact that these two origins have a common super-class, i.e., they are similar in something. Note, that AMBR makes such inferences only if the whole paths of the markers involve only relevant agents.

Structural Correspondences. Like marker passing, this mechanism creates hypotheses between agents. The difference is that the former is sensitive to semantic justifications, whereas the latter – to propositional ones. There are different kinds of structural correspondences – if two relations are mapped, their arguments should also be mapped; if two instances are mapped, their respective concepts and situations should be mapped.

Because several mechanisms create hypotheses independently, it is possible for some of them to be duplicated, or some of them to be contradictory. A special procedure, attached to each agent monitories the hypotheses in which the respective agent is involved and establishes inhibitory or excitatory links between them.

The Constraint Satisfaction Network. The Constraint Satisfaction Network (CSN) consists of hypotheses for correspondences and is interconnected with the main one. Each hypothesis receives activation from its arguments and from its justifications. It is also inhibited from its competitors (responding to the pressure for one-to-one mapping). Thus, CSN simultaneously reflects the semantic, pragmatic, and structural pressures of the analogy–making task. Due to the CSN, the global behavior of the system emerges from the local interactions between agents. However, it is important to note that no time is spent waiting for the CSN to settle in order to read out the 'solution' from the activation pattern. This allows cognition to be viewed as a continuous process, without breaks between the given tasks.

Rating and Promotion. Because at some moment the system should finish its work, each agent rates its competing hypotheses at regular time intervals. If one of them holds for a long time as a leader, it is promoted to a winner.

Some of the instance-agents (the elements of the target situation) are authorized to use the rating mechanism. The purpose of this mechanism is to monitor all hypotheses that involve the agent and to send *promotion incentives* to those that emerge as stable and unambiguous leaders.

Each authorized instance keeps a data structure, called *rating table*. The *individual ratings* for each registered mature hypothesis are stored in this table. Individual ratings are just a numbers that characterize the relative success of the respective hypothesis – how long, how recently, and how strongly has it been a leader, according to its rivals. The instance-agents periodically (in a fixed time interval and whenever a hypothesis registration request come) adjust the individual ratings. The rating of the current leader increases, whereas those of the other hypotheses decrease. The amount of the change is proportional to the difference between the activation levels of the leader and its closest competitors.

When the individual rating of some hypothesis exceeds a certain threshold, the respective instance-agent sends to it a symbolic structure, called *promotion incentive*. In addition, it eliminates all looser hypotheses. When a hypothesis agent receives such message, it transforms itself to winner hypothesis.

Note, that there is not any central executor that monitors the CSN and decides whether the network is relaxed enough. Instead, some hypotheses become winners locally and in asynchrony. This allows blending between episodes to occur, or unique solutions to be found (of course, sometimes useless solutions are also proposed by the system).

Instantiation. There are two mechanisms for instantiation - skolemization and transfer. The former augments the descriptions of the retrieved episodes based on semantic and structural information. The latter adds new information to the target situation. These mechanisms ensure tolerance towards the lack of information and make the reformulation of the task possible. However, in the current paper we describe a novel instantiation mechanism for creating anticipation-agents.

Imagine the following situation: Let the green cube in the target situation be mapped onto the red cylinder from an old situation in memory. Let the system recalls that the bone was behind the red cylinder. This is a reason to anticipate that now the bone should be behind the green cube. A new instance of the relation 'behind' should be created and should be connected to its arguments – 'bone' and 'green cube'. This will be an example of anticipation-agent.

The exact algorithm for creating anticipation agents is the following: each agent that wins in the mapping competition, checks whether it is an argument of a relation. If yes, the agent informs the relation it is part of. Thus, each relation monitors whether all of its arguments are mapped with winners. If this is the case, the relation decides whether to instantiate a self-copy procedure. If it lacks any promising hypotheses, it starts the process of self-copy. If there are promising hypotheses then the mapping is correct and these hypotheses may be promoted at a later moment, thus there is no need to create a new instance.

The just born agent starts its life cycle as anticipation agent. Some of the anticipation agents serve to direct the attention. For example, the relation 'same color', connecting two red cylinders from a base situation can instantiate itself. Thus, a new anticipation-agent would connect the respective correspondences of the two cylinders. It is now the responsibility of the perceptual system to check whether the anticipation is correct or not.

An interesting result emerges, however, when the perceptual system cannot check directly the new prediction, for example, when the relation 'behind' is transferred. In such a case, multiple new anticipations emerge. In particular, the causal relation from the base that represents the knowledge "The bone was behind the red cylinder and that caused AIBO to move to it and it successfully found the bone" would instantiate itself using the same mechanism. Analogously, an action-agent "move to the green cube" would also be created.

Finally, the new casual relation would recognize that this certain movement would cause reaching the goal and would activate the corresponding procedural knowledge for actual movement.

3.3 Anticipation by Analogy

We have designed new mechanisms for creation of anticipations. AIBO transfers some relations and objects from the past episodes to the current situation. This transferred knowledge is assumed to be an expectation. Again, following the main principles of the DUAL architectures, all instantiation operations are performed locally and only looking at the system "from above", the whole pattern of new anticipation-agents could be viewed as a certain anticipated state of the world.

Note, however, that all mechanisms in AMBR work in parallel and thus, the instantiation mechanisms influence other cognitive processes. In particular, we assume that the anticipation-agents play an important role for controlling the attention and in turn the processes of retrieval and mapping.

4 Simulation Results

Suppose the AIBO robot faces several objects in a room. It must predict where (behind which object) the bone is hidden and then to go to the chosen object. Here, we present the simulated version of the real world scenario.

4.1 Mapping Between Close Situations

In the first simulation the mechanisms for transfer and skolemization were switched off in order to check the mapping process. The AIBO robot had four past episodes encoded in its memory (**Fig. 3**).

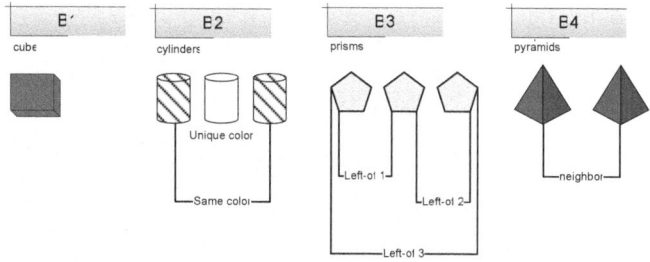

Fig. 3. Old episodes in the memory of the robot, used in the first simulation

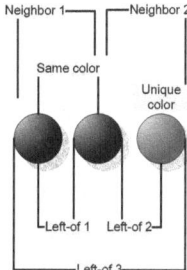

Fig. 4. The target situation, used in the first simulation

The specific number, color and shape of the objects varied across the episodes. The relations between the objects also varied. In all six runs of the simulation one and the same target situation (see **Fig. 4**) was given to the model but different aspects of it were attached to the INPUT node, thus simulating different attention biases. The representation of the target included DUAL-agents representing the objects (OBJECT1, OBJECT2, OBJECT3), as well as some of their properties (BLACK1, BLACK2, GREEN1, BALL1, BALL2, BALL3). Other DUAL-agents represented object properties (e.g. COLOR-OF1, SHAPE-OF1, etc.). Finally, there were relations that involved one or more objects, e.g. LEFT_OF, SAME-COLOR1, NEIGHBOR1, etc. (see **Fig. 4**).

The task of the robot, in this simulation, was to establish a mapping between the target and certain base situation. The target situation (more precisely, some aspects of the target situation) was used for recalling old episodes from memory. These old episodes were gradually and partially retrieved; gradual mappings between various target and base elements emerged; the mapping in turn was influencing the retrieval process. Thus, implicitly the memorized bases competed with each other for the best mapping with the target. It is important to note, however, that all these mappings emerged from local interactions only, without any central mechanism that would calculate the best match among all episodes available in the long term memory.

The results of the simulation are systematized in Table 1.

Table 1. The results from the first situation

Run	Relations at the INPUT	Winner base situation
1	The three balls	B2
2	The three colors	B4
3	The three left-of relations	B3
4	The two neighbor relations and the three colors	B4
5	Only the concept 'green'	B2
6	The relations 'same-color' and 'unique-color'	B2

As shown in Table 1, by varying only the attention bias, the model retrieved different past situations for further analogy making. Note, that this result was achieved by AMBR, testing it in extreme conditions: all old episodes are very similar. They all involve simple geometrical objects and relations that are relatively close semanticaly. Typically, analogy-making models (SME, ACME, LISA, as well as AMBR in previous simulations) are used and tested with dissimilar episodes in memory.

4.2 Single Run of the Model

After all additional mechanisms for instantiation and moving were added, four new bases were designed, including a bone hidden behind a specific object in each of them (see **Fig. 5**). The target consisted only of three red cubes and a specific task to find the bone was also represented and attached to the GOAL list. Again certain properties of the cubes, namely their colors, were highlighted by attaching them to the INPUT node. The simulated AIBO robot was supposed to use the target to retrieve an old episode and to establish mappings between the elements of the two situations. In addition, the position of the bone in the base situation had to be transferred in order to predict the place of the hidden bone in the target situation. Finally, a motor command had to be sent in order to execute the movement. Again, all these operations emerged only from the local interactions of a large number of microagents. Each DUAL-agent just made its specific job with a speed, proportional to its relevance to the task, and the observed behavior was a result of all these small micro-mechanisms simultaneously at play.

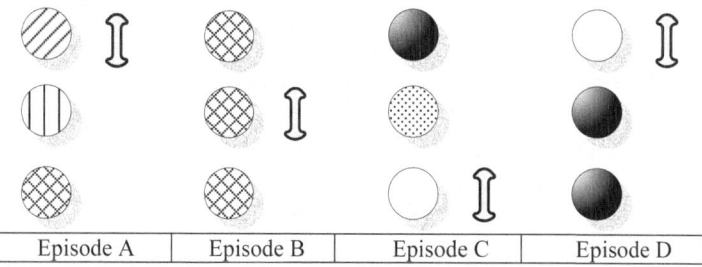

| Episode A | Episode B | Episode C | Episode D |

Fig. 5. Old episodes in the memory of the robot used in the second and third simulations

In Figure 6, part of the log from the run of the model is presented:

```
T=52.10, #<PR SAME-COLOR-SIT-003> received in SAME-COLOR-SIT-
003<-->SAME-COLOR-THREEGREEN.
T=56.20, #<PR SIT-SIT-003> received in SIT-SIT-003<-->SIT-
THREEGREEN.
T=56.90, #<PR FIND-SIT-003> received in FIND-SIT-003<-->FIND-
THREEGREEN.
T=61.00, #<PR AIBO-I-SIT-003> received in AIBO-I-SIT-003<--
>AIBO-I-THREEGREEN.
T=69.20, #<PR INITST-SIT-003> received in INITST-SIT-003<--
>INITST-THREEGREEN.
T=69.30, #<PR BONE-SIT-003> received in BONE-SIT-003<-->BONE-
THREEGREEN.
T=119.00, #<PR IN-FRONT-OF-SIT-003> received in IN-FRONT-OF-
SIT-003<-->IN-FRONT-OF-THREEGREEN.
T=290.20, #<PR MIDLE-CUBE-SIT-003> received in MIDLE-CUBE-
SIT-003<-->MIDLE-CYLINDER-THREEGREEN.
Time: 294.600: Just created agent: ANTICIP-MOVE-THREEGREEN-
Time: 295.400: Just created agent: ANTICIP-BEHIND-THREEGREEN-
Time: 303.400: Just created agent: ANTICIP-CAUSE-THREEGREEN-
THE ACTION ANTICIP-MOVE-THREEGREEN- IS EXECUTING!!
T=314.60, #<PR RED-2-SIT-003> received in RED-2-SIT-003<--
>GREEN-2-THREEGREEN.
```

Fig. 6. Part of the script of a single run of the system

The first several lines are reports for establishing winner hypotheses between certain target elements and elements from the four bases. The first winner connected two 'same-color' relations – one from the target and one from episode B, named 'ThreeGreen' (see **Fig. 5**). The second hypothesis was established between the two situation-agents and thus resulted in additional massive retrieval of all of the elements of situation 'ThreeGreen'. Soon after, many relations and objects found their corresponding elements (FIND, AIBO, INITIAL-STATE, BONE, IN-FRONT-OF, MIDLE-CUBE). Note that the bone in the past episode was hidden behind the middle cylinder, which has just found its corresponding element. The system did not wait a full mapping to be established. Instead, the instantiation mechanisms immediately begun their work and soon some anticipation agents were created. Thus, gradually, in parallel with the processes of retrieval and mapping, the anticipated state emerged in

AIBO's memory system. Namely, it anticipated that the bone was hidden behind the middle cube (because the current situation was analogical to the retrieved episode of three balls with same color and a bone behind the middle one). Moreover, AIBO inferred that if it would move to the middle cube, it would achieve its goal. As a consequence, the respective action was executed. Note, that the execution of the action did not stop the process. The establishment of additional mappings could continue even after achieving the goal.

4.3   Statistical Results from Many Runs of the System

At the end, a set of eight target situation was designed, varying the shape and the color of the objects involved (Figure 7).

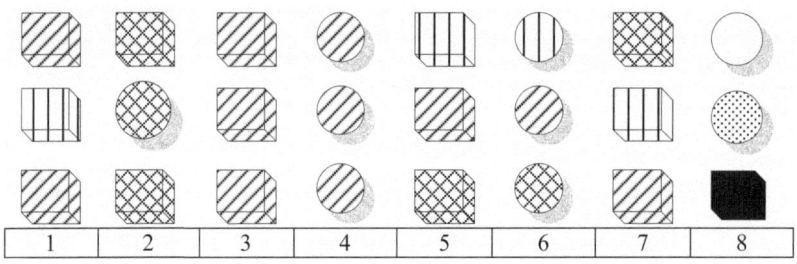

Fig. 7. The eighth target situations used in the third simulation

In order to achieve statistical results, a special tool for creating variants of the knowledge base was used. More precisely, 50 combinations of top-down links from the concepts to the particular instances were randomly created, thus simulating 50 different variants of the core memory. Each of the eight targets was run against each of the knowledge-base variants in two different conditions – focussing the attention of the system on the colors or on the shapes of the objects, respectively. Thus, two distributions of the preferred past episode were achieved (see **Fig. 8**).

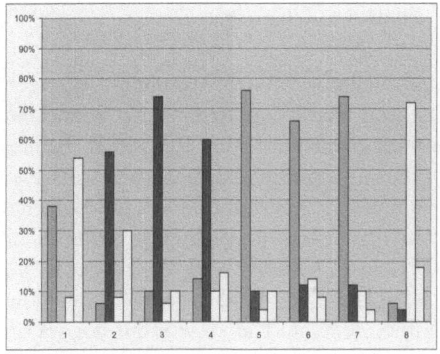 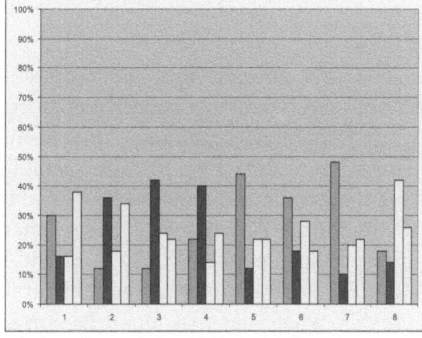

Fig. 8. Comparing the statistical data when the attention was focused on color (left panel) or shape (right panel)

It is shown that when the attention is focused on the shapes only, the base was chosen almost randomly. This is because in all base situations the shapes were equal and thus, the shape was actually irrelevant to the task. AMBR successfully responded to this irrelevance, assuming all bases as almost equally good candidates for analogy.

However, this was not the case, if the color was in the focus of attention.

Clear tendencies to prefer certain bases depending on the target can be observed. At the same time the model was flexible enough, allowing multiple solutions of the problem to be found depending on context (top-down links). These results reconfirm the context-sensitive behavior of AMBR observed previously in other domains [12].

Consider for example the retrieved bases for the second target problem (two green cubes and a green ball in between). The preferred base was B (three green balls) mainly because of the identical colors of all three objects and the fact that attention is concentrated on color. However, quite often an interesting analogy between the second target and the episode D (one white ball and two black balls) was produced. In this case the system detected that the relation `same-form` in the target could be mapped onto the relation `same-color` in the base. The system mapped these two relations and this in turn helped for the retrieval of episode D.

5 Conclusion

A new approach towards building anticipatory agents has been presented, namely building predictions using analogy with a single past episode.

The AMBR model was extended with new agent-types and new mechanisms for building anticipatory-agents based on the existing mechanisms of skolemisation and transfer. These new mechanism were successfully tested in simulation experiments. Additionally, the role of the attention bias for the choice of a base for analogy was explored. Thus, new pressures, which a future fully fledged perceptual system should account for, were defined.

These simulations showed that the newly proposed mechanism can be seriously considered as an important part of any advanced cognitive system with anticipatory capabilities.

The presented work is just a small step in a long-term project. Designing perceptual capabilities would allow for automatic encoding of new situations and thus, the behavior of real robots instead of simulated ones could be tested. Bottom-up and top-down mechanisms for locating the attention should be carefully modeled and their influence on the robot's behavior further investigated.

References

1. Baldassarre G. (2002). Planning with Neural Networks and Reinforcement Learning. PhD Thesis. Colchester - UK: Computer Science Department, University of Essex.
2. Baldassarre G. (2003). Forward and Bidirectional Planning Based on Reinforcement Learning and Neural Networks in a Simulated Robot. In Butz M., Sigaud O., Gérard P. (Eds.), Adaptive Behaviour in Anticipatory Learning Systems, Springer Verlag, Berlin, Heidelberg, pp. 179-200.

3. Brooks, R.A., (1991). How to build complete creatures rather than isolated cognitive simulators, in K. VanLehn (ed.), Architectures for Intelligence, pp. 225-239, Lawrence Erlbaum Assosiates, Hillsdale, NJ.
4. Butz, M. V., Goldberg, D. E., & Stolzmann, W. (2002). The anticipatory classifier system and genetic generalization. Natural Computing, 1, pp. 427-467.
5. Butz, M.V., & Goldberg, D.E. (2003). Generalized state values in an anticipatory learning classifier system. In Butz M., Sigaud O., Gérard P. (Eds.), Anticipatory behavior in adaptive learning systems, Springer Verlag, Berlin, Heidelberg, Germany, pp. 282-301.
6. Gentner, D. (1983). Structure-mapping: A theoretical framework for analogy. Cognitive Science, 7, 155-170.
7. Hofstadter, D. R. (1995). Fluid Concepts and Creative Analogies: Computer Models of the Fundamental Mechanisms of Thought, NY: Basic Books.
8. Holyoak K. & Thagard P. (1989). Analogical mapping by constraint satisfaction. CognitiveScience, 13, 295-355.
9. Hummel, J. & Holyoak, K. (1997). Distributed representation of structure: A theory of analogical access and mapping. Psychological Review, 104, 427-466.
10. Kokinov, B. (1994a). A hybrid model of reasoning by analogy. In K. Holyoak & J. Barnden (Eds.), Advances in connectionist and neural computation theory: Vol. 2. Analogical connections (pp. 247-318). Norwood, NJ: Ablex
11. Kokinov, B. (1994b). The DUAL cognitive architecture: A hybrid multi-agent approach. Proceedings of the Eleventh European Conference of Artificial Intelligence (ECAI-94). London: John Wiley & Sons, Ltd.
12. Kokinov, B., Grinberg, M. (2001). Simulating Context Effects in Problem Solving with AMBR. In: Akman, V., Thomason, R., Bouquet, P. (eds.) Modeling and Using Context. Lecture Notes in Computer Science (Lecture Notes in Artificial Intelligence), vol. 1775, Springer Verlag.
13. Kokinov, B. Petrov, A. (2000). Dynamic Extension of Episode Representation in Analogy-Making in AMBR. In: Proceedings of the 22nd Annual Conference of the Cognitive Science Society. Erlbaum, Hillsdale, NJ.
14. Kokinov, B., Petrov, A. (2001). Integration of Memory and Reasoning in Analogy-Making: The AMBR Model. In: Gentner, D., Holyoak, K., Kokinov, B. (eds.). The Analogical Mind: Perspectives from Cognitive Science, Cambridge, MA: MIT Press
15. Petrov, A. & Kokinov, B. (1999). Processing symbols at variable speed in DUAL: Connectionist activation as power supply. In Proceedings of the Sixteenth International Joint Conference on Artificial Intelligence (IJCAI-99). San Francisco, CA: Morgan Kaufman, p. 846-851.
16. Pomerleau, D. (1989). "ALVINN: An Autonomous Land Vehicle. In a Neural Network," Advances in Neural Information Processing Systems 1, Morgan Kaufmann
17. Stolzmann, W. (1998). Anticipatory classifier systems. Genetic Programming 1998: Proceedings of the Third Annual Conference, 658-664.
18. Sutton, R.S. (1990). Integrated architectures for learning, planning, and reacting based on approximating dynamic programming. In Proceeding of the Seventh International Conference on Machine Learning, pp. 216-224. San Mateo, Ca.: Morgan Kaufmann.

Classification of Skewed and Homogenous Document Corpora with Class-Based and Corpus-Based Keywords

Arzucan Özgür and Tunga Güngör

Boğaziçi University, Computer Engineering Department, Bebek,
34342 İstanbul, Turkey
{ozgurarz,gungort}@boun.edu.tr

Abstract. In this paper, we examine the performance of the two policies for keyword selection over standard document corpora of varying properties. While in corpus-based policy a single set of keywords is selected for all classes globally, in class-based policy a distinct set of keywords is selected for each class locally. We use SVM as the learning method and perform experiments with boolean and tf-idf weighting. In contrast to the common belief, we show that using keywords instead of all words generally yields better performance and tf-idf weighting does not always outperform boolean weighting. Our results reveal that corpus-based approach performs better for large number of keywords while class-based approach performs better for small number of keywords. In skewed datasets, class-based keyword selection performs consistently better than corpus-based approach in terms of macro-averaged F-measure. In homogenous datasets, performances of class-based and corpus-based approaches are similar except for small number of keywords.

1 Introduction

The amount of electronic text information available such as Web pages, digital libraries, and email messages is increasing rapidly. As a result, the challenge of extracting relevant knowledge increases as well. The need for tools that enable people find, filter, and manage these resources has grown. Thus, automatic categorization of text document collections has become an important research issue.

SVM is one of the most successful text categorization methods [1, 2, 3]. It was designed for solving two-class pattern recognition problems [4]. The problem is to find the decision surface that separates the positive and negative training examples of a category with maximum margin. SVM can be used to learn linear or non-linear decision functions. Pilot experiments to compare the performance of various classification algorithms including linear SVM, SVM with polynomial kernel of various degrees, SVM with RBF kernel with different variances, k-nearest neighbor algorithm and Naive Bayes technique have been performed [5]. In these experiments, SVM with linear kernel was consistently the best performer. These results confirm the results of previous studies [1, 2, 3]. Thus, in this study we use SVM with linear kernel as the classification technique. For our experiments, we use the *SVMlight* system [6], which has been commonly used in previous studies [1, 2, 3].

C. Freksa, M. Kohlhase, and K. Schill (Eds.): KI 2006, LNAI 4314, pp. 91 – 101, 2007.

Keyword selection can be implemented in two alternative ways. In the first one, which we name as *corpus-based keyword selection*, a common keyword set for all classes that reflects the most important words in all documents is selected. In the alternative approach, named as *class-based keyword selection*, the keyword selection process is performed separately for each class. In this way, the most important words specific to each class are determined and a different set of keywords is used for each class.

Most previous studies focus on keyword selection metrics such as chi-square, information gain, odds ratio, probability ratio, document frequency, and binormal separation [3, 7, 8]. They use either the class-based or the corpus-based approach. In SVM-based text categorization, generally all available words in the document set are used instead of limiting to a set of keywords [1, 2, 5, 9]. In some studies, it was stated that using all the words leads to the best performance and using keywords is unsuccessful with SVM [3, 9, 10]. An interesting study by Forman covers the keyword selection metrics for text classification using SVM [3]. While this study makes extensive use of class-based keywords, it naturally does not cover some of the important points. The main focus of the study is on the keyword selection metrics; and there does not exist a comparison of the class-based and corpus-based keyword selection approaches. In [9], Debole and Sebastiani focus on supervised term weighting approaches and report their results both for class-based keyword selection, which they name as local policy and corpus-based keyword selection, which they call global policy. They use Reuters-21578 in their study, which is a highly skewed corpus. Different from our findings, they report that global keyword selection performs better than local keyword selection and SVM performs best when all the words are used. In [11], Özgür *et al.*, compare class-based and corpus-based keyword selection. However, they use a single dataset, Reuters-21578, and do not study the effect of these keyword selection approaches for document corpora of varying class distributions.

The aim of this paper is to evaluate the use of keywords for SVM-based text categorization and examine how class-based and corpus-based keyword selection approaches perform for datasets with varying class distribution properties. We use six standard document corpora in our study. Classic3 is a homogenous corpus, where all the classes are nearly equally well represented in the training set. Reuters-21578 and Wap corpora are highly skewed. A few of the classes are prevalent in the training set, while some classes are represented with very few documents. Hitech, LA1, and Reviews are neither homogenous nor highly skewed. Our results reveal that using keywords in SVM-based text categorization instead of using all the available words generally leads to better performance. We show that when corpus-based keyword selection is used for highly skewed datasets, less prevalent classes are represented poorly and macro-averaged F-measure performance drops down. In this case, class-based keyword selection is preferable. In homogenous datasets, although class-based approach performs better for small number of keywords, corpus-based approach performs slightly better or similar for large number of keywords. We perform our experiments with the two most commonly used term weighting approaches, boolean and tf-idf weighting. Surprisingly, we find that tf-idf weighting does not always outperform boolean weighting. As the keyword selection metric, we use total tf-idf scores of each term. In this way, keyword selection and term weighting phases are reduced to a single phase since tf-idf is also used for term weighting. This reduces the overall time of term weighting and keyword selection.

The paper is organized as follows: Section 2 discusses the document representation and Section 3 gives an overview of the keyword selection approaches. In Section 4, we describe the six standard datasets we used in the experiments, our experimental methodology, and the results we have obtained. We conclude in Section 5.

2 Document Representation

In our study, documents are represented by the vector-space model. In this model, each document is represented as a vector **d**, where each dimension stands for a distinct term in the term space of the document collection. We use the bag-of-words representation. To obtain the document vectors, each document is parsed, non-alphabetic characters and mark-up tags are discarded, case-folding is performed, and stop words are eliminated. We use the list of 571 stop words used in the Smart system [12]. We stem the words by using Porter's Stemming Algorithm [13], which is commonly used for word stemming in English. Each document is represented as $\mathbf{d}=(w_1, w_2, \ldots, w_n)$, where, w_i is the weight of ith term of document **d**.

We use boolean and tf-idf weighting schemes which are most commonly used in the literature. In boolean weighting, the weight of a term is considered to be 1 if the term appears in the document and it is considered to be 0 if the term does not appear in the document. tf-idf weighting scheme is defined as follows:

$$w_i = tf_i \cdot \log\left(\frac{n}{n_i}\right).$$
(1)

where tf_i is the raw frequency of term i in document d, n is the total number of documents in the corpus and n_i is the number of documents in the corpus where term i appears. Tf-idf weighting approach weights the frequency of a term in a document with a factor that discounts its importance if it appears in most of the documents, as in this case the term is assumed to have little discriminating power. Also, in order to account for documents of different lengths we normalize each document vector so that it is of unit length. Previous studies report that tf-idf weighting performs better than boolean weighting [14]. On the other hand, boolean weighting has the advantages of being very simple and requiring less memory. This is especially important in the high dimensional text domain. In the case of scarce memory resources, less memory requirement also leads to less classification time. Interestingly, we found that boolean approach does not always perform worse than tf-idf approach.

3 Keyword Selection

Most previous studies that apply SVM to text categorization use all the words in the document collection without any attempt to identify the important keywords [1, 2, 9]. On the other hand, there are various remarkable studies on keyword selection for text categorization in the literature [3, 7, 8]. As stated above, these studies mainly focus on keyword selection metrics and employ either the corpus-based or the class-based keyword selection approach, and do not use standard datasets. In addition, most studies do not use SVM as the classification algorithm. For instance, in [7] kNN and LLSF are used, and in [8] Naive Bayes is used. Later studies reveal that SVM performs consistently better than these classification algorithms [1, 2, 3].

In this study, rather than focusing on keyword selection metrics, we focus on the two keyword selection approaches, corpus-based keyword selection and class-based keyword selection. These two approaches have not been studied extensively together in the literature. In [9], Debole and Sebastiani perform experiments for both of the approaches. However their study is not extensive in this aspect since their main focus is on supervised term weighting methods and they use only the Reuters-21578 dataset. In contrast to our findings, they report that corpus-based keyword selection performs better than class-based keyword selection and SVM performs best when all the words are used. In [11], Özgür *et al.*, compare class-based and corpus-based keyword selection. However, they use a single dataset, Reuters-21578, and do not study the effect of these keyword selection approaches for document corpora of varying class distributions. In this study, we compare these keyword selection approaches with the alternative method of using all words without any keyword selection. We evaluate the performance of these approaches over datasets with varying class size distributions, i.e. homogenous, skewed, and highly skewed.

We use total tf-idf scores of terms as the keyword selection metric. Although it has not been used as a keyword selection metric in the literature, it has the advantage of leading to the reduction of keyword selection and term weighting phases into a single phase, when tf-idf is also used for term weighting. Our results show that it performs well, since in contrast to the previous studies we could obtain performances better than the approach where all the available words are used with SVM-based text categorization. In corpus-based keyword selection approach, terms that achieve the highest total tf-idf score in the overall corpus are selected as the keywords. To obtain the total tf-idf score of a term, the tf-idf weights of that term in each document are summed. This approach favors the prevailing classes and gives penalty to classes with small number of training documents in document corpora where there is high skew. In the class-based keyword selection approach, on the other hand, distinct keywords are selected for each class. The total tf-idf score of a term is calculated separately for each class. To obtain the total tf-idf score of a term for a specific class, the tf-idf weights of that term in only the documents that belong to that class are summed. This approach gives equal weight to each class in the keyword selection phase. So, less prevailing classes are not penalized.

4   Experiment Results

4.1   Document Data Sets

In our experiments we used six standard document corpora, widely used in automatic text organization research. The contents of these document sets, after preprocessing as described in Section 2, is summarized in Table 1. Classic3 data set contains 1,398

CRANFIELD documents from aeronautical system papers, 1,033 MEDLINE documents from medical journals, and 1,460 CISI documents from information retrieval papers. This dataset is homogenous since all the classes are represented equally well in the training set. This data set is relatively easy, because the classes are disjoint from each other.

Table 1. Summary description of document sets

Data set	# of documents	# of classes	# of terms
Classic3	3,891	3	10,930
Hitech	2,300	6	18,867
LA1	3,204	6	25,024
Reviews	4,069	5	31,325
Reuters-21578	12,902	90	20,307
Wap	1,560	20	8,064

The Hitech, LA1, and Reviews [15] datasets are neither highly skewed nor homogenous. They are very high dimensional compared to the number of documents in the training sets. The Hitech data set was derived from the San Jose Mercury newspaper articles, which are delivered as part of the TREC collection [16]. The classes of this document corpora are computers, electronics, health, medical, research, and technology. LA1 data set consists of documents from Los Angeles Times newspaper, used in TREC-5 [16]. The categories correspond to the desk of the paper that each article appeared. The data set consists of documents from entertainment, financial, foreign, metro, national, and sports desks. Reviews data set contains articles from San Jose Mercury Newspaper, that are distributed as part of the TREC collection TIPSTER vol. 3 [16]. The classes of this document corpora are food, movie, music, radio, and restaurant.

The documents in Reuters-21578 v1.0 document collection [17], which is considered as the standard benchmark for automatic document organization systems, have been collected from Reuters newswire in 1987. This corpus consists of 21,578 documents. 135 different categories have been assigned to the documents. The maximum number of categories assigned to a document is 14 and the mean is 1.24. This dataset is highly skewed. For instance, the "earnings" category is assigned to 2,709 training documents, but 75 categories are assigned to less than 10 training documents. 21 categories are not assigned to any training documents. 7 categories contain only one training document and many categories overlap with each other such as grain, wheat, and corn.

Wap data set consists of 1,560 web pages from Yahoo! subject hierarchy collected and classified into 20 different classes for the WebACE project [18]. This dataset is also highly skewed. Minimum class size is 5, maximum class size is 341, and average class size is 78. Many categories of Wap are close to each other.

In order to divide the Reuters-21578 corpus into training and test sets, mostly the modified Apte (ModApte) split has been used [17]. With this split the training set consists of 9,603 documents and the test set consists of 3,299 documents. For our results to be comparable with the results of other studies, we also used this splitting method. We also removed the classes that do not exist both in the training set and in the test set, remaining with 90 classes out of 135. For the other data sets, we used the initial 2/3 of the documents as the training set and the remaining 1/3 as the test set. Below we report the results for the test sets of the corpora.

4.2 Results and Discussion

Tables 2 and 3 display, respectively, the micro-averaged and macro-averaged F-measure results, for boolean and tf-idf document representations using all words and using keywords ranging in number from 10 to 2000. Bool (cl), tf-idf (cl), and tf-idf (co) stand for class-based approach with boolean weighting, class-based approach with tf-idf weighting, and corpus-based approach with tf-idf weighting, respectively. Micro-averaged F-measure gives equal weight to each document and therefore it tends to be dominated by the classifier's performance on common categories. Macro-averaged F-measure gives equal weight to each category regardless of its frequency and thus it is influenced more by the classifier's performance on rare categories.

In the following discussion, it is assumed that tf-idf weighting is used unless it is stated otherwise. When we examine Classic3 dataset, whose class distribution is homogenous, we observe that micro-averaged and macro-averaged F-measure results are similar. Also, there is not much performance difference among class-based keyword selection and corpus-based keyword selection. For instance, in the case of 30 keywords, both achieve 90% success in terms of micro-averaged F-measure and

Table 2. Micro-averaged F-measure Results

	# words	10	30	50	70	100	200	300	400	500	1000	1500	2000	All
Classic3	bool (cl)	0.824	0.892	0.930	0.937	0.936	0.940	0.942	0.947	0.953	0.955	0.961	0.965	0.981
	tf-idf (co)	0.768	0.900	0.901	0.926	0.937	0.956	0.963	0.971	0.981	0.988	0.992	0.992	0.994
	tf-idf (cl)	0.845	0.900	0.944	0.948	0.950	0.959	0.955	0.957	0.960	0.964	0.965	0.971	0.994
Hitech	bool (cl)	0.543	0.534	0.533	0.551	0.567	0.593	0.587	0.578	0.569	0.585	0.578	0.586	0.581
	tf-idf (co)	0.377	0.518	0.539	0.586	0.603	0.606	0.614	0.627	0.623	0.643	0.647	0.659	0.649
	tf-idf (cl)	0.553	0.597	0.619	0.628	0.629	0.625	0.631	0.622	0.644	0.621	0.618	0.627	0.649
LA1	bool (cl)	0.598	0.685	0.699	0.718	0.730	0.739	0.759	0.764	0.783	0.790	0.791	0.791	0.797
	tf-idf (co)	0.467	0.648	0.723	0.754	0.766	0.793	0.806	0.816	0.816	0.817	0.824	0.833	0.841
	tf-idf (cl)	0.634	0.731	0.761	0.773	0.784	0.789	0.801	0.809	0.807	0.814	0.812	0.815	0.841
Reviews	bool (cl)	0.800	0.834	0.844	0.846	0.858	0.882	0.891	0.893	0.898	0.900	0.903	0.903	0.915
	tf-idf (co)	0.778	0.862	0.869	0.886	0.894	0.934	0.939	0.942	0.944	0.943	0.937	0.936	0.941
	tf-idf (cl)	0.843	0.867	0.891	0.901	0.901	0.906	0.912	0.914	0.918	0.926	0.924	0.920	0.941
Reuters 21578	bool (cl)	0.738	0.780	0.802	0.802	0.806	0.811	0.819	0.823	0.821	0.820	0.818	0.818	0.817
	tf-idf (co)	0.425	0.543	0.628	0.671	0.697	0.761	0.786	0.804	0.813	0.845	0.859	0.861	0.857
	tf-idf (cl)	0.780	0.814	0.831	0.833	0.838	0.838	0.839	0.842	0.848	0.854	0.853	0.855	0,857
Wap	bool (cl)	0.650	0.674	0.713	0.715	0.713	0.721	0.736	0.757	0.754	0.758	0.766	0.762	0.759
	tf-idf (co)	0.135	0.496	0.585	0.607	0.655	0.691	0.716	0.723	0.721	0.740	0.749	0.743	0.752
	tf-idf (cl)	0.688	0.736	0.748	0.746	0.751	0.736	0.728	0.726	0.722	0.746	0.741	0.747	0.752

Table 3. Macro-averaged F-measure Results

	# words	10	30	50	70	100	200	300	400	500	1000	1500	2000	All
Classic3	bool (cl)	0.811	0.880	0.930	0.938	0.935	0.938	0.939	0.946	0.951	0.953	0.959	0.965	0.980
	tf-idf (co)	0.769	0.886	0.899	0.925	0.936	0.953	0.962	0.971	0.980	0.989	0.992	0.992	0.994
	tf-idf (cl)	0.829	0.886	0.941	0.946	0.949	0.957	0.955	0.956	0.959	0.964	0.964	0.970	0.994
Hitech	bool (cl)	0.397	0.413	0.421	0.434	0.453	0.469	0.472	0.465	0.446	0.451	0.461	0.456	0.433
	tf-idf (co)	0.230	0.371	0.466	0.497	0.507	0.505	0.508	0.538	0.530	0.538	0.582	0.598	0.558
	tf-idf (cl)	0.489	0.555	0.577	0.571	0.574	0.565	0.565	0.570	0.589	0.567	0.549	0.561	0.558
LA1	bool (cl)	0.476	0.590	0.620	0.656	0.662	0.670	0.686	0.696	0.713	0.729	0.726	0.728	0.729
	tf-idf (co)	0.284	0.530	0.628	0.674	0.692	0.715	0.738	0.745	0.752	0.748	0.752	0.765	0.777
	tf-idf (cl)	0.552	0.674	0.706	0.712	0.727	0.735	0.745	0.760	0.755	0.762	0.756	0.764	0.777
Reviews	bool (cl)	0.767	0.814	0.829	0.830	0.843	0.871	0.881	0.879	0.881	0.885	0.884	0.885	0.874
	tf-idf (co)	0.558	0.692	0.693	0.710	0.720	0.931	0.928	0.933	0.939	0.939	0.935	0.932	0.928
	tf-idf (cl)	0.847	0.864	0.894	0.903	0.904	0.904	0.911	0.912	0.916	0.916	0.912	0.906	0.928
Reuters 21578	bool (cl)	0.481	0.469	0.472	0.466	0.443	0.398	0.384	0.385	0.377	0.349	0.332	0.328	0.294
	tf-idf (co)	0.010	0.030	0.051	0.082	0.091	0.162	0.207	0.242	0.263	0.373	0.425	0.431	0.439
	tf-idf (cl)	0.500	0.515	0.519	0.510	0.508	0.511	0.492	0.494	0.494	0.498	0.492	0.492	0.439
Wap	bool (cl)	0.442	0.453	0.482	0.502	0.502	0.480	0.476	0.490	0.488	0.488	0.482	0.477	0.448
	tf-idf (co)	0.092	0.208	0.306	0.321	0.350	0.412	0.416	0.435	0.442	0.455	0.468	0.455	0.450
	tf-idf (cl)	0.550	0.593	0.590	0.550	0.533	0.507	0.505	0.497	0.495	0.509	0.477	0.482	0.450

88.6% success in terms of macro-averaged F-measure. However, class-based approach converges faster than corpus-based approach and thus performs better for small number of keywords (200 keywords and less). As number of keywords increases performance tends to increase. Although all words approach (10930 words) achieves the highest performance of 99.4%, tf-idf corpus-based approach achieves a very close performance of 99.2% with 1500 keywords. Boolean class-based approach does not perform much worse than the tf-idf class-based approach and it performs generally better than tf-idf corpus-based approach for 100 and less keywords.

Hitech, LA1, and Reviews datasets have neither homogenous nor highly skewed class distributions. Micro-averaged and macro-averaged F-measure results of Reviews dataset are similar to each other. However, macro-averaged F-measure results are considerably less than micro-averaged F-measure results for Hitech and LA1 datasets. When we examine the results on the Hitech dataset, we observe that for 300 and less keywords class-based approach achieves better micro-averaged F-measure performance than corpus-based approach and for 1000 and less keywords it achieves better macro-averaged F-measure performance. On the other hand, corpus-based approach achieves the highest performance for 2000 keywords, i.e. 65.9% micro-averaged and 59.8% macro-averaged F-measure performance. These results are higher than the all words approach (18867 words), which achieves 64.9% and 55.8% micro-averaged and macro-averaged F-measure results, respectively. In terms of macro-averaged F-measure performance, class-based approach with 50 and more keywords and corpus-based approach with 1500 and 2000 keywords achieve better results than the all words approach. Boolean class-based approach with 200 keywords achieves higher F-measure performance than boolean all words approach. Although boolean class-based approach performs worse than tf-idf class-based approach, it performs better than tf-idf corpus-based approach for 10 and 30 keywords.

Over LA1 dataset, class-based approach performs better than corpus-based approach for 100 and less keywords in terms of micro-averaged F-measure. Macro-averaged F-measure results of class-based approach are generally higher than that of the corpus-based approach. Only for 2000 keywords, corpus-based approach achieves slightly better macro-averaged F-measure performance than class-based approach (76.5% versus 76.4%). All words approach achieves the best performance of 84.1% micro-averaged and 77.7% macro-averaged F-measure. The closest performance to these results is achieved by the corpus-based approach with 2000 keywords, 83.3% micro-averaged and 76.5% macro-averaged F-measure. Boolean class-based approach performs worse than tf-idf class-based approach, but it performs better than tf-idf corpus-based approach for 10 and 30 keywords.

Over Reviews dataset, tf-idf corpus-based approach achieves the highest micro-averaged (94.4%) and macro-averaged (93.9%) F-measure performance with 500 keywords. These results are even higher than the all words approach (31325 words), which achieves 94.1% micro-averaged and 92.8% macro-averaged F-measure performance. For 100 and less keywords class-based approach achieves higher performance than corpus-based approach both in terms of micro-averaged and macro-averaged F-measure. There is a gap between macro-averaged F-measure results. For instance, while class-based approach achieves 90.3% macro-averaged performance for 70 keywords, corpus-based approach achieves only 71.0% performance. Even boolean class-based approach performs better than tf-idf corpus-based approach in terms of macro-averaged F-measure for 100 and less keywords.

Reuters-21578 and Wap datasets have highly skewed class distributions. Thus, there is a large gap between micro-averaged and macro-averaged F-measure results. For both datasets, we can conclude that class-based keyword selection achieves consistently higher macro-averaged F-measure performance than corpus-based approach. The high skew in the distribution of the classes in the datasets affects the macro-averaged F-measure values in a negative way because macro-average gives equal weight to each class instead of each document and documents of rare classes tend to be more misclassified. By this way, the average of correct classifications of classes drops dramatically for datasets having many rare classes. Class-based keyword selection is observed to be very useful for this skewness. For instance, in Reuters-21578 dataset, with even a small portion of words (50-100-200), class-based tf-idf method reaches 50% success which is far better than the 43.9% success of tf-idf with all words. In Wap dataset, class-based approach with 30 keywords achieves the highest performance in terms of macro-averaged F-measure (59.3%), which is considerably higher than the macro-averaged F-measure performance of all words approach (45.0%). Also, tf-idf class based approach for small number of keywords (100 keywords and less) achieves better or similar performance compared to the case where all words are used. Rare classes are characterized in a successful way with class-based keyword selection, because every class has its own keywords for the categorization problem. Corpus-based approach shows worse results because most of the keywords are selected from prevailing classes, which prevents rare classes to be represented fairly by their keywords. In text categorization, most of the learning takes place with a small but crucial portion of keywords for a class [19]. Class-based keyword selection, by definition, focuses on this small portion; on the other hand, corpus-based approach finds general keywords concerning all classes. So, with few keywords, class-based approach achieves much more success by finding more crucial class keywords. Corpus-based approach is not successful with that small portion, but has a steeper learning curve. For instance, for the Reuters-21578 dataset, it leads to the peak micro-averaged F-measure value of our study (86.1%) with 2000 corpus-based keywords, which exceeds the success scores of recent studies with standard usage of Reuters-21578 [1, 20].

Boolean class-based approach generally performs worse than tf-idf class-based approach for all number of keywords. This is an expected result, since it does not take into account term frequencies and inverse document frequencies. However, surprisingly, for Wap dataset, for 300 and more keywords, boolean approach achieves higher micro-averaged F-measure performance than tf-idf class-based and corpus-based approaches. Also, boolean all words approach performs better than tf-idf all words approach in terms of micro-averaged F-measure and performs similar in terms of macro-averaged F-measure. In addition, boolean approach achieves the highest micro-averaged F-measure performance in the overall for 2000 keywords (76.2%). Thus, in this case boolean approach may be preferred to tf-idf approach since it is simpler and needs less memory and time.

5   Conclusion

In this paper, we investigated the use of keywords in text categorization with SVM. Unlike previous studies that focus on keyword selection metrics, we studied the performance of the two approaches for keyword selection, corpus-based approach and class-based approach, over datasets of varying class distribution properties. We used six standard document corpora and both boolean and tf-idf weighting schemes.

In text categorization literature, generally all of the words in the documents were used for categorization with SVM. Keyword selection was not performed in most of the studies; even in some studies, keyword selection was stated to be unsuccessful with SVM [3, 9, 10]. In contrast to these studies, we observed that keyword selection generally improves the performance of SVM. This is quite important since there is considerable gain in terms of classification time and memory when small number of keywords is used.

For all datasets (homogenous, skewed, and highly skewed) class-based approach performs better than corpus-based approach for small number of keywords (generally 100 and less keywords) in terms of micro-averaged F-measure. Corpus-based approach generally achieves higher micro-averaged F-measure performance for larger number of keywords. There is not much difference between micro-averaged and macro-averaged F-measure values and between class-based and corpus-based approaches in homogenous datasets. On the other hand, for skewed and highly skewed datasets, there is a gap between micro-averaged and macro-averaged F-measure results. In highly skewed datasets, class-based keyword selection approach performs consistently better than corpus-based approach and the approach where all words are used, in terms of macro-averaged F-measure. In the corpus-based approach, the keywords tend to be selected from the prevailing classes. Rare classes are not represented well by these keywords. However, in the class-based approach, rare classes are represented equally well as the prevailing classes because each class is represented with its own keywords for the categorization problem. Therefore, class-based keyword selection approach should be preferred to corpus-based approach for highly skewed datasets. It should also be preferred when small number of keywords will be used due to space and time limitations.

When we compare the tf-idf and boolean weighting approaches, surprisingly we see that boolean approach is not always worse than tf-idf approach although it is simpler. It can be preferred to tf-idf approach especially in cases where there are limited space resources.

Acknowledgement

This work was supported by the Bo azi̧ci University Research Fund under the grant number 05A103. The authors would like to thank Levent Özgür for helpful discussions.

References

1. Yang, Y., Liu, X.: A Re-examination of Text Categorization Methods. In Proceedings of SIGIR-99, 22nd ACM International Conference on Research and Development in Information Retrieval. Berkeley (1996)
2. Joachims, T.: Text Categorization with Support Vector Machines: Learning with Many Relevant Features. In: European Conference on Machine Learning (ECML) (1998)
3. Forman, G.: An Extensive Empirical Study of Feature Selection Metrics for Text Classification. Journal of Machine Learning Research 3 (2003) 1289–1305
4. Burges, C.J.C.: A Tutorial on Support Vector Machines for Pattern Recognition. Data Mining and Knowledge Discovery 2(2) (1998) 121–0167
5. Özgür, A.: Supervised and Unsupervised Machine Learning Techniques for Text Document Categorization. MS Thesis, Bo azi̧ci University, Istanbul (2004)
6. Joachims, T.: Advances in Kernel Methods-Support Vector Learning. Chapter Making Large-Scale SVM Learning Practical. MIT (1999)
7. Yang, Y., Pedersen, J.O.: A Comparative Study on Feature Selection in Text Categorization. In: Proceedings of the 14th International Conference on Machine Learning (1997) 412–420
8. Mladenic, D., Grobelnic, M.: Feature Selection for Unbalanced Class Distribution and Naive Bayes. In: Proceedings of the 16th International Conference on Machine Learning (1999) 258–267
9. Debole, F., Sebastiani, F.: Supervised Term Weighting for Automated Text Categorization. In: Proceedings of SAC-03, 18th ACM Symposium on Applied Computing. ACM Press (2003) 784–788
10. Aizawa, A.: Linguistic Techniques to Improve the Performance of Automatic Text Categorization. In: Proceedings of 6th Natural Language Processing Pacific Rim Symposium. Tokyo (2001) 307–314
11. Özgür, A., Özgür, L., Güngör T.:Text Categorization with Class-Based and Corpus-Based Keyword Selection. In: Proceedings of ISCIS'05. Lecture Notes in Computer Science 3733. Springer Verlag (2005) 607–616

12. ftp://ftp.cs.cornell.edu/pub/smart/ (2004)
13. Porter, M. F.: An Algorithm for Suffix Stripping. Program 14 (1980) 130–137
14. Salton, G., Buckley, C.: Term Weighting Approaches in Automatic Text Retrieval. Information Processing and Management 24(5) (1988) 513–523
15. Karypis G.: Cluto 2.0 Clustering Toolkit. http://www.users.cs.umn.edu/~karypis/cluto (2004)
16. TREC. Text Retrieval Conference. http://trec.nist.gov (1999)
17. Lewis, D.D.: Reuters-21578 Document Corpus V1.0. http://kdd.ics.uci.edu/databases/reuters21578/reuters21578.html
18. Han, E-H.S., Boley, D., Gini, M., Gross, R., Hastings, K., Karypis, G., Kumar, V., Mobasher, B., Moore, J.: WebAce: A Web Agent for Document Categorization and Exploration. In: Proceedings of the 2nd International Conference on Autonomous Agents (1998)
19. Özgür, L., Güngör, T., Gürgen, F.: Adaptive Anti-Spam Filtering for Agglutinative Languages. A Special Case for Turkish. Pattern Recognition Letters 25(16) (2004) 1819–1831
20. Sebastiani, F.: Machine Learning in Automated Text Categorization. ACM Computing Surveys 34(5) (2002) 1–47

Learning an Ensemble of Semantic Parsers for Building Dialog-Based Natural Language Interfaces

Lappoon R. Tang

Department of Computer Sciences
University of Texas at Brownsville
Brownsville, TX 78520, U.S.A.
lappoon.tang@utb.edu

Abstract. Building or learning semantic parsers has been an interesting approach for creating natural language interfaces (NLI's) for databases. Recently, the problem of imperfect precision in an NLI has been brought up as an NLI that might answer a question incorrectly can render it unstable, if not useless. In this paper, an approach based on ensemble learning is proposed to trivially address the problem of unreliability in an NLI due to imperfect precision in the semantic parser in a way that also allows the recall of the NLI to be improved. Experimental results in two real world domains suggested that such an approach can be promising.

1 Introduction

Semantic parsing refers to the process of mapping a sentence to its meaning representation [1]. Using a machine learning approach to the task was shown to be effective in making natural language interfaces (NLI's) portable across domains [2,3]. Unfortunately, the NLI's learned might be unreliable; it might deliver incorrect information to a user since the precision of the learned semantic parser might not be perfect [4]. However, as it was mentioned in [4], a user prefers reliable user interfaces over intelligent but unreliable ones.

The core of the problem is that a clarification dialog was never initiated to verify that the system indeed correctly interpreted the user question – a feature implied necessary for constructing reliable (learning based) NLI's [4]. A "dialog" here refers to a brief interaction between the user and the NLI for the purpose of confirming the correctness of the NLI's interpretation of the user question. Hence, our work here is somewhat unrelated to existing approaches in building dialog interfaces [5,6]. Incorporating such dialoguing capabilities into an NLI can *trivially* guarantee that incorrect information is never delivered to a user. Instead of returning one interpretation, one can do better by allowing multiple interpretations to be returned from which a user can choose a correct one or else he ignores the NLI altogether. This can be achieved by learning an ensemble of semantic parsers for the NLI.

While learning an ensemble of parsers has been shown to be effective in probabilistic (syntactic) parsing [7], to the author's knowledge, it has not been attempted

C. Freksa, M. Kohlhase, and K. Schill (Eds.): KI 2006, LNAI 4314, pp. 102–112, 2007.
© Springer-Verlag Berlin Heidelberg 2007

on learning semantic parsers. This has motivated an investigation on the effectiveness of using a bagging approach to learning semantic parsers for creating a reliable NLI *and* improving the NLI's recall at the same time. In particular, a variant of bagging is being explored here in which *multiple learners* with different language biases are used on different resamplings of the training data to learn the parser ensemble for enhancing ensemble diversity. Also, while traditionally in bagging one creates random samples of data with replacement, random samples here are created without replacement (i.e. the different samples are mutually disjoint). Random samples are so created because the training data may already have duplicated examples (e.g. the same question was asked by two different users of the NLI). This paper presents an ensemble approach on learning semantic parsers the author calls CHILLE (CHILL using an Ensemble) which is a generalization of an earlier approach CHILL [2] that was demonstrated to be effective for learning semantic parsers. CHILLE is demonstrated to be effective in two real world domains.

The rest of the paper is organized as follows. Section 2 will provide a brief background on bagging classifiers and on inductive logic programming [8]. The approach CHILLE on learning an ensemble of semantic parsers for building dialog based NLI's is presented in Section 3. Experimental results are presented in Section 4 followed by a discussion on related work in Section 5. Finally, conclusions and future work are presented in Section 6.

2 Background

2.1 Bagging Classifiers

Bagging [9] (i.e. "bootstrap aggregating") is a machine learning technique that creates an ensemble of classifiers $\{C_k\}$ (instead of traditionally a single classifier) from replicate data sets $\{D_k\}$ randomly drawn *with replacement* from a bootstrap distribution of training data D; each classifier $C_i \in \{C_k\}$ is learned from the data sample $D_i \in \{D_k\}$. A new instance x (drawn from the same distribution underlying D) is classified using the ensemble $\{C_k\}$ by averaging the numerical values produced by the classifiers in $\{C_k\}$ if the label for x is numerical. Otherwise, voting is performed among the classifiers given x where the label with the highest vote is returned.

It has been demonstrated to be an effective approach in learning when the underlying learning procedure is unstable (i.e. a small perturbation in the training data would produce a large deviation in classification accuracy). While bagging is helpful for reducing biases among classifiers [7], the technique is being exploited here to learn a *diverse* ensemble of semantic parsers for returning a set of interpretations (of the user question) that has a high probability of containing a correct interpretation.

2.2 Inductive Logic Programming

Inductive Logic Programming (ILP) is a subfield in AI which concerns learning a first order Horn theory that explains a given set of training data (i.e. observations) and also generalizes to novel cases; it is at the intersection of machine learning and logic programming. The problem is defined as follows. Given a set of examples $\xi = \xi^+ \cup \xi^-$ consisting of positive and negative examples of a target concept, and background knowledge B, find an hypothesis $H \in \mathcal{L}$ (the language of hypotheses) such that it is consistent with the training data. In practice, one finds a hypothesis that is sufficiently accurate as there might be noise in training data. Due to the use of a more expressive first-order formalism, ILP techniques are proven to be more effective in tackling problems that require learning relational knowledge than traditional propositional approaches [10]. ILP techniques are employed for the task of semantic parser acquisition as subtle relations among objects in a parse state might be useful for disambiguation and they can be effectively learned via ILP methods.

There are two traditional approaches in the design of ILP learning algorithms: top-down and bottom-up. In the former, one builds a clause in a general-to-specific order where the search usually starts with the most general clause and successively specializes it with background predicates according to some search heuristic. A representative example of this approach would be the FOIL algorithm [10,11]. In the latter, the search begins at the other end of the space where it starts with the most specific hypothesis, the set of examples, and constructs clauses in a specific-to-general order by generalizing the more specific clauses. A representative example of this approach would be the GOLEM algorithm [12].

Two ILP learners will be used in CHILLE. The first one is CHILLIN [13] that was originally employed in CHILL. The second is COCKTAIL which combines clauses constructed by MFOIL [14] and those by CHILLIN through a theory evaluation metric based on the minimum description length principle [15]. COCKTAIL was proven to be more effective than CHILLIN and MFOIL on the task of semantic parser acquisition [3]. Readers are recommended to refer to the literature for more details.

3 Ensemble Learning for Semantic Parsing

3.1 The CHILL Architecture for Semantic Parser Induction

The system CHILL can learn to map a question in a natural language (e.g. English) to its target meaning representation (e.g. a Prolog logic query). Given a corpus of training data D (each is a pair of user question and its meaning representation), a lexicon $Lex(D)$, and an ILP learner L, CHILL learns control rules that specialize an overly general parser (generated directly from the corpus) to produce a specialized parser P. The algorithm is outlined below:

Chill(D, $Lex(D)$, L, P):

1. Parser Operator Generation: Given D and $Lex(D)$, generate a set of parsing operators O_x for each sentence $x \in D$ such that O_x is complete with respect to x (i.e. it can produce a correct parse for x). The set $S_D = \bigcup_{x \in D} O_x$ forms the overly general parser for the corpus D; it produces many spurious parses for the sentences in D.

2. Example Analysis: Using the target meaning representation $q(x)$ for the sentence $x \in D$, identify derivations in O_x that produce spurious parses for x to generate positive and negative examples for each operator $o \in O_x$.
3. Control Rule Induction: Use the learner L to induce a set of control rules R_o for each $o \in S_D$ given the positive and negative examples for o.
4. Parser Specialization: For each $o \in S_D$, incorporate the control rules R_o into o (as its preconditions) to produce the specialized parser P.

3.2   The CHILLE Algorithm

CHILLE generalizes CHILL by allowing the use of a *set* of ILP learners $Ls = \{L_k\}$ to learn an ensemble of N ($N \geq |Ls|$) parsers $E = \{P_j\}$ from different sets of training data $\{D_m\}$ sampled from the given corpus D. However, as we mentioned before, a variant of bagging is used here that samples training data *without* replacement since the corpus may already contain duplicated sentences. Besides, this allows one to potentially maximize the *variety* of sentences given to learn a parser. For the sake of simplicity, CHILLE is outlined as a randomized algorithm as follows:

ChillE(N, D, $Lex(D)$, Ls, E):

1. $E = \emptyset$
2. Generate N samples of training data $\{D_m\}$ by randomly drawing training data from the corpus D.
3. Randomly draw a learner $L \in Ls$ and a sample of data $D_i \in \{D_m\}$.
4. Learn a parser P by running Chill(D_i, $Lex(D)$, L, P).
5. Add P to E.
6. Repeat 3 to 5 until N parsers have been learned.

There are two possible ways to use the parser ensemble to create the NLI: 1) perform voting among the ensemble on a novel user question and select the most common interpretation, or 2) return all the queries produced by the ensemble (paraphrased in natural language) and let the user choose a correct one (if any). These schemes will be discussed below:

Voting: Given a novel sentence s, a set of queries Q is produced by applying each parser in the ensemble to s. Suppose $Q' \subseteq Q$ is the set of queries that are actually executable (since some parsers in the ensemble might fail to produce a successful parse due to problems in learning). One can return the query $q \in Q'$ whose answer set has the highest vote in Q'. If there is a tie, one can select the $q_L \in Q'$ produced by the semantic parser that was learned by the best learner $L \in Ls$ (a learner is better if the parsers produced by it have a higher recall on average).

Returning a Query Set: Instead of returning only the "best" query from the ensemble by voting, one can consider returning a practically small but probable set of queries and let the user select a correct one. Not only will this increase the probability that a correct interpretation can be found, and, hence, improve the recall of the NLI, but it will also ensure that the NLI does not deliver incorrect information to the user.

Suppose the ensemble has N reasonably independent parsers such that the expected error rate of parser i is e_i. The expected accuracy R (i.e recall) of the entire ensemble is, approximately, as follows:

$$R \approx 1 - \prod_i e_i \qquad (1)$$

If all the learners in Ls have similar generalization performance and are relatively stable, then Equation 1 can be reduced to:

$$R \approx 1 - \hat{e}^N \qquad (2)$$

where $\hat{e}$ is the average error rate of a parser in the ensemble. Obviously, if $\hat{e}$ remains relatively constant for increasing N, the error rate of the entire ensemble will converge to arbitrarily small value, thus, R will converge to nearly perfect. While this is only an analysis of the ideal situation in which the parsers in the ensemble are relatively independent of each other, it does shed light on the theoretical optimal performance of such an approach, and, hence, provides a basis of the (ideal) goals for which one should strive when employing such an ensemble approach to the task.

In practice, this ideal is limited by at least two factors: 1) the response time of the NLI would increase proportionally to N (thus one has to practically bound N), and 2) the level of mutual independence of the parsers in the ensemble (i.e. ensemble diversity). One purpose of this paper is to explore the trade-off between efficiency and accuracy in using an ensemble approach on building reliable dialog-based NLI's and see if the approach can scale up to large data sets.

4   Experimental Evaluation

4.1   Domains and Corpora

Two different domains are used for experimentation. The first one is the United States Geography domain. The database contains about 800 facts implemented in Prolog as relational tables containing basic information about the U.S. states like population, area, capital city, neighboring states, and so on. The second domain consists of a set of 1000 computer-related job postings, such as job announcements, from the USENET newsgroup `austin.jobs`. Information from these job postings is extracted to create a database which contains the following types of information: 1) the job title, 2) the company, 3) the recruiter, 4) the location, 5) the salary, 6) the languages and platforms used, and 7) required or desired years of experience and degrees [16].

The U.S. Geography domain has a corpus of 1000 sentences (Geo1000) in which 250 are collected from undergraduate students in the CS department at the University of Texas at Austin and the rest from real users of the Web interface "powered" by CHILL[1]. The job database information system has a corpus

---

[1] `www.cs.utexas.edu/users/ml/geo.html`

of 640 sentences (Jobs640); 400 of which are artificially made using a simple grammar that generates certain obvious types of questions people may ask and the other 240 are obtained from undergraduate students or the Web interface. Both corpora are available at www.cs.utexas.edu/users/ml/nldata.html.

4.2   Experimental Design

Comparison on performance will be conducted among five types of NLI's using the two corpora: 1) non-dialog based NLI's built by CHILL [3], 2) dialog based NLI's built by CHILLE using voting, 3) dialog based NLI's built by CHILLE that return a query set, and 4) PRECISE (a non-learning NLI) [4], and 5) Microsoft English Query (EQ) (also a non-learning NLI) [4,17]. In CHILL, the learner used was COCKTAIL (as it performed the best among the choices) while in CHILLE both COCKTAIL and CHILLIN were used. The average test time *per trial* (i.e. the time taken to evaluate the performance of a parser on test questions) for the systems except that of PRECISE and EQ (as results are not available) are also reported here for a comparison on trade-off between efficiency and accuracy among the systems. Notice that comparison of performance is done on NLI's instead of on semantic parsers for two reasons: 1) while a semantic parser may be imperfect in precision, a carefully designed NLI using the same semantic parser can nonetheless be made to delivery only correct information, and 2) not all NLI's are learning based and/or employing semantic parsers for processing semantics of sentences. The test time of an NLI's semantic parser is used to give us an idea of the NLI's approximate response time; such an estimate should be reasonable since the time required in parsing (i.e. the parser's efficiency) plus the time required in executing the query produced by the parser (i.e. efficiency in information retrieval) is usually the bottleneck of the response time of the NLI.

The experiments were conducted using 10-fold cross validation. In each test, the recall (a.k.a. accuracy) and the precision of the NLI's will be computed along with their test time. Recall and precision are defined as

$$Recall = \frac{\text{\# of correctly answered questions}}{\text{\# of test questions}} \tag{3}$$

$$Precision = \frac{\text{\# of correctly answered questions}}{\text{\# of questions answered by the NLI}}. \tag{4}$$

Precision is defined as the number of questions answered correctly by an NLI divided by the number of questions for which the NLI produced an answer because sometimes a semantic parser might not produce an executable query for a sentence and hence not all questions are answered by the NLI.

In other words, for all the NLI's created by CHILL or CHILLE, recall is the same as the number of correct logic queries produced divided by the total number of sentences in the test set. If a parser ensemble is used and an entire query set is returned, the parser ensemble is considered correct if one of the parsers produced a correct logic query. Precision is the number of correct logic queries produced divided by the number of sentences in the test set from which the parser (or parser ensemble) produced a query (i.e. a successful parse). Please note that a

Table 1. Results in the U.S. Geography domain. Combinations of ensemble size N and % of training data M are listed as (N x M). QS means a query set is returned. V means voting is performed in the ensemble. CorInfo means a user is guaranteed that the information delivered is correct.

NLI \ Corpora	Geo1000			
	Recall	Precision	CorInfo	Test Time (seconds)
CHILL	81.70±1.28	83.64±1.43	×	55.82±24.50
CHILLE (2x100%) (QS)	86.00±1.52	87.50± 1.67	√	229.25±40.91
CHILLE (2x100%) (V)	82.40±1.63	84.01± 1.74	×	233.11±39.20
CHILLE (2x50%) (QS)	86.8±1.89	87.69± 2.07	√	209.43±51.00
CHILLE (2x50%) (V)	80.40±2.48	81.46±2.31	×	199.81±44.62
CHILLE (4x25%) (QS)	**88.44±2.60**	89.23±2.47	√	**343.19±93.39**
CHILLE (4x25%) (V)	76.78±3.60	77.64±3.55	×	294.68±69.47
CHILLE (6x16.67%) (QS)	88.80±1.98	89.42±1.76	√	444.44±132.08
CHILLE (6x16.67%) (V)	75.10±3.07	75.80±3.20	×	344.38±89.17
PRECISE	77.50	100.00	√	n/a
EQ	≈ 57.50	≈ 82.00	×	n/a

logic query is considered correct here if it produces the same answer set as that of the target logic query for the test question.

In information extraction, recall is usually defined as $|I|/|C|$ where $I = R \cap C$, R is the set of retrieved documents (from executing the user query), and C is the set of correct documents. Precision is usually defined as $|I|/|R|$. Here, in semantic parsing, recall is defined as $|I|/|C|$ where I is the intersection of the set S of logic queries produced by the parser (or parser ensemble) in parsing the set of test sentences and C which is the set of correct logic queries, one for each sentence, in the given set of test sentences. Precision is defined as $|I|/|S|$. In other words, recall is simply the fraction of test sentences that were correctly parsed by the learned parser (or parser ensemble), and precision is the probability that the logic query produced by the learned parser (or parser ensemble) from parsing a test sentence is correct.

4.3   Discussion of Results

For all the experiments performed, a beam size of four, a significant threshold of 6.64 (99% level of significance), and a parameter $m = 10$ (the m-estimate [18]) were used for mFOIL. The best four clauses (by coverage) found by CHILLIN were used. In order to study trade off between efficiency and accuracy in an NLI constructed using an ensemble, various combinations of the ensemble size and the percentage of training data in a random sample given to learn each parser in the ensemble were explored. The results of the various NLI's in the U.S. Geography domain and the job posting domain are shown in Table 1 and Table 2 respectively.

The recall for PRECISE in the U.S. Geography domain was taken from [19] while in the job posting domain it was obtained by very carefully inspecting the bar chart results reported in [4], and likewise for the results on EQ. All confidence intervals in the results were computed at the 95% confidence level.

Table 2. Results in the job posting domain. Combinations of ensemble size N and % of training data M are listed as $(N \times M)$. QS means a query set is returned. V means voting is performed in the ensemble. CorInfo means a user is guaranteed that the information delivered is correct.

NLI \ Corpora	Jobs640			
	Recall	Precision	CorInfo	Test Time (seconds)
CHILL	84.53±2.35	86.89±2.71	×	33.23±4.27
CHILLE (2x100%) (QS)	87.66±1.67	88.93±1.99	√	100.81±6.10
CHILLE (2x100%) (V)	85.47±2.50	86.73±2.93	×	101.21±6.23
CHILLE (2x50%) (QS)	87.03±2.09	87.59±2.19	√	86.08±4.28
CHILLE (2x50%) (V)	84.22±1.96	84.76±2.11	×	86.45±4.38
CHILLE (4x25%) (QS)	88.91±2.06	89.48±2.26	√	129.48±3.99
CHILLE (4x25%) (V)	82.50±2.99	83.04±3.16	×	129.57±4.37
CHILLE (6x16.67%) (QS)	**90.00±2.29**	90.57±2.29	√	**170.85±8.04**
CHILLE (6x16.67%) (V)	81.56±2.36	82.08±2.33	×	169.81±8.39
PRECISE	≈ 87.50	100.00	√	n/a
EQ	≈ 47.5	≈ 75.00	×	n/a

As it was mentioned before, when bagging the parsers in CHILLE, the different training samples were not allowed to overlap with each other. For example, in the 2x50% setup, the two random samples form a partition of the training data. It was found consistently that maximizing diversity *among random samples* by using less but different data per sample led to getting better trade off in performance than using more data per sample (but with overlapping among the samples) given the same ensemble size. For example, in all combinations attempted, it is evident that the 2x100% combination (two parsers learned with the entire set of data by differen learners) did not provide a dramatic gain in performance in the two domains to warrant the significantly increase in response time.

In the U.S. Geography domain, CHILLE outperformed CHILL, PRECISE, and EQ in two different combinations: 1) 2x50%, and 2) 4x25%. The response time (test time per trial divided by number of test questions) for the former was roughly 2.09 seconds per question (s/q), and the latter was 3.43 s/q. Although the combination 6x16.67% (i.e. 1/6 of training data) performed best in recall, it was not significantly different from that of 4x25% but the increase in NLI response time is not worth the trivial gain in accuracy. Overall, the 4x25% combination represents the best trade off in efficiency and accuracy as it significantly outperformed CHILL in both recall and precision, dramatically outperformed both PRECISE and EQ, and required a tolerable amount of response time. It is arguable that the increase in response time is worth the wide margin of performance gain produced as CPU speed is still being improved, and, hence, increase in response time with ensemble size could only be a lesser issue over time.[2]

In the job posting domain, CHILLE using the combination 6x16.67% performed the best in recall – significantly outperformed CHILL and EQ, and slightly

---

[2] Besides, there are other ways to further improve response time like using a more efficient language like C++ for implementation instead of the current choice Prolog.

outperformed PRECISE. The response time of this combination was 2.67 s/q which is slightly more than that of 4x25% (2.02 s/q); a faster machine will further minimize the gap. So, CHILLE with 6x16.67% represents the best trade off between efficiency and accuracy.

In both domains, using a query set significantly outperformed voting without significant loss in response time and hence made a better approach in both domains. In other words, returning a query set is a feasible approach if voting were considered the "accepted" approach. Although increasing the (training) sample size for each parser might improve recall for voting, it would be done at the expense of ensemble diversity (a critical factor in ensemble accuracy) and response time.[3]

In all domains, only the dialog based NLI's created using CHILLE with returning a query set and those created using PRECISE were reliable. Of course, one could create dialog based NLI's using CHILLE with voting, and, hence, trivially ensure reliability as the reliability of an NLI depends only on if it is dialog-based or not here. Since traditionally bagging relies on voting to give a single output, it is, therefore, interesting to compare the performance of an NLI created using purely such a traditional approach and that of an NLI created by combining bagging and a dialog-based approach. Alternatively, one can think of the results of non dialog-based NLI's created by bagging with voting as "intermediate points" between the two extremes of using a purely non dialog-based and non ensemble-based approach and a purely dialog-based and ensemble-based approach in a "continuum" of NLI building approaches.

Our results demonstrated that ensuring reliability in an NLI doesn't have to come at the cost of abandoning a learning approach to the problem – a change of NLI framework can solve the problem. Also, using advanced machine learning techniques (e.g. ensemble learning) may open a door for breakthrough in building highly accurate NLI's. In fact, the best recall of the NLI's created using CHILLE with returning a query set are the best reported results in the literature in the attempted domains.

Using a higher amount of data per random sample will lead to learning a bigger semantic parser – one with more parsing operators. Since using a bigger parser will increase NLI response time, it is interesting to see if one can manage response time by keeping the amount of data per sample fairly constant *without* losing accuracy. The experimental results are suggesting that this could be a possibility, and, hence, the CHILLE approach could possibly scale up to larger corpora.

5   Related Work

PRECISE is (non-learning) based NLI that maps English questions to SQL [4]. Its approach has the advantage that it can recognize in advance whether it will be able to correctly answer a question or not (i.e. whether the question is

---

[3] Therefore, the author suspects that using a combination like 4x80% in voting would not dramatically increase performance due to the mediocre results obtained in 2x100% in *voting* but this has yet to be experimentally confirmed.

"semantically tractable" or not); hence, it never delivers incorrect information back to a user. However, it cannot utilize available training data to improve its own performance – a hallmark of learning systems. CHILLE addresses the problem of imperfect precision in a different way that allows it to potentially handle a wider variety of questions and incorporate latest advances in machine learning.

Another related system on the task is called SCISSOR [19]. SCISSOR is an improvement over CHILL in that it integrates syntactic and semantic processing in a mutually beneficial way. Also, since SCISSOR employs a statistical parsing framework, it is more robust than a purely rule-based deterministic approach like CHILL. CHILLE addresses the issue of robustness in parsing in a different way – by learning multiple semantic parsers, it reduces the generalization bias of a learned semantic parser in way similar to how bagging helps to reduce generalization bias of a single classifier.

6   Conclusion and Future Work

CHILLE is a novel approach that generalizes a previous approach called CHILL on learning semantic parsers. By learning an ensemble of semantic parsers, it allows returning a set of interpretations of a user question to 1) facilitate building dialog-based NLI's that guarantee their reliability as a user can choose a correct interpretation from the set if one exists, and 2) at the same time improve the recall of NLI's in a way that allows setting optimal trade off between efficiency and accuracy for a given domain based on the NLI builder's preference. Experimental results in two different real world domains suggested that such an approach can tackle the problem of imperfect precision in the semantic parser (or parsers) of an NLI and utilize available training data for continual improvement in system performance.

Since one source of "semantic intractability" [4] is "incompleteness" of a learned parser with respect to a novel sentence (e.g. the parser might be lacking a parsing operator required for successfully parsing the question), the author will also investigate the issue of making a learned parser complete with respect to a novel sentence provided that the lexicon has all the necessary information for constructing a correct meaning representation for the sentence.

Acknowledgments

Part of this research was supported by the National Science Foundation under grant IRI-9704943.

References

1. Allen, J.F.: Natural Language Understanding (2nd Ed.). Benjamin/Cummings, Menlo Park, CA (1995)
2. Zelle, J.M., Mooney, R.J.: Learning to parse database queries using inductive logic programming. In: Proceedings of the Thirteenth National Conference on Artificial Intelligence (AAAI-96), Portland, OR (1996) 1050–1055

3. Tang, L.R., Mooney, R.J.: Using multiple clause constructors in inductive logic programming for semantic parsing. Lecture Notes in Computer Science **2167** (2001) 466+

4. Popescu, A.M., Etzioni, O., Kautz, H.: Towards a theory of natural language interfaces to databases. In: Proceedings of the 2003 International Conference on Intelligent User Interfaces (IUI-2003), Miami, FL, ACM (2003) 149–157

5. Boye, J., Wirén, M.: Negotiative spoken-dialogue interfaces to databases. In: Proceedings Diabruck (7th workshop on the semantics and pragmatics of dialogue), Wallerfangen, Germany (2003)

6. Thompson, C., Goker, M.: Learning to suggest: The adaptive place advisor. In: Proceedings of the 2000 AAAI Spring Symposium on Adaptive User Interfaces, Menlo Park, CA (2000)

7. Henderson, J., Brill, E.: Bagging and boosting a treebank parser. In: Proceedings of NAACL 2000, Seattle, WA (2000) 34–41

8. Muggleton, S.H., Raedt, L.D.: Inductive logic programming: Theory and methods. Journal of Logic Programming **19** (1994) 629–679

9. Breiman, L.: Bagging predictors. Machine Learning **24**(2) (1996) 123–140

10. Quinlan, J.R.: Learning logical definitions from relations. Machine Learning **5**(3) (1990) 239–266

11. Cameron-Jones, R.M., Quinlan, J.R.: Efficient top-down induction of logic programs. SIGART Bulletin **5**(1) (1994) 33–42

12. Muggleton, S., Feng, C.: Efficient induction of logic programs. In Muggleton, S., ed.: Inductive Logic Programming. Academic Press, New York (1992) 281–297

13. Zelle, J.M., Mooney, R.J.: Combining top-down and bottom-up methods in inductive logic programming. In: Proceedings of the Eleventh International Conference on Machine Learning (ICML-94), New Brunswick, NJ (1994) 343–351

14. Džeroski, S.: Handling noise in inductive logic programming. Master's thesis, Faculty of Electrical Engineering and Computer Science, University of Ljubljana (1991)

15. Rissanen, J.: Modeling by shortest data description. Automatica **14** (1978) 465–471

16. Califf, M.E., Mooney, R.J.: Relational learning of pattern-match rules for information extraction. In: Proceedings of the Sixteenth National Conference on Artificial Intelligence (AAAI-99), Orlando, FL (1999) 328–334

17. Blum, A.: Microsoft english query 7.5: Automatic extraction of semantics from relational databases and OLAP cubes. In: Proceedings of 25th International Conference on Very Large Data Bases (VLDB'99). (1999) 247–248

18. Cestnik, B.: Estimating probabilities: A crucial task in machine learning. In: Proceedings of the Ninth European Conference on Artificial Intelligence, Stockholm, Sweden (1990) 147–149

19. Ge, R., Mooney, R.J.: A statistical semantic parser that integrates syntax and semantics. In: Proceedings of the Ninth Conference on Computational Natural Language Learning. (2005) 9–16

Game-Theoretic Agent Programming in Golog Under Partial Observability

Alberto Finzi[1,2] and Thomas Lukasiewicz[2,1]

[1] Institut für Informationssysteme, Technische Universität Wien
Favoritenstraße 9-11, A-1040 Vienna, Austria
[2] Dipartimento di Informatica e Sistemistica, Università di Roma "La Sapienza"
Via Salaria 113, I-00198 Rome, Italy
{finzi,lukasiewicz}@dis.uniroma1.it

Abstract. We present the agent programming language POGTGolog, which integrates explicit agent programming in Golog with game-theoretic multi-agent planning in partially observable stochastic games. It deals with the case of one team of cooperative agents under partial observability, where the agents may have different initial belief states and not necessarily the same rewards. POGTGolog allows for specifying a partial control program in a high-level logical language, which is then completed by an interpreter in an optimal way. To this end, we define a formal semantics of POGTGolog programs in terms of Nash equilibria, and we specify a POGTGolog interpreter that computes one of these Nash equilibria. We illustrate the usefulness of POGTGolog along a rugby scenario.

1 Introduction

During the recent years, the development of controllers for autonomous agents has become increasingly important in AI. One way of designing such controllers is the programming approach, where a control program is specified through a language based on high-level actions as primitives. Another way is the planning approach, where goals or reward functions are specified and the agent is given a planning ability to achieve a goal or to maximize a reward function. An integration of both approaches has recently been proposed through the seminal language DTGolog [3], which integrates explicit agent programming in Golog [16] with decision-theoretic planning in (fully observable) Markov decision processes (MDPs) [15]. It allows for partially specifying a control program in a high-level language as well as for optimally filling in missing details through decision-theoretic planning, and it can thus be seen as a decision-theoretic extension to Golog, where choices left to the agent are made by maximizing expected utility. From a different perspective, it can also be seen as a formalism that gives advice to a decision-theoretic planner, since it naturally constrains the search space.

DTGolog has several other nice features, since it is closely related to first-order extensions of decision-theoretic planning (see especially [2,19,8]), which allow for (i) compactly representing decision-theoretic planning problems without explicitly referring to atomic states and state transitions, (ii) exploiting such compact representations for efficiently solving large-scale problems, and (iii) nice properties such as *modularity*

C. Freksa, M. Kohlhase, and K. Schill (Eds.): KI 2006, LNAI 4314, pp. 113–127, 2007.

(parts of the specification can be easily added, removed, or modified) and *elaboration tolerance* (solutions can be easily reused for similar problems with few or no extra cost).

As a serious drawback, however, DTGolog is designed only for the single-agent framework. That is, the model of the world essentially consists of a single agent that we control by a DTGolog program and the environment summarized in "nature". But there are many applications where we encounter multiple agents, which may compete against each other, or which may also cooperate with each other. For example, in *robotic soccer*, we have two competing teams of agents, where each team consists of cooperating agents. Here, the optimal actions of one agent generally depend on the actions of all the other ("enemy" and "friend") agents. In particular, there is a bidirected dependence between the actions of two agents, which generally makes it inappropriate to model enemies and friends of the agent that we control simply as a part of "nature". As an example for an important cooperative domain, in *robotic rescue*, mobile agents may be used in the emergency area to acquire new detailed information (such as the locations of injured people in the emergency area) or to perform certain rescue operations. In general, acquiring information as well as performing rescue operations involves several and different rescue elements (agents and/or teams of agents), which cannot effectively handle the rescue situation on their own. Only the cooperative work among all the rescue elements may solve it. Since most of the rescue tasks involve a certain level of risk for humans (depending on the type of rescue situation), mobile agents can play a major role in rescue situations, especially teams of cooperative heterogeneous mobile agents.

This is the motivation behind GTGolog [5], which is a generalization of DTGolog that integrates agent programming in Golog with game-theoretic multi-agent planning in (fully observable) stochastic games [14], also called Markov games [17,11].

Example 1.1. Consider a rugby player a_1, who is deciding his next $n > 0$ moves and wants to cooperate with a team mate a_2. He has to deliberate about if and when it is worth to pass the ball. His options can be encoded by the following GTGolog program:

proc $step(n)$
if $(haveBall(a_1) \land n > 0)$ **then**
    πx $(\pi y$ (**choice**$(a_1 : moveTo(x) \mid passTo(a_2))$ $\|$
        choice$(a_2 : moveTo(y) \mid receive(a_1))))$;
$step(n{-}1)$
end.

This program encodes that while a_1 is the ball owner and $n > 0$, the two agents a_1 and a_2 perform a parallel action choice in which a_1 (resp., a_2) can either go somewhere or pass (resp., receive) the ball. Here, the preconditions and effects of the actions are to be formally specified in a suitable action theory. Given this high-level program and the action theory for a_1 and a_2, the program interpreter then fills in the best moves for a_1 and a_2, reasoning about all the possible interactions between the two agents.

Another crucial aspect of real-world environments, however, is that they are typically only partially observable, due to noisy and inaccurate sensors, or because some relevant parts of the environment simply cannot be sensed. For example, especially in the robotic rescue domain described above, every agent has generally only a very partial view on the environment. However, both DT- and GTGolog assume full observability, and have not been generalized to the partially observable case so far.

In this paper, we try to fill this gap. We present the agent programming language POGTGolog, which extends GTGolog and thus also DTGolog by partial observability. The main contributions of this paper can be summarized as follows:

- We present the agent programming language POGTGolog, which integrates explicit agent programming in Golog with game-theoretic multi-agent planning in partially observable stochastic games [9]. POGTGolog allows for modeling one team of cooperative agents under partial observability, where the agents may have different initial belief states and not necessarily the same rewards (and thus in some sense the team does not necessarily have to be homogeneous).
- POGTGolog allows for specifying a partial control program in a high-level language, which is then completed in an optimal way. To this end, we associate with every POGTGolog program a set of (finite-horizon) policies, which are possible (finite-horizon) instantiations of the program where missing details are filled in. We then define a semantics of a POGTGolog program in terms of Nash equilibria, which are optimal policies (that is, optimal instantiations) of the program.
- We define a POGTGolog interpreter and show that it computes a Nash equilibrium of POGTGolog programs. We also prove that POGTGolog programs can represent partially observable stochastic games, and that the POGTGolog interpreter can be used to compute one of their (finite-horizon) Nash equilibria. Furthermore, we illustrate the usefulness of the POGTGolog approach along a rugby scenario.

2 Preliminaries

In this section, we first recall the main concepts of the situation calculus and of the agent programming language Golog; for further details see especially [16]. We then recall the basics of normal form games and of partially observable stochastic games (POSGs).

2.1 The Situation Calculus

The situation calculus [12,16] is a first-order language for representing dynamically changing worlds. Its main ingredients are *actions*, *situations*, and *fluents*. An *action* is a first-order term of the form $a(u_1, \ldots, u_n)$, where the function symbol a is its *name* and the u_i's are its *arguments*. All changes to the world are the result of actions. For example, the action $moveTo(r, x, y)$ may stand for moving the agent r to the position (x, y). A *situation* is a first-order term encoding a sequence of actions. It is either a constant symbol or of the form $do(a, s)$, where do is a distinguished binary function symbol, a is an action, and s is a situation. The constant symbol S_0 is the *initial situation* and represents the empty sequence, while $do(a, s)$ encodes the sequence obtained from executing a after the sequence of s. For example, the situation $do(moveTo(r, 1, 2), do(moveTo(r, 3, 4), S_0))$ stands for executing $moveTo(r, 1, 2)$ after executing $moveTo(r, 3, 4)$ in the initial situation S_0. We write $Poss(a, s)$, where $Poss$ is a distinguished binary predicate symbol, to denote that the action a is possible to execute in the situation s. A *(relational) fluent* represents a world or agent property that may change when executing an action. It is a predicate symbol whose most right argument is a situation. For example, $at(r, x, y, s)$ may express that the agent r is at the position (x, y) in the situation s. In the situation calculus, a dynamic domain is represented by a *basic action theory* $AT = (\Sigma, \mathcal{D}_{una}, \mathcal{D}_{S_0}, \mathcal{D}_{ssa}, \mathcal{D}_{ap})$, where:

- Σ is the set of (domain-independent) foundational axioms for situations [16].
- $\mathcal{D}_{una}$ is the set of unique names axioms for actions, which express that different actions are interpreted in a different way.
- $\mathcal{D}_{S_0}$ is a set of first-order formulas describing the initial state of the domain (represented by S_0). For example, $at(r, 1, 2, S_0) \wedge at(r', 3, 4, S_0)$ may express that the agents r and r' are initially at the positions $(1, 2)$ and $(3, 4)$, respectively.
- $\mathcal{D}_{ssa}$ is the set of *successor state axioms* [16]. For each fluent $F(\boldsymbol{x}, s)$, it contains an axiom of the form $F(\boldsymbol{x}, do(a, s)) \equiv \Phi_F(\boldsymbol{x}, a, s)$, where $\Phi_F(\boldsymbol{x}, a, s)$ is a formula with free variables among $\boldsymbol{x}, a, s$. These axioms specify the truth of the fluent F in the next situation $do(a, s)$ in terms of the current situation s, and are a solution to the frame problem (for deterministic actions). For example, the axiom $at(r, x, y, do(a, s)) \equiv a = moveTo(r, x, y) \vee (at(r, x, y, s) \wedge \neg \exists x', y' (a = move$-$To(r, x', y')))$ may express that the agent r is at the position (x, y) in the situation $do(a, s)$ iff either r moves to (x, y) in the situation s, or r is already at the position (x, y) and does not move away in s.
- $\mathcal{D}_{ap}$ is the set of *action precondition axioms*. For each action a, it contains an axiom of the form $Poss(a(\boldsymbol{x}), s) \equiv \Pi(\boldsymbol{x}, s)$, which characterizes the preconditions of the action a. For example, $Poss(moveTo(r, x, y), s) \equiv \neg \exists r' \, at(r', x, y, s)$ may express that it is possible to move the agent r to the position (x, y) in the situation s iff no other agent r' is at (x, y) in s.

We use the concurrent version of the situation calculus [16], which is an extension of the standard situation calculus by concurrent actions. A *concurrent action* c is a set of standard actions, which are concurrently executed when c is executed.

2.2 Golog

Golog is an agent programming language that is based on the situation calculus. It allows for constructing *programs* from primitive actions that are defined in a basic action theory AT, where standard (and not so standard) Algol-like control constructs can be used. More precisely, *programs* p in Golog have one of the following forms (where c is a primitive action, ϕ is a *condition*, which is obtained from a situation calculus formula over fluents by suppressing all situation arguments, $p, p_1, p_2, \ldots, p_n$ are programs, $P_1, \ldots, P_n$ are procedure names, and $x, \boldsymbol{x}_1, \ldots, \boldsymbol{x}_n$ are arguments):

1. *Primitive action*: c. Do c.
2. *Test action*: $\phi?$. Test the truth of ϕ in the current situation.
3. *Sequence*: $[p_1; p_2]$. Do p_1 followed by p_2.
4. *Nondeterministic choice of two programs*: $(p_1 \mid p_2)$. Do either p_1 or p_2.
5. *Nondeterministic choice of program argument*: $\pi x \, (p(x))$. Do any $p(x)$.
6. *Nondeterministic iteration*: p^*. Do p zero or more times.
7. *Conditional*: **if** ϕ **then** p_1 **else** p_2. If ϕ is true, then do p_1 else do p_2.

8. *While-loop*: **while** ϕ **do** p. While ϕ is true in the current situation, do p.

9. *Procedures*: **proc** $P_1(\boldsymbol{x}_1)$ p_1 **end** ; ... ; **proc** $P_n(\boldsymbol{x}_n)$ p_n **end** ; p.

For example, the Golog program **while** $\neg at(r, 1, 2)$ **do** $\pi x, y \, (moveTo(r, x, y))$ stands for "while the agent r is not at the position $(1, 2)$, move r to a nondeterministically chosen position (x, y)". Golog has a declarative formal semantics, which is defined in the situation calculus. Given a Golog program p, its execution is represented by a situation calculus formula $Do(p, s, s')$, which encodes that the situation s' can be reached by executing the program p in the situation s.

2.3 Normal Form Games

Normal form games from classical game theory [18] describe the possible actions of $n \geq 2$ agents and the rewards that the agents receive when they simultaneously execute one action each. For example, in *two-finger Morra*, two players E and O simultaneously show one or two fingers. Let f be the total numbers of fingers shown. If f is odd, then O gets f dollars from E, and if f is even, then E gets f dollars from O. More formally, a *normal form game* $G = (I, (A_i)_{i \in I}, (R_i)_{i \in I})$ consists of a set of *agents* $I = \{1, \ldots, n\}$ with $n \geq 2$, a nonempty finite set of *actions* A_i for each agent $i \in I$, and a *reward function* $R_i \colon A \to \mathbf{R}$ for each agent $i \in I$, which associates with every *joint action* $a \in A = \times_{i \in I} A_i$ a reward $R_i(a)$ to agent i. If $n = 2$, then G is called a *two-player normal form game* (or simply *matrix game*). If additionally $R_1 = -R_2$, then G is a *zero-sum matrix game*; we then often omit R_2 and abbreviate R_1 by R.

The behavior of the agents in a normal form game is expressed through the notions of pure and mixed strategies, which specify which of its actions an agent should execute and which of its actions an agent should execute with which probability, respectively. For example, in two-finger Morra, a pure strategy for player E (or O) is to show two fingers, and a mixed strategy for player E (or O) is to show one finger with the probability $7/12$ and two fingers with the probability $5/12$. Formally, a *pure strategy* for agent $i \in I$ is any action $a_i \in A_i$. A *pure strategy profile* is any joint action $a \in A$. If the agents play a, then the *reward* to agent $i \in I$ is given by $R_i(a)$. A *mixed strategy* for agent $i \in I$ is any probability distribution π_i over its set of actions A_i. A *mixed strategy profile* $\pi = (\pi_i)_{i \in I}$ consists of a mixed strategy π_i for each agent $i \in I$. If the agents play π, then the *expected reward* to agent $i \in I$, denoted $\mathbf{E}[R_i(a) \,|\, \pi]$ (or $R_i(\pi)$), is defined as $\sum_{a = (a_j)_{j \in I} \in A} R_i(a) \cdot \Pi_{j \in I} \pi_j(a_j)$.

Towards optimal behavior of the agents in a normal form game, we are especially interested in mixed strategy profiles π, called Nash equilibria, where no agent has the incentive to deviate from its part, once the other agents play their parts. Formally, given a normal form game $G = (I, (A_i)_{i \in I}, (R_i)_{i \in I})$, a mixed strategy profile $\pi = (\pi_i)_{i \in I}$ is a *Nash equilibrium* of G iff for every agent $i \in I$, it holds that $R_i(\pi_i' \circ \pi_{-i}) \leq R_i(\pi)$ for every mixed strategy π_i', where $\pi_i' \circ \pi_{-i}$ is obtained from π by replacing π_i by π_i'. For example, in two-finger Morra, the mixed strategy profile where each player shows one finger resp. two fingers with the probability $7/12$ resp. $5/12$ is a Nash equilibrium. Every normal form game G has at least one Nash equilibrium among its mixed (but not necessarily pure) strategy profiles, and many have multiple Nash equilibria. In the two-player case, they can be computed by linear complementary programming and linear programming in the general and the zero-sum case, respectively. A *Nash selection function f* associates with every normal form game G a unique Nash equilibrium $f(G)$. The expected reward to agent $i \in I$ under $f(G)$ is denoted by $v_f^i(G)$.

2.4 Partially Observable Stochastic Games

Partially observable stochastic games [9] generalize normal form games, partially observable Markov decision processes (POMDPs) [10], and decentralized POMDPs [7,13]. A partially observable stochastic game consists of a set of states S, a normal form game for each state $s \in S$, a set of joint observations of the agents O, and a transition function that associates with every state $s \in S$ and joint action of the agents $a \in A$ a probability distribution on all combinations of next states $s' \in S$ and joint observations $o \in O$. Formally, a *partially observable stochastic game (POSG)* $G = (I, S, (A_i)_{i \in I}, (O_i)_{i \in I}, P, (R_i)_{i \in I})$ consists of a set of *agents* $I = \{1, \ldots, n\}$, $n \geq 2$, a nonempty finite set of *states* S, two nonempty finite sets of *actions* A_i and *observations* O_i for each $i \in I$, a transition function $P \colon S \times A \to PD(S \times O)$, which associates with every state $s \in S$ and joint action $a \in A = \times_{i \in I} A_i$ a probability distribution over $S \times O$, where $O = \times_{i \in I} O_i$, and a *reward function* $R_i \colon S \times A \to \mathbf{R}$ for each $i \in I$, which associates with every state $s \in S$ and joint action $a \in A$ a *reward* $R_i(s, a)$ to i.

Since the actual state $s \in S$ of the POSG G is not fully observable, every agent $i \in I$ has a belief state b_i that associates with every state $s \in S$ the belief of agent i about s being the actual state. A *belief state* $b = (b_i)_{i \in I}$ of G consists of a probability function b_i over S for each agent $i \in I$. The POSG G then defines probabilistic transitions between belief states as follows. The new belief state $b^{a,o} = (b_i^{a,o})_{i \in I}$ after executing the joint action $a \in A$ in $b = (b_i)_{i \in I}$ and jointly observing $o \in O$ is given by $b_i^{a,o}(s') = \sum_{s \in S} P(s', o \mid s, a) \cdot b_i(s) / P_b(b_i^{a,o} \mid b_i, a)$, where $P_b(b_i^{a,o} \mid b_i, a) = \sum_{s' \in S} \sum_{s \in S} P(s', o \mid s, a) \cdot b_i(s)$ is the probability of observing o after executing a in b_i.

The notions of finite-horizon pure (resp., mixed) policies and their rewards (resp., expected rewards) can now be defined as usual using the above probabilistic transitions between belief states. Informally, given a finite horizon $H \geq 0$, a pure (resp., mixed) time-dependent policy associates with every belief state b of G and number of steps to go $h \in \{0, \ldots, H\}$ a pure (resp., mixed) normal form game strategy.

Finally, the notion of a finite-horizon Nash equilibrium for a POSG G is then defined as follows. A policy π is a *Nash equilibrium* of G under a belief state b iff for every agent $i \in I$, it holds that $G_i(H, b, \pi_i' \circ \pi_{-i}) \leq G_i(H, b, \pi_i \circ \pi_{-i})$ for all policies π_i', where $G_i(H, b, \alpha)$ denotes the *H-step reward* to agent $i \in I$ under an initial belief state $b = (b_i)_{i \in I}$ and the policy α. A policy π is a *Nash equilibrium* of G iff it is a Nash equilibrium of G under every belief state b.

3 Partially Observable GTGolog (POGTGolog)

In this section, we present the agent programming language POGTGolog, which is a generalization of GTGolog [5] that allows for partial observability. We first describe the domain theory and the syntax and semantics of POGTGolog programs.

We focus on the case of one team of cooperative agents under partial observability, where the agents may have different initial belief states and not necessarily the same rewards (and so may also be heterogeneous). We assume that (i) each agent knows the initial local belief state of every other agent, and (ii) after each action execution, each agent can observe the actions of every other agent and receives their local observations.

3.1 Domain Theory

POGTGolog programs are interpreted relative to a domain theory, which extends a basic action theory by stochastic actions, reward functions, and utility functions. Formally, in addition to a basic action theory AT, a *domain theory* $DT = (AT, ST, OT)$ consists of a *stochastic theory* ST and an *optimization theory* OT, which are both defined below.

We assume a team $I = \{1, \ldots, n\}$ consisting of $n \geq 2$ cooperative agents $1, \ldots, n$. The finite nonempty set of primitive actions A is partitioned into nonempty sets of primitive actions $A_1, \ldots, A_n$ of agents $1, \ldots, n$, respectively. A *single-agent action* of agent $i \in I$ (resp., *multi-agent action*) is any concurrent action over A_i (resp., A). We assume a finite nonempty set of observations O, which is partitioned into nonempty sets of observations $O_1, \ldots, O_n$ of agents $1, \ldots, n$, respectively. A *single-agent observation* of agent $i \in I$ is any $o_i \in O_i$. A *multi-agent observation* is any $o \in \bigtimes_{i \in I} O_i$.

A *stochastic theory* ST is a set of axioms that define stochastic actions. We represent stochastic actions through a finite set of deterministic actions, as usual [6,3]. When a stochastic action is executed, then with a certain probability, "nature" executes exactly one of its deterministic actions and produces exactly one possible observation. We use the predicate $stochastic(c, s, n, o, \mu)$ to encode that when executing the stochastic action c in the situation s, "nature" chooses the deterministic action n producing the observation o with the probability μ. Here, for every stochastic action c and situation s, the set of all (n, o, μ) such that $stochastic(c, s, n, o, \mu)$ is a probability function on the set of all deterministic components n and observations o of c in s. We also use the notation $prob(c, s, n, o)$ to denote the probability μ such that $stochastic(c, s, n, o, \mu)$. We assume that c and all its nature choices n have the same preconditions. A stochastic action c is indirectly represented by providing a *successor state axiom* for every associated nature choice n. The stochastic action c is *executable* in a situation s with observation o, denoted $Poss(c_o, s)$, iff $prob(c, s, n, o) > 0$ for some n.

The *optimization theory* OT specifies a reward and a utility function. The former associates with every situation s and multi-agent action c, a reward to every agent $i \in I$, denoted $reward(i, c, s)$. The utility function maps every reward and success probability to a real-valued utility $utility(v, pr)$. We assume $utility(v, 1) = v$ and $utility(v, 0) = 0$ for all v. An example is $utility(v, pr) = v \cdot pr$. The utility function suitably mediates between the agent reward and the failure of actions due to unsatisfied preconditions.

Example 3.1 (Rugby Domain). Consider the following rugby domain, which is inspired by the soccer domain in [11]. The rugby field consists of 22 rectangles, which are divided into a 4×5 grid of 20 squares and two goal rectangles (see Fig. 1). We assume a team of two agents $a = \{a_1, a_2\}$ against a (static) team of two agents $o = \{o_1, o_2\}$, where a_1 and o_1 are the *captains* of a and o, respectively. Each agent occupies a square and is able to do one of the following actions on each turn: $N, S, E, W, stand,$

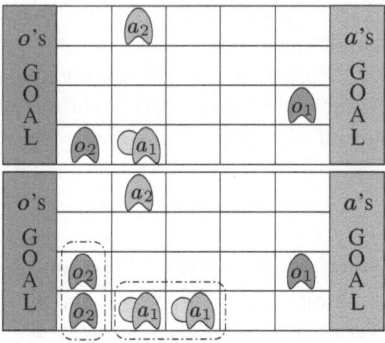

Fig. 1. Rugby Domain: Initial belief states of a_1 and a_2, respectively

$passTo(\alpha)$, and *receive* (move up, move down, move right, move left, no move, pass, and receive the ball, respectively). The ball is represented by an oval and also occupies a square. An agent is a *ball owner* iff it occupies the same square as the ball. The ball follows the moves of the ball owner, and we have a goal when the ball owner steps into the adversary goal. An agent can also pass the ball to another agent of the same team, but this is possible only if the receiving agent is not closer to the opposing end of the field than the ball, otherwise, an offside fault is called by the referee, and the ball possession goes to the captain of the opposing team. When the ball owner goes into the square occupied by the other agent, if the other agent stands, possession of ball changes. Thus, a good defensive maneuver is to stand where the other agent wants to go.

We define the domain theory $DT = (AT, ST, OT)$ as follows. Concerning the basic action theory AT, we assume the deterministic action $move(\alpha, m)$ (encoding that agent α executes m), where $\alpha \in \boldsymbol{a} \cup \boldsymbol{o}$, $m \in \{N, S, E, W, stand, passTo(\alpha'), receive\}$, and α' is a team mate of α, and the fluents $at(\alpha, x, y, s)$ (encoding that agent α is at position (x, y) in situation s) and $haveBall(\alpha, s)$ (encoding that agent α has the ball in situation s). They are defined by the following successor state axioms:

$$at(\alpha, x, y, do(c, s)) \equiv \exists x', y' \, (at(\alpha, x', y', s) \ \wedge \exists m \, (move(\alpha, m) \in c \ \wedge$$
$$((m = stand \vee m = receive \vee \exists\beta \, (m = passTo(\beta))) \wedge x = x' \wedge y = y') \vee$$
$$(m = N \wedge x = x' \wedge y = y'+1) \vee (m = S \wedge x = x' \wedge y = y'-1) \vee$$
$$(m = E \wedge x = x'+1 \wedge y = y') \vee (m = W \wedge x = x'-1 \wedge y = y')));$$
$$haveBall(\alpha, do(c, s)) \equiv \exists\beta \, (haveBall(\beta, s) \wedge (\alpha = \beta \wedge \neg\exists\beta' \, (cngBall(\beta', c, s) \vee$$
$$rcvBall(\beta', c, s))) \vee (\alpha \neq \beta \wedge (cngBall(\alpha, c, s) \vee rcvBall(\alpha, c, s)))).$$

Here, $cngBall(\alpha, c, s)$ is true iff the ball possession changes to α after an action c in s (in the case of either an adversary block or an offside ball passage). The predicate $rcvBall(\alpha, c, s)$ is true iff agent α receives the ball from the ball owner or is in offside.

As for the stochastic theory ST, we assume the stochastic action $moveS(\alpha, m)$, which represents agent α's attempt in doing m among $N, S, E, W, stand, passTo(\beta)$, and *receive*. It can either succeed, and then the deterministic action $move(\alpha, m)$ is executed, or it can fail, and then the deterministic action $move(\alpha, stand)$ (that is, no change) is executed. Furthermore, after each execution of $moveS(\alpha, m)$, agent α can observe the presence of a team mate α' in the direction of the movement, given that agent α' is visible, that is, not covered by another agent:

$$stochastic(\{moveS(\alpha, m)\}, s, \{a\}, \{obs(\beta, out)\}, \mu) \equiv$$
$$\exists \mu_1, \mu_2 \, ((a = move(\alpha, m) \wedge (out = succ \wedge \mu_1 = 0.8 \vee out = fail \wedge \mu_1 = 0.1) \vee$$
$$a = move(\alpha, stand) \wedge (out = succ \wedge \mu_1 = 0.01 \vee out = fail \wedge \mu_1 = 0.09)) \wedge$$
$$(visible(\alpha, \alpha', a, s) \wedge (\beta = \alpha' \wedge \mu_2 = 0.7 \vee \beta = none \wedge \mu_2 = 0.1) \vee$$
$$\neg visible(\alpha, \alpha', a, s) \wedge (\beta = \alpha' \wedge \mu_2 = 0 \vee \beta = none \wedge \mu_2 = 0.2)) \wedge \mu = \mu_1 \cdot \mu_2) \, ;$$
$$stochastic(\{moveS(\alpha, m), moveS(\alpha', m')\}, s, \{a_\alpha, a_{\alpha'}\}, \{o_\alpha, o_{\alpha'}\}, \mu) \equiv$$
$$\exists \mu_1, \mu_2 \, (stochastic(\{moveS(\alpha, m)\}, s, \{a_\alpha\}, \{o_\alpha\}, \mu_1) \wedge$$
$$stochastic(\{moveS(\alpha', m')\}, s, \{a_{\alpha'}\}, \{o_{\alpha'}\}, \mu_2) \wedge \mu = \mu_1 \cdot \mu_2) \, .$$

Here, $visible(\alpha, \alpha', a, s)$ is true if α can observe α' after the execution of a in s. The stochastic action $moveS(\alpha, m)$ is associated with the observations $obs(\beta, out)$, where $\beta \in \{\alpha', none\}$ and $r \in \{succ, fail\}$. That is, after the execution of the action $move(\alpha, m)$, agent α can observe both whether its team mate α' is present or not (first argument) and the success or failure of the action (second argument). Note that we assume that $obs(\alpha', out)$ has the probability zero, if α' is not visible. Notice also that in the last axiom, we assume the independence of the observations.

As for the optimization theory OT, the reward function for the agents is defined by:

$$reward(\alpha, c, s) = r \equiv \exists \alpha'(goal(\alpha', do(c, s)) \wedge (\alpha' \in \mathbf{a} \wedge r = M \vee$$
$$\alpha' \in \mathbf{o} \wedge r = -M)) \vee \neg \exists \alpha' \, (goal(\alpha', do(c, s)) \wedge evalTeamPos(c, r, s)) \, .$$

Here, the reward of agent α is very high (that is, M stands for a "big" integer), if a team mate scores a goal. Otherwise, the reward depends on $evalTeamPos(c, r, s)$, that is, the position of its team relative to the adversary team as well as the ball possession.

3.2 Belief States

We next introduce belief states over situations, and define the semantics of actions in terms of transitions between belief states. A *belief state (over situations)* has the form $b = (b_i)_{i \in I}$, where every b_i is a set of pairs (s, μ) consisting of a standard situation s and a real $\mu \in (0, 1]$ such that all μ sum up to 1. Informally, every b_i represents the belief of agent $i \in I$ expressed as a probability distribution over ordinary situations. The *probability* of a fluent formula $\phi(s)$ in $b = (b_i)_{i \in I}$, denoted $\phi(b)$, is the probability vector $pr = (pr_i)_{i \in I}$, where every pr_i with $i \in I$ is the sum of all μ such that $\phi(s)$ is true and $(s, \mu) \in b_i$. Similarly, $reward(c, b)$ denotes the vector $r = (r_i)_{i \in I}$, where every r_i with $i \in I$ is the sum of all $reward(i, c, s) \cdot \mu$ such that $(s, \mu) \in b_i$.

Given a deterministic action c and a belief state $b = (b_i)_{i \in I}$, the *successor belief state* after executing c in b, denoted $do(c, b)$, is the belief state $b' = (b'_i)_{i \in I}$, where $b'_i = \{(do(c, s), \mu / Poss(c, b)) \mid (s, \mu) \in b_i, Poss(c, s)\}$ for every $i \in I$. Given a stochastic action c, an observation o of c, and a belief state $b = (b_i)_{i \in I}$, the *successor belief state* after executing c in b and observing o, denoted $do(c_o, b)$, is the belief state $b' = (b'_i)_{i \in I}$, where b'_i is obtained from all pairs $(do(n, s), \mu \cdot \mu')$ such that $(s, \mu) \in b_i$, $Poss(c, s)$, and $\mu' = prob(c, s, n, o) > 0$ by normalizing the probabilities to sum up to 1.

The probability of observing o after executing the stochastic action c in the belief state $b = (b_i)_{i \in I}$, denoted $prob(c, b, o)$, is the vector $pr = (pr_i)_{i \in I}$, where every pr_i with $i \in I$ is the sum of all $\mu \cdot \mu'$ such that $(s, \mu) \in b_i$ and $\mu' = prob(c, s, n, o) > 0$.

Example 3.2 (Rugby Domain cont'd). Consider the following scenario relative to the domain theory of Example 3.1 (see Fig. 1). We focus only on controlling the members of the team a, which cooperate to score a goal against the (static) team o. The captain a_1 of a has a complete view of the situation, and its belief state b_{a_1} is shown in Fig. 1, upper part: There is only the situation s_1 with probability 1 such that $at(a_1, 2, 1, s_1)$, $at(a_2, 2, 4, s_1)$, $at(o_2, 1, 1, s_1)$, $at(o_1, 5, 2, s_1)$, and $haveBall(a_1, s_1)$ are true. That is, the captain o_1 of o is very close to the goal of a. From the perspective of a_1, the goal seems quite done: a_1 can pass to a_2, which has a paved way towards the goal. But a_1 has to cooperate with a_2, whose vision of the situation is more confused and expressed by the belief state b_{a_2} in Fig. 1, lower part: a_2 could be either at (a) $(1, 1)$ or at (b) $(1, 2)$, and a_1 could be either at (c) $(2, 1)$ or at (d) $(3, 1)$. Hence, a_2's belief state may e.g. be given by $b_{a_2} = \{(s_{a,c}, 0.5), (s_{a,d}, 0.3), (s_{b,c}, 0.1), (s_{b,d}, 0.1)\}$.

3.3 Syntax

Given the actions specified by a domain theory DT, a *program p* in POGTGolog has one of the following forms (where α is a multi-agent action, ϕ is a condition, p, p_1, p_2 are programs, $a_{i,1}, \ldots, a_{i,n_i}$ are actions of agents $i \in I$, and $J \subseteq I$ with $|J| \geq 2$):

1. *Deterministic or stochastic action:* α. Do α.
2. *Nondeterministic action choice of agent $i \in I$:* **choice**$(i: a_{i,1}| \cdots |a_{i,n_i})$.
   Do an optimal action (for agent $i \in I$) among $a_{i,1}, \ldots, a_{i,n_i}$.
3. *Nondeterministic joint action choice:* $\|_{j \in J}$**choice**$(j : a_{j,1}| \cdots |j : a_{j,n_j})$.
   Do any action $\|_{j \in J} a_{j,i_j}$ with an optimal probability $\pi = \Pi_{j \in J} \pi_{j,i_j}$.
4. *Test action:* ϕ?. Test the truth of ϕ in the current situation.
5. *Action sequence:* $[p_1; p_2]$. Do p_1 followed by p_2.
6. *Nondeterministic choice of two programs:* $(p_1 | p_2)$. Do p_1 or p_2.
7. *Nondeterministic choice of an argument:* $\pi x \, (p(x))$. Do any $p(x)$.
8. *Nondeterministic iteration:* $p^\star$. Do p zero or more times.
9. *Conditionals:* **if** ϕ **then** p_1 **else** p_2.
10. *While-loops:* **while** ϕ **do** p.
11. *Procedures, including recursion.*

Hence, compared to Golog, we now also have multi-agent actions and stochastic actions (instead of only primitive resp. deterministic actions). Furthermore, we now additionally have different kinds of nondeterministic action choices for the agents in 2 and 3, where one or any subset of the agents in I can choose among a finite set of single-agent actions. The formal semantics of 2 and 3 is defined in such a way that an optimal action is chosen for the agents (see Section 3.4). As usual, the sequence operator ";" is associative (for example, $[[p_1; p_2]; p_3]$ and $[p_1; [p_2; p_3]]$ have the same semantics), and we often use "$p_1; p_2$" to abbreviate "$[p_1; p_2]$" when there is no danger of confusion.

Example 3.3 (Rugby Domain cont'd). Consider again the scenario (and its belief states b_{a_1} and b_{a_2}) of Example 3.2 relative to the domain theory of Example 3.1 (see Fig. 1). Both agents a_1 and a_2 have to decide when (and if) it is worth to pass the ball, considering that if a_1 tries to pass while a_2 is in offside (for example, in $s_{a,d}$ or $s_{b,d}$), then the ball goes to the captain o_1 of the adversary team o, which is in a very good position to score a goal. The subsequent POGTGolog program, denoted $schema$, represents a way of acting of a_1 and a_2 in this scenario, where the agents a_1 and a_2 have two possible chances to coordinate themselves in order to pass the ball, and thereafter both of them have to run towards the goal (with or without the ball).

> **choice**$(a_1:\ moveS(a_1, E) \mid moveS(a_1, stand) \mid moveS(a_1, passTo(a_2))) \parallel$
>   **choice**$(a_2:\ moveS(a_2, S) \mid moveS(a_2, E) \mid moveS(a_2, receive))$;
> **choice**$(a_1:\ moveS(a_1, E) \mid moveS(a_1, stand) \mid moveS(a_1, passTo(a_2))) \parallel$
>   **choice**$(a_2:\ moveS(a_2, E) \mid moveS(a_2, receive))$;
> $\{moveS(a_1, E), moveS(a_2, E)\}$;
> $\{moveS(a_1, E), moveS(a_2, E)\}$.

3.4 Semantics

We now define the formal semantics of POGTGolog programs p relative to a domain theory $DT = (AT, ST, OT)$ in terms of Nash equilibria. We first associate with every POGTGolog program p, a belief state b, and horizon $H \geq 0$, a set of executable H-step policies π along with their expected utility U_i to every agent $i \in I$. We then define the notion of a Nash equilibrium to characterize a subset of optimal such policies, which is the natural semantics of a POGTGolog program relative to a domain theory.

Intuitively, given a horizon $H \geq 0$, an H-step policy π of a POGTGolog program p relative to a domain theory is obtained from the H-horizon part of p by replacing every single-agent choice by a single action, and every multi-agent choice by a collection of probability distributions, one over the actions of each agent. Formally, for every POGT-Golog program p, we define the *nil-terminated variant* of p, denoted $\widehat{p}$, by $\widehat{p} = [p_1; \widehat{p}_2]$, if $p = [p_1; p_2]$, and $\widehat{p} = [p; nil]$, otherwise. Given a POGTGolog program p relative to a domain theory DT, a horizon $H \geq 0$, and a start belief state b, we say that π is an H-step *policy* of p in b iff $DT \models G(\widehat{p}, b, H, \pi, \langle v, pr \rangle)$, where $v = (v_i)_{i \in I}$ and $pr = (pr_i)_{i \in I}$. The *expected H-step utility* of π in b to $i \in I$, denoted $U_i(H, b, \pi)$, is $utility(v_i, pr_i)$. Here, we define the macro $G(\widehat{p}, b, h, \pi, \langle v, pr \rangle)$ by induction as follows:

- Null program ($\widehat{p} = nil$) or zero horizon ($h = 0$):

$$G(\widehat{p}, b, h, \pi, \langle v, pr \rangle) =_{def} \pi = stop \land v = \mathbf{0} \land pr = \mathbf{1}.$$

  Intuitively, p ends when it is null or at the horizon end.

- First program action c is deterministic (resp., stochastic with observation):

$$G([c; p'], b, h, \pi, \langle v, pr \rangle) =_{def}$$
$$(Poss(c, b) = \mathbf{0} \land \pi = stop \land v = \mathbf{0} \land pr = \mathbf{1}) \lor$$
$$(Poss(c, b) > \mathbf{0} \land \exists \pi', v', pr'\, (G(p', do(c, b), h-1, \pi', \langle v', pr' \rangle) \land$$
$$\pi = c; \pi' \land v = v' + reward(c, b) \land pr = pr' \cdot Poss(c, b)).$$

  Here, $(s_i)_{i \in I}$ op $(t_i)_{i \in I} = (s_i$ op $t_i)_{i \in I}$ for op $\in \{+, \cdot\}$. Informally, suppose that $\widehat{p} = [c; p']$, where c is a deterministic action (resp., stochastic action with observation). If c is not executable in the belief state b, then p has only the policy $\pi = stop$ along with the expected reward $v = 0$ and the success probability $pr = \mathbf{0}$. Otherwise, the optimal execution of $[c; p']$ in the belief state b depends on that one of p in $do(c, b)$. Observe that c is executable in b with the probability $Poss(c, b)$, which affects the overall success probability pr.

- Stochastic first program action c (choice of nature):

$$G([c\,;p'], b, h, \pi, \langle v, pr\rangle) =_{def}$$
$$\exists \pi_q, v_q, pr_q \, (\textstyle\bigwedge_{q=1}^{l} G([c_{o_q}\,;p'], b, h, c_{o_q}\,;\pi_q, \langle v_q, pr_q\rangle) \wedge$$
$$\pi = c_{o_q}\,;\textbf{for } q=1 \textbf{ to } l \textbf{ do if } o_q \textbf{ then } \pi_q \wedge$$
$$v = \textstyle\sum_{q=1}^{l} v_q \cdot prob(c, b, o_q) \wedge pr = \textstyle\sum_{q=1}^{l} pr_q \cdot prob(c, b, o_q))\,.$$

Here, $o_1, \ldots, o_l$ are the possible observations. The generated policy is a conditional plan in which every such observation o_q is considered.

- Nondeterministic first program action (choice of agent $i \in I$):

$$G([\textbf{choice}(i\colon a_1|\cdots|a_n)\,;p'], b, h, \pi, \langle v, pr\rangle) =_{def}$$
$$\exists \pi_q, v_q, pr_q, k \, (\textstyle\bigwedge_{q=1}^{n} G([a_q\,;p'], b, h, a_q\,;\pi_q, \langle v_q, pr_q\rangle) \wedge$$
$$k \in \{1, \ldots, n\} \wedge \pi = a_k\,;\textbf{for } q=1 \textbf{ to } n \textbf{ do if } \psi_q \textbf{ then } \pi_q \wedge$$
$$v = v_k \wedge pr = pr_k)\,.$$

Here, the ψ_q's denote conditions that the other agents in I test to observe j's choice.

- Nondeterministic first program action (joint choice of the agents in J):

$$G([\,\|_{j\in J}\textbf{choice}(j\colon a_{j,1}|\cdots|a_{j,n_j})\,;p'], b, h, \pi, \langle v, pr\rangle) =_{def}$$
$$\exists \pi_a, v_a, pr_a, \pi_a \, (\textstyle\bigwedge_{a\in A} G([\bigcup_{j\in J} a_j\,;p'], b, h, \bigcup_{j\in J} a_j\,;\pi_a, \langle v_a, pr_a\rangle) \wedge$$
$$\textstyle\bigwedge_{j\in J} (\pi_j \in PD(\{a_{j,1}, \ldots, a_{j,n_j}\})) \wedge$$
$$\pi = \|_{j\in J}\pi_j\,;\textbf{for each } a \in A \textbf{ do if } \phi_a \textbf{ then } \pi_a \wedge$$
$$v = \textstyle\sum_{a\in A} v_a \cdot \Pi_{j\in J}\pi_j(a_j) \wedge pr = \textstyle\sum_{a\in A} pr_a \cdot \Pi_{j\in J}\pi_j(a_j)\,.$$

Here, $A = \times_{j\in J}\{a_{j,1}, \ldots, a_{j,n_j}\}$, and each π_j with $j \in J$ is a probability distribution over $\{a_{j,1}, \ldots, a_{j,n_j}\}$. Informally, we compute the joint policy for each possible combination of actions $a \in A$. The conditions ϕ_a with $a \in A$ are to observe what the agents have actually executed.

- Nondeterministic choice of two programs:

$$G([(p_1\,|\,p_2)\,;p'], b, h, \pi, \langle v, pr\rangle) =_{def}$$
$$\exists \pi_q, v_q, pr_q, k \, (\textstyle\bigwedge_{q\in\{1,2\}} G([p_q\,;p'], b, h, \pi_q, \langle v_q, pr_q\rangle) \wedge$$
$$k \in \{1, 2\} \wedge \pi = \pi_k \wedge v = v_k \wedge pr = pr_k)\,.$$

- Test action:

$$G([\phi?\,;p'], b, h, \pi, \langle v, pr\rangle) =_{def} (\phi[b] = \mathbf{0} \wedge \pi = stop \wedge v = \mathbf{0} \wedge pr = 0) \vee$$
$$\exists pr'(\phi[b] > \mathbf{0} \wedge G(p', b, h, \pi, \langle v, pr'\rangle) \wedge pr = pr' \cdot \phi[b])\,.$$

Informally, let $p = [\phi?\,;p']$. If ϕ is false in b, then p has only the policy $\pi = stop$ along with the expected reward $v = 0$ and the success probability $pr = 0$. Otherwise, π is a policy of p with the expected reward v and success probability $pr' \cdot \phi[b]$ iff π is a policy of p' with the expected reward v and success probability pr'.

- The macro G is naturally extended to nondeterministic choices of action arguments, nondeterministic iterations, conditionals, while-loops, and procedures.

We are now ready to define the notion of a Nash equilibrium as follows. An H-step policy of a POGTGolog program p in a belief state b is an H-step Nash equilibrium of p in b iff, for every agent $i \in I$, it holds that $U_i(H, b, \pi) \leq U_i(H, b, \pi')$ for all H-step policies π' of p in b obtained from π by modifying only actions of agent i.

Example 3.4 (Rugby Domain cont'd). Consider again the scenario (and its belief states b_{a_1} and b_{a_2}) of Example 3.2 relative to the domain theory of Example 3.1 (see Fig. 1). Assuming the horizon $H = 4$, a 4-step policy π of the POGTGolog program *schema* of Example 3.3 is given by $DT \models G([schema; nil], (b_{a_1}, b_{a_2}), 4, \pi, \langle(v_1, v_2), (pr_1, pr_2)\rangle)$. For agent a_1, an optimal way of acting is to pass the ball as soon as possible, which can be encoded by the following (pure) 4-step policy $\pi_{a_1} = c; \pi_{a_1}^1$, where $c = \{moveS(a_1, passTo(a_2)), moveS(a_2, receive)\}$, and $\pi_{a_1}^1$ is an optimal 3-step policy of *schema'* in the belief state $(do(c, b_{a_1}), do(c, b_{a_2}))$. Here, *schema'* is obtained from *schema* by removing the first nondeterministic joint action choice. The policy π_{a_1} gives to agent a_2 three $moveS(a_2, E)$ attempts to achieve the touch-line. From the standpoint of a_2, instead, it is worth to do a $moveS(a_2, S)$ to observe if agent a_1 is aligned, trying to minimize the likelihood of a wrong passage. In this case, a_1 has to delay the passage waiting for the move of a_2. The resulting (pure) 4-step policy π_{a_2} is more favorable to a_2's belief state:

$$\pi_{a_2} = c; \text{ if } obs(a_1, succ) \text{ then } \pi_{a_2}^{1,o_1}$$
$$\text{else if } obs(a_1, fail) \text{ then } \pi_{a_2}^{1,o_2}$$
$$\text{else if } obs(none, succ) \text{ then } \pi_{a_2}^{2,o_3}$$
$$\text{else if } obs(none, fail) \text{ then } \pi_{a_2}^{2,o_4},$$

where $c = \{moveS(a_1, S), moveS(a_2, stand)\}$ and $\pi_{a_2}^{k,o_i}$ is an optimal 3-step policy of *schema'*, when observing o_i after the execution of c from (b_{a_1}, b_{a_2}). Given this conflict of opinions, an optimal compromise for both a_1 and a_2 is a Nash equilibrium.

4   A POGTGolog Interpreter

In this section, we define an interpreter for POGTGolog programs relative to domain theories and provide optimality and representation results.

We define an interpreter for POGTGolog programs p relative to a domain theory DT by specifying the macro $DoG(\hat{p}, b, H, \pi, \langle v, pr \rangle)$, which takes as input the *nil*-terminated variant $\hat{p}$ of a POGTGolog program p, a belief state $b = (b_i)_{i \in I}$, and a finite horizon $H \geq 0$, and which computes as output an optimal H-step policy π and the vectors $v = (v_i)_{i \in I}$ and $pr = (pr_i)_{i \in I}$, respectively, where v_i is the expected H-step reward of π to i, and $pr_i \in [0, 1]$ is the H-step success probability of π for i.

We define the macro $DoG(\hat{p}, b, h, \pi, \langle v, pr \rangle)$ in nearly the same way as the macro $G(\hat{p}, b, h, \pi, \langle v, pr \rangle)$ in Section 3.4, except for the following modifications:

- Nondeterministic first program action (choice of agent $i \in I$): The characterization of DoG is obtained from the one of G by replacing the condition "$k \in \{1, \ldots, n\}$" by the condition "$k = \text{argmax}_{q \in \{1, \ldots, n\}} utility(v_{q,i}, pr_{q,i})$", where $v_q = (v_{q,i})_{i \in I}$ and $pr_q = (pr_{q,i})_{i \in I}$. Informally, given the possible actions $a_1, \ldots, a_n$ for agent $i \in I$, we select an optimal one for i, that is, one with greatest $utility(v_{q,i}, pr_{q,i})$.
- Nondeterministic first program action (joint choice of the agents in J): The characterization of DoG is obtained from the one of G by replacing "$\bigwedge_{j \in J}(\pi_j \in PD(\{a_{j,1}, \ldots, a_{j,n_j}\}))$" by "$(\pi_j)_{j \in J} = selectNash(\{utility(v_a, pr_a)|_J \mid a \in A\})$", where $utility((s_i)_{i \in I}, (t_i)_{i \in I}) = (utility(s_i, t_i))_{i \in I}$, and $s|_J$ is the restriction of s to J, for $s = (s_i)_{i \in I}$ and $J \subseteq I$. Informally, we compute a local Nash equilibrium

.

$(\pi_j)_{j \in J}$ from a normal form game using the Nash selection function $selectNash$. Note that we assume that all agents have the same Nash selection functions, and thus they automatically select a common unique Nash equilibrium.

- Nondeterministic choice of two programs: The characterization of DoG is obtained from the one of G by replacing "$k \in \{1, 2\}$". by "$k = \mathrm{argmax}_{q \in \{1,2\}} \; utility(v_{q,j}, pr_{q,j})$". Informally, given two possible programs p_1 and p_2, we select an optimal one for agent j, that is, one with greatest $utility(v_{q,j}, pr_{q,j})$.

The following theorem shows the important result that the macro DoG is optimal in the sense that, for every horizon $H \geq 0$, among the set of all H-step policies π of a POGTGolog program p relative to a domain theory DT in a belief state b, it computes an H-step Nash equilibrium and its expected H-step utility.

Theorem 4.1. *Let* $DT = (AT, ST, OT)$ *be a domain theory, and let* p *be a POGT-Golog program relative to* DT. *Let* b *be a belief state (over situations), let* $H \geq 0$ *be a horizon, and let* $DT \models DoG(\hat{p}, b, H, \pi, \langle v, pr \rangle)$. *Then,* π *is an* H-*step Nash equilibrium of* p *in* b, *and* $utility(v_i, pr_i)$ *is its expected* H-*step utility to agent* $i \in I$.

The next theorem shows that, given any horizon $H \geq 0$, every POSG can be encoded as a program p in POGTGolog, such that DoG computes one of its H-step Nash equilibria and its expected H-step reward.

Theorem 4.2. *Let* $G = (I, Z, (A_i)_{i \in I}, (O_i)_{i \in I}, P, (R_i)_{i \in I})$ *be a POSG, let* $H \geq 0$ *be a horizon, and let* b_0 *be a belief state of* G. *Then, there exists a domain theory* $DT = (AT, ST, OT)$, *and a set of POGTGolog programs* $\{\hat{p}^h \mid h \in \{0, \ldots, H\}\}$ *relative to* DT *such that* $\delta = (\delta_i)_{i \in I}$ *is an* H-*step Nash equilibrium for* G, *where every* $(\delta_i(b, h))_{i \in I} = (\pi_i)_{i \in I}$ *is given by* $DT \models DoG(\hat{p}^h, B_b, h+1, \|_{i \in I} \pi_i ; \pi', \langle v, pr \rangle)$, *for every belief state* b *reachable from* b_0 *and every* $h \in \{0, \ldots, H\}$, *where* B_b *is the belief state over situations associated with the belief state* b *of* G. *Furthermore, the expected* H-*step reward* $G_i(H, b, \delta)$ *to agent* $i \in I$ *is given by* $utility(v_i, pr_i)$, *where* $DT \models DoG(\hat{p}^H, B_b, H+1, \pi, \langle v, pr \rangle)$, *for every belief state* b *reachable from* b_0.

5 Summary and Outlook

We have presented the agent programming language POGTGolog, which combines explicit agent programming in Golog with game-theoretic multi-agent planning in POSGs, and which allows for modeling one team of cooperative agents under partial observability, where the agents may have different initial belief states and not necessarily the same rewards. It allows for specifying a partial control program in a high-level logical language, which is then completed by an interpreter in an optimal way. To this end, we have defined a formal semantics of POGTGolog programs in terms of Nash equilibria, and specified a POGTGolog interpreter that computes one of these Nash equilibria. We have illustrated the usefulness of this approach along a rugby scenario.

An interesting topic of future research is to generalize POGTGolog to the case in which we can give up the assumption that every agent knows the initial local belief states of all the other agents, their locally executed actions, and their local observations.

This may, for example, be achieved by explicit communication between the agents or by independence assumptions between the local actions and observations of different agents. A further direction of future research is to generalize POGTGolog to the case of two competing teams of cooperative agents under partially observability.

Acknowledgments. This work was supported by the Austrian Science Fund Project P18146-N04 and by a Heisenberg Professorship of the German Research Foundation (DFG). We thank the reviewers for their comments, which helped to improve this work.

References

1. F. Bacchus, J. Y. Halpern, and H. J. Levesque. Reasoning about noisy sensors and effectors in the situation calculus. *Artif. Intell.*, 111:171–208, 1999.
2. C. Boutilier, R. Reiter, and B. Price. Symbolic dynamic programming for first-order MDPs. In *Proceedings IJCAI-2001*, pp. 690–700.
3. C. Boutilier, R. Reiter, M. Soutchanski, and S. Thrun. Decision-theoretic, high-level agent programming in the situation calculus. In *Proceedings AAAI-2000*, pp. 355–362.
4. A. Ferrein, C. Fritz, and G. Lakemeyer. Using Golog for deliberation and team coordination in robotic soccer. *Künstliche Intelligenz*, 1:24–43, 2005.
5. A. Finzi and T. Lukasiewicz. Game-theoretic agent programming in Golog. In *Proceedings ECAI-2004*, pp. 23–27.
6. A. Finzi and F. Pirri. Combining probabilities, failures and safety in robot control. In *Proceedings IJCAI-2001*, pp. 1331–1336.
7. C. V. Goldman and S. Zilberstein. Decentralized control of cooperative systems: Categorization and complexity analysis. *J. Artif. Intell. Res.*, 22:143–174, 2004.
8. C. Guestrin, D. Koller, C. Gearhart, and N. Kanodia. Generalizing plans to new environments in relational MDPs. In *Proceedings IJCAI-2003*, pp. 1003–1010.
9. E. A. Hansen, D. S. Bernstein, and S. Zilberstein. Dynamic programming for partially observable stochastic games. In *Proceedings AAAI-2004*, pp. 709–715.
10. L. Pack Kaelbling, M. L. Littman, and A. R. Cassandra. Planning and acting in partially observable stochastic domains. *Artif. Intell.*, 101(1–2):99–134, 1998.
11. M. L. Littman. Markov games as a framework for multi-agent reinforcement learning. In *Proceedings ICML-1994*, pp. 157–163.
12. J. McCarthy and P. J. Hayes. Some philosophical problems from the standpoint of Artificial Intelligence. In *Machine Intelligence* 4, pp. 463–502. Edinburgh University Press, 1969.
13. R. Nair, M. Tambe, M. Yokoo, D. V. Pynadath, and S. Marsella. Taming decentralized POMDPs: Towards efficient policy computation for multiagent settings. In *Proceedings IJCAI-2003*, pp. 705–711. 2003.
14. G. Owen. *Game Theory: Second Edition*. Academic Press, 1982.
15. M. L. Puterman. *Markov Decision Processes: Discrete Stochastic Dynamic Programming*. Wiley, 1994.
16. R. Reiter. *Knowledge in Action: Logical Foundations for Specifying and Implementing Dynamical Systems*. MIT Press, 2001.
17. J. van der Wal. *Stochastic Dynamic Programming*, volume 139 of *Mathematical Centre Tracts*. Morgan Kaufmann, 1981.
18. J. von Neumann and O. Morgenstern. *The Theory of Games and Economic Behavior*. Princeton University Press, 1947.
19. S. W. Yoon, A. Fern, and B. Givan. Inductive policy selection for first-order MDPs. In *Proceedings UAI-2002*, pp. 569–576.

Finding Models for Blocked 3-SAT Problems in Linear Time by Systematical Refinement of a Sub-model

Gábor Kusper

Eszterházy Károly College
Department of Information Technology
1. sqr. Eszterházy , Eger 3300 Hungary
gkusper@sztech.ektf.hu
http://sztech.ektf.hu/~gkusper

Abstract. We report a polynomial time SAT problem instance, the Blocked SAT problem. A blocked clause set, an instance of the Blocked SAT problem, contains only blocked clauses. A close is blocked (for resolution) if it has a literal on which no resolution is possible in the clause set. We know from work of O. Kullmann that a blocked clause can be added or deleted from a clause set without changing its satisfiability. Hence, any blocked clause set is satisfiable, but it is not clear how to find a satisfying assignment for it. We introduce the Blocked SAT Solver algorithm, which provides a model for Blocked SAT problems in linear time, if we know at least one blocked literal per clause. To collect these information polynomial time is needed in general. We show that in case of 3-SAT we can collect these information in linear time. This means that the Blocked 3-SAT problem is a linear time problem. We also discuss how to use blocked clauses if the whole clause set is not blocked.

1 Introduction

Propositional Satisfiability is the problem of determining, for a formula of the propositional calculus, if there is an assignment of truth values to its variables for which that formula evaluates the true. By SAT we mean the problem of propositional satisfiability for formulae in conjunctive normal form (CNF).

SAT is the first, and one of the simplest, of the many problems which have been shown to be NP-complete [Coo71]. It is dual of propositional theorem proving, and many practical NP-hard problems may be transformed efficiently to SAT. Thus, a good SAT algorithm would likely have considerable utility. It seems improbable that a polynomial time algorithm will be found for the general SAT problem but we know that there are restricted SAT problems that can be solved in polynomial time. So a "good" SAT algorithm should check the input SAT instance first whether it is an instance of such a restricted SAT problem. In this work we introduce the Blocked SAT problem, which is solvable in polynomial time. We list some polynomial time solvable restricted SAT problems:

C. Freksa, M. Kohlhase, and K. Schill (Eds.): KI 2006, LNAI 4314, pp. 128–142, 2007.
© Springer-Verlag Berlin Heidelberg 2007

1. The restriction of SAT to instances where all clauses have length k is denoted by k-SAT. Of special interest are *2-SAT* and *3-SAT*: 3 is the smallest value of k for which k-SAT is **NP**-complete, while 2-SAT is solvable in linear time [EIS76, APT79].

2. *Horn SAT* is the restriction to instances where each clause has at most one un-negated variable. Horn SAT is solvable in linear time [DG84, Scu90], as are a number of generalizations such as *renamable Horn SAT* [Lew78, Asp80], *extended Horn SAT* [CH91] and *q-Horn SAT* [BHS94, BCH+94].

3. The hierarchy of *tractable* satisfiability problems [DE92], which is based on Horn SAT and 2-SAT, is solvable in polynomial time. An instance on the k level of the hierarchy is solvable in $O(nk + 1)$ time.

4. *Nested SAT*, in which there is a linear ordering on the variables and no two clauses overlap with respect to the interval defined by the variables they contain [Knu90].

5. SAT in which no variable appears more than twice. All such problems are satisfiable if they contain no unit clauses [Tov84].

6. *r,r-SAT*, where r,s-SAT is the class of problems in which every clause has exactly r literals and every variable has at most s occurrences. All r,r-SAT problems are satisfiable in polynomial time [Tov84].

7. A formula is *SLUR* (Single Lookahead Unit Resolution) *solvable* if, for all possible sequences of selected variables, algorithm SLUR does not give up. Algorithm SLUR is a nondeterministic algorithm based on unit propagation. It eventually gives up the search if it starts with, or creates, an unsatisfiable formula with no unit clauses. The class of SLUR solvable formulae was developed as a generalization including Horn SAT, renamable Horn SAT, extended Horn SAT, and the class of CC-balanced formulae [SAF+95].

8. *Resolution-Free* SAT Problem, where every resolution results in a tautologous clause, is solvable in linear time [Kus05].

The Blocked SAT problem is also a restriction of SAT to instances where each clause of the clause set is blocked, i.e., we have at least one blocked literal in each clause. It is a generalization of the Resolution-Free SAT problem, where all literals are blocked.

A clause is blocked in a clause set if it has a literal on which no resolution is possible in that clause set, i.e., it is blocked for resolution. The notion of blocked clause was introduced by O. Kullmann in [Kul99a, Kul99b]. He studied blocked clauses because we wanted to add new short clauses to the input clause set to improve the worst case time complexity of his algorithm. He proved that a blocked clause can be added or deleted from a clause set without changing its satisfiability.

Based on that idea it is easy to show that any blocked clause set is satisfiable, as if we remove all blocked clauses then we obtain the trivially satisfiable clause set. But this process do not show how to find a model for blocked clause sets.

The Blocked SAT Solver algorithm provides a model for Blocked SAT problems in linear time, if we know at least one blocked literal per clauses. It uses heavily the notion of sub-model [Kus02, Kus05].

A sub-model is a partial assignment created, by definition, by the negation of a resolution-mate. A resolution-mate is obtained from a clause, the generator clause, by negating one of its literals, the generator literal.

The algorithm exploits the fact that if a sub-model is generated form a blocked clause (generator clause) using one of its blocked literal as the generator one (generator literal), then it is a model for those clauses in the set which contain either positively or negatively the generator literal. Moreover, it satisfies from the rest the ones which differ from it. (The clause A differs from the clause B if it has a literal a, i.e., $a \in A$, which occurs in B negatively, i.e., $\bar{a} \in B$.) Which means it satisfies almost the whole clause set! Only those clauses are not satisfied which do not contain the generator literal neither positively nor negatively and do not differ from the generator clause. Here comes the trick. If we have such a clause (no-occurrence clause), then the union of it and the generator clause is a clause. This union will be our new generator clause. It contains an unused blocked liter. This literal will our next generator literal (keeping also the older ones). We generate a new sub-model from the generator clause and generator literals which will satisfy all the clauses which was satisfied by the old one and which satisfies also many more clauses. This process is called as the refinement of the sub-model. After finitely many refinement steps the sub-model will satisfy the whole clause set.

It is not necessary that from the beginning all clauses are blocked, but only that the (confluent) reduction process of eliminating blocked clauses finally eliminates all clauses.

We also discuss how to collect the information whether a literal is blocked or not. We show that in case of 3-SAT we can collect these information in linear time by using cubic memory space.

We introduce the data structure called NLC, Number of Literal Combinations. We create NLC by reading each clause only once. For every subset of every clause we increase the corresponding counter in NLC by one. Afterwards we read again the clause set and for every literal in every clause we calculate the number of possible resolution partners minus the number of blocking clauses. For example if the clause is $\{a, b, c\}$ and the literal is a than this number is

$$NLC[\bar{a}] - NLC[\bar{a}, \bar{b}] - NLC[\bar{a}, \bar{c}] + NLC[\bar{a}, \bar{b}, \bar{c}].$$

If this number is zero then this literal is blocked in the input clause set.

In case of 3-SAT we need cubic memory space to store NLC. We read and write it 7 times (3 times for one length subsets of the clause, 3 times for two length subsets of the clause and 1 time for the clause itself) per clauses in the first loop. In the second loop we read it 4 times per literal (see the example above). Both steps are linear in the number of literals. This means, since the Blocked SAT Solver is a linear time method if we know at least one blocked literal per clauses, that the Blocked 3-SAT problem is a linear time problem.

An input clause set is rarely blocked but during the work of a general SAT solver algorithm we may encounter a blocked clause set. Any general SAT solver uses some simplification steps: resolution, unit-propagation, removing subsumed clauses, etc. The fewer clauses (or literals) are in a clause set the more likely that it is blocked. This means that soon or later a general SAT solver encounters a blocked clause set. In this case it is worth switching to the Blocked SAT Solver algorithm since it is polynomial. The simplification steps may update the NLC data structure which makes it easier to decide whether an immediate clause set is blocked or not.

We also discuss how to use blocked clauses if the whole clause set is not blocked. We introduce two lemmas, the Blocked Clear Clause Rule and the Independent Blocked Clause Rule, to describe cases where the answer is true.

2   Definitions

We use the well known set based representation of SAT. The abbreviation "iff" means "if and only if".

Let V be a finite *set of Boolean variables*. The *negation of a variable* v is denoted by $\overline{v}$. Given a set U, we denote $\overline{U} := \{\overline{u} \mid u \in U\}$ and call the *negation of the set* U.

Literals are members of the set $W := V \cup \overline{V}$. *Positive literals* are the members of the set V. *Negative literals* are their negations. If w denotes a negative literal $\overline{v}$, then $\overline{w}$ denotes the positive literal v.

Clauses and *assignments* are finite sets of literals that do not contain any literal together with its negation simultaneously. A clause is interpreted as a disjunction of its literals. An assignment is interpreted as a conjunction of its literals. A *clause set* is a finite set of clauses. A clause set is interpreted as a conjunction of its clauses.

We use the following constants: n is the number of variable, and m is the number of clauses in the input clause set.

If C is a clause and $|C| = k$, then we say that C is a *k-clause*. Special cases are *unit clauses* or *units* which are 1-clauses, and *clear* or *total clauses* which are n-clauses. Note that any unit clause is a clause and an assignment at the same time. The clause set CC is the set of all clear clauses.

$$CC := \{C \mid Clause(C) \land |C| = n\}.$$

A clause C is *subsumed by the clause set* S, denoted by $C \supseteq\in S$, iff

$$C \supseteq\in S : \Longleftrightarrow Clause(C) \land ClauseSet(S) \land \underset{B\in S}{\exists} B \subseteq C.$$

A clause C is *entailed by the clause set* S, denoted by $C \supseteq\in_{CC} S$, iff

$$C \supseteq\in_{CC} S : \Longleftrightarrow Clause(C) \land ClauseSet(S) \land \underset{\substack{D\in CC \\ C\subseteq D}}{\forall} \underset{B\in S}{\exists} B \subseteq D.$$

A clause C is *independent in clause set* S iff it is not entailed by S.

If A and B are clauses then we define the *clause difference* of them, denoted by $diff(A, B)$, as

$$diff(A, B) := A \cap \overline{B}.$$

If $diff(A, B) \neq \emptyset$ then we say that A *differs from* B.

Resolution can be performed on two clauses iff they differ only in one variable. If resolution can be performed then the *resolvent*, denoted by $Res(A, B)$, is

$$Res(A, B) := (A \cup B) \setminus (diff(A, B) \cup diff(B, A)).$$

If S is a clause set and A is an assignment, then we can do *hyper-unit propagation*, for short *HUP*, by A on S, denoted by $HUP(S, A)$, as follows:

$$HUP(S, A) := \{C \setminus \{\overline{A}\} \mid C \in S \wedge C \cap A = \emptyset\}.$$

A *literal* $c \in C$ is *blocked* in the clause C and in the clause set S, denoted by $Blck(c, C, S)$, iff

$$Blck(c, C, S) : \Longleftrightarrow \underset{\substack{B \in S \\ \overline{c} \in B}}{\forall} \; \underset{\substack{b \in B \\ b \neq \overline{c}}}{\exists} \; \overline{b} \in C,$$

A *clause* C is *blocked* in the clause set S, denoted by $Blck(C, S)$, iff a blocked literal exists in it. A *clause set* S is *blocked*, denoted by $Blck(S)$, iff all clauses in it are blocked.

If C is a clause and A is an assignment then we say that the sub-model generated by C and A, denoted by $sm(C, A)$, is

$$sm(C, A) = \begin{cases} \overline{(C \setminus A)} \cup A & , if C \neq \emptyset; \\ \emptyset & , otherwise. \end{cases}$$

A partial assignment I is a *model for a clause set* S, denoted by $M(S, I)$, iff $HUP(S, I)$ is the empty clause set.

3   The Blocked SAT Problem

In this section we introduce the Blocked SAT problem which is a restriction of SAT to instances where each clause of the clause set is blocked.

A clause is blocked in a clause set if it has a literal on which no resolution is possible in that clause set. The notion of blocked clause was introduced in [Kul99a, Kul99b]. A blocked clause can be added or deleted from a clause set without changing its satisfiability. From this it is easy to show that any blocked clause set is satisfiable, as if we remove all blocked clauses then we obtain the trivially satisfiable clause set. But this process do not show how to find a model for blocked clause sets.

Now we recall the formal definition of the blocked clause set:

$$Blck(S) : \Longleftrightarrow \underset{A \in S}{\forall} \; \underset{a \in A}{\exists} \; \underset{\substack{B \in S \\ \overline{a} \in B}}{\forall} \; \underset{\substack{b \in B \\ b \neq \overline{a}}}{\exists} \; \overline{b} \in A.$$

Since the definition has 2 quantifiers on clauses and 2 quantifiers on literals, in these clauses we need $O(n^2 m^2)$ time to decide whether the clause set is blocked.

We give three examples for blocked clause sets. Notation is explained below.

1st	2nd	3rd
(+)+ x	(-)+ x x	+ +(-)
x +(+)	x(+)+ x	+(-)+
(-)- x	x -(-)x	(+)- -
x -(-)	x x -(+)	(-)+ +
		-(+)-
		- -(+)
+ - +	- - + +	+ + +
- + -	x + - x	- - -

In the example we used the literal matrix representation of clause sets where rows corresponds to clauses and columns to variable. The symbol + means positive literal occurrence, - means negative literal occurrence and x means no literal occurrence. Under each example we list all models for it. Blocked literals are in brackets.

The Blocked SAT problem is a very restrictive subset of the general SAT problem, but it can be solved in polynomial time. During the work of a general SAT solver immediate blocked clause sets are frequent.

4   The Blocked SAT Solver Algorithm

In this section we introduce the blocked SAT Solver algorithm which solves the Blocked SAT problem in polynomial time. To be more precise we present a linear time version which uses $O(4m)$ basic set operations. But this version assumes that we already know about at least one blocked literal per clause. However, in general, polynomial ($O(n^2m^2)$) time is needed to collect that information.

Algorithm 1 (Blocked SAT Solver)
BlockedSATSolver(S, Z)
input: clause set S that is non-empty and blocked,
output: assignment Z, a model for S.

  1 **START**

  2    $(A, B) = (\emptyset, \emptyset)$;

  3   *// A is the <u>generator clause</u>, B is the set of <u>generator literals</u>.*

  4   **for each** $C \in S$ **do**

  5   *// If C and A differ then sm(A, B) satisfies C, see Lemma 1.*

  6   **if** $(diff(A, C) = \emptyset)$ **then**

  7   *// Otherwise, C is a <u>no-occurrence clause</u>.*

  8   *// We have to consider it.*

  9        $A := A \cup C$;

  10   *// This is the new <u>generator clause</u>.*

  11   **if** $(B \cap C = \emptyset)$ **then**

```
12                  //  In this case we have to refine B
13                  //  to gain C ∩ sm(A, B) ≠ ∅.
14              Let c ∈ C be a blocked literal in C, S;
15              B := B ∪ {c};
16                  //  This is the new set of generator literals.
17          fi
18      fi
19   od
20   Z := sm(A, B);
21 HALT
```

The main idea of the algorithm is the following. If the input clause set is blocked then if we generate a sub-model from a blocked clause (generator clause, in the algorithm variable A) and from a blocked literal (generator literal(s), in the algorithm variable B) then we obtain a model for those clauses from the clause set which contain the generator literal or the negation of it. But there might be a clause (no-occurrence clause, in the algorithm variable C) which does not contain the generator literal either positively or negatively. This means that the union of the no-occurrence clause and the generator clause is a clause. This clause will be our new generator clause. Since the no-occurrence clause is blocked too, we can select a blocked literal from it and add to the generator literals. We do this only if it is necessary, i.e., if the sub-model does not satisfies the no-occurrence clause. The new sub-model is generated from the new generator clause and from the new generator literals. This step is the refinement step. One can show that this new sub-model is still a model for the clauses which are satisfied by the old one (see the 2nd auxiliary lemma). Hence, we obtain a model for the input clause set by performing refinement steps while we read its clauses one after the other.

It is not necessary that from the beginning all clauses are blocked, but only that the (confluent) reduction process of eliminating blocked clauses finally eliminates all clauses.

We see that this algorithm uses each clause only once and in an iteration it does 4 basic set operations in the worst case.

First we give an example on how the Blocked SAT Solver algorithm works. The variables S, C, A, c, B are the variables from the algorithm. Blocked literals are in brackets in columns S and C. The abbreviation "unch." means unchanged.

S	C	A	{c}	B	sm(A, B)
+(-)x x	+(-)x x	+-xx	x-xx	x-xx	--xx
(+)x + x					
- x(-)x					
x x -(+)					
unch.	(+)x + x	+-+x	+xxx	+-xx	+--x
unch.	- x(-)x	unch.		unch.	+--x
unch.	x x -(+)	unch.		unch.	+--x

The input clause set, S, is non-empty and blocked. So the precondition is fulfilled. We take the first clause and one of its blocked literal. We generate a sub-model from them. This will be the base of the refinement. We read the next clause, which is a no-occurrence clause. We refine the sub-model. We read the next clause, but it is satisfied, so we do not refine the sub-model. The same is true for the lest clause. The last sub-model as we expected is indeed a model for the input clause set.

The following lemmas are needed to show that the Blocked SAT Solver algorithm always finds a model for a blocked clause set. The first one states that if we have a blocked clause set S and its subset G which contains only clauses that do not differ form each other then there is an algorithm which constructs B that has $sm(\bigcup G, B)$ is a model for the clauses which differ from $\bigcup G$.

Lemma 1 (Auxiliary Lemma 1 for Blocked SAT Solver)
Assume S is a non-empty, blocked clause set. Assume $G \subseteq S$, G is non-empty and for all $E, F \in G$ we have $diff(E, F) = \emptyset$. Assume $A = \bigcup G$. Assume B is a clause constructed by the following piece of pseudo-code:

1    $(B, i) := (\emptyset, 1)$;

2   **for each** $E \in G$ **do**

3   **if** $(B \cap E = \emptyset)$ **then**

4   **Let** $a_i \in E$ be a blocked literal in E, S;

5        $(B, i) := (B \cup \{a_i\}, i + 1)$;

6   **fi**

7   **od**

Assume $C \in S$ and $diff(A, C) \neq \emptyset$. Then $C \cap sm(A, B) \neq \emptyset$.

Proof by Contradiction: Assume $C \cap sm(A, B) = \emptyset$. We show that this assumption leads to a contradiction. From this assumption we know, by definition of sub-model, that $diff(A, C) \subseteq B$. Let $k = |B|$. Note that we know, from the construction of B, that $B = \{a_1..a_k\}$. Let $i \in \{1..k\}$ be the largest index such that $a_i \in diff(A, C)$ and $a_i \in B$. (Such $i \in \{1..k\}$ exists, because $diff(A, C) \neq \emptyset$ and $diff(A, C) \subseteq B$.) Then we know, from the construction of B, that there is a clause $E \in G$ which has $(B \setminus \{a_i..a_k\}) \cap E = \emptyset$ and $a_i \in E$ and $a_i \in E$ is blocked in E, S. Since $a_i \in diff(A, C)$ we know, by definition of clause difference, that $\overline{a_i} \in C$. From this and from that $a_i \in E$ is blocked in E, S and from $C \in S$ we know, by definition of blocked literal, that for some $\overline{c} \in C$ we have $\overline{c} \neq \overline{a_i}$ and $c \in E$. From this and from $A = \bigcup G$ we know that $c \in A$. Hence, $c \in diff(A, C)$.

From $c \in E$ and from $(B \setminus \{a_i..a_k\}) \cap E = \emptyset$ we know that $c \notin \{a_1..a_{i-1}\}$. From $\overline{c} \neq \overline{a_i}$, i.e., from $c \neq a_i$ and from $c \in diff(A, C)$ and from the fact that $i \in \{1..k\}$ is the largest index such that $a_i \in diff(A, C)$ and $a_i \in B$ we know that $c \notin \{a_i..a_k\}$. Hence, $c \notin B$.

But this is a contradiction, because $c \in diff(A, C)$ and $diff(A, C)$ is a subset of B. Hence, $C \cap sm(A, B) \neq \emptyset$.

The main idea of this proof is that in the worst-case A and C differ only in blocked literals from B, i.e., the sub-model does not satisfy C. But the construction of B makes sure that this cannot happen.

This lemma makes sure that $sm(A, B)$ satisfies C provided that it is not a no-occurrence clause.

If we look at the Blocked SAT Solver algorithm we see that it does the same as the piece of the pseudo-code in this auxiliary lemma. The only difference is that in the lemma we already have the set of clauses which does not differ from each other (this is the clause set G), while in the algorithm we collect these clauses during the computation. In both cases A is the union of these clauses.

Remember, we enlarge A and B only if we encounter a no-occurrence clause C in order to refine the sub-model, which means that the new sub-model will satisfies C.

The second auxiliary lemma states that this new sub-model still satisfies those clauses which were satisfied by the old one.

Lemma 2 (Auxiliary Lemma 2 for Blocked SAT Solver). *Let S be a non-empty, blocked clause set. Let A be a clause. Let $B \subseteq A$. Let $C \in S$ such that $C \subseteq A$ and $B \cap C = \emptyset$. Let $c \in C$ be a blocked literal in C, S. Let $D \in S$ such that $D \cap sm(A, B) \neq \emptyset$. Then $D \cap sm(A, B \cup \{c\}) \neq \emptyset$.*

Proof: If $\bar{c} \notin D$ then $D \cap sm(A, B \cup \{c\}) \neq \emptyset$ follows from $D \cap sm(A, B) \neq \emptyset$. Assume $\bar{c} \in D$. Then since $c \in C$ is blocked in C, S, by definition of blocked literal, we know that for some $d \in D$ we have $d \neq \bar{c}$ and $\bar{d} \in C$. Since $C \subseteq A$ we know that $\bar{d} \in A$ and from $B \cap C = \emptyset$ we know that $\bar{d} \notin B$. Therefore, by definition of sub-model, $\bar{d} \in sm(A, B)$. From $\bar{d} \neq c$ we know, by definition of sub-model, that $\bar{d} \in sm(A, B \cup \{c\})$. Hence, $D \cap sm(A, B \cup \{c\}) \neq \emptyset$.

Now we show that the Blocked SAT Solver algorithm solves the Blocked SAT problem. We use "{}" to mark formulae that are True at the respective points of algorithm, in order to prove the correctness of the algorithm in the Hoare calculus. The Hoare calculus makes sure that if we have a so-called "while program" and we can arrive from the precondition to the postcondition using the rules of the calculus, then for each input which fulfills the precondition the computed output fulfills the postcondition if the computation terminates. For more details about Hoare calculus please consult [LS87].

For better readability we use the infix version of hyper-unit propagation, which is denoted by an asterisk (*). This means that instead of $HUP(S, A)$ we use $S * A$. This variant of Blocked SAT Solver uses a new variable T in order to be able to give the invariant of the algorithm. We know that Block SAT Solver considers each clause in the input claus set. In T we collect the visited clauses. The essence of the invariant is that $sm(A, B)$ is a model of the visited clauses.

In Hoare calculus we will use the following invariant and auxiliary formulae:

$$Inv :\Leftrightarrow Blck(S) \wedge S \neq \emptyset \wedge T * sm(A, B) = \emptyset.$$

$$If1 :\Leftrightarrow Inv \wedge C \in (S \setminus T).$$

$$If2 :\Leftrightarrow If1 \wedge C \subseteq A.$$

$$IT2 :\Leftrightarrow If2 \wedge B \cap C = \emptyset.$$

Algorithm 2 (Blocked SAT Solver)

BlockedSATSolver(S, Z)
input: clause set S that is non-empty and blocked,
output: assignment Z, a model for S.

1   **START**

2   // $\{Blck(S) \wedge S \neq \emptyset\}$, *precondition*

3   // $\{Blck(S) \wedge S \neq \emptyset \wedge \emptyset * sm(\emptyset, \emptyset) = \emptyset\}$

4   $(A, B, T) := (\emptyset, \emptyset, \emptyset);$

5   // $\{Blck(S) \wedge S \neq \emptyset \wedge T * sm(A, B) = \emptyset\}$, *by assignment axiom*

6   // $\{\{Inv\}$, *loop invariant*

7   **while** $(T \neq S)$ **do**

8   // $\{\{Inv \wedge T \neq S\}$, *by while rule*

9   // $\{\{Inv \wedge T \neq S\}$

10   **Let** $C \in S \setminus T$;

11   // $\{\{Inv \wedge C \in S \setminus T\}$

12   // $\{\{If1\}$

13   **if** $(diff(A, C) = \emptyset)$ **then**

14   // $\{\{If1 \wedge diff(A, C) = \emptyset\}$, *by if-then rule*

15   // $\{\{If1 \wedge T * sm(A \cup C, B) = \emptyset\}$

16      $A := A \cup C;$

17   // $\{\{If1 \wedge C \subseteq A \wedge T * sm(A, B) = \emptyset\}$, *by assignment axiom*

18   // $\{\{If2\}$, *it is the same*

19   **if** $(B \cap C = \emptyset)$ **then**

20   // $\{\{If2 \wedge B \cap C = \emptyset\}$, *by if-then rule*

21   // $\{\{IT2\}$, *it is the same*

22   **Let** $c \in C$ be a blocked literal in C, S;

23   // $\{\{IT2 \wedge Blck(c, C, S)\}$

24   // $\{\{IT2 \wedge T * sm(A, B \cup \{c\}) = \emptyset\}$, *by Lemma 2*

25          $B := B \cup \{c\};$

26   // $\{\{If2 \wedge B \cap C \neq \emptyset \wedge T * sm(A, B) = \emptyset\}$, *by assign. axiom*

27   // $\{\{If2 \wedge B \cap C \neq \emptyset\}$

28   **fi**

29   // $\{\{If2 \wedge B \cap C \neq \emptyset\}$, *by if-else rule*

30   // $\{\{If2 \wedge C \cap sm(A, B) \neq \emptyset\}$

31   **else**

32   // $\{\{If1 \wedge diff(A,C) \neq \emptyset\}$, by if-else rule

33   // $\{\{If1 \wedge C \cap sm(A,B) \neq \emptyset\}$, by Lemma 1

34   **fi**

35   // $\{\{If1 \wedge C \cap sm(A,B) \neq \emptyset\}$

36   // $\{\{Inv \wedge C \in S \setminus T \wedge T \cup \{C\} * sm(A,B) = \emptyset\}$, logical consequence

37    $T := T \cup \{C\}$;

38   // $\{\{Inv \wedge T * sm(A,B) = \emptyset\}$, by assignment axiom

39   // $\{\{Inv\}$, it is the same

40   **od**

41   // $\{\{Inv \wedge T * sm(A,B) = \emptyset \wedge T = S\}$, by while rule

42   // $\{\{S * sm(A,B) = \emptyset\}$, logical consequence

43    $Z := sm(A,B)$;

44   // $\{\{S * Z = \emptyset\}$, by assignment axiom

45   // $\{\{M(S,Z)\}$, postcondition

46   **HALT**

Now we see that from the precondition (the clause set is non-empty and blocked) the postcondition (the computed assignment is a model for the clause set) follows if we follow the steps of the algorithm.

The only question is whether the algorithm always terminates or sometimes runs in an infinite loop. But it terminates always, because it does nothing else but uses each clause from the clause set one after the other and clause sets are finite sets.

Theorem 3 (Correctness of the Blocked SAT Solver)
Let S be a non-empty and blocked clause set. Then after execution of Blocked SAT Solver(S, I), I is a model for S.

Proof: From Algorithm 2, and from the fact that it terminates we obtain that Blocked SAT Solver terminates for every non-empty and blocked clause set and gives back a model for it.

Before we discuss the worst case time complexity of the algorithm we show how to decide in linear time whether a 3-SAT problem is blocked or not.

We use a special data structure called NLC, Number of Literal Combinations. For every subset of every clause we increase the corresponding counter in NLC by one. This means that NLC contains 3 arrays, a one dimensional (1D) with $2n$ entries for counting 1 length subsets, a 2D one with $2n * 2n$ entries for counting 2 length subsets and a 3D one with $2n * 2n * 2n$ entries for counting the clauses.

We use literals as indices in square brackets. If we give one literal in the square brackets then we access the 1D array, by 2 literals the 2D one and by 3 literals the 3D one.

Since $NLC[x, y]$ is the same as $NLC[y, x]$, we assume that the indices are written in alphabetical order. Note that this means that the 2D and 3D arrays contain only zeros below the diagonal.

We initialize NLC by filling in it by zeros. Afterwards, we use it in two loops.

1. In the first loop we read the clauses of the input clause set one after the other. For all subsets of all clauses we increase the corresponding entry in NLC by one. For example if the input clause is $\{a, b, c\}$ then we increase these entries:

$$NLC[a], NLC[b], NLC[c], NLC[a, b], NLC[a, c], NLC[b, c], NLC[a, b, c].$$

This means that we do 7 read and write steps (to increase one entry, we have to read its value first, add one, and write it back) for all clauses, i.e., the first loop is an $O(2 * 7m)$ time method.

Before we go to the second loop we have to investigate the 2D and 3D arrays further. We see that they contain a lots of zeros. In the 3D one we have at most only m non zero entries. In the 2D one we have at most only $3m$ non zero entries. Hence, it is worth using hash tables instead of arrays. We assume that we can initialize these hash tables in $O(nm)$ time. Therefore, the initialization of NLC takes $O(nm + 2n)$ write steps.

2. In the second loop we read again the clauses of the input clause set one after the other. For each literal in each clause we compute the number of possible resolution partners minus the number of blocking clauses. Let us assume that the actual clause is $A = \{a, b, c\}$ and the actual literal is a. Then the possible resolution partners are those clauses that contain $\bar{a}$. We know the number of those closes, it is stored in $NLC[\bar{a}]$. We recall the definition of blocked literal:

$$Blck(a, A, S) : \iff \mathop{\forall}_{\substack{B \in S \\ \bar{a} \in B}} \mathop{\exists}_{\substack{\bar{x} \in B \\ \bar{x} \neq \bar{a}}} x \in A,$$

This means that in those clauses which are possible resolution partners, there is an other literal $\bar{x}$ such that x occurs in A. In this case x is either b or c. So we obtain the number of blocking clauses as $NLC[\bar{a}, \bar{b}] + NLC[\bar{a}, \bar{c}]$. But if we do so, then we count the clauses, which contains $\bar{a}$ and $\bar{b}$ at the same time, twice. This means that we have to subtract $NLC[\bar{a}, \bar{b}, \bar{c}]$ from the number of blocking clauses. Hence, the number of possible resolution partners minus the number of blocking clauses is:

$$NLC[\bar{a}] - NLC[\bar{a}, \bar{b}] - NLC[\bar{a}, \bar{c}] + NLC[\bar{a}, \bar{b}, \bar{c}].$$

If this number is zero then it means that all possible resolution partners are at the same time blocking clauses, i.e., the literal a in the clause $\{a, b, c\}$ is blocked in the input clause set.

In the second loop we read NLC 4 times for each literal, i.e., it is a $O(4nm)$ time method in the worst case.

The overall worst case time complexity of the usage of NLC is $O(nm + 2n + 2 * 7m + 4nm) = O(5nm + 14m + 2n) = O(5nm)$.

We can also use NLC for the general SAT problem. Then we need $O(2^k)$ memory space and $O(2 * (2^k - 1)m)$ time for the first loop, and $O(2^{k-1}nm)$ time for the second loop in the worst case, where k is the length of largest clause in the input clause set.

Now we can prove that Blocked SAT Solver is a linear time algorithm if at least one blocked literal per clause is known or the input clause set is a 3-SAT problem, otherwise it is a quadratic time algorithm.

Theorem 4 (Complexity of the Blocked SAT Solver)

Let S be a non-empty and blocked clause set. The worst-case time complexity of Blocked SAT Solver(S, I) is $O(4nm)$, if at least one blocked liter per clause is known; $O(9nm)$ if S is a 3-SAT problem; $O(n^2m^2)$, otherwise.

> Proof: If at least one blocked literal per clause is known then in the worst-case we need $O(4m)$ basic set operations, because there is only one for loop on clauses of the set and in one iteration in the worst-case we perform 4 basic set operation. Since any basic set operation can be performed in $O(n)$ time, the worst-case time complexity of Blocked SAT Solver is $O(4nm)$, i.e., linear in the number of literals.
>
> Otherwise, if the S is a 3-SAT problem then by using NLC data structure we need $O(5nm)$ time in the worst case to provide the necessary information for the Blocked SAT Solver algorithm. This means that we need $O(9nm)$ time altogether which is linear in the number of literals. Otherwise, in the worst-case we need $O(n^2m^2)$ time, see the previous section, to provide the necessary information for the Blocked SAT Solver algorithm. Since $O(n^2m^2)$ dominates $O(4nm)$ the worst-case time complexity of Blocked SAT Solver is $O(n^2m^2)$.

5   Blocked Clause Rules

If the clause set is not blocked but it contains blocked clauses then it is a question whether we can use the blocked clauses to simplify the clause set or not? We also discuss this question briefly.

The Blocked Clear Clause Rule states that if a clause set contains only clear clauses (i.e., full clauses) and one of them is blocked, then the sub-model generated from this blocked one is a model. This is a very rare case but it serves as a basis for the next rule.

Lemma 3 (Blocked Clear Clause Rule). *Let S be a clause set. Let $C \in S$ be a blocked and clear clause. Let $a \in C$ be a blocked literal C, S. Then $sm(C, \{\, a\})$ is a model for S, provided that*
(a) either S is a subset of CC,
(b) or C is not subsumed by $S \setminus \{C\}$.

> Proof:
> (a) To show this, by definition of model, it suffices to show that for an arbitrary but fixed $B \in S$ we have that $B \cap sm(C, \{\, a\})$ is not empty. Since S is a subset of CC we know that B is a clear clause. Hence, there are two cases, either $a \in B$ or $\bar{a} \in B$.

In case $a \in B$ we have, by definition of sub-model, that $a \in sm(C, \{a\})$. Hence, $B \cap sm(C, \{a\})$ is not empty.

In case $\bar{a} \in B$, since $a \in C$ is blocked in C, S we know, by definition of blocked literal, that for some $b \in B$ we have $b \neq \bar{a}$ and $\bar{b} \in C$. From this, by definition of sub-model, we know that $b \in sm(C, \{a\})$. Hence, $B \cap sm(C, \{a\})$ is not empty.

Hence, if S is a subset of CC, then $sm(C, \{a\})$ is a model for S.

(b) To show this, by definition of model, it suffices to show that for an arbitrary but fixed $B \in S$ we have that $B \cap sm(C, \{a\})$ is not empty. Since C is not subsumed by $S \setminus \{C\}$ we know, by definition of subsumption, that $B \nsubseteq C$. From this, since C is a clear clause we know that for some $b \in B$ we have $\bar{b} \in C$. There are two cases, either $\bar{b} = a$ or $\bar{b} \neq a$.

In the first case we have $\bar{b} = a$, i.e., $\bar{a} \in B$. From this since $a \in C$ is blocked in C, S we know, by definition of blocked literal, that for some $d \in B$ we have that $d \neq \bar{a}$ and $\bar{d} \in C$. From this, by definition of sub-model, we know that $d \in sm(C, \{a\})$. Hence, $B \cap sm(C, \{a\})$ is not empty.

In the second case we have $\bar{b} \neq a$. From this and from $\bar{b} \in C$ we know, by definition of sub-model, that $b \in sm(C, \{a\})$. Hence, $B \cap sm(C, \{a\})$ is not empty.

Hence, If C is not subsumed by $S \setminus \{C\}$, then $sm(C, \{a\})$ is a model for S.

The Independent Blocked Clause Rule states that if a clause set contains an independent blocked clause, then it is satisfiable and a sub-model generated from this clause is a part of a model. This situation occurs quite often, but checking independent-ness is expensive.

Lemma 4 (Independent Blocked Clause Rule). *Let S be a clause set. Let $A \in S$ be blocked in S and independent in $S \setminus \{A\}$. Let $a \in A$ be a blocked literal in A, S. Then there is a model M for S such that $sm(A, \{a\}) \subseteq M$, i.e., $HUP(S, sm(A, \{a\}))$ is satisfiable.*

Proof: We know that A is independent in $S \setminus \{A\}$. Hence, by definition of independent, we know that there is a clear clause C that is subsumed by A and not subsumed by any other clause in S. Since $A \subseteq C$ we know that $sm(A, \{a\}) \subseteq sm(C, \{a\})$. Hence, it suffices to show that $sm(C, \{a\})$ is a model for S. To show this, by definition of model, it suffices to show that for an arbitrary but fixed $B \in S$ we have that $B \cap sm(C, \{a\})$ is not empty. The remaining part of the proof is the same as the proof of the (b) variant of the Blocked Clear Clause Rule.

Hence, $B \cap sm(C, \{a\})$ is not empty. Hence, there is a model M for S such that $sm(A, \{a\}) \subseteq M$.

This proof is traced back to the proof of Blocked Clear Clause Rule. We can do this because we know that there is a clear clause which is blocked and not entailed by $S \setminus \{A\}$. Note that for clear clauses the notion of subsumed and entailed are the same.

The proof shows that if we perform an independent clause check and we find a clear clause which is subsumed by only one clause, then we know the whole model $(sm(C, \{ a\}))$ and not only a part of the model $(sm(A, \{ a\}))$.

References

[APT79] B. Aspvall, M. F. Plass, and R. E. Tarjan. A Linear-Time Algorithm for Testing the Truth of Certain Quantified Boolean Formulas. Information Processing Letters, 8(3):121–132, 1979.

[Asp80] B. Aspvall. Recognizing Disguised NR(1) Instances of the Satisfiability Problem. J. of Algorithms, 1:97–103, 1980.

[BHS94] E. Boros, P. L. Hammer, and X. Sun. Recognition of q-Horn Formulae in Linear Time. Discrete Applied Mathematics, 55:1–13, 1994.

[BCH+94] E. Boros, Y. Crama, P. L. Hammer, and M. Saks. A Complexity Index for Satisfiability Problems. SIAM J. on Computing, 23:45–49, 1994.

[CH91] V. Chandru and J. Hooker. Extended Horn Sets in Propositional Logic. J. of the ACM, 38(1):205–221, 1991.

[Coo71] S. A. Cook. The Complexity of Theorem-Proving Procedures. Proceedings of the 3rd ACM Symposium on Theory of Computing, 151–158, 1971.

[DE92] M. Dalal and D. W. Etherington. A Hierarchy of Tractable Satisfiability Problems. Information Processing Letters, 44:173–180, 1992.

[DG84] W. F. Dowling and J. H. Gallier. Linear-Time Algorithms for Testing the Satisfiability of Propositional Horn Formulae. J. of Logic Programming, 1(3):267–284, 1984.

[EIS76] S. Even, A. Itai, and A. Shamir. On the Complexity of Timetable and Multi-Commodity Flow Problems. SIAM J. on Computing, 5(4):691–703, 1976.

[Knu90] D. E. Knuth. Nested Satisfiability. Acta Informatica, 28:1–6, 1990.

[Kul99a] O. Kullmann. New methods for 3-SAT decision and worst-case analysis. Theoretical Computer Science, 223(1-2):1–72, 1999.

[Kul99b] O. Kullmann. On a Generalization of Extended Resolution. Discrete Applied Mathematics, 96-97(1-3):149–176, 1999.

[Kus02] G. Kusper. Solving the SAT Problem by Hyper-Unit Propagation. RISC Technical Report 02-02, 1–18, University Linz, Austria, 2002.

[Kus05] G. Kusper. Solving the Resolution-Free SAT Problem by Hyper-Unit Propagation in Linear Time. Annals of Mathematics and Artificial Intelligence, 43(1-4):129–136, 2005.

[Lew78] H. R. Lewis. Renaming a set of clauses as a Horn set. J. of the Association for Computing Machinery, 25:134–135, 1978.

[LS87] J. Loeckx and K. Sieber. The Foundations of Program Verification. Wiley-Teubner, second edition, 1987.

[SAF+95] J. S. Schlipf, F. Annexstein, J. Franco, and R. P. Swaminathan. On finding solutions for extended Horn formulas. Information Processing Letters, 54:133–137, 1995.

[Scu90] M. G. Scutella. A Note on Dowling and Gallier's Top-Down Algorithm for Propositional Horn Satisfiability. J. of Logic Programming, 8(3):265–273, 1990.

[Tov84] C. A. Tovey. A Simplified NP-complete Satisfiability Problem. Discrete Applied Mathematics, 8:85–89, 1984.

Towards the Computation of Stable Probabilistic Model Semantics

Emad Saad

College of Computer Science and Information Technology
Abu Dhabi University
Abu Dhabi, U.A.E.
emad.saad@adu.ac.ae

Abstract. In [22], a stable model semantics extension of the language of hybrid probabilistic logic programs [21] with non-monotonic negation, normal hybrid probabilistic programs (NHPP), has been developed by introducing the notion of stable probabilistic model semantics. It has been shown in [22] that the stable probabilistic model semantics is a natural extension of the stable model semantics for normal logic programs and the language of normal logic programs is a subset of the language NHPP. This suggests that efficient algorithms and implementations for computing the stable probabilistic model semantics for NHPP can be developed by extending the efficient algorithms and implementation for computing the stable model semantics for normal logic programs, e.g., SMODELS [17]. In this paper, we explore an algorithm for computing the stable probabilistic model semantics for NHPP along with its auxiliary functions. The algorithm we develop is based on the SMODELS [17] algorithms. We show the soundness and completeness of the proposed algorithm. We provide the necessary conditions that these auxiliary functions have to satisfy to guarantee the soundness and completeness of the proposed algorithm. This algorithm is the first to develop for studying computational methods for computing the stable probabilistic models semantics for hybrid probabilistic logic programs with non-monotonic negation.

1 Introduction

Hybrid Probabilistic Programs (HPP) [5] is a probabilistic logic programming framework that enables the user to *explicitly* encode his/her knowledge about the type of dependencies existing between the probabilistic events being described by the programs. HPP generalizes the *probabilistic annotated logic programming framework*, originally proposed in [15] and further extended in [16]. In this paper we study the problem of automating the probabilistic reasoning under the stable probabilistic model (p-model) semantics proposed in [22]. Stable p-model semantics is the first formalism to study non-monotonic negation in hybrid probabilistic programs originally proposed in [5] and further modified and extended in [21]. Stable p-model semantics [22] generalizes both the stable model semantics for normal logic programs [10] and the semantics of hybrid probabilistic logic programs introduced in [21]. The idea in [21] comes upon observing that commonsense

C. Freksa, M. Kohlhase, and K. Schill (Eds.): KI 2006, LNAI 4314, pp. 143–158, 2007.
© Springer-Verlag Berlin Heidelberg 2007

reasoning about probabilities relies on how *likely* (probable) are the various events to occur, rather than how precise our knowledge about these probabilities is [5]. Hybrid probabilistic programs (HPP) [21] is a probabilistic logic programming framework that enables the user to *explicitly* encode his/her knowledge about the type of dependencies existing between the probabilistic events being described by the programs. Moreover, it has the ability to encode the user's knowledge about how to combine the probabilities of the same event derived from different rules. HPP semantics [21] intuitively, captures the commonsense probabilistic reasoning according to how likely are the various events to occur. In addition, HPP subsumes Lakshmanan and Sadri's [11] probabilistic implication-based framework as well as it is a natural extension of classical logic programming.

In [22], we extended the language of HPP [21] to support non-monotonic negation. In addition, we defined two alternative semantics for the extended language; the stable probabilistic model semantics and the probabilistic well-founded semantics and studied their relationships. We showed that the stable probabilistic model semantics and the probabilistic well-founded semantics generalize the stable model semantics [10] and the well-founded semantics [9] for normal logic programs, and they reduce to the semantics of HPP [21] in the absence of non-monotonic negation. An important result is that the relationship between the stable probabilistic model semantics and the probabilistic well-founded semantics *preserves* the relationship between the stable model semantics and the well-founded semantics for normal logic programs [9].

Since the stable p-model semantics naturally generalize its classical counterpart, hence, this suggests that efficient algorithms and implementations can be developed by *extending* the existing efficient algorithms and implementations developed for computing the stable models for normal logic programs, such as SMODELS [17]. (In [21], we have presented an algorithm for computing the least fixpoint for HPP without non-monotonic negation, that extends the Dowling-Gallier algorithm for computing the satisfiability of a set of Horn formulae [7], which is the ground base for developing the various auxiliary functions in SMODELS.) In this paper, we provide an algorithm for computing the stable p-model semantics for normal hybrid probabilistic programs [22] based on the decision procedure of SMODELS [17] along with its auxiliary functions. We provide the necessary conditions that these auxiliary functions have to satisfy to guarantee the soundness and completeness of the proposed algorithm. We present formal definitions for these auxiliary functions and show that they satisfy the necessary conditions for the soundness and completeness of the proposed algorithm. In this paper, we focus on the the computation of the stable probabilistic model semantics. Motivating examples and extensive comparisons between stable probabilistic model semantics and other related work can be found in [22].

2   Normal Hybrid Probabilistic Programs

In the following subsections, we present the syntax and semantics of the Normal Hybrid Probabilistic Programs (NHPP) as presented in [22]. The notions of probabilistic strategies, annotations, and hybrid basic formulae, which are defined below, were first introduced in [5].

2.1 Probabilistic Strategies

Let $C[0,1]$ denote the set of all closed intervals in $[0,1]$. In the context of HPP, probabilities are assigned to primitive events (atoms) and compound events (conjunctions or disjunctions of atoms) as intervals in $C[0,1]$. Let $[a_1, b_1], [a_2, b_2] \in C[0,1]$. Then the *truth order* asserts that $[a_1, b_1] \leq_t [a_2, b_2]$ iff $a_1 \leq a_2$ and $b_1 \leq b_2$. The set $C[0,1]$ and the relation $\leq_t$ form a complete lattice. In particular, the join ($\oplus_t$) operation is defined as $[a_1, b_1] \oplus_t [a_2, b_2] = [max\{a_1, a_2\}, max\{b_1, b_2\}]$ and the meet ($\otimes_t$) is defined as $[a_1, b_1] \otimes_t [a_2, b_2] = [min\{a_1, a_2\}, min\{b_1, b_2\}]$ w.r.t. $\leq_t$. The type of dependency among the primitive events within a compound event is described by *probabilistic strategies*, which are explicitly selected by the user. We call ρ, a pair of functions $\langle c, md \rangle$, a probabilistic strategy (p-strategy), where $c : C[0,1] \times C[0,1] \rightarrow C[0,1]$, the *probabilistic composition function*, which is *commutative, associative, monotonic* w.r.t. $\leq_t$, and meets the following *separation* criteria: there are two functions c_1, c_2 such that $c([a_1, b_1], [a_2, b_2]) = [c_1(a_1, a_2), c_2(b_1, b_2)]$. Whereas, $md : C[0,1] \rightarrow C[0,1]$ is the *maximal interval function*. The maximal interval function md of a certain p-strategy returns an estimate of the probability range of a primitive event, A, from the probability range of a compound event that contains A. The composition function c returns the probability range of a conjunction (disjunction) of two events given the ranges of its constituents. For convenience, given a multiset of probability intervals $M = \{\![a_1, b_1], \ldots, [a_n, b_n]\!\}$, we use cM to denote $c([a_1, b_1], c([a_2, b_2], \ldots, c([a_{n-1}, b_{n-1}], [a_n, b_n])) \ldots)$. According to the type of combination among events, p-strategies are classified into *conjunctive* p-strategies and *disjunctive* p-strategies. Conjunctive (disjunctive) p-strategies are employed to compose events belonging to a conjunctive (disjunctive) formula (please see [5,21] for the formal definitions).

2.2 Language Syntax

In this subsection, we describe the syntax of NHPP. Let L be an arbitrary first-order language with finitely many predicate symbols, constants, and infinitely many variables. Function symbols are disallowed. In addition, let $S = S_{conj} \cup S_{disj}$ be an arbitrary set of p-strategies, where S_{conj} (S_{disj}) is the set of all conjunctive (disjunctive) p-strategies in S. The Herbrand base of L is denoted by B_L. An *annotation* denotes a probability interval and it is represented by $[\alpha_1, \alpha_2]$, where α_1, α_2 are called annotation items. An *annotation item* is either a constant in $[0,1]$, a variable (*annotation variable*) ranging over $[0,1]$, or $f(\alpha_1, \ldots, \alpha_n)$ (called *annotation function*) where f is a representation of a computable total function $f : ([0,1])^n \rightarrow [0,1]$ and $\alpha_1, \ldots, \alpha_n$ are annotation items. The building blocks of the language of NHPP are *hybrid basic formulae*. Let us consider a collection of atoms $A_1, \ldots, A_n$, a conjunctive p-strategy ρ, and a disjunctive p-strategy ρ'. Then $A_1 \wedge_\rho \ldots \wedge_\rho A_n$ and $A_1 \vee_{\rho'} \ldots \vee_{\rho'} A_n$ are called *hybrid basic formulae*, and $bf_S(B_L)$ is the set of all ground hybrid basic formulae formed using distinct atoms from B_L and p-strategies from S. An annotated hybrid basic formula is an expression of the form $F : \mu$ where F is a hybrid basic formula and μ is an annotation. A *hybrid literal* is an annotated hybrid basic formula $F : \mu$ (positive annotated hybrid basic formula or positive hybrid literal) or the negation of an annotated hybrid basic formula $not\ (F : \mu)$ (negative annotated hybrid basic formula or negative hybrid literal).

Definition 1 (Rules). *A normal hybrid probabilistic rule (nh-rule) is an expression of the form*
$$A : \mu \leftarrow F_1 : \mu_1, \ldots, F_n : \mu_n, not\ (G_1 : \mu_{n+1}), \ldots, not\ (G_m : \mu_{n+m})$$
where A is an atom, $F_1, \ldots, F_n, G_1, \ldots, G_m$ are hybrid basic formulae, and μ, μ_i $(1 \leq i \leq m + n)$ are annotations.

A hybrid probabilistic rule (h-rule) is an nh-rule where $m = 0$—i.e., there are no negative hybrid literals.

The intuitive meaning of an nh-rule, in Definition 1, is that, if for each $F_i : \mu_i$, the probability interval of F_i is at least μ_i and for each $not\ (G_j : \mu_j)$, it is not *provable* that the probability interval of G_j is at least μ_j, then the probability interval of A is μ.

Definition 2 (Programs). *A normal hybrid probabilistic program over S (nh-program) is a pair $P = \langle R, \tau \rangle$, where R is a finite set of nh-rules with p-strategies from S, and τ is a mapping $\tau : B_L \rightarrow S_{disj}$. A hybrid probabilistic program (h-program) is an nh-program where all the rules are h-rules.*

The mapping τ in the above definition associates to each atomic hybrid basic formula A a disjunctive p-strategy that will be employed to combine the probability intervals obtained from different rules having A in their heads. An nh-program is ground if no variables appear in any of its rules.

Example 1 ([22]). Consider an insurance company which determines the premium categories, by calculating the risk factor according to a genetic test for cancer and the family history for this disease. Assume that customers who have a family history of the disease have a probability of developing cancer with at least 92%. The insurance company will assign high premiums to the customers who have family history of the disease and tested positive as long as their risk conditions are unchanged. Risk conditions can be changed by taking specific medications. This situation can be represented by the following nh-rules:

$risk(X) : [0.9, 1] \qquad\qquad\qquad \leftarrow (test(X) \wedge_{pc} history(X)) : [0.60, 0.75],$
$\qquad\qquad\qquad\qquad\qquad\qquad\qquad\quad not\ changeRisk(X)[0.8, 1]$

$risk(X) : [0, 0.1] \qquad\qquad\qquad\ \leftarrow (test(X) \wedge_{pc} history(X)) : [0.60, 0.75],$
$\qquad\qquad\qquad\qquad\qquad\qquad\qquad\quad changeRisk(X) : [0.8, 1]$

$changeRisk(X) : [0.9, 1] \qquad\quad \leftarrow medicine(X, Med) : [0.65, 1]$
$highPremium(X) : [1, 1] \qquad\quad\ \leftarrow risk(X) : [0.9, 1]$
$lowPremium(X) : [1, 1] \qquad\qquad \leftarrow risk(X) : [0, 0.1]$
$test(sam) : [0.92, 1] \qquad\qquad\qquad \leftarrow$
$history(sam) : [0.95, 1] \qquad\qquad\quad\ \leftarrow$
$medicine(sam, medication) : [0.98, 1] \leftarrow$

and the mapping τ assigns ncd to $risk(sam)$ and an arbitrary disjunctive p-strategy [5,21] to the other hybrid basic formulae. The ncd denotes the disjunctive negative correlation p-strategy, which is defined as: $c_{ncd}([a_1, b_1], [a_2, b_2]) = [\min(1, a_1 + a_2), \min(1, b_1 + b_2)]$. The first nh-rule asserts that the risk factor is at least 90% whenever the cancer genetic test for a customer is positive and that customer has a family history of cancer with probability between 60% and 75%, and it is not provable that his risk conditions have changed with probability at least 80%. Observe that $test$ and $history$ events are conjoined according to the *positive correlation* p-strategy (denoted by $\wedge_{pc}$) where $c_{pcc}([a_1, b_1], [a_2, b_2]) = [\min(a_1, a_2), \min(b_1, b_2)]$. The second rule says that the risk factor is at most 10% whenever the customer risk conditions are changed, even though the person tested positive and have a family history of the disease with probability between 60% and 75%. The third nh-rule describes the change of the risk conditions of a customer with probability at least 90% if a medication for the disease becomes available with probability at least 65%. The fourth and fifth nh-rules assert that definite high premium and low premium are considered whenever the probability of risk factors are at least 90% and at most 10% respectively. The last three nh-rules represent the facts available about a specific customer named sam.

2.3 Satisfaction and Models

In this subsection, we review the declarative semantics of nh-programs [22]. The notion of a probabilistic model (p-model) is based on *hybrid formula functions* defined below.

Definition 3. *A hybrid formula function is a mapping* $h : bf_S(B_L) \rightarrow C[0, 1]$ *that satisfies the following conditions:*

- *Commutativity:* $h(G_1 *_\rho G_2) = h(G_2 *_\rho G_1)$, $* \in \{\wedge, \vee\}$, $\rho \in S$
- *Composition:* $c_\rho(h(G_1), h(G_2)) \leq_t h(G_1 *_\rho G_2)$, $* \in \{\wedge, \vee\}$, $\rho \in S$
- *Decomposition. For any hybrid basic formula* F, $\rho \in S$, *and* $G \in bf_S(B_L)$: $md_\rho(h(F *_\rho G)) \leq_t h(F)$.

The notion of truth order can be extended to hybrid formula functions. Given hybrid formula functions h_1 and h_2, we say $(h_1 \leq_t h_2) \Leftrightarrow (\forall F \in bf_S(B_L) : h_1(F) \leq_t h_2(F))$. The set of all hybrid formula functions, HFF, and the truth order $\leq_t$ form a complete lattice. The meet $\otimes_t$ and the join $\oplus_t$ operations are defined respectively as: for all $F \in bf_S(B_L)$, $(h_1 \otimes_t h_2)(F) = h_1(F) \otimes_t h_2(F)$ and $(h_1 \oplus_t h_2)(F) = h_1(F) \oplus_t h_2(F)$. We say that a probabilistic interpretation (p-interpretation) of an nh-program P is a hybrid formula function.

Definition 4 (Probabilistic Satisfaction). *Let* $P = \langle R, \tau \rangle$ *be a ground nh-program,* h *be a p-interpretation, and*
$r \equiv A : \mu \leftarrow F_1 : \mu_1, \ldots, F_n : \mu_n, not\,(G_1 : \beta_1), \ldots, not\,(G_m : \beta_m) \in R$. *Then*

- h *satisfies* $F_i : \mu_i$ *(denoted by* $h \models F_i : \mu_i$*) iff* $\mu_i \leq_t h(F_i)$.
- h *satisfies* $not\,(G_j : \beta_j)$ *(denoted by* $h \models not\,(G_j : \beta_j)$*) iff* $\beta_j \not\leq_t h(G_j)$.

- h satisfies $Body \equiv F_1 : \mu_1, \ldots, F_n : \mu_n, not\ (G_1 : \beta_1), \ldots, not\ (G_m : \beta_m)$ (denoted by $h \models Body$) iff $\forall (1 \leq i \leq n), h \models F_i : \mu_i$ and $\forall (1 \leq j \leq m), h \models not\ (G_j : \beta_j)$.
- h satisfies $A : \mu \leftarrow Body$ iff $h \models A : \mu$ or h does not satisfy $Body$.
- h satisfies P iff h satisfies every nh-rule in R and for every atomic formula $A \in bf_S(B_L)$, $c_{\tau(A)}\{\!\{\mu | A : \mu \leftarrow Body \in R \text{ and } h \models Body\}\!\} \leq_t h(A)$.

Definition 5 (Models). Let P be an nh-program. A probabilistic model (p-model) of P is a probabilistic interpretation of P that satisfies P.

Proposition 1. Let P be an h-program. $h_P = \otimes_t \{h | h$ is a p-model of $P \}$ is the least p-model of P.

Associated with each h-program P, is an operator, T_P, called the *fixpoint operator*, which maps probabilistic interpretations to probabilistic interpretations.

Definition 6. Let $P = \langle R, \tau \rangle$ be a ground h-program and h be a total probabilistic interpretation. The fixpoint operator T_P is a mapping $T_P : HFF \rightarrow HFF$ which is defined as follows:

1. if A is an atom, $T_P(h)(A) = c_{\tau(A)} M_A$ where $M_A = \{\!\{\mu | A : \mu \leftarrow Body \in R$ such that $h \models Body\}\!\}$ and $M_A \neq \emptyset$. If $M_A = \emptyset$, then $T_P(h)(A) = [0,0]$
2. $T_P(h)(G_1 \wedge_\rho G_2) = c_\rho(T_P(h)(G_1), T_P(h)(G_2))$ where $(G_1 \wedge_\rho G_2) \in bf_S(B_L)$
3. $T_P(h)(G_1 \vee_{\rho'} G_2) = c_{\rho'}(T_P(h)(G_1), T_P(h)(G_2))$ where $(G_1 \vee_{\rho'} G_2) \in bf_S(B_L)$.

Proposition 2. Let P be an h-program. Then, $h_P = lfp(T_P)$.

3   Stable Probabilistic Model Semantics

In this section we introduce the notion of *stable probabilistic models (sp-models)*, which extends the notion of stable models for classical logic programming [10]. The semantics is defined in two steps. First, we guess a p-model h for a certain nh-program P, then we define the notion of the probabilistic reduct of P with respect to h—which is an h-program. Second, we determine whether h is a stable p-model for P or not by employing the fixpoint operator of the probabilistic reduct to verify whether h is its least p-model. It must be noted that every h-program has a unique least p-model [21].

Definition 7 (Probabilistic Reduct). Let $P = \langle R, \tau \rangle$ be a ground nh-program and h be a probabilistic interpretation. The probabilistic reduct P^h of P w.r.t. h is $P^h = \langle R^h, \tau \rangle$ where:

$$R^h = \left\{ A : \mu \leftarrow F_1 : \mu_1, \ldots, F_n : \mu_n \ \left| \ \begin{array}{l} A : \mu \leftarrow F_1 : \mu_1, \ldots, F_n : \mu_n, \\ \quad not\ (G_1 : \beta_1), \ldots, not\ (G_m : \beta_m) \in R \text{ and} \\ \forall (1 \leq j \leq m),\ \beta_j \not\leq_t h(G_j) \end{array} \right. \right\}$$

The probabilistic reduct P^h is an h-program. For any $not\ (G_j : \beta_j)$ in the body of $r \in R$ with $\beta_j \not\leq_t h(G_j)$ is simply satisfied by h, and $not\ (G_j : \beta_j)$ is removed from the body of r. If $\beta_j \leq_t h(G_j)$ then the body of r is not satisfied and r is trivially ignored.

Definition 8 (Stable Probabilistic Model). *A probabilistic interpretation h is a stable p-model of an nh-program P if h is the least p-model of P^h.*

Example 2. It is easy to verify that the only stable p-model of the program in Example 1 is given by:

$h(risk(sam))$	$= [0, 0.1]$	$h(changeRisk(sam))$	$= [0.9, 1]$
$h(highPremium(sam))$	$= [0, 0]$	$h(lowPremium(sam))$	$= [1, 1]$
$h(test(sam))$	$= [0.92, 1]$	$h(history(sam))$	$= [0.95, 1]$
$h(medicine(sam, medication))$	$= [0.98, 1]$	$h(test(sam) \wedge_{pc} history(sam))$	$= [0.92, 1]$

Example 3. Consider the following nh-program $P = \langle R, \tau \rangle$ where R is

$$a : [0.89, 0.91] \leftarrow not\ (b : [0.3, 0.4])$$
$$b : [0.55, 0.60] \leftarrow not\ (a : [0.7, 0.75])$$
$$c : [0.2, 0.3] \quad \leftarrow d : [0.1, 0.15]$$
$$d : [0.1, 0.2] \quad \leftarrow not\ (e : [0.1, 0.3])$$

and $\tau(a) = \tau(b) = \tau(c) = \tau(d) = \pi$ where π is any arbitrary disjunctive p-strategy. This nh-program has two stable p-models h_1 and h_2 where $h_1(a) = [0.89, 0.91], h_1(b) = [0, 0], h_1(c) = [0.2, 0.3], h_1(d) = [0.1, 0.2], h_1(e) = [0, 0]$ and $h_2(a) = [0, 0], h_2(b) = [0.55, 0.60], h_2(c) = [0.2, 0.3], h_2(d) = [0.1, 0.2], h_2(e) = [0, 0]$. Since, for example, h_1 can be verified as a stable p-model because the probabilistic reduct of P w.r.t. h_1 contains the h-rules:

$$a : [0.89, 0.91] \leftarrow$$
$$c : [0.2, 0.3] \quad \leftarrow d : [0.1, 0.15]$$
$$d : [0.1, 0.2] \quad \leftarrow$$

and $lfp(T_{P^{h_1}}) = h_1$.

Theorem 1. *Every h-program P has a unique stable p-model h iff h is the least p-model of P.*

Let us show that the stable probabilistic model semantics generalizes the stable model semantics of normal logic programs [10]. A normal logic program P can be represented as an nh-program $P' = \langle R, \tau \rangle$ where each normal rule

$$a \leftarrow b_1, \ldots, b_n, not\ c_1, \ldots, not\ c_m \in P$$

can be encoded, in R, as an nh-rule of the form

$$a : [1, 1] \leftarrow b_1 : [1, 1], \ldots, b_n : [1, 1], not\ (c_1 : [1, 1]), \ldots, not\ (c_m : [1, 1])$$

where $a, b_1, \ldots, b_n, c_1, \ldots, c_m$ are atomic hybrid basic formulae and $[1, 1]$ represents the truth value *true*. τ is any arbitrary assignment of disjunctive p-strategies. We call the class of nh-programs that consists only of nh-rules of the above form as $NHPP_1$.

Proposition 3. *Let P be a normal logic program. Then S' is a stable model of P iff h is a stable p-model of $P' \in NHPP_1$ that corresponds to P where $h(a) = [1,1]$ iff $a \in S'$ and $h(b) = [0,0]$ iff $b \in B_L \setminus S'$.*

4   An Algorithm for Computing Stable P-Models

In this section, we develop an algorithm for computing the stable p-models for an nh-program, which is based on SMODELS algorithm [17] for computing the stable model semantics for normal logic programs. The algorithm we develop constructs a stable p-model *incrementally*. It takes a ground nh-program P and a partial hybrid formula function (partial p-interpretation) h as inputs and returns true if h can be extended to a stable p-model (which is a total hybrid formula function) for P. Otherwise, it returns false. In the following we provide definitions and notions that we use throughout the rest of the paper. Let h be a probabilistic interpretation, then $dom(h) \subseteq bf_S(B_L)$ denotes the domain of h ($dom(h) \subsetneq bf_S(B_L)$ if h is a partial probabilistic interpretation). We use $negdom(h)$ to denote the set $\{F \mid F \in dom(h), h(F) = [0,0]\}$. We also define $posdom(h) = dom(h) \backslash negdom(h)$. Given a hybrid formula function h, $Pos(h)$ and $Neg(h)$ denote the following mappings:

- $Pos(h)(F) = h(F) \ \forall F \in dom(h)$ such that $h(F) \neq [0,0]$.
- $Neg(h)(F) = h(F) \ \forall F \in dom(h)$ such that $h(F) = [0,0]$.

We will describe each hybrid formula function using its graph. In other words, if h is a hybrid formula function, then h can be represented as the set $\{(A, \mu) | A \in dom(h)$ and $\mu = h(A)\}$. More conveniently, we use $A : \mu$ to denote (A, μ). Thus, we will frequently refer to h as a *set of annotated hybrid basic formulae*. Furthermore, if P is a ground nh-program, we consider $bf_S(B_L)$ as the set of all distinct ground hybrid basic formulae that appear in P, denoted by $Formulae(P)$. If h is a partial or total hybrid formula function, then $dom(h) = posdom(h) \cup negdom(h)$ and $h = Pos(h) \cup Neg(h)$ viewing h as a set of annotated hybrid basic formulae.

Definition 9. *A set of annotated hybrid basic formulae h (a hybrid formula function) is said to* cover *a set of annotated hybrid basic formulae g (a hybrid formula function) if $dom(g) \subseteq dom(h)$ and $\forall F \in dom(g), g(F) \leq_t h(F)$.*

Definition 10. *A set of annotated hybrid basic formulae g (a hybrid formula function) is said to* agree with *a set of annotated hybrid basic formulae h (a hybrid formula function) if the following conditions hold:*

- *$posdom(h) \subseteq posdom(g)$,*
- *$negdom(h) \subseteq negdom(g)$, and*
- *$\forall F \in dom(h), h(F) \leq_t g(F)$.*

Definition 10 is closely related to the definition of the well-founded order defined in [22].

Definition 11 ([22]). *Let P be an nh-program, H_P be the set of all partial hybrid formula functions of P, and $h_1, h_2 \in H_P$. We define the following partial order ($\leq_w$) on H_P: $h_1 \leq_w h_2$ iff $posdom(h_1) \subseteq posdom(h_2)$, $negdom(h_1) \subseteq negdom(h_2)$, and $\forall F \in dom(h_1), h_1(F) \leq_t h_2(F)$.*

The notion of *cover* can also be applied to sets of hybrid basic formulae. If h and g are two hybrid formula functions, then $dom(h)$ covers $dom(g)$ if $dom(g) \subseteq dom(h)$. Moreover, a hybrid basic formula F is said to be covered by $dom(h)$ if $F \in dom(h)$.

Definition 12. *Let $P = \langle R, \tau \rangle$ be a ground nh-program and h be a total hybrid formula function, then we define the operator F_P as a mapping $F_P : HFF \to HFF$ where $F_P(h) = lfp(T_{P^h})$.*

Lemma 1. *Let $P = \langle R, \tau \rangle$ be a ground nh-program and h be a total hybrid formula function, then h is a stable p-model of P iff $F_P(h) = h$.*

Lemma 2. *The function F_P is anti-monotone with respect to $\leq_t$.*

Now, we describe an algorithm for computing the stable p-model semantics for an nh-program along with the auxiliary functions. In addition, we present the necessary conditions that these auxiliary functions have to satisfy to guarantee the soundness and completeness of stable p-model semantics computation algorithm. Figure 1 describes a decision procedure for determining whether an nh-program P has a stable p-model or not. The function $spmodels(P, h)$ computes one stable p-model for P, however, it can be modified to compute all the stable p-models of P. It returns true if there is a stable p-model for P agreeing with the set of annotated hybrid basic formulae h (a hybrid formula function), otherwise it returns false. It takes a ground nh-program P and a partial hybrid formula function (a set of annotated hybrid basic formulae) h as an input. The set h represents the partially computed stable p-model.

The function $spmodels(P, h)$ calls two functions: $pexpand(P, h)$ and $pconflict(P, h)$. The function $pexpand(P, h)$ expands the set of annotated hybrid basic formulae h by the functions $PAtleast(P, h)$ and $PAtmost(P, h)$, whereas the function $pconflict(P, h)$ discovers the conflicts. The function $pconflict(P, h)$ determines whether h is a hybrid formula function (partial or total) that could be extended to a stable p-model of P, by checking that each hybrid basic formula defined in h is assigned exactly one probability interval and h satisfies P. To guarantee the soundness and completeness of $spmodels(P, h)$ we present the conditions E1-E2 and C1-C2 required for designing $pexpand(P, h)$ and $pconflict(P, h)$. Let $h' = pexpand(P, h)$ we assume that:

E1: $posdom(h) \subseteq posdom(h'), negdom(h) \subseteq negdom(h')$, and for all $F \in dom(h)$,
$\quad h(F) \leq_t h'(F)$, and
E2: every stable p-model of P that agrees with h agrees also with h'.

In addition, we assume that $pconflict(P, h)$ satisfies the following conditions

C1: if $dom(h)$ covers $Formulae(P)$ and there is no stable p-model that agrees with h, then $pconflict(P, h)$ returns true, and

C2: if $pconflict(P, h)$ returns true, then there is no stable p-model of P that agrees with h.

The function $spmodels(P, h)$ starts by expanding the partially computed stable p-model h (line 2). Condition E1 ensures that h is really extended and E2 guarantees that no stable p-model is lost. Then a test for checking a conflict is performed. Condition C1 ensures that if $Formulae(P)$ is covered and there is a conflict, the conflict is detected (lines 3 and 4). Condition C2 guarantee that if there is a conflict, then there is no stable p-model agreeing with h'. If there is no conflict (lines 5 and 6) and $dom(h')$ covers $Formulae(P)$, then $spmodels(P, h)$ returns true with h' is a stable p-model of P. If there is $x \in Formulae(P)$

```
1: function spmodels(P, h)
2:   h' := pexpand(P, h)
3:   if pconflict(P, h') then
4:     return false
5:   else if dom(h') covers Formulae(P) then
6:     return true
7:   else
8:     take some x ∈ Formulae(P) not covered by dom(h')
9:     if spmodels(P, h' ∪ {x : [0, 0]}) then
10:      return true
11:    else
12:      take x : [a, b] ∈ lfp(F_P^2) or gfp(F_P^2)
13:      return spmodels(P, h' ∪ {x : [a, b]})
14:    end if
15: end if
```

```
1: function pexpand(P, h)
2: repeat
3:    h' := h
4:    h := PAtleast(P, h)
5:    h := h ∪ {F : [0, 0]|F ∈ Formulae(P) and F : [0, 0] ∈ PAtmost(P, h)}
6: until h' = h
7: return h
```

```
1: function pconflict(P, h)
2: { Precondition: h = expand(P, h)}
3: if posdom(h) ∩ negdom(h) ≠ ∅ then
4:    return true
5: else if dom(h) covers Formulae(P) and h does not satisfy P then
6:    return true
7: else
8:    return false
9: end if
```

Fig. 1. A decision procedure for the stable p-model semantics

not covered by $dom(h')$ (line 8), then either $x : [0,0]$ belongs to the partially computed stable p-model (line 9) or there is some constant annotation $[a,b]$ such that $x : [a,b]$, with $[a,b] \neq [0,0]$, belongs to the partially computed stable p-model (line 12 and 13). The two cases are handled by backtracking. In the first case we extend h' by $\{x : [0,0]\}$, but if $spmodels(P, h' \cup \{x : [0,0]\})$ returns false, then $x : [0,0]$ does not belong to the computed stable p-model. Hence, $spmodels(P, h)$ returns what $spmodels(P, h' \cup \{x : [a,b]\})$ returns, where $x : [a,b] \in lfp(F_P^2)$ or $gfp(F_P^2)$. Since F_P is antimonotone, F_P^2 is monotone and its least fixpoint and greatest fixpoint limit the fixpoints of F_P [12]. Therefore, because of the possibility of having multiple nh-rules in P with x in their heads with different annotations, we select $[a,b]$ that is guaranteed to be in the computed stable p-model. This is achieved by selecting $x : [a,b]$ such that $x : [a,b] \in lfp(F_P^2)$ or $x : [a,b] \in gfp(F_P^2)$. This is because any stable p-model is a fixpoint of the operator F_P. The following theorem proves the correctness of decision procedure described in Figure 1.

Theorem 2. *Let P be an nh-program and h be a hybrid formula function. Then, there is a stable p-model of P agreeing with h if and only if $spmodels(P, h)$ returns true.*

Proof. The proof of this theorem is similar to the proof of a corresponding result presented in [17]. The proof proceeds as follows. Let $NC(P, h) = Formulae(P) \setminus dom(h)$ be the set of hybrid basic formulae that is in $Formulae(P)$ but not covered by $dom(h)$. We prove the theorem by induction on $NC(P, h)$. Assume that $NC(P, h) = \emptyset$ which implies that $dom(h)$ covers $Formulae(P)$. Then, $h' = pexpand(P, h)$ and by E1 $dom(h')$ covers $Formulae(P)$ as well and $spmodels(P, h)$ returns true if and only if $pconflict(P, h')$ returns false. By E2, C1, and C2 $pconflict(P, h')$ returns false exactly when there is a stable p-model agreeing with h.

Assume that $NC(P, h) \neq \emptyset$. If $pconflict(P, h')$ returns true, then $spmodels(P, h)$ returns false. Hence, there is no stable p-model agreeing with h by the conditions E2 and C2. However, if $pconflict(P, h')$ returns false and $dom(h')$ covers $Formulae(P)$, then $spmodels(P, h)$ returns true. Therefore, there is a stable p-model that agrees with h due to the conditions E2 and C1. Otherwise, since $spmodels(P, h)$ returns true and $dom(h')$ still not covers $Formulae(P)$ and since the size of $NC(P, h' \cup \{x.[0,0]\}) = NC(P, h' \cup \{x.[a,b]\}) \subset NC(P, h')$ then by inductive hypothesis together with E1 and E2 we have that that either $spmodels(P, h' \cup \{x.[0,0]\})$ or $spmodels(P, h' \cup \{x.[a,b]\})$ returns true if and only if there is a stable p-model agreeing with h.   ∎

5   PAtleast(P,h) and PAtmost(P,h)

In this subsection we provide foundations for computing the functions $PAtleast(P, h)$ and $PAtmost(P, h)$. The function $PAtleast(P, h)$ enlarges the partially computed stable p-model h by adding annotated hybrid basic formulae and/or monotonically increasing the annotations associated to the hybrid basic formulae that already exist in the partially computed stable p-model h. The function $PAtleast(P, h)$ computes the least fixpoint of the operator D_P, which is a variation of probabilistic well-founded operator W_P defined in [22]. We say that an nh-program globally satisfies $F : \nu \ (not \ (G : \beta))$ if the nh-program as a whole provides evidence for satisfying $F : \nu \ (not \ (G : \beta))$.

Definition 13 (Global Satisfaction). *Let P be an nh-program and $F : \nu \ (not$ $(G : \beta))$ be a positive (negative) hybrid literal. We say that $F : \nu \ (not \ (G : \beta))$ is globally satisfied by P if every minimal probabilistic interpretation that satisfies P also satisfies $F : \nu \ (not \ (G : \beta))$.*

Let $P = \langle R, \tau \rangle$ be an nh-program and g be a stable p-model of P agreeing with the set of annotated hybrid basic formulae h and H_P is the set of all partial p-interpretations of P. Then we define $PAtleast(P, h)$ to be the least fixpoint of the operator $D_P : H_P \to H_P$ defined as follows:

1. For each atom A we have that $D_P(h)(A) = c_{\tau(A)} \ M_A$, where $M_A \neq \emptyset$ contains the probability intervals μ obtained from the nh-rules $A : \mu \leftarrow Body \in R$, such that h satisfies $Body$, and for each negative hybrid literal $not \ (G_j : \beta_j)$ in $Body$ we have that P globally satisfies $not \ (G_j : \beta_j)$.
2. For each atom A we have that $D_P(h)(A) = [0, 0]$ if for each nh-rule $r \in R$ such that A appears in its head, h does not satisfy some hybrid literal $F : \nu$ or $not \ (G : \beta)$ in the body of r and P does not globally satisfy $F : \nu$.
3. $D_P(h)(G_1 \wedge_\rho G_2) = c_\rho(D_P(h)(G_1), D_P(h)(G_2))$ where $(G_1 \wedge_\rho G_2) \in bf_S(B_L)$ and for each atom A in $(G_1 \wedge_\rho G_2)$, A is defined in $D_P(h)$.
4. $D_P(h)(G_1 \vee_{\rho'} G_2) = c_{\rho'}(D_P(h)(G_1), D_P(h)(G_2))$ where $(G_1 \vee_{\rho'} G_2) \in bf_S(B_L)$ and for each atom A in $(G_1 \vee_{\rho'} G_2)$, A is defined in $D_P(h)$.

Example 4. Consider the following nh-program P

$$a : [0.9, 1] \quad \leftarrow b : [0.7, 0.8], not \ (c : [0.5, 0.55])$$
$$d : [0.9, 1] \quad \leftarrow not \ (a : [0.9, 1])$$
$$e : [0.2, 0.35] \leftarrow not \ (b : [0.7, 0.8])$$

We will compute $h = PAtleast(P, \emptyset)$. Since b and c do not appear in heads of any nh-rules in P, $b : [0, 0] \in h$ and $c : [0, 0] \in h$ by 2 in the above definition. In addition, $a : [0, 0] \in h$ by 2 as well since the first nh-rule is not satisfied due to $b : [0.7, 0.8]$ in the nh-rule because $[0.7, 0.8] \not\leq_t [0, 0]$. Obviously, $d : [0.9, 1]$ and $e : [0.2, 0.35]$ are in h by 1 in the above definition. Hence, $PAtleast(P, \emptyset) = \{a : [0, 0], b : [0, 0]c : [0, 0], d : [0.9, 1], e : [0.2, 0.35]\}$ which is the least fixpoint of D_P.

Lemma 3. *The function $PAtleast(P, h)$ is monotonic with respect to $\leq_w$ in its second argument.*

Note that $D_P(h)$ is a variation of the probabilistic well-founded operator W_P defined in [22]. This implies that $PAtleast(P, h) = lfp(D_P(h)) = lfp(W_P)$. Therefore, according to Theorem 4 of [22], g is a stable p-model of P if g is a fixpoint of W_P, and hence a fixpoint of D_P which in turn a fixpoint of $PAtleast(P, h)$. This implies that g is a stable p-model of P iff $g = W_P(g) = D_P(g) = PAtleast(P, g)$.

Proposition 4. *If g is a stable p-model of an nh-program P that agrees with the partial hybrid formula function h, then g agrees with $PAtleast(P, h)$.*

Furthermore, we can bound a stable p-model from above by defining the function $PAtmost(P, h)$. The function $PAtmost(P, h)$ computes the least fixpoint of P^h, the probabilistic reduct of P with respect to h (defined below). The idea is to extend the set of annotated hybrid basic formulae h which corresponds to the partially computed stable p-model by adding annotated hybrid basic formulae of the form $x : [0, 0]$ without violating condition E2. We can add $x : [0, 0]$ to the set h if $x : [0, 0] \in PAtmost(P, h) = lfp(T_{P^h})$. However, a different notion of probabilistic reduct from the one defined in Definition 7 is needed in this context as defined below.

Definition 14. *Let $P = \langle R, \tau \rangle$ be a ground nh-program, h be a partial hybrid formula function, and*

$$r \equiv A : \mu \leftarrow F_1 : \mu_1, \ldots, F_n : \mu_n, not \ (G_1 : \beta_1), \ldots, not \ (G_m : \beta_m) \in R.$$

Then the probabilistic reduct P^h of P with respect to h is $P^h = \langle R^h, \tau \rangle$ where R^h is the set of h-rules obtained from R by:

- *deleting every nh-rule r in R where there is a $not \ (G_j : \beta_j)$ in the body of r such that $\beta_j \leq_t h(G_j)$,*
- *deleting every $not \ (G_k : \beta_k)$ from the body of the remaining nh-rules.*

The notion of reduct in the above definition is a generalization of the notion of reduct in Definition 7, to cope with partial hybrid formula functions. For total hybrid formula functions both notions of reduct coincides. In addition, $PAtmost(P, h)$ is a total hybrid formula function. Consequently, if g is a stable p-model of an nh-program P, then $g = PAtmost(P, g)$. It is worth noting that, from the definition of the probabilistic reduct with respect to partial hybrid formula function h, h can be extended to a total hybrid formula function and we still get the same probabilistic reduct. This is achieved by adding to h the annotated hybrid basic formulae $F : [0, 0]$ such that $F \in bf_S(B_L) \setminus dom(h)$. This means, if $h_1 \leq_w h_2$, then $h_1 \leq_t h_2$ as well, after extending both h_1 and h_2 to total hybrid formula functions by adding $F : [0, 0]$ such that $F \in bf_S(B_L) \setminus dom(h_1)$ to h_1 and $F : [0, 0]$ such that $F \in bf_S(B_L) \setminus dom(h_2)$ to h_2 respectively.

Lemma 4. *The function $PAtmost(P, h)$ is anti-monotone in its second argument.*

The above lemma shows that the function $PAtmost(P, h)$ is anti-monotone with respect to $\leq_t$. This is because given $h_1 \leq_w h_2$, then we also get $h_1 \leq_t h_2$, which implies that $PAtmost(P, h_2) \leq_t PAtmost(P, h_1)$.

Proposition 5. *Let g be a stable p-model of P that agrees with h. Then $g \leq_t PAtmost(P, h)$.*

Corollary 1. *The function $pexpand(P, h)$ satisfies conditions E1 and E2*

Corollary 2. *The function conflict(P, h) satisfies conditions C2*

It follows that *pexpand(P, h)* satisfies conditions E1 and E2. The function *conflict(P, h)* obviously fulfills C2, and the next proposition shows that C1 also holds.

Proposition 6. *If* $h = pexpand(P, h)$, *dom(h) covers Formulae(P), and posdom(h) $\cap$ negdom(h) = $\emptyset$ and h satisfies P, then h is a stable p-model of P.*

Example 5. Consider the following nh-program P

$$a : [0.45, 0.55] \leftarrow c : [1, 1], not\ (b : [0.7, 0.95])$$
$$b : [0.7, 0.95] \leftarrow c : [1, 1], not\ (a : [0.45, 0.55])$$
$$c : [1, 1] \leftarrow not\ (a : [0.45, 0.55])$$

We use the decision procedure *spmodels* to determine whether P has a stable p-model or not and return it if exist. The $lfp(F_P^2)$ is the empty set and $gfp(F_P^2)$ is

$$\{a : [0.45, 0.55], b : [0.7.0.95], c : [1, 1]\}.$$

First*pexpand(P, $\emptyset$)* returns $\emptyset$ and *pconflict(P, $\emptyset$)* returns false. Since *Formulae(P)* = $\{a, b, c\}$ is not covered by $\emptyset$, we choose either a, b, or c in order to proceed. Let us take b, then *spmodels(P, $\{b : [0, 0]\}$)* is executed. *pexpand(P, $\{b : [0, 0]\}$)* returns $\{b : [0, 0]\}$. Then *pconflict(P, $\{b : [0, 0]\}$)* returns false. Since *Formulae(P)* is not covered by $\{b\}$, we choose either a or c in order to proceed. Let us take a, then

$$spmodels(P, \{a : [0, 0], b : [0, 0]\})$$

is executed. *pexpand(P, $\{a : [0, 0], b : [0, 0]\}$)* returns

$$\{a : [0, 0], b : [0, 0], c : [1, 1], b : [0.7, 0.95], a : [0.45, 0.55]\}.$$

Then *pconflict(P, $\{a : [0, 0], b : [0, 0], c : [1, 1], b : [0.7, 0.95], a : [0.45, 0.55]\}$)* returns true. Then we backtrack and execute *spmodels(P, $\{a : [0.45, 0.55], b : [0, 0]\}$)*.

pexpand(P, $\{a : [0.45, 0.55], b : [0, 0]\}$) returns $\{a : [0.45, 0.55], b : [0, 0], a : [0, 0],$ c.[0, 0]\}.

Then *pconflict(P, $\{a : [0.45, 0.55], b : [0, 0], a : [0, 0], c.[0, 0]\}$)* returns true. Finally, we backtrack and execute *spmodels(P, $\{b : [0.7, 0.95]\}$)*. *pexpand(P, $\{b : [0.7, 0.95]\}$)* returns $\{b : [0.7, 0.95]\}$. Then *pconflict(P, $\{b : [0.7, 0.95]\}$)* returns false. Since *Formulae(P)* is not covered by $\{b\}$, we choose either a or c in order to proceed. Let us take a, then *spmodels(P, $\{a : [0, 0], b : [0.7, 0.95]\}$)* is executed. *pexpand(P, $\{a : [0, 0], b : [0.7, 0.95]\}$)* returns $\{a : [0, 0], b : [0.7, 0.95], c : [1, 1]\}$. Then

$$pconflict(P, \{a : [0, 0], b : [0.7, 0.95], c : [1, 1]\})$$

returns false and *spmodels(P, $\{a : [0, 0], b : [0.7, 0.95]\}$)* returns true as well as *spmodels(P, $\{b : [0.7, 0.95]\}$)* and *spmodels(P, $\emptyset$)* having $\{a : [0, 0], b : [0.7, 0.95], c : [1, 1]\}$ as a stable p-model of P.

6   Conclusions

In this work, we have proposed an algorithm for computing the stable probabilistic model semantics [22]. The proposed algorithm is a modification of the decision procedure of SMODELS [17], a state-of-the-art system for computing the stable model semantics of normal logic programs. We have described the modified algorithm, along with its auxiliary functions, and we have provided the necessary conditions that these auxiliary functions have to satisfy to guarantee the soundness and completeness of the proposed algorithm. We have presented formal definitions and algorithms for these auxiliary functions and shown that they satisfy the necessary conditions for the soundness and completeness of the proposed algorithm.

As future work, we plan to provide an implementation of these algorithms, and investigate applications of the resulting framework in the context of knowledge representation and reasoning in presence of uncertainty (e.g., probabilistic planning).

References

1. C. Baral et al. Probabilistic reasoning with answer sets. *In 7th International Conference on Logic Programming and Nonmonotonic Reasoning*, Springer Verlag, 2004.

2. C. Bell, A. Nerode, R. Ng, V. S. Subrahmanian. Mixed integer programming methods for computing Nonmonotonic Deductive Databases. *Journal of ACM*, 41(6): 1178-1215, 1994.

3. W. D. Chen, D. S. Warren. Computation of stable models and its integration with logical query processing. *IEEE Transaction on Knowledge and Data Engineering*, 8(5): 742-757, 1996.

4. P. Cholewinski, V. Marek, M. Truszczynski, A. Mikitiuk. Computing with default logic. *Artificial Intelligence*, 112(1-2): 105-146, 1999.

5. A. Dekhtyar and V. S. Subrahmanian. Hybrid probabilistic program. *Journal of Logic Programming*, 43(3): 187-250, 2000.

6. M. Dekhtyar, A. Dekhtyar, and V. S. Subrahmanian. Hybrid probabilistic programs: algorithms and complexity. *In Uncertainty in Artificial Intelligence Conference*, pages 160-169, 1999.

7. W. F. Dowling and J. H. Gallier. Linear-time algorithms for testing the satisfiability of propositional Horn formulae. *Journal of Logic Programming*, 1(3): 267-284, 1984.

8. A. Van. Gelder. The alternating fixpoint of logic programs with negation. *Journal of Computer and System Sciences*, 47(1):185-221, 1993.

9. A. Van. Gelder, K. A. Ross, and J. S. Schlipf. The Well-founded semantics for general logic programs. *Journal of ACM*, 38(3):620-650, 1991.

10. M. Gelfond and V. Lifschitz. The stable model semantics for logic programming. In R. Kowalski and K. Bowen, editors, *In Fifth International Conference and Symposium on Logic Programming*, 1070-1080, 1988.

11. L. V.S. Lakshmanan and F. Sadri. On a theory of probabilistic deductive databases. *Journal of Theory and Practice of Logic Programming*, 1(1):5-42, January 2001.

12. V. Lifschitz. Foundations of logic programming. *In Principles of Knowledge Representation*, 69-127, CSLI Publications, 1996.

13. J. J. Lu and S. M. Leach. Computing annotated logic programs. *In International Conference on Logic Programming*, Pascal van Hentenryck, editor, MIT press, Cambridge, MA, 1994.
14. W. Marek and M. Truszczynski. Autoepistemic logic. *Journal of ACM*, 38(3):588–619, 1991.
15. R. T. Ng and V. S. Subrahmanian. Probabilistic logic programming. *Information and Computation*, 101(2):150-201, December 1992.
16. R. T. Ng and V. S. Subrahmanian. Stable semantics for probabilistic deductive databases. *Information and Computation*, 110(1):42-83, 1994.
17. I. Niemela and P. Simons. Efficient implementation of the well-founded and stable model semantics. *In Joint International Conference and Symposium on Logic Programming*, 289-303, 1996.
18. I. Niemela, P. Simons, T. Soininen. Stable model semantics of weight constraint rules. *In Fifth International Conference on Logic Programming and Nonmonotonic Reasoning*, 317-331, 1999.
19. R. Reiter. A logic for default reasoning. *Artificial Intelligence*, 13(1-2):81-132, 1980.
20. E. Saad. *Hybrid probabilistic programs with non-monotonic negation: semantics and algorithms*. Ph.D. thesis, New Mexico State University, May 2005.
21. E. Saad and E. Pontelli. Towards a more practical hybrid probabilistic logic programming framework. *In Practical Aspects of Declarative Languages*. Springer Verlag, 2005.
22. E. Saad and E. Pontelli. Hybrid probabilistic logic programming with non-monotoic negation. In *Twenty First International Conference on Logic Programming*. Springer Verlag, 2005.
23. V. S. Subrahmanian, D. S. Nau, C. Vago. wfs + branch and bound = stable models. *IEEE Transaction on Knowledge and Data Engineering*, 7(3): 362-377, 1995.
24. J. Vennekens, S. Verbaeten, and M. Bruynooghe. Logic programs with annotated disjunctions. *In International Workshop on Nonmonotonic Reasoning*, 2004.

DiaWOz-II – A Tool for Wizard-of-Oz Experiments in Mathematics*

Christoph Benzmüller[1], Helmut Horacek[1], Ivana Kruijff-Korbayová[2],
Henri Lesourd[1], Marvin Schiller[1], and Magdalena Wolska[2]

[1] Dept. of Computer Science
[2] Dept. of Computational Linguistics
Saarland University, Germany
{chris,horacek,henri,schiller}@ags.uni-sb.de,
{korbay,magda}@coli.uni-sb.de

Abstract. We present DiaWOz-II, a configurable software environment
for Wizard-of-Oz studies in mathematics and engineering. Its interface
is based on a structural *wysiwyg* editor which allows the input of complex mathematical formulae. This allows the collection of dialog corpora
consisting of natural language interleaved with non-trivial mathematical expressions, which is not offered by other Wizard-of-Oz tools in the
field. We illustrate the application of DiaWOz-II in an empirical study
on tutorial dialogs about mathematical proofs, summarize our experience with DiaWOz-II and briefly present some preliminary observations
on the collected dialogs.

Keywords: Dialog systems, natural language dialog in mathematics,
tutoring systems, Wizard-of-Oz experiments.

1 Introduction

For the development of natural language dialog systems, experiments in the
Wizard-of-Oz (WOZ) paradigm are a valued source of dialog corpora.[1]

Existing environments for WOZ experiments, even those for the domain of
mathematics tutoring, generally operate in domains that either require only
simple mathematical formulae (with operators like + and ×), or they separate
the mathematical objects (geometric figures or equations) from the tutorial dialog (such as in the Wooz tutor [2], for example). In this paper we present our
WOZ environment DiaWOz-II which, in contrast to that, enables the collection
of dialogs where natural language text is interleaved with mathematical notation, as is typical for (informal) mathematical proofs. The interface components

---

* This work has been funded by the DFG Collaborative Research Center on Resource-
Adaptive Cognitive Processes, SFB 378 (http://www.coli.uni-saarland.de/
projects/sfb378/).

[1] A Wizard-of-Oz experiment [1] serves to test the usability of a hypothetical software
system. The system is (partially) simulated by a human expert, the *wizard*. Typically,
a mediator software partially implements the functionality of the simulated system.

C. Freksa, M. Kohlhase, and K. Schill (Eds.): KI 2006, LNAI 4314, pp. 159–173, 2007.

of DiaWOz-II are based on the *what-you-see-is-what-you-get* scientific text editor TEX$_{\text{MACS}}$[2] [3]. DiaWOz-II provides one interaction window for the user and one for the wizard, together with additional windows displaying instructions and domain material for the user, and additional notes and pre-formulated text fragments for the wizard. All of these windows allow for copying freely from one to the other. Furthermore, our DiaWOz-II allows the wizard to annotate user dialog turns with their categorization. DiaWOz-II is also connected to a spell-checker for checking both the user's and the wizard's utterances.

This paper is organized as follows: In Sect. 2 we motivate the design of our system. In Sect. 3.1 we describe the TEX$_{\text{MACS}}$ *wysiwyg* editor, on which the interface of DiaWOz-II is based. The DiaWOz-II system is discussed in detail in Sect. 3. In Sect. 4 we discuss the application of DiaWOz-II in a recently completed series of experiments. Section 5 concludes the paper.

2   Design Aspects

General Requirements for WOZ Tools. We list some general requirements we considered in the development of DiaWOz-II:

Plausibility and Comfort. For WOZ experiments, it is crucial to maintain the user's belief that he is interacting with a fully artificial system. Therefore, the software that mediates between wizard and student should enable the wizard to conceal his human identity. This is not a trivial pursuit, since it is common sense that "people are flexible, computers are rigid (or consistent), people are slow at typewriting, computer output is fast" [4]. Thus, the WOZ tool is required to enable the wizard to respond to the participant quickly and comfortably and in a plausible way.

Suitability for Book-keeping. The main goal of WOZ experiments is the analysis of the interactions between the subjects and the simulated system. Therefore, the WOZ tool is required to record the dialogs using a representation format suitable for further processing and analysis.

Flexibility and Simplicity. The WOZ tool should be easily adjustable, so that it can be used under different experimental conditions and in different domains. Adjustments to the software should not significantly add to the complexity of carrying out a series of experiments, a process which by itself poses enough challenges.

Tool Integration. The WOZ tool should support the integration of other software components, for example, modules that already realize single parts of the simulated overall system.

Specific Requirements for DiaWOz-II. DiaWOz-II has been developed for application in the DIALOG project [5], which investigates the use of natural language dialog for teaching mathematical proofs. The particular research foci of the DIALOG project are natural language analysis, domain reasoning for mathematics, and tutorial aspects of mathematics tutoring.

---

[2] www.texmacs.org

In 2003, we carried out a first empirical study [6] in the WOZ paradigm in which we collected a corpus of tutorial dialogs on mathematical proofs in German. The study concentrated on the comparison between three tutoring strategies, namely the *Socratic*, *didactic* and the *minimal feedback* strategies. For this purpose, we developed the DiaWoZ [7] environment, the predecessor of DiaWOz-II. DiaWoZ supports complex dialog specifications, which were needed in order to specify a particular hinting algorithm used in the *Socratic* tutoring condition. DiaWoZ allows keyboard-to-keyboard interaction between the wizard and the student. The interfaces consist mainly of a text window with the dialog history and a menu bar providing mathematical symbols. Furthermore, the wizard can assign dialog state transitions and speech act categories to student turns w.r.t. the underlying dialog model. The DiaWoZ interface allowed free mixing of natural language text with mathematical symbols. Still, there was room for improvement with respect to the *plausibility and comfort* criterion postulated above. For example, the experiment participants suggested the use of the keyboard instead of the mouse for inserting mathematical symbols.

The first study motivated a second series of experiments [8], which we briefly describe in Sect. 4. In contrast to the first study, the more recent study imposes less constraints on the wizards' tutoring and assumes a rather simple dialog model. However, in comparison to the first study, the second study is more focused on linguistic phenomena and mathematical domain reasoning in tutorial dialogs and the interplay between these two.

Related Work. A variety of WOZ tools and dialog system toolkits already exist. Examples are the simulation environment ARNE [4], the SUEDE prototyping tool for speech user interfaces [9] and MD-WOZ [10].

In the domain of mathematics, a WOZ simulation of the ALPS environment [11] and the Wooz tutor [2] should be mentioned. In the case of ALPS, the Synthetic Interview (SI) method is used, i.e. the student formulates free-form questions in a chat window, and receives a video clip with an answer. In the ALPS system, these video clips are pre-recorded, stored in a database, and retrieved as answers for the questions from the user, whereas in the WOZ simulation of ALPS, the wizard's responses are spontaneous. The ALPS tutor is designed to be an algebra tutor. Typical problems in the domain of ALPS are for example the computation of area and perimeter of geometric figures.

The Wooz tutor is also a tool for keyboard-to-keyboard interaction in the domain of algebra. It offers a chat window displaying the tutorial dialog, a dedicated window displaying the problem statement and a dedicated editor for editing equations. A typical problem given to the participants is "please factor $11x^2 - 11x + 6$".

The interfaces of these two systems are not intended for mixing natural language input with the mathematical notation employed for proving theorems, which we investigate in the DIALOG project. For our dialog system we aim for an interface that allows flexible and easy input for mathematical formulae and natural language text. This requirement is addressed by the interface in DiaWOz-II.

3   The DiaWOz-II System

We decided to build a new WOZ tool rather than trying to improve the existing DiaWoZ system. An important motivation was to use TeX$_{\text{MACS}}$ [3] as a platform for the new system in order to benefit from its typesetting abilities, its configurable GUI and its event-handling as a building block for the creation of a more lightweight software.

DiaWOz-II is realized as a classical client-server architecture, and consists of a server and two client interfaces for the student and the tutor respectively. The architecture allows keyboard-to-keyboard interaction between the student and the tutor. Furthermore, the server fulfills other central functions, namely the recording of the interaction in a log-file, controlling turn-taking between the dialog participants, and providing an interface to a spell-checker. We first describe TeX$_{\text{MACS}}$ and its role in DiaWOz-II before we further elaborate on each of these aspects in turn.

3.1   TeX$_{\text{MACS}}$

TeX$_{\text{MACS}}$ is a scientific text editor with strong support for mathematical typesetting which is inspired by TeX and *GNU emacs*. The internal representation of a TeX$_{\text{MACS}}$-document is well organized in a tree-like structure. TeX$_{\text{MACS}}$ provides two alternative editing modes: (i) a *wysiwyg interface* that allows to directly manipulate the typeset document and (ii) a *source mode* that provides a view of the internal document representation in the underlying, structured TeX$_{\text{MACS}}$ markup language. This language supports the definition of *macros*, which are generally easy to read and understand. It is also worth noting that the standard TeX$_{\text{MACS}}$ markup language inherits many usual LaTeX constructs, in such a way that for LaTeX-literate persons, starting to use TeX$_{\text{MACS}}$ is usually straightforward. Thus, extending the markup (namely, defining new kinds of tags together with how these newly defined tags must be typeset) can be done in a very convenient way using macros. For more sophisticated behavior, for example, the implementation of an interactive application, one can use *Scheme*, the standard TeX$_{\text{MACS}}$ scripting language.

TeX$_{\text{MACS}}$ fulfills the *plausibility and comfort* requirement introduced in Sect. 2 by offering various advanced modes of input for mathematical symbols, and in particular it enables LaTeX commands. Using TeX$_{\text{MACS}}$ also fulfills the *flexibility and simplicity* requirement, since it can be reconfigured with little effort.

The TeX$_{\text{MACS}}$ editor has already been adapted as an interface to a diversity of external tools, most of which are computer algebra systems. However, using TeX$_{\text{MACS}}$ as an interface for a (simulated) tutoring system is novel.

3.2   TeX$_{\text{MACS}}$ as Base Component of DiaWOz-II

A TeX$_{\text{MACS}}$ application as employed in DiaWOz-II has the overall structure shown in Fig. 1. Such an application consists of (i) a set of TeX$_{\text{MACS}}$ *macros* which implement the *visualization* of the different parts of the user interface (i.e. what are their shapes, their locations, the text attributes (e.g. color, font, ...), etc.), and

(2) a set of *Scheme scripts*, which implement the mechanism which interprets the *events* (e.g., a mouse click, a key press, etc.) and *modifies* the interface accordingly.

Macros. A very basic example of a T_EX_{MACS} macro that can be used to turn a part of the document into *italics underlined* text is (cf. [12] for more details on the macro language):

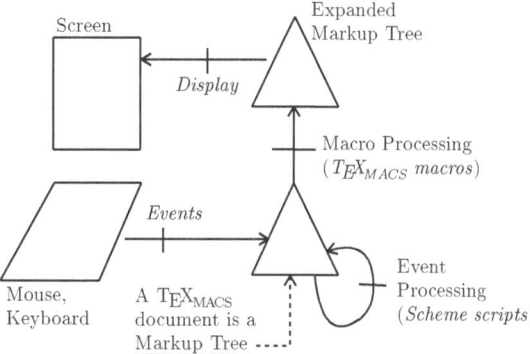

Fig. 1. Structure of a T_EX_{MACS} application

```
<underlined-italics|x> => <with|font-shape|italic|<underline|<arg|x>>>
```

The left-hand side of this expression defines the use of the macro (i.e., the *non-expanded* markup, as it can be found in a T_EX_{MACS} document file) and the right-hand side its expansion. Given this macro definition, the T_EX_{MACS} markup fragment   `<underlined-italics|This is italics underlined text.>` is first rewritten by the macro processor as   `<with|font-shape|italic|` `<underline|This is italics underlined text.>>`   and then displayed in T_EX_{MACS} as *This is italics underlined text.*

Processing the Markup Using *Scheme*. The event processor can be extended by plugins written as *Scheme* scripts. These scripts can manipulate the internal markup tree that represents the user interface, typically as a reaction to an event (e.g., mouse, keyboard, network, etc.). As a reaction to the changes in the markup, the macros are reevaluated, and the display is then updated.

3.3   Student and Wizard Interfaces

The dialog system simulated by DiaWOz-II is presented to the student as a window, referred to as the interaction window. It consists of menu bars and a text field, as shown in Fig. 2. The dialog history and the prompt for the current input are displayed in the same text field, separated by a horizontal bar at the bottom in Fig. 2. The utterances from the tutor and the student are displayed in different colors for better readability. The student can send messages by pressing the "absenden" (submit) button. Upon submitting, the message becomes part of the dialog history. The answers by the tutor are accompanied by an acoustic signal.

In a second window, which is independent of the interaction window, supplementary study material with mathematical concepts and definitions is displayed.

The wizard's interface, as shown in Fig. 3, is conceptually similar to the student's interface. In addition, the wizard is asked to categorize each student turn w.r.t. three dimensions: correctness, granularity and relevance; the wizard fills out the fields of a small table referring to the three dimensions by making

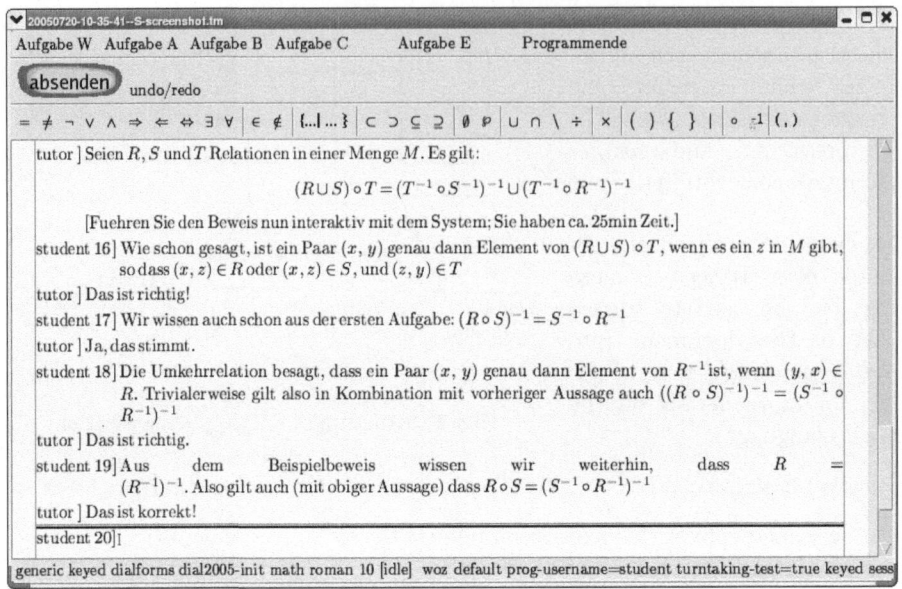

Fig. 2. Interaction window of the student interface

choices in pull-down menus, or directly by typing. The wizard's button for sending messages is only enabled once all the fields have been filled. If the student's utterance does not represent a mathematical statement the wizard fills in default values (N/A).

We now turn in more detail to the methods for inserting mathematical symbols in DiaWOz-II, which are made available by TEX$_{MACS}$. Mathematical symbols (e.g., $\emptyset$) can be created by using LaTeX commands (e.g., \emptyset) or by using additional commands defined when designing the interface (e.g., the command \emptyset in German language, i.e. \leeremenge). These commands are also made available in the menu bar. DiaWOz-II also allows for structured commands, e.g. commands that create pairs of brackets for pair ($\square$, $\square$) and for set notation $\{\square | \square\}$. An example is the macro *paar* (German for *pair*):

```
<paar|left|right> => ( <arg|left> , <arg|right> )
```

Invoking \paar with the arguments x and y yields the formula (x, y). The two arguments need not necessarily be provided when invoking the macro, their respective placeholders can also be filled in interactively and modified later. Macros can be nested, and most importantly, they avoid missing parentheses when the user writes expressions using the pair notation. The set of macros provided with DiaWOz-II can be easily extended with further TEX$_{MACS}$ macros.

TEX$_{MACS}$ furthermore makes it possible to distinguish between mathematical symbols created via the menu bar and via LaTeX commands, even if they appear to be the same at the typesetting level.

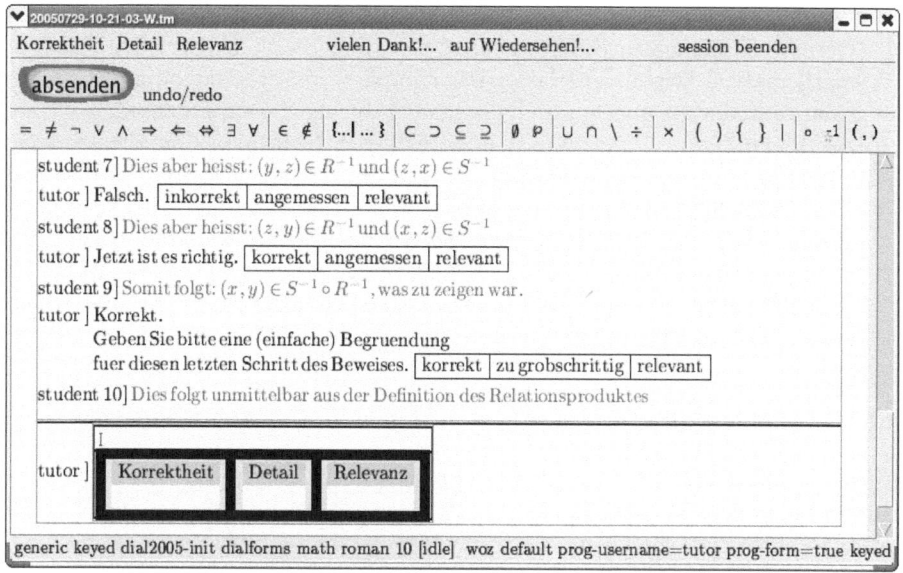

Fig. 3. The interaction window of the wizard interface

Using structured building blocks for constructing mathematical formulae via macros is similar to the MATHS TILES approach [13]. MATHS TILES are graphical tiles that can contain text, diagrammatic shapes and *sockets*, which are place-holders where other MATHS TILES can be inserted to form composite objects. TEX$_{\text{MACS}}$ has the advantage over MATHS TILES that it already includes by default a large set of macros for constructing formulae, such as a large number of macros that represent LAT$_{\text{E}}$X commands.

3.4   The Server

The central capabilities of DiaWOz-II reside in the server. Its main task is to pass the dialog contributions back and forth between the student and the wizard interface. Furthermore, it provides the following other central services:

Log-file Mechanism. All dialogs are recorded in a log-file in DiaWOz-II. The log-file format is based on the representation format of TEX$_{\text{MACS}}$, which is a structured, extensible and open document format. Naturally, the annotations performed by the wizard for each student turn are also stored in the log-file.

Spell-Checking. Spelling mistakes by the wizard can be a giveaway of human simulation. Therefore, our server (optionally) integrates a spell-checker. If spell-checking is activated, a message from the wizard is only passed on by the server if it passes the spell-checker, otherwise the wizard is asked to correct the message. The student's input is also spell-checked. Messages exceeding a threshold of spelling errors are refused (i.e. not passed on to the wizard). The underlying rationale is that it would be implausible that an automated system could deal with such misspelled input.

We currently employ the spell-checker GNU Aspell[3] with the standard German dictionary provided with Aspell together with an extra dictionary of mathematical jargon. The latter was compiled from the introductory mathematics materials and gradually extended during the experiments.

Turn-Taking Control. DiaWOz-II imposes strict turn-taking on the student: once the student makes a turn, the sending of new messages is disabled (i.e., the dedicated button for "sending" is deactivated and displayed in a darker shade) until the tutor provides a response. Without this constraint, it might become unclear to which turn of the student an answer from the wizard belongs. However, the tutor is allowed to barge in at any time, which enables him to offer support or prompt if the student appears to be inactive.

3.5   Implementation

Figure 4 illustrates the architecture of DiaWOz-II. In order to customize the client interfaces, we have

- adapted the menu bars and buttons to the needs of our application and
- restricted the editing facilities so that the student can only type in a designated text area with all other T$_E$X$_{MACS}$ functionalities disabled (for example, inserting an image, or editing the dialog history).

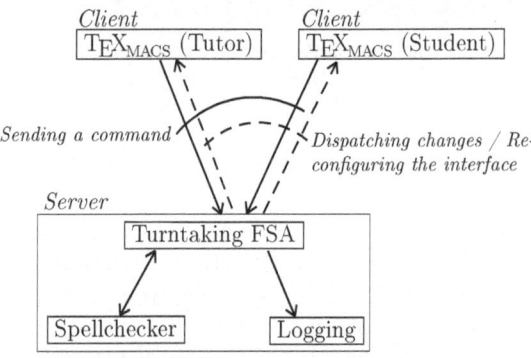

Fig. 4. System architecture

On the server side, turn-taking is controlled by a finite-state automaton. A message received by the server is written to the log-file and sent to a spell-checker. If it passes, it is broadcast to the clients. If it does not pass, it is sent back to the sender for correction. Disallowing the student from sending new messages until the wizard makes a turn is technically realized by server messages to the student's client to reconfigure the client's interface (i.e. enable/disable the interface's elements according to the current state).

The combination of macros and *Scheme* provided in TeXmacs has turned out to be very useful for our development of DiaWOz-II. In particular, the amount of code we wrote (a dozen of *Scheme* files of approximately 100 Kb in total) is relatively small considering the implemented functionality, and it remained manageable over time (as opposed to the previous version of DiaWoZ that consisted of about 200 Kb Java code spread among 70 files). The environment enabled also

---

[3] http://aspell.sourceforge.net/

people who are not professional software developers to participate in developing the system. Thus, TeXmacs has proven to be a good choice for our WOZ software, both from the point of view of the level of functionality it offers (word processing with LaTeX-like mathematical typesetting in a customizable editor) as well as from the point of view of prototyping and extending the software. The combination of the *Scheme* programming language with the large set of features already provided by TeXmacs allows for a lightweight, inclusive software development process.

4   An Empirical Study Using DiaWOz-II

Exploiting the DiaWOz-II system, we carried out a series of experiments in July 2005. In this study (see [8]), we collected a corpus of tutorial dialogs in German on mathematical proofs in the domain of binary relations. The collected data serves to investigate linguistic phenomena related to the mixing of mathematical formulae and natural language, underspecification phenomena, qualitative aspects of proof steps and mutual dependencies between natural language analysis and non-trivial mathematical domain reasoning.

4.1   Method

Thirty-seven students from Saarland University participated in the experiments. They were instructed to solve proof exercises collaboratively with a computer system that was described to them as a natural language dialog system on mathematics. This system was simulated with the DiaWOz-II software and four experts[4], who took the role of the wizard in turn (the set-up is shown in Fig. 5).

Fig. 5. An experiment in progress: The participant (left) and the wizard, experimenter and research assistant in the control room (right)

The wizards were given general instructions on the *Socratic* style of tutoring (cf. [14]), which is characterized by the use of questions to elicit information from

---

[4] The experts consisted of the lecturer of a course *Foundations of Mathematics*, a maths teacher, and two maths graduates with teaching experience.

the student. The tutors were instructed to reject utterances outside the mathematical domain and to respond in a uniform manner. Apart from that, the wizards were not restricted in the verbalization of their answers to the students. This allowed us to investigate the use of mathematical language without possibly influencing it by a-priori restrictions, even if more restrictions might have contributed to making the simulated system appear even more machine-like. In addition to the interaction window of DiaWOz-II, the tutors were provided with a second TeXmacs window in which they could save text and formulae for re-use.

The exercises were taken from the domain of relations, and were centered around the concepts of relation composition and relation inverse. Because of the advanced character of the exercises, the participants were required to have taken part in at least one mathematics course at university level. First, the subjects were required to fill out a questionnaire, asking about previous experiences with dialog systems and mathematics background. Subjects were also given study material with the mathematical definitions that were required to solve the exercises which was studied for approximately 25 minutes. The materials were handed out on paper and were also available as a TeXmacs document on the screen. This helped to achieve a uniform (and thus plausible) appearance of the system. Prior to the tutoring session, the students received a short introduction to the interface, during which the different modes of input for mathematical symbols – as menu items, as LaTeX commands or via commands in German language – and the copy & paste facility were demonstrated.

The largest part of the two-hour experimental session was allotted to the interaction between the student and the simulated system. In addition to the log-files recorded by DiaWOz-II, screen recordings were made. Furthermore, the participants were encouraged to "think aloud" and they were audio-recorded and filmed. This comprehensive collection of data not only documents the text of the tutorial dialogs, but also allows us to analyze how the participants used the interface and the study material.

At the end of the experimental session, the participants were required to fill out a second questionnaire asking about their verdict on the usability of the system, how difficult they found the exercises, and suggestions for improvements of the system.

4.2   Discussion

The experiments resulted in a large and diverse corpus of dialogs. During a session, a participant made on average 24 dialog turns, excluding those that were rejected for bad spelling. We briefly discuss how DiaWOz-II fulfilled its role, how the participants coped with the interface. Furthermore, the collected data hints at a potential influence of the interface features in combination with the reading material on the resulting tutorial dialogs.

Observations from the Corpus. An example of two dialog fragments from the experiment is given in Fig. 6. These dialogs were obtained under two different modes of presentation of the study material: formal (FM) vs. verbose (VM). Note

S33: Nach Aufgabe W ist
$(S \circ (S \cup R)^{-1})^{-1} = ((S \cup R)^{-1})^{-1} \circ S^{-1}$
By Exercise W
$(S \circ (S \cup R)^{-1})^{-1} = ((S \cup R)^{-1})^{-1} \circ S^{-1}$
holds
T34: Das ist richtig! *That is correct!*
S34: Dies ist nach Theorem 1 gleich
$(S \cup R) \circ S^{-1}$
This is by Theorem 1 equal to
$(S \cup R) \circ S^{-1}$
T35: Das ist auch richtig!
That is also correct!
S35: Ein Element (a,b) ist genau dann
in dieser Menge, wenn es ein $z \in M$ gibt
mit $(a,z) \in S \cup R$ und $(z,b) \in S^{-1}$
An element (a,b) is in this set exactly
when there is a $z \in M$ with $(a,z) \in S \cup R$
und $(z,b) \in S^{-1}$
T36: Das ist korrekt! *That is correct!*
S36: Also $(b,z) \in S$. *Therefore $(b,z) \in S$.*
T37: Auch das ist korrekt!
That is also correct!
S37: DAmit [sic] kann aber nicht die
geforderte Gleichheit bewiesen werden.
But with this the given equality cannot be
proven.
T38: Das haben Sie gut erkannt! Was
fehlt Ihnen dazu?
You have made a good observation! What
is missing?

S25: $(R \cup S) \circ S = \{(x,y) \mid \exists z(z \in M \land$
$((x,z) \in R \lor (x,z) \in S) \land (z,y) \in S\}$
T28: Diese Beziehung hat Relevanz fuer
die Aufgabe. Sie muessen sich also etwas
dabei gedacht haben, als Sie diese ver-
muteten. Koennen Sie begruenden, wie
sie mit der Aufgabe zusammenhaengt?
This relation is relevant for the exercise.
You must have considered something as
you proposed it. Can you justify how it is
related to the exercise?
T29: Ihre Formel ist richtig.
Your formula is correct.
T30: Koennen Sie die Behauptung unter
zusaetzlichen Voraussetzungen beweisen?
Can you prove the expression under
additional assumptions?
S26: $\{(x,y) \mid \exists z(z \in M \land ((x,z) \in R \lor$
$(x,z) \in S) \land (z,y) \in S\} = \{(x,y) \mid \exists z(z$
$\in M \land (y,z) \in S \land ((x,z) \in R \lor (x,z) \in$
$S))\} \Leftrightarrow ((y,z) \in S \land (z,y) \in S)\}$
T31: Auf der rechten Seite ist z nicht
spezifiziert. Meinten Sie vielleicht $\Leftrightarrow$ oder
etwas Aehnliches [sic] statt $\land$?
On the right side z is not specified.
Do you perhaps mean $\Leftrightarrow$ or something
similar instead of $\land$?

Fig. 6. Excerpts of dialogs in the two conditions: VM-group (left) and FM-group
(right). English translations are given in italics. S- and T- indicate student and wizard
turns, respectively.

that the dialogs clearly differ in the employed mathematical style and that in
Fig. 6 (right), the mathematical operations performed by the student can be
characterized as term rewriting steps, i.e. a subformula of a term is replaced by
an equivalent subformula. Also note that in Fig. 6 (right), the student uses no
natural language. Even though all subjects were informed before the interaction
that the system can handle a combination of natural language and formula input,
we observed great variations in the amount of natural language used by the
subjects.

Corpus analysis reveals differences in the use of natural language and mathematical expressions that was at least partially influenced by the mode of presentation of the study material. The group presented with the verbose material
tended to use more natural language than the formal material group and the

dialog turns of the VM-subjects contained more, but shorter, mathematical expressions. The formal material group tended to use more and longer formulas overall, and less natural language. More details on the differences in language production between the two conditions can be found in [15].

The copy & paste facilities provided by DiaWOz-II allowed copying definitions from the study material into the dialog contributions, and allowed copying previously uttered formulae for constructing new formulae. We observed that many subjects constructed larger and larger formulae with several levels of nesting. No such phenomenon was observed in the first study [6]. Even though the predecessor DiaWoZ software used in this study allowed copy & paste, this feature was not explained to the users and discovered only by some. Furthermore, in the first study the introduction material was only presented on paper, so that students could not copy from there as was possible in the second study. Another difference is the mathematical domain itself - the proofs concerning relations in the second experiment series require considerably longer formulae than those concerning naive set theory in the first experiment.

Usability of DiaWOz-II. The students were required to fill out post-experiment questionnaires, which among other things asked questions about the interface.

Students were asked if they had problems while using the interface, and to qualify their answer by a rating on a five-point scale between one (no problems) and five (great problems). Their ratings[5] (median 2, average 2.14, standard deviation 0.85) indicate that the participants generally had little trouble using the DiaWOz-II interface.

Even though a direct comparison between DiaWoZ and DiaWOz-II would require an experiment on its own (the two reported experiments involved different mathematical domains and different requirements imposed on the participants), these ratings are not far from those obtained in the first series of experiments with DiaWoZ. There, students had also been asked the same question, where they indicated a rating of 1.59 on average and a median of 1.

A small number of participants commented to the experimenter that they suspect a human teacher. However, comments by other subjects indicated that these were convinced of having interacted with an automated system.

Participants were asked to give comments about the system in general and the interface in particular, which are summarized in Table 1. The fact that the input facilities of DiaWOz-II were positively mentioned by numerous participants can be contrasted with the first series of experiments, where eight of the seventeen participants complained that the sole input method for mathematical symbols via the menu bar required constant switching between the mouse and the keyboard for inputting mathematical formulae.

A serious criticism concerned the speed of the system. This refers to two aspects: (1) the fact that the students had to wait for the answers from the

---

[5] The ratings from thirty-six participants are distributed as follows: A rating of 1 was assigned by 7 participants, a rating of 2 by 21 participants, a rating of 3 by 4 participants and a rating of 4 by 4 participants. No participant gave a rating of 5.

Table 1. Most frequent comments on the DiaWOz-II interface (number of participants indicated in brackets)

Positive Comments	
– Variety of formula input methods[1] (7)	– Interface is simple to use/clear (5)
– LaTeX commands available[1] (6)	– Questions can be formulated in NL (4)
– Math symbols in menu[1] (5)	
[1] In total, 20 subjects mentioned at least one positive aspect w.r.t. to formula input.	
Negative Comments	
– TeX$_{MACS}$-specific problems (14)	– Interface delay (10)
– Bad screen size/font size (8)	– Sending messages not via return key (6)
– No direct keyboard shortcuts for math symbols available (3)	

system, and (2) the behavior of the interface itself. The waiting times consisted in the time spent by the tutor to read the dialog contributions from the students and to write an answer (even with the help of pre-formulated answers), but also the message-passing between the client, the server and the spell-checker. An important fact was that the wizards were sometimes challenged by the size of formulae created by the students, which made checking them particularly time-consuming. The insufficient speed attributed to the system's interface refers to a small but noticeable delay when typing symbols in DiaWOz-II. This delay is not experienced when using a standard TeXmacs, but results from the extra mechanism that protects the dialog history from being edited mentioned above. Another criticism concerns the window layout. For the experiment we used a screen capturing software and a low screen resolution to save disk space, which was commented on negatively by the subjects.

In summary, the questionnaires show that the input methods for mathematical text available in DiaWOz-II were well received by many participants, but that other mainly technical difficulties remain. A possible improvement proposed by some of the participants is an option for the user to withdraw a message after it is sent, in case the user himself becomes aware of a minor error and wants to correct it himself.

5 Conclusion

We have presented DiaWOz-II, our mediator software for WOZ experiments based on the *wysiwyg* editor TeX$_{MACS}$. DiaWOz-II allows various modes of input for mathematical symbols, such as LaTeX commands, customized commands and menu items, and editing facilities that allow for the creation of complex formulae. Furthermore, DiaWOz-II inherits high quality typesetting from TeX$_{MACS}$. One purpose of this paper is to advocate DiaWOz-II to the AI community for similar WOZ studies in domains such as engineering, physics, economics, etc. where mathematical input in combination with natural language plays a crucial role.

We also briefly addressed the set-up and some results of a series of experiments conducted with DiaWOz-II. The corpus we obtained is important to guide our research in the DIALOG project. It is currently being evaluated and can be obtained from http://www.ags.uni-sb.de/~dialog (see [8] for a preliminary analysis). We have observed that the capabilities of DiaWOz-II for editing and copying mathematical formulae introduced artifacts into some of the tutorial dialogs that we collected, which we did not observe in the previous, similar experiment. These manifest themselves in a term-rewriting style of proving mathematical theorems leading to unnecessarily large and nested formulae. This hints at the importance of incorporating didactic knowledge into tutoring systems in our field (as simulated by DiaWOz-II) which prevent students from abusing such a system's features in a technology-driven manner, and to help the students to use these features purposefully and with moderation.

As a part of our ongoing work, we are combining the dialog specification mechanism from DiaWoZ with the DiaWOz-II system to obtain an environment that reflects our expertise gained with both systems. The DiaWOz-II system can be downloaded from http://www.ags.uni-sb.de/~dialog/diawoz2.

Acknowledgments. We would like to thank all of the members of the Dialog team for their input and comments on initial drafts of this paper, and of course for their contributions to DiaWOz-II and the experiments.

References

1. Fraser, N.M., Gilbert, G.N.: Simulating speech systems. Computer Speech and Language (5) (1991) 81–99
2. Kim, J.H., Glass, M.: Evaluating dialogue schemata with the Wizard of Oz computer-assisted algebra tutor. In: Intelligent Tutoring Systems. (2004) 358–367
3. Hoeven, J.v.d.: GNU TeXmacs: A free, structured, wysiwyg and technical text editor. In Flipo, D., ed.: Le document au XXI-ième siècle. Volume 39-40., Metz (2001) 39–50 Actes du congrès GUTenberg.
4. Dahlbäck, N., Jönsson, A., Ahrenberg, L.: Wizard of Oz studies – Why and how. Knowledge-Based Systems **6**(4) (1993) 258–266
5. Benzmüller, C., Fiedler, A., Gabsdil, M., Horacek, H., Kruijff-Korbayová, I., Pinkal, M., Siekmann, J., Tsovaltzi, D., Vo, B.Q., Wolska, M.: Tutorial dialogs on mathematical proofs. In: Proceedings of the IJCAI Workshop on Knowledge Representation and Automated Reasoning for E-Learning Systems, Acapulco (2003) 12–22
6. Benzmüller, C., Fiedler, A., Gabsdil, M., Horacek, H., Kruijff-Korbayová, I., Pinkal, M., Siekmann, J., Tsovaltzi, D., Vo, B.Q., Wolska, M.: A Wizard of Oz experiment for tutorial dialogues in mathematics. In: Proceedings of AI in Education (AIED 2003) Workshop on Advanced Technologies for Mathematics Education, Sydney, Australia (2003) 471–481
7. Fiedler, A., Gabsdil, M., Horacek, H.: A tool for supporting progressive refinement of wizard-of-oz experiments in natural language. In Lester, J.C., Vicari, R.M., Paraguaçu, F., eds.: Intelligent Tutoring Systems — 7th International Conference (ITS 2004). Number 3220 in LNCS, Springer (2004) 325–335

8. Benzmüller, C., Horacek, H., Lesourd, H., Kruijff-Korbayová, I., Schiller, M., Wolska, M.: A corpus of tutorial dialogs on theorem proving; the influence of the presentation of the study-material. In: Proceedings of International Conference on Language Resources and Evaluation (LREC 2006), Genoa, Italy, ELDA (2006) To Appear.

9. Klemmer, S.R., Sinha, A.K., Chen, J., Landay, J.A., Aboobaker, N., Wang, A.: Suede: a wizard of oz prototyping tool for speech user interfaces. In: UIST. (2000) 1–10

10. Munteanu, C., Boldea, M.: MDWOZ: A Wizard of Oz environment for dialog systems development. In: Proceedings 2nd International Conference on Language Resources and Evaluation, Athens, Greece (2000) 1579–82

11. Anthony, L., Corbett, A.T., Wagner, A.Z., Stevens, S.M., Koedinger, K.R.: Student question-asking patterns in an intelligent algebra tutor. In Lester, J.C., Vicari, R.M., Paraguaçu, F., eds.: Intelligent Tutoring Systems. Volume 3220 of Lecture Notes in Computer Science., Springer (2004) 455–467

12. Hoeven, J.v.d., et al.: The TeXmacs manual. http://www.texmacs.org/tmweb/manual/web-manual.en.html (1999-2006)

13. Billingsley, W., Robinson, P.: Towards an intelligent online textbook for discrete mathematics. In: Proceedings of the 2005 International Conference on Active Media Technology, Takamatsu, Japan (2005) 291 – 296

14. Rosé, C.P., Moore, J.D., VanLehn, K., Albritton, D.: A comparative evaluation of socratic versus didactic tutoring. In: 23rd Annual Conference of the Cognitive Science Society, Edinburgh, Scotland (2001)

15. Wolska, M., Kruijff-Korbayová, I.: Factors influencing input styles in tutoring systems: the case of the study-material presentation format. In: Proceedings of the ECAI-06 Workshop on Language-enabled Educational Technology. (2006) To Appear.

Applications of Automated Reasoning

Ulrich Furbach and Claudia Obermaier

Universität Koblenz-Landau
D56070 Koblenz, Germany
{uli,obermaie}@uni-koblenz.de

Abstract. This paper offers an informal overview and discussion on first order predicate logic reasoning systems together with a description of applications which are carried out in the Artificial Intelligence Research Group of the University in Koblenz. Furthermore the technique of knowledge compilation is shortly introduced.

1   Introduction

Automated theorem proving systems have made increasing progress during the last decades. There was even a prominent open problem, the Robbins problem, which has puzzled logicians since 1930, which was solved by the Automated Reasoner EQP for first order equational logic, developed at Argonne National Laboratory [McC97]. Propositional reasoning systems are very successful in soft- and hardware verification, where the length of formulae which can be processed has grown by orders of magnitude over the last 10 years; today it is very well possible to solve real world verification tasks from hardware design.

In knowledge representation there was a shift from graphic oriented systems like KL-One in the beginning of the 90s towards concept languages or description logic, as it is called nowadays. For the processing of description logics the most commonly used algorithms are basically tableau calculi, which reached a very sophisticated level, allowing the use of description logics for numerous interesting applications (see e.g. [BCM+03]). Because of the close relationship between description logic and modal logic, the fact emerged, that in many cases description logic systems are the more powerful modal logic provers ([Mas99]). This is of importance because modal logic is a decidable fragment of first order predicate logic and thus it plays an important role in computer science.

In this paper we want to demonstrate, that automated reasoning systems are very well ready for real world applications. We are dealing with first order predicate logic systems, accepting its semi-decidability and taking advantage from its higher descriptive power. We are aware that there are many applications of propositional and even higher order interactive reasoning systems; in this paper, however, we want to focus on own experiences and therefore we concert this presentation mainly about first order automated reasoning.

In the following section we briefly depict the state of the art in the development of first order high performance theorem proving, while in the main part we then focus on applications we carried out in the AI Research Group of University of

C. Freksa, M. Kohlhase, and K. Schill (Eds.): KI 2006, LNAI 4314, pp. 174–187, 2007.
© Springer-Verlag Berlin Heidelberg 2007

Koblenz and in wizAI GmbH, a spin-off of this research group. In a last section we will introduce some aspects of knowledge compilation.

2   State of the Art in Automated Deduction

In this section we will use a small toy example to clarify and to discuss some aspects of automated reasoning systems. Given the knowledge base from Figure 1(a), most systems start by transforming this set of formulae into a set of clauses form Figure 1(b). This is astonishing because there are lot of arguments against this transformation: most important that the structure of the formulae gets lost despite equivalence transformation. This structure might mirror some properties of the domain which is modeled, and which can possibly be used to control the navigation through the search space while proving a theorem based on this formula. If the reasoning system allows the user to control the proof by interaction, it might be helpful to retain the structure, in order to facilitate navigation for the user. To our knowledge, there are very few systems which directly work with the original formula; two of them are used in a program verification context, where user interaction is often helpful ([BHOS96, ABB$^+$02]). Most high performance theorem proving systems for predicate logic use clause normal form (e.g. [Wei97, Sch04, RV02, Wer03])

$symptom(s) \leftarrow$ $symptom(s)$

$cause(c_1) \vee cause(c_2) \leftarrow symptom(s)$ $cause(c_1) \vee cause(c_2) \vee \neg symptom(s)$

$treatment(t_0) \leftarrow cause(c_1)$ $treatment(t_0) \vee \neg cause(c_1)$

$treatment(t_1) \leftarrow cause(c_1)$ $treatment(t_1) \vee \neg cause(c_1)$

$treatment(t_0) \leftarrow cause(c_2)$ $treatment(t_0) \vee \neg cause(c_2)$

$treatment(t_2) \leftarrow cause(c_2)$ $treatment(t_2) \vee \neg cause(c_2)$

(a) Knowledge base KB   (b) Set of clauses

Fig. 1. Knowledge base KB and corresponding set of clauses

Linear Deduction. Most textbooks on Artificial Intelligence (AI) present a resolution calculus to reason about knowledge bases (see e.g. [PMG97] or an overview on different textbooks [Fur03]). In the 70s of the previous century this was indeed the main approach to process logical formulae in AI and a very common understanding was at that time that goal oriented linear deduction should be used to prove logical consequences from a set of formulae. Assume for example the knowledge base KB from Figure 1 together with the task to prove that there is a treatment given that special situation; the existence of a treatment can be modeled by the additional formula $\exists X treatment(X)$, which is called a goal. Altogether we have to prove $KB \models \exists X treatment(X)$. Since resolution is a refutational calculus, we have to negate the goal and after a slight equivalent preserving transformation we get the clause set $KB \wedge \neg treatment(X)$, where all variables are implicitly universal quantified. In order to find a refutation of this, it seems to be very natural to start with the goal and to work "backwards" until one reaches the empty clause $\square$, indicating that the clause set is unsatisfiable and hence the goal logically follows from the knowledge base. This would lead to a sequence of resolvents $\neg treatment(X), \neg cause(c_1), cause(c_2) \vee \neg symptom(s), cause(c_2), treatment(t_2), \square$. One of the first calculi advocating this goal oriented linear approach are model elimination ([Lov68]) and linear resolution ([KK71]) and a model elimination theorem prover SETHEO even won the CASC competition (which will be discussed later). It was much later in the 90s where we really understood that these calculi are not based on resolution – they are much closer to tableau calculi, at least if one takes the treatment of variables as a discriminating parameter[1].

One appealing aspect of linear refutation is that it is very close to the concept of logic programming: one starts from a call of the program and then works through a sequence of intermediate computations until $\square$ is found. The answer to such a computation can be constructed from the unifiers used during the inference steps. In [BF97] it is shown how the logic programming paradigm can be used not only for Horn clause programming, but also for full first order logic by means of a variant of model elimination. Another argument for this goal oriented search for a refutation usually was its higher potential in search space pruning. We intentionally use past tense, because nowadays most high performance theorem provers are working in a saturation based manner, which is explained in the next section.

Saturation. Assume again the proof task $KB \wedge \neg treatment(X)$ from the previous discussion. Instead of assigning some of the clauses, e.g. the goal clause, a particular importance, we just take the clause set as it is given by this task. If we further assume that we have resolution as the inference rule at hand, we simply add new resolvents to this set. Starting with the initial set $S = KB \cup \{\neg treatment(X)\}$ one can derive by this the new set $S' = S \cup \{cause(c_2) \vee \neg symptom(s) \vee treatment(t_0)\}$. Such an extension is done until the set contains the empty clause or (in special cases) until there are no new clauses to derive, i.e. the set is saturated.

There are some issues to solve if one tries to do saturation based proving. The amount of clauses which are generated from a given set of clauses, can increase dramatically. Therefore it is mandatory to avoid generating in some sense useless clauses and to get rid of redundant clauses. A very powerful technique to this end is the use of term ordering to control the generation of new clauses. For an overview of this technique in the context of resolution the reader is pointed towards [BG01]; theorem provers which are successful with this techniques are, among others, Otter, Spass and Vampire ([McC90, Wei97, RV02]). The use of ordering for controlling the generation of new clauses has another advantage: it also helps in handling equality. If the formulae to be handled by the reasoning system contain an equality predicate, there are basically two different methods

---

[1] In resolution variables are treated as being implicitly universal quantified, whereas in tableau calculi they usually are rigid, i.e. placeholders for a yet unknown constant.

to handle this. Either one adds axioms to the set of clauses describing the usual properties of equational logic or an additional inference rule, like paramodulation, is used to handle the equations. The latter approach also raises the problem of generating too many new clauses, which, however, can be controlled as well by term ordering.

It is not only resolution which is available as an inference rule in saturation based theorem proving. If the entire formula which is given as the proof task, is transformed into an equivalent formula in equational logic, superposition together with ordering restriction can be used. This approach is followed in the theorem prover E ([Sch04]) or in the Waldmeister-system ([HL02]).

It is no doubt that the systems employing saturation techniques nowadays belong to the most successful high performance first order systems; this aspect will be discussed in more detail below.

Tableaux. Although introduced more or less at the same time as the resolution principle (1950 – 1960), there was only evidence in the 1980th that tableau calculi offer an alternative approach with very interesting properties. These are in particular, that parts of the history of the current proof attempt are coded into the proof object and that the variables are treated rigidly. We will explain this on a special form of tableaux, the so called *Hypertableaux*, which are introduced in [BFN96] and which is used in KRHyper, a theorem prover, which is the basis throughout our applications. The calculus is a clause normal form tableau calculus and hence we start constructing a tableau from a given set of clauses, which are regarded as implications (negative literals are the premises, positive literals the conclusion). In our example from Figure 1 there is one single fact, i.e. implication with empty premise. Hence we construct the tableau consisting of one node, namely $symptom(s)$. The only inference rule works as follows: we take a branch from the tableau and a clause from the clause set; if all literals from the premise of the clause are contained in the branch (in the case of variables it is slightly more complicated), then the branch can be extended by the literal in the conclusion. If there is more than one literal in the conclusion, the branch is split; if there is no conclusion in the clause, the branch is closed. A clause set is unsatisfiable, if a tree constructed by this method only contains closed branches. An interesting property o f this method can be seen if one omits the goal clause $\neg treatment(X)$, which is a clause without positive literal, i.e. without conclusion. The tableau from Figure 2 is an exhausted (i.e. maximal) tableau which can be constructed from the clauses in our example. In such a case we not only have a proof object, containing information from the proof search, we also can read two models of the clause set, namely the atoms from each of the two branches. Hence tableaux are also very helpful for constructing models for satisfiable clauses. This is of particular importance in a non-monotonic setting, where minimal models have to be computed as a basis of a closed world assumption; an overview of such approaches can be found in [DFN01].

Tableau methods are also the main mechanism for the design of description logic systems, which are gaining increasing importance in the design of the Semantic Web project. A drawback of tableau calculi is the handling of equality;

$$\neg tr(X)$$
$$\neg ca(c_1)$$
$$ca(c_2) \vee \neg sy(s)$$
$$ca(c_2)$$
$$tr(t_2)$$
$$\bot$$

$$ca(c_2) \vee \neg sy(s) \vee tr(t_0)$$
$$ca(c_2) \vee \neg sy(s) \vee tr(t_1)$$
$$ca(c_2) \vee tr(t_1)$$
$$\dots$$

$$sy(s)$$

$$ca(c_1) \quad ca(c_2)$$
$$| \qquad |$$
$$tr(t_0) \quad tr(t_2)$$
$$| \qquad |$$
$$tr(t_1) \quad tr(t_0)$$

(a) Linear resolution   (b) Saturation   (c) Hypertableau

Fig. 2. Different calculi – predicates are abbreviated by the first two letters

the variables in a tableau have to be substituted simultaneously in the entire tableau during a unification, which is necessary in an extension step with first oder clauses. This makes the handling of equality very difficult, and, indeed, there are no high performance tableau proofers which are also dealing with equality in a way comparable with saturation based systems.[2]

Empirical Aspects. Two important achievements in automated reasoning research are the commonly used benchmark suite TPTP ([SS98]) and the CASC-competition ([PSS02]). The TPTP (Thousands of Problems for Theorem Provers) problem library is a library of test problems for automated theorem proving (ATP) systems. Currently the TPTP contains 7000 test problems with a large variety in complexity and difficulties. These problems are grouped into domains, like lattice theory, hardware creation and verification and many others. Besides the problem library, the TPTP contains a utility to convert the problems to existing ATP formats; it offers conversions to nearly all systems and thus facilitates the use of the library. The principal motivation for the TPTP project is to move the testing and evaluation of ATP systems from the previous ad hoc situation onto a firm footing. This goal is certainly reached, and, even more, the TPTP idea led to the CASC competition, which is held annually during a deduction conference.

CASC evaluates the performance of sound, fully automatic, classical first order ATP systems. The evaluation is in terms of the number of problems solved and the average runtime for successful solutions. The problems are chosen from the TPTP Problem Library and they are presented together with a specified time limit for each solution attempt. Although there might be the danger that system designers try to tune their provers towards the event and the possible problem set (the TPTP), there are certainly a number of advantages:

- It turned out that different calculi and systems are winning in different problem classes.
- The systems are becoming increasingly robust. They have to run fully automated, to be invoked from batch, such that their developers have no chance to interfere during the entire competition.

---

[2] In the Hyper tableau calculus the situation is different, because we have universal variables; efficiently equality handling is in development right now.

- The progress of the field becomes transparent, by having the winners from the previous year participate, even if a new version of the system is also an entry into the current competition.

As said above, one way to present the success of automated theorem proving is to refer to TPTP and CASC. However, it is time to point out that applications, of course, are another important measure of success. We experienced that model generation deduction offers a very flexible way to use automated systems in applications and embedded systems. This is what we will exemplify in the following section.

3   Applications

In this section we will focus on application projects we worked through the recent years in the AI Research Group (AGKI) of Koblenz University and in wizAI GmbH, which is a spin-off of the research group. When researchers talk about applications, this can have very different semantics; some mean the application of a theoretical tool or method within the own field, e.g. using a theorem prover for knowledge representation purposes in Artificial Intelligence research. Others mean that a problem for which there was known no solution can be solved by means of the research carried out; e.g. the solution of an open mathematical problem by an automated reasoning system, mentioned in the introduction. In this paper we offer a different understanding of 'application': we have a reasoning system, the KRHyper, based on hyper tableau; this system has been developed during many years, it is tested in various contexts and we assume that it is a very reliable and flexible tool. And this is exactly what we are benefitting from in other projects; we use this tool as part of the software developing process. It is used to quickly and safely solve subproblems during the software engineering process. Of course the problems could have been solved differently by programming it from scratch. By the use of our KRHyper the solution can be achieved quicker, easier to test and more flexible to allow modification in case the requirements of the project change, which is a very likely the case in commercial projects.

We used KRHyper in the following larger projects:

- Together with Dresdner Bank we developed a prototype of a knowledge management system, which is used for early discovery of reputational risks caused by decisions and statements from own bank divisions (for details see [FGHT+04]). This is presumably the only software system in a major bank, where an automated theorem prover is running its kernel.
- In a PhD-project, which was aiming at the intelligent processing of XML database queries, it turned out, that KRHyper could be used to transform incomplete queries into queries which can be processed efficiently by the underlying database system (details can be found in [BFGHK04])
- In RoboCup we are working towards the use of logic in the simulation league. Until now, we have been working on a soccer team which was programmed in large parts by the use of logic programming techniques. KRHyper was

used to check formal properties of the team, i.e. the multi-agent system. Recently we changed the focus of the project, which is carried out in the DFG Special Focus Program 1125 "RoboCup"; because of the mixture of real valued computation and logical reasoning we are using hybrid automata for model checking of properties ([Hen96]).

- The Living Book project was carried out over several years; funded by the German Ministry of Research and Education and by the European Comission. We developed a system which allows the development and use of intelligent personalised textbooks via the internet. This project will be discussed in more detail below.

- The Spatial Metro project is an ongoing project carried out together with the city of Koblenz and with two of her twin cities, Norwich and Rouen. It is financed by the European Commission and the State Government of Rheinland-Pfalz. The goal of our part of the project is to use AI techniques for efficient guidance of tourists in the city. This project will be discussed in more detail below.

Living Books. Living Book is a project which was carried out during several years aiming at the development of personalized intelligent books. Intelligent in the sense, that a user is able to work and interact with her book, which is maintained on a central server. The book also contains interactive systems, which can be used for exercises and practice. For access to some books published in this project the reader is referred to http://www.in2math.de; in this paper we want to concentrate on the underlying technique, the Slicing Book Technique. By this technique a document, say, a mathematics text book, is separated once as a preparatory step into a number of small units, such as definitions, theorems, proofs, etc. The purpose of the sliced book then is to enable authors, teachers and students to produce personalized teaching or learning materials based on a selective assembly of units. Once a reader is entering the portal of the book in the web, she can login with her account and gets the entry page of the book. There it is possible to select parts of the book from the table of contents and to specify preferences, e.g. to include all prerequisites necessary for the understanding of the selected units or to include all parts were the contents of the selected units is used – such a view is depicted in Figure 3.

Once the user has specified the current view of the book, the system has to provide the appropriate units and compose them in order to receive a final pdf-document. This task is depicted in Figure 4 for the general case, where the user can even select from various books. Assume she is asking for an overview of the notion of "Normal Forms" by selecting the appropriate parts. In addition the user has some preferences, like preferring formal notations or explanations by examples, which the system already knows about the user.

The slices or units, whose collection constitutes the books basically contain LaTex-code. This is connected with appropriate meta data, like the relations according to the prerequisite and refers relation, meta-data stating the type of the unit (example, proof, theorem and things like this) or ontologies which allow the combination of different keyword systems. All this data belonging to the users query are put together and stated as proof task, i.e. a logical formula for the KRHyper system. KRHyper computes a model of the given set of clauses; it is important to note that the formula contains all the slices of the books in a certain representation. From the model for the given query the system can extract the identification numbers of the slices, put together the LaTex parts and generate a pdf document, which can be presented to the user.

There are some lessons we learned from this application: the KRHyper system must be able to process very many, i.e. ten thousands of slices efficiently and it needs non-monotonic negation in order to deal with closed-world assumptions. Another important property is that it must be possible to process description logic parts of the task. For details of all this the reader is referred to [BFGHS04].

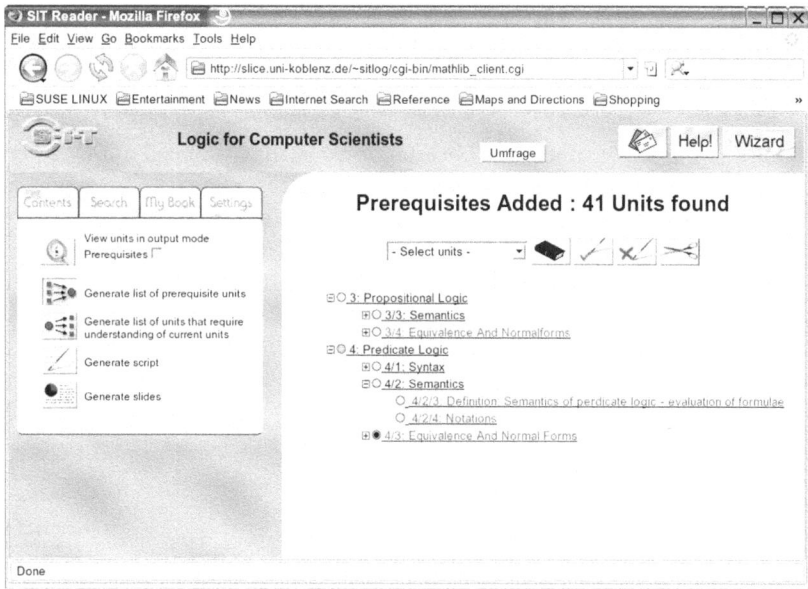

Fig. 3. A personalized view of Living Books

Spatial Metro. One goal of this European Commission project is the use of AI techniques for efficient guidance of tourists in a city. For these possible tours within a city the metaphor 'spatial metro' is used. The points of interest in a city are depicted in the form of a metro map: according to the type of these points we can have different 'metro lines'; for example there may be a monument line, a shopping and a culture line. Figure 5 shows two of these lines together with the points of interest they contain.

Each of these points of interests is equipped with a bluetooth access point, which is able to send information about this location. This can be information about buildings, history, a map or even latest offers from a shop. If a tourist is reaching the area of this access point his mobile phone or his PDA can connect with this access point and present the information. Two aspect are of importance: this connection and hence the service is for free, no phone or WLAN fees have to be paid for and, more interesting (at least for this paper), the information which is offered by the access point is processed and filtered by the users phone. For this the user was able to edit a special profile on his phone, which contains preferences and other private information. This information is kept secure within her phone and is compared with the information (and its meta data), such that only those information which are of interest for the user are presented.

The comparison and processing of the information offered by the access point is done by KRHyper. For this we re-implemented the theorem prover in Java ME in order to get it running on a smart phone; presumably its the first theorem prover for first order predicate calculus running on a mobile phone (if the phone is not in use it can be used to solve TPTP-problems); for more information see [KS05b]; more about the entire approach can be found in [KS05a].

The lessons we learned until now from this project: Firstly, implementation language matters! Our KRHyper system is implemented in Ocaml, mainly because this was the Ph.D. student's favorite language; when we tried to get KRHyper running on a smart phone, it became obvious that we need a JAVA version and hence a re-implementation became necessary. Maybe such a porting could have been taken into account from the very beginning. The second lesson is more on the project design, concerning the willingness of users to download a piece of software, i.e. the reasoning machinery, on their mobile phones. In a field study we carried out, it turned out, that users are rather reluctant to do this. In a second phase of the project we are working to get rid of this bottleneck.

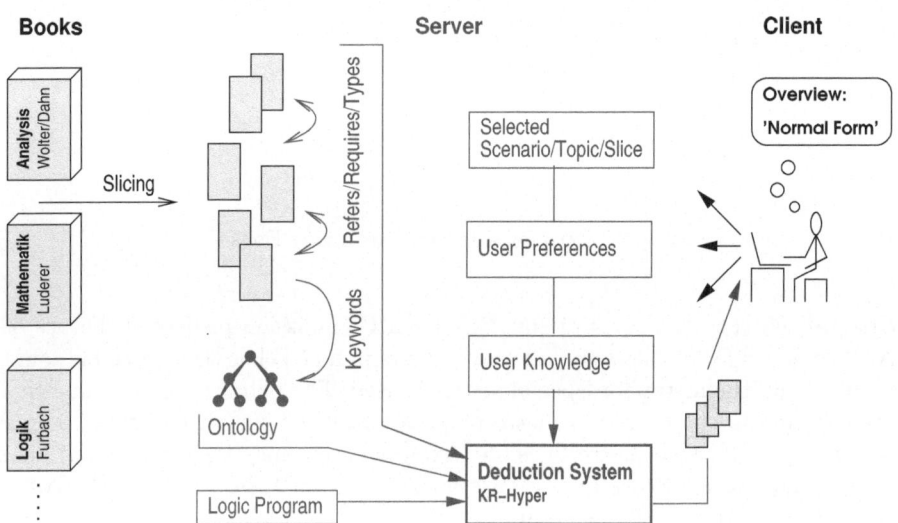

Fig. 4. The reasoning part of Living Books

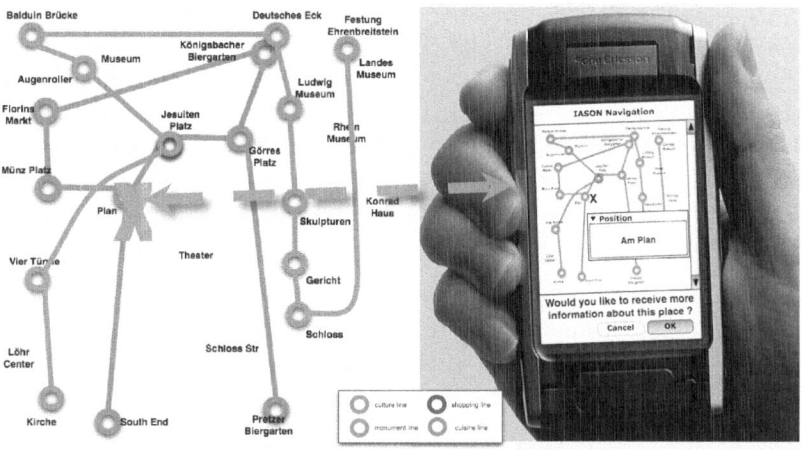

Fig. 5. A tourist guide on your mobile

4   Knowledge Compilation

In practice we are very often confronted with the following task: given a knowledge base, we want to answer a set of queries from that knowledge base. For example in diagnosis of electrical circuits, the system description of the correctly functioning circuit is used for various different queries. The naive approach to solve this problem would be to answer all the queries independently. But this would cause an exponential complexity for each query. That is why a new approach called knowledge compilation evolved. The basic idea of knowledge compilation is to precompile the knowledge base into a special form. This precompilation step is very costly (meaning of exponential complexity) but has to be performed only once. After that precompilation, some types of queries can be answered in polynomial or even linear time. Usually the formula in the target language of that precompilation has lots of other very nice properties such as the possibility of projecting the formula onto a set of atoms in linear time. Since the costly precompilation only has to be performed once, its complexity is relativized. There is a huge number of target languages for the mentioned precompilation. A rather new target language for knowledge compilation is DNNF. We will now take a closer look at this normal form. In the following, the term formula always means propositional logic formula.

Decomposable Negation Normalform. DNNF is short for decomposable negation normal form and is a special normal form developed in [Dar01]. A formula is in DNNF, if it is in negation normal form (NNF) and additionally satisfies the decomposability property. This property means that for any conjunction which occurs in the formula, the conjuncts do not share atoms. As an example take the set of clauses $F = \{\{a \lor b\}, \{c \lor \neg b\}\}$. This clause set is not in DNNF, because the atom b occurs in both clauses and as usual, the clauses of the set are combined by conjunction.

DNNF has the very nice property, that satisfiability can be decided in time, which is linear to the length of the DNNF. This is a direct consequence of the decomposability property. Because of this property it is possible to perform the satisfiability test on each subformula independently. Another very interesting feature of formulae in DNNF is the possibility to check the minimal cardinality in linear time. The next very interesting feature of formulae in DNNF ist the possibility to project a DNNF on a set of atoms in linear time. A list of other important features can be found in [Dar01].

Compilation of Propositional Logic Formulae into DNNF. A very naive approach is to use Shannon's rule to transform a formula into DNNF. As an example we will transform the clauseset of our previous example F into DNNF. F is not in DNNF, because the variable b occurs in two different clauses. Now we transform F into DNNF by using Shannon's rule:

$$dnnf(F) = b \wedge F_{|b=true} \vee \neg b \wedge F_{|b=false}$$
$$= (b \wedge c) \vee (\neg b \wedge a)$$

In this example the transformation is very short. But one can easily imagine that the compilation of bigger formulae gets a lot more complicated. Because in huge sets of clauses, as occuring in practice, usually a great deal of atoms are shared between different clauses. That is why a variety of algorithms for the compilation into DNNF were developed. Many of these algorithms are based on DPLL ([Dar04],[Dar01]).

Weaving Projection into the Computation of DNNF. Quite often, we are only interested in a special part of our knowledge base F. Meaning for example that we are only interested in the values of a set of atoms S. Hence we want to *project* the knowledge base on this set of atoms S. Let Σ be the set of all propositional atoms occuring in our knowledge base F. Then the projection of F on the atoms in S is dual to *forgetting* all the atoms included in $\Sigma \setminus S$. Given that the projection of a DNNF on a set of atoms is linear, the common procedure is to transform the knowledge base into DNNF and to project on S afterwards. In [Wer06] it is suggested to weave in projection into the precompilation step. It is shown that this leads to an exponential saving of space.

Let's take a closer look at the technique of weaving in projection into the DNNF transformation. As mentioned above, a DPLL based algorithm is used to transform formulae into DNNF. To use this algorithm, the knowledge base is required to be given in CNF. During the computation of the DNNF, it is possible to use a number of rules to get rid of the atoms which are supposed to be forgotten. If one of the atoms we want to forget is pure, we can use pure literal elimination to get rid of this atom. The Isol* rule is another possibility to remove atoms which are supposed to be forgotten. This rule is related to resolution. The application of these rules not only removes atoms which we want to forget, but can also have a positive effect on the transformation into DNNF. Atoms which are shared only between clauses which are removed by this rules do not violate the decomposability property after the application of these rules.

Although we only used a very small example to explain the knowledge compilation approach, it should be clear that such techniques can be very helpful in realistic applications, as they are described in this paper. Currently we are working on separating subproblems where knowledge compilation can be applied.

5 Conclusion

In this paper we gave a very rough overview on first order predicate logic systems and its most recent developments. We then depicted some applications and we tried to demonstrate that automated reasoning systems are a valuable tool to be used in important parts in the application systems. Of course, even if the reasoner is fully developed and tested there is a considerable amount of work to be done for its integration and the design of the appropriate interfaces.

As a last remark concerning the development of commercial real world applications, we want to point out that it is a long way from an academic prototype system towards a real product. There is a lot of manpower to invest, but on the other hand, academic research and development can also benefit from this.

References

[ABB+02]   Wolfgang Ahrendt, Thomas Baar, Bernhard Beckert, Martin Giese, Reiner Hähnle, Wolfram Menzel, Wojciech Mostowski, and Peter H. Schmitt. The KeY System: Integrating Object-Oriented Design and Formal Methods. In *Fundamental Approaches to Software Engineering. 5th International Conference, FASE 2002*, LNCS 2306, pages 327–330. Springer, 2002.

[BCM+03]   Franz Baader, Diego Calvanese, Deborah L. McGuinness, Daniele Nardi, and Peter F. Patel-Schneider, editors. *The Description Logic Handbook: Theory, Implementation, and Applications*. Cambridge University Press, 2003.

[BF97]   Peter Baumgartner and Ulrich Furbach. Calculi for Disjunctive Logic Programming. In Jan Maluszynski, editor, *Logic Programming - Proceedings of the 1997 International Symposium*, New York, 1997. The MIT Press.

[BFGHK04]   Peter Baumgartner, Ulrich Furbach, Margret Groß-Hardt, and Thomas Kleemann. Model Based Deduction for Database Schema Reasoning. In *KI 2004*, volume 3238 of *LNCS*, pages 168–182. Springer Verlag, Berlin, Heidelberg, New-York, 2004.

[BFGHS04]   Peter Baumgartner, Ulrich Furbach, Margret Groß-Hardt, and Alex Sinner. Living Book - Deduction, Slicing, and Interaction. *J. Autom. Reasoning*, 32(3):259–286, 2004.

[BFN96]   Peter Baumgartner, Ulrich Furbach, and Ilkka Niemelä. Hyper Tableaux. In José Júlio Alferes, Luís Moniz Pereira, and Ewa Orlowska, editors, *JELIA*, volume 1126 of *LNCS*, pages 1–17. Springer, 1996.

[BG01]   Leo Bachmair and Harald Ganzinger. Resolution Theorem Proving. In Robinson and Voronkov [RV01], pages 19–99.

[BHOS96]   Bernhard Beckert, Reiner Hähnle, Peter Oel, and Martin Sulzmann. The Tableau-based Theorem Prover $_3T^A P$ Version 4.0. In Michael A. McRobbie and John K. Slaney, editors, *CADE*, volume 1104 of *LNCS*, pages 303–307. Springer, 1996.

[Dar01]   Adnan Darwiche. Decomposable Negation Normal Form. *Journal of the ACM*, 48(4), 2001.

[Dar04]   Adnan Darwiche. New Advances in Compiling CNF into Decomposable Negation Normal Form. In *Proceedings of the 16th Eureopean Conference on Artificial Intelligence, ECAI'2004*, pages 328–332, 2004.

[DFN01]   Jürgen Dix, Ulrich Furbach, and Ilkka Niemelä. Nonmonotonic Reasoning: Towards Efficient Calculi and Implementations. In Robinson and Voronkov [RV01], pages 1241–1354.

[FGHT⁺04]   Ulrich Furbach, Margret Groß-Hardt, Bernd Thomas, Tobias Weller, and Alexander Wolf. Issues Management: Erkennen und Beherrschen von kommunikativen Risiken und Chancen. Fachberichte Informatik 2–2004, Universität Koblenz-Landau, Institut für Informatik,Universitätsstr. 1, D-56070 Koblenz, 2004.

[Fur03]   Ulrich Furbach. AI – A Multiple Book Review. *Artificial Intelligence*, 145(1-2):245 – 252, 2003.

[Hen96]   Thomas A. Henzinger. The Theory of Hybrid Automata. In *Proceedings of the IEEE Symposium on Logic in Computer Science (LICS 1996)*, pages 278–292, 1996.

[HL02]   Thomas Hillenbrand and Bernd Löchner. The Next WALDMEISTER Loop. In Andrei Voronkov, editor, *CADE*, volume 2392 of *LNCS*, pages 486–500. Springer, 2002.

[KK71]   R. A. Kowalski and D. Kuehner. Linear Resolution with Selection Function. *Artificial Intelligence*, 2:227–260, 1971.

[KS05a]   Thomas Kleemann and Alex Sinner. Decision Support for Personalization on Mobile Devices. In *Proceedings of the 21st International Conference, ICLP 2005*, pages 404–406, 2005.

[KS05b]   Thomas Kleemann and Alex Sinner. Krhyper - in your Pocket, System Description. In Robert Nieuwenhuis, editor, *CADE*, volume 3632 of *LNCS*, pages 452–458. Springer, 2005.

[Lov68]   D. Loveland. Mechanical Theorem Proving by Model Elimination. *JACM*, 15(2), 1968.

[Mas99]   Fabio Massacci. Design and Results of the Tableaux-99 Non-classical (Modal) Systems Comparison. In Neil V. Murray, editor, *TABLEAUX*, volume 1617 of *LNCS*, pages 14–18. Springer, 1999.

[McC90]   William McCune. Otter 2.0. In Mark E. Stickel, editor, *CADE*, volume 449 of *LNCS*, pages 663–664. Springer, 1990.

[McC97]   William McCune. Solution of the Robbins Problem. *J. Autom. Reasoning*, 19(3):263–276, 1997.

[PMG97]   David Poole, Alan Mackworth, and Randy Goebel. *Computational Intelligence: A Logical Approach*. Oxford University Press, 1997.

[PSS02]   F. Pelletier, G. Sutcliffe, and C. Suttner. The Development of CASC. *AI Communications*, 15(2-3):79–90, 2002.

[RV01]   John Alan Robinson and Andrei Voronkov, editors. *Handbook of Automated Reasoning (in 2 volumes)*. Elsevier and MIT Press, 2001.

[RV02]   Alexandre Riazanov and Andrei Voronkov. The design and implementation of VAMPIRE. *AI Commun.*, 15(2-3):91–110, 2002.

[Sch04]   Stephan Schulz. System description: E 0.81. In David A. Basin and Michaël Rusinowitch, editors, *IJCAR*, volume 3097 of *LNCS*, pages 223–228. Springer, 2004.

[SS98]   G. Sutcliffe and C. Suttner. The TPTP Problem Library: CNF Release v1.2.1. *Journal of Automated Reasoning*, 21(2):177–203, 1998.

[Wei97]   Christoph Weidenbach. Spass - version 0.49. *Journal of Automated Reasoning*, 18(2):247–252, 1997.

[Wer03]   Christoph Wernhard. System Description: KRHyper. Fachberichte Informatik 14–2003, Universität Koblenz-Landau, Institut für Informatik, Universitätsstr. 1, D-56070 Koblenz, 2003.

[Wer06]   Christoph Wernhard. Tableaux Between Proving, Projection and Compilation. Technical report, Universität Koblenz-Landau, 2006. In preparation.

On the Scalability of
Description Logic Instance Retrieval

Ralf Möller[1], Volker Haarslev[2], and Michael Wessel[1]

[1] Hamburg University of Technology
[2] Concordia University, Montreal

Abstract. Although description logic systems can adequately be used
for representing and reasoning about incomplete information (e.g., for
John we know he is French or Italian), in practical applications it can
be assumed that (only) for some tasks the expressivity of description
logics really comes into play whereas for building complete applications,
it is often necessary to effectively solve instance retrieval problems with
respect to largely deterministic knowledge. In this paper we present and
analyze the main results we have found about how to contribute to this
kind of scalability problem. We assume familiarity with description logics
in general and tableau provers in particular.

1 Introduction

Although description logics (DLs) are becoming more and more expressive (e.g.,
[14]), our experience has been that it is only for some tasks that the expressivity
of description logics really comes into play; for many applications, it is necessary
to be able to deal with largely deterministic knowledge very effectively. Thus, in
practice, description logic systems offering high expressivity must also be able
to handle large bulks of data descriptions (Aboxes with concepts and role assertions) which are largely deterministic. Users expect that DL systems scale w.r.t.
these practical needs. In our view there are two kinds of scalability problems:
scalability w.r.t. large sets of data descriptions (data description scalability) and
scalability w.r.t. high expressivity, which might only be important for small parts
of the data descriptions (expressivity scalability).

In the literature, the data description scalability problem has been tackled
from different perspectives. We see two main approaches, the layered approach
and the integrated approach. In the layered approach the goal is to use databases
for storing and accessing data, and exploit description logic ontologies for convenient query formulation. The main idea is to support ontology-based query translation to relational query languages (SQL, datalog). See, e.g., [21,9] (DLDB), [3]
(Instance Store), or [4] (DL-Lite). We notice that these approaches are only
applicable if reduced expressivity does not matter. Despite the most appealing
argument of reusing database technology (in particular services for persistent
data), at the current state of the art it is not clear how expressivity can be
increased to, e.g., $\mathcal{SHIQ}$ without losing the applicability of database transformation approaches. Hence, while data description scalability is achieved, it is

C. Freksa, M. Kohlhase, and K. Schill (Eds.): KI 2006, LNAI 4314, pp. 188–201, 2007.

not clear how to extend these approaches to achieve expressivity scalability (at least for some parts of the data descriptions).

Tableau-based DL systems are now widely used in practical applications because these systems are quite successful w.r.t. the expressivity scalability problem. Therefore, for investigating solutions to both problems, the expressivity and the data description scalability problem, we pursue the integrated approach that considers query answering with a tableau-based description logic system augmented with new techniques inspired from database systems. For the time being we ignore the problems associated with persistency and investigate specific knowledge bases (see below).

The contribution presents and analyzes the main results we have found about how to start solving the scalability problem with tableau-based prover systems given large sets of data descriptions for a large number of individuals. Note that we do not discuss query answering speed of a particular system but investigate the effect of optimization techniques that could be exploited by any (tableau-based) DL inference system that already exists or might be built. Since DLs are very popular now, and tableau-based systems have been extensively studied in the literature (see [2] for references), we assume the reader is familiar with DLs in general and tableau-based decision procedures in particular (see, e.g., [1]).

2   Lehigh University Benchmark

In order to investigate the data description scalability problem, we use the Lehigh University BenchMark (LUBM, [8,9]). LUBM queries are conjunctive queries referencing concept, role, and individual names from the Tbox. A query language tailored to description logic applications that can express these queries is described in [20].[1] The language is called nRQL and supports a restricted form of conjunctive queries (variables are only bound to individuals mentioned in the Abox and not to "anonymous" individuals that denote objects from the domain that provably must exist). Although some work on standard conjunctive queries is published [5,17,7], to the best of the authors' knowledge, (efficient) algorithms for answering full conjunctive queries for expressive description logics such as $\mathcal{SHIQ}$ [16] are not known.

Below, LUBM queries 9 and 12 are shown in order to illustrate LUBM queries – note that $'www.University0.edu'$ is an individual and $subOrganizationOf$ is a transitive role. Please refer to [8,9] for more information about the LUBM queries.

$$Q9 : ans(x, y, z) \leftarrow Student(x), Faculty(y), Course(z),$$
$$advisor(x, y), takesCourse(x, z), teacherOf(y, z)$$
$$Q12 : ans(x, y) \leftarrow Chair(x), Department(y), memberOf(x, y),$$
$$subOrganizationOf(y,' www.University0.edu')$$

---

[1] In the notation for queries used in this paper we assume that different variables may have the same bindings.

In order to investigate the data description scalability problem, we used a TBox for LUBM with inverse and transitive roles as well as domain and range restrictions but no number restrictions, value restrictions, or disjunctions (after GCI absorption). Among other axioms, the LUBM TBox contains axioms that express necessary *and* sufficient conditions for some concept names. For instance, there is an axiom $Chair \doteq Person \sqcap \exists headOf.Department$. For evaluating optimization techniques for query answering we consider runtimes for a whole query set (queries 1 to 14 in the LUBM case).

3   Optimization Techniques

If the queries mentioned in the previous section are answered in a naive way by evaluating subqueries in the sequence of syntactic notation, acceptable answering times can hardly be achieved. Determining all bindings for a variable (with a so-called generator) is much more costly than verifying a particular binding (with a tester). Treating the one-place predicates *Student*, *Faculty*, and *Course* as generators for bindings for corresponding variables results in combinatorial explosion (cross product computation). Optimization techniques are required that provide for efficient query answering in the average case.

3.1   Query Optimization

The optimization techniques that we investigated are inspired by database join optimizations, and exploit the fact that there are few *Faculties* but many *Students* in the data descriptions. For instance, in case of query Q9 from LUBM, the idea is to use *Faculty* as a generator for bindings for y and then generate the bindings for z following the role *teacherOf*. The heuristics applied here is that the average cardinality of a set of role fillers is rather small. For the given z bindings we apply the predicate *Course* as a tester (rather than as a generator as in the naive approach). Given the remaining bindings for z, bindings for x can be established via the inverse of *takesCourse*. These x bindings are then filtered with the tester *Student*.

If z was not mentioned in the head, i.e., in set of variables for which bindings are to be computed, and the tester *Course* was not used, there would be no need to generate bindings for z at all. One could just check for the existence of a *takesCourse* role filler for bindings w.r.t. x.

In the second example, query Q12, the constant $'www.University0.edu'$ is mentioned. Starting from this individual the inverse of *subOrganizationOf* is applied as a generator for bindings for y which are filtered with the tester *Department*. With the inverse of *memberOf*, bindings for x are computed which are then filtered with *Chair*. Since for the concept *Chair* sufficient conditions are declared in the TBox, instance retrieval reasoning is required if *Chair* is a generator. Thus, it is advantageous that *Chair* is applied as a tester (and only instance tests are performed).

For efficiently answering queries, a query execution plan is determined by a cost-based optimization component (c.f., [6, p. 787ff.]) which orders query atoms such that queries can be answered effectively. For computing a total order relation on query atoms with respect to a given set of data descriptions (assertions in an ABox), we need information about the number of instances of concept and role names. An estimate for this information can be computed in a preprocessing step by considering given data descriptions, or could be obtained by examining the result set of previously answered queries. We assume that ABox realization is too costly (takes about 6 minutes for LUBM with one university, excluding the initial Abox consistency test), so this alternative is ruled out.

3.2   Indexing by Exploiting Told and Taxonomical Information

In many practical applications that we encountered, data descriptions often directly indicate (some of the) concept names of which an individual is an instance. Therefore, in a preprocessing step, it is useful to compute an index that maps concept names to sets of individuals which are their instances. In a practical implementation this index might be realized with some form of hashtable.

Classifying the TBox yields the set of ancestors for each concept name, and if an individual i is an instance of a concept name A due to explicit data descriptions, it is also an instance of the ancestors of A. This information can be made accessible by an index that maps concept names to instances. The index is organized in such a way that retrieving the instances of a concept name A, or one of its ancestors, requires (almost) constant time. The index is particularly useful to provide bindings for variables if, despite all optimization attempts for deriving query execution plans, concept names must be used as generators. In addition, the index is used to estimate the cardinality of concept extensions. The estimates are used to compute an order relation for query atoms. The smaller the cardinality of a concept or a set of role fillers is assumed to be, the more priority is given to the query atom. Optimizing LUBM query $Q9$ with the techniques discussed above yields the following query execution plan (denoted as a query, substeps to be read from left to right).

$$Q9' : ans(x, y, z) \leftarrow Faculty(y), teacherOf(y, z), Course(z),$$
$$advisor^{-1}(y, x), Student(x), takesCourse(x, z)$$

Using this kind of rewriting, queries can be answered much more efficiently.

If the TBox contains only GCIs of the form $A \sqsubseteq A_1 \sqcap \ldots \sqcap A_n$, i.e., if the TBox forms a hierarchy, the index-based retrieval discussed in this section is complete (see [3]). However, this is not the case for LUBM. In LUBM, besides domain and range restrictions, axioms are also of the form $A \doteq A_1 \sqcap A_2 \sqcap \ldots \sqcap A_k \sqcap \exists R_1.B_1 \sqcap \ldots \sqcap \exists R_m.B_m$ (actually, $m = 1$). If sufficient conditions with exists restrictions are specified as in the case of $Chair$, optimization is much more complex. In LUBM data descriptions, no individual is explicitly declared as a $Chair$ and, therefore, reasoning is required, which is known to be rather costly. If $Chair$ is used as a generator and not as a tester such as in the simple query $ans(x) \leftarrow Chair(x)$, optimization is even more important. The idea to optimize instance retrieval is to detect an additional number of obvious instances using further incomplete tests, and, in addition, to determine obvious non-instances. We first present the latter technique and continue with the former afterwards.

3.3 Obvious Non-instances: Exploiting Information from One Completion

The detection of "obvious" non-instances of a given concept C can be implemented using a model merging operator defined for so-called individual pseudo models (aka pmodels) as defined in [10]. Since these techniques have already been published, we just sketch the main idea here for the sake of completeness. The central idea is to compute a pmodel from a completion that is derived by the tableau prover.

For instance, in the DL $\mathcal{ALC}$ a pseudo model for an individual i mentioned in a consistent initial A-box A w.r.t. a Tbox T is defined as follows. Since A is consistent, there exists a set of completions $\mathcal{C}$ of A. Let $A' \in \mathcal{C}$. An *individual pseudo model* M for an individual i in A is defined as the tuple $\langle M^D, M^{\neg D}, M^{\exists}, M^{\forall} \rangle$ w.r.t. A' and A using the following definition.

$$M^D = \{D \mid i : D \in A', D \text{ is a concept name}\}$$
$$M^{\neg D} = \{D \mid i : \neg D \in A', D \text{ is a concept name}\}$$
$$M^{\exists} = \{R \mid i : \exists R.C \in A'\} \cup \{R \mid (i,j) : R \in A\}$$
$$M^{\forall} = \{R \mid i : \forall R.C \in A'\}$$

Note the distinction between the initial A-box A and its completion A'. It is important that all restrictions for a certain individual are "reflected" in the pmodel. The idea of model merging is that there is a simple sound but incomplete test for showing that adding the assertion $i : \neg C$ to the ABox will not lead to a clash (see [10] for details) and, hence, i is not an instance of the query concept C. Let MS be a set of pmodels. The pmodel merging test is: $atoms\_mergable(MS) \wedge roles\_mergable(MS)$ where $atoms\_mergable$ tests for a possible primitive clash between pairs of pseudo models. It is applied to a set of pseudo models MS and returns *false* if there exists a pair $\{M_1, M_2\} \subseteq MS$ with $(M_1^D \cap M_2^{\neg D}) \neq \emptyset$ or $(M_1^{\neg D} \cap M_2^D) \neq \emptyset$. Otherwise it returns *true*.

The algorithm $roles\_mergable$ tests for a possible role interaction between pairs of pseudo models. It is applied to a set of pseudo models MS and returns *false* if there exists a pair $\{M_1, M_2\} \subseteq MS$ with $(M_1^{\exists} \cap M_2^{\forall}) \neq \emptyset$ or $(M_1^{\forall} \cap M_2^{\exists}) \neq \emptyset$. Otherwise it returns *true*. The reader is referred to [11] for the proof of the soundness of this technique and for further details.

It should be emphasized that the complete set of data structures for a particular completion is not maintained by a DL reasoner. The pmodels provide for an appropriate excerpt of a completion needed to determine non-instances.

3.4 Obvious Instances: Exploiting Information from the Precompletion

Another central optimization technique to ensure data description scalability as it is required for LUBM is to also find "obvious" instances with minimum effort.

Given an initial ABox consistency test and a completion one can consider all deterministic restrictions, i.e., one considers only those completion data structures (from now on called constraints) for which there are no choice points in the tableau proof (in other words, consider only those constraints that do not have dependency information attached). These constraints constitute a so-called precompletion.[2] Note that in a precompletion, no constraints are violated because we assume that the precompletion is computed from an existing completion.

Given the precompletion constraints, for each individual i, an approximation of the most-specific concept (MSC) is computed as follows (the approximation is called MSC'). For all constraints representing role assertions of the form $(i, j) :$ R (or $(j, i) : R$) add constraints of the form $i : \exists R.\top$ (or $i : \exists R^{-1}.\top$). Afterwards, constraints for a certain individual i are collected into a set $\{i : C_1, \ldots, i : C_n\}$. Then, $MSC'(i) := C_1 \sqcap \ldots \sqcap C_n$. Now, if $MSC'(i)$ is subsumed by the query concept C, then i must be an instance of C. In the case of LUBM many of the assertions lead to deterministic constraints in the tableau proof which, in turn, results in the fact that for many instances of a query concept C (e.g., *Faculty* as in query $Q9$) the instance problem is decided with a subsumption test based on the MSC' of each individual. Subsumption tests are known to be fast due to caching and model merging [13]. The more precisely $MSC'(i)$ approximates $MSC(i)$, the more often an individual can be determined to be an obvious instance of a query concept. Obviously, it might be possible to determine obvious instances by directly considering the precompletion data structures. However, at this implementation level a presentation would be too detailed. The main point is that, due to our findings, the crude approximation with MSC' suffices to solve many instance tests in LUBM.

If query atoms are used as testers, in LUBM it is the case that in a large number of cases the test for obvious non-instances or the test for obvious instances determines the result. However, for some individuals i and query concepts C both tests do not determine whether i is an instance of C (e.g., this is the case for *Chair*). Since both of these "cheap" tests are incomplete, for some individuals i a refutational ABox consistency test resulting from adding the claim $i : \neg C$ (refutational instance test) must be decided with a sound and complete tableau prover. For some concepts C, the set of candidates is quite large. Considering the volume of assertions in LUBM (see below for details), it is easy to see that the refutational instance test should not start from the initial, unprocessed ABox in order to ensure scalability.

For large ABoxes and many repetitive instance tests it is a waste of resources to "expand" the very same initial constraints over and over again. Therefore, the precompletion resulting from the initial ABox consistency test is used as a starting point for refutational instance tests. The tableau prover keeps the precompletion in memory. All deterministic constraints are expanded, so if some

---

[2] Cardinality measures for concept names, required for determining optimized query execution plans, could be made more precise if cardinality information was computed by considering a precompletion. However, in the case of LUBM this did not result in better query execution plans.

constraint is added, only a limited amount of work is to be done. To understand the impact of refutation-based instance tests on the data description scalability problem, a more low-level analysis on tableau provers architectures is required.

3.5   Index Structures for Optimizing Tableau Provers

Tableau provers are fast w.r.t. backtracking, blocking, caching and the like. But not fast enough if applied in a naive way. If a constraint $i : \neg C$ is added to a precompletion, the tableau prover must be able to very effectively determine related constraints for i that already have been processed. Rather than using linear search through lists of constraints, index structures are required for bulk data descriptions.

First of all, it is relatively easy to classify various types of constraints (for exists restrictions, value restrictions, atomic restrictions, negated atomic restrictions, etc.) and access them effectively according to their type. We call the corresponding data structure an active record of constraint sets (one set for each kind of constraint). For implementing a tableau prover, the question for an appropriate data structure for these sets arises. Since ABoxes are not models, (dependency-directed) backtracking cannot be avoided in general. In this case, indexing the set of "relevant" constraints in order to provide algorithms for checking if an item is an element of a set or list (element problem) is all but easy. Indexing requires hashtables (or trees), but backtracking requires either frequent copying of index structures (i.e., hashtables) or frequent insertion and deletion operations concerning hashtables. Both operations are known to be costly.

Practical experiments with LUBM and the DL system RacerPro (see below for a detailed evaluation) indicate that the following approach is advantageous in the average case. For frequent updates of the search space structures during a tableau proof, we found that simple lists for different kinds of constraints are most efficient, thus we have an active record of lists of constraints. New constraints are added to the head of the corresponding list, a very fast operation. During backtracking, the head is chopped off with minimum effort. The list representation is used if there are few constraints, and the element problem can be decided efficiently. However, if these lists of constraints get large, performance decreases due to linear search. Therefore, if some list from the active record of constraints gets longer than a certain threshold, the record is restructured and the list elements are entered into an appropriate index structure (hashtables with individuals as keys). Afterwards the tableau prover continues with a new record of empty lists as the active record. The pair of previous record of lists and associated hashtable is called a generation. From now on, new constraints are added to the new active record of constraints and the list(s) of the first generation are no longer used. For the element problem the lists from the active record are examined first (linear search over small lists) and then, in addition, the hashtable from the first generation is searched (almost linear search). If a list from the active record gets too large again, a new generation is created. Thus, in general we have a sequence of such generations, which are then considered for the element test in the obvious way. If backtracking occurs, the lists of the appropriate generation are installed again as the active record of lists. This way of dealing with the current search state allows for a functional implementation style of the tableau prover which we prefer for debugging purposes. However, one might also use a destructive way to manage constraints during backtracking. Obviously, all (deterministic) constraints from the initial Abox can be stored in a hashtable. In any case, the main point here is that tableau provers need an individual-based index to efficiently find all constraints in which an individual is involved. In the evaluation of other optimization techniques (see below) we presuppose that a tableau prover is equipped with this technology, and thus we can assume that each refutational instance test is rather fast.

3.6   Transforming Sufficient Conditions into Conjunctive Queries

Up to now we can detect obvious instances based on told and taxonomical information (almost constant time, see Section 3.2) as well as information extracted from the precompletion (linear time w.r.t. the number of remaining candidate individuals and a very fast test, see Section 3.4). Known non-instances can be determined with model merging techniques applied to individual pmodels (also a linear process w.r.t. the number of remaining candidate individuals but with a very fast test, see Section 3.3). However, there might still be some candidates left. Using the results in [10] it is possible to use dependency-directed instance retrieval and binary partitioning. Our findings suggest that in the case of LUBM, for example for the concept *Chair*, the remaining refutational tableau proofs are very fast. However, for *Chair* a considerable number of candidates remain since there are many *Persons* in LUBM. In application scenarios such as those we investigate with LUBM we have 200,000 individuals and more (see the evaluation below) with many *Persons*. Even if each single instance test lasts only a few dozen microseconds, query answering will be too slow, and hence additional techniques should be applied to solve the data description scalability problem.

The central insight for another optimization technique is that in the presence of sufficient conditions for concept names given in the Tbox, query atoms that refer to names might be transformed. Let us consider the query $ans(x) \leftarrow Chair(x)$. For *Chair*, sufficient conditions are given as part of the TBox (see above). Thus, in principle, we are looking for instances of the concept $Person \sqcap \exists headOf.Department$. The key to optimizing query answering becomes apparent if we transform the definition of *Chair* into a conjunctive query and derive the optimized version $Q15'$

$Q15 : ans(x) \leftarrow Person(x), headOf(x, y), Department(y)$
$Q15' : ans(x) \leftarrow Department(y), headOf^{-1}(y, x), Person(x)$

Because there exist fewer *Departments* than *Persons* in LUBM, search for bindings for x is substantially more focused in $Q15'$ (which is the result of automatic query optimization, see above). In addition, in LUBM, the extension of *Department* can be determined with simple index-based tests only (only hierarchies are involved) and thus the heuristics of the query optimizer produce optimal results. With the *Chair* example one can easily see that the standard approach for instance retrieval can be optimized dramatically with rewriting concept query atoms if certain conditions are met.

Algorithm 1. $rewrite(tbox, concept, var)$:

if $meta\_constraints(tbox) \neq \emptyset \vee definition(concept) = \top$ **then**
    return $(concept(var))$
else
    $\{atom_1, \ldots, atom_n\} :=$
        $rewrite\_0(tbox, concept, definition(tbox, concept), var, \{\})$
    return $(atom_1, \ldots, atom_n)$

Algorithm 2. $rewrite\_0(tbox, concept, var, exp)$:

if $definition(concept) = \top \vee concept \in exp$ **then**
    return $\{concept(var)\}$
else
    ;; catch installs a marker to which the control flow can be thrown
    catch $not\_rewritable$
        $rewrite\_1(tbox, concept, definition(tbox, concept), var, \{concept\} \cup exp)$

Algorithm 3. $rewrite\_1(tbox, concept\_name, definition, var, exp)$:

if $(definition = A)$ where A is an atomic concept **then**
    return $rewrite\_0(tbox, definition, var, \{definition\} \cup exp)$
else
    if $(definition = \exists R.C)$ **then**
        $filler\_var := fresh\_variable()$
        return $\{R(var, filler\_var)\} \cup rewrite\_0(tbox, C, filler\_var, exp)$
    else
        if $(definition = C_1 \sqcap \ldots \sqcap C_n)$ **then**
            return $rewrite\_1(tbox, concept\_name, C_1, var, exp)$
                $\cup \ldots \cup$
                $rewrite\_1(tbox, concept\_name, C_n, var, exp)$
        else
            ;; throw the control flow out of rewrite_1 recursion
            ;; back to the call to rewrite_1 in rewrite_0 and return {concept_name(var)}
            throw $not\_rewritable \{concept\_name(var)\}$

The rewriting algorithm is defined in Algorithms 1, 2, and 3. Every concept query atom $C(x)$ in a conjunctive query used is replaced with $rewrite(query\_tbox, C, x)$ (and afterwards, the query is optimized).

Some auxiliary functions are used. The function $definition(C)$ returns sufficient conditions for a concept name C (the result is a concept), and the function $meta\_constraints(tbox)$ indicates whether there are some meta constraints left after GCI transformation (see [15], the result is a set of concepts). In addition, we use a function $fresh\_variable$ that generates a new variable that was not used before.

If there is no specific definition or there are meta constraints, rewriting is not applied (see Algorithm 1). It is easy to see that the rewriting approach is sound. However, it is complete only under specific conditions, which can be automatically detected. If we consider the Tbox $T = \{D \doteq \exists R.C\}$, the Abox $A = \{i : \exists R.C\}$ and the query $ans(x) \leftarrow D(x)$, then due to the algorithm presented above the query will be rewritten as $ans(x) \leftarrow R(x, y), C(y)$. For variable bindings, the query language nRQL (see above) considers only those individuals that are explicitly mentioned in the Abox. Thus i will not be part of the result set because there is no binding for y in the Abox A. Examining the LUBM Tbox and Abox it becomes clear that in this case for every $\exists R.C$ that is applicable to an individual i in a tableau proof there already exist constraints $(i, j) : R$ and $j : C$ in the original Abox (LUBM was derived from a database scenario). However, even if this is not the case, the technique can be employed under some circumstances.

Usually, in order to construct a model (or a completion to be more precise), tableau provers create a new individual for each constraint of the form $i : \exists R.C$ and add corresponding concept and role assertions. These newly created individuals are called anonymous individuals. Let us assume, during the initial Abox consistency test a completion is found. As we have discussed above, a precompletion is computed by removing all constraints that depend on a choice point. If there is no such constraint, the precompletion is identical to the completion that the tableau prover computed. Then, assuming that blocking is postponed to fit the query with the largest nesting depth, the set of bindings for variables is extended to the anonymous individuals found in the precompletion. The rewriting

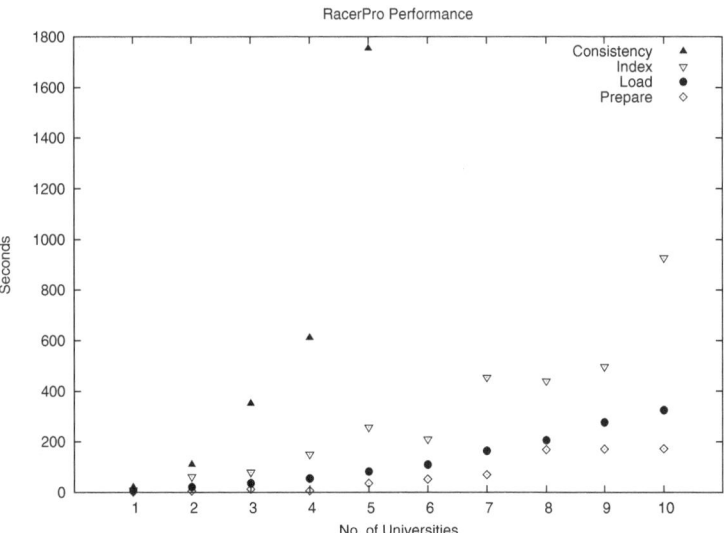

Fig. 1. Runtimes for loading, preparation, abox consistency checking and indexing

technique for concept query atoms is applicable (i.e., is complete) under these conditions. Even if the rewriting technique is not complete (i.e., s.th. is removed from the completion in order to derive the precompletion), it can be employed to reduce the set of candidates for binary partitioning techniques that can speed-up this process considerably in the average case (c.f., [10]).

The transformation approach discussed in this section is reminiscent of an early transformation approach discussed in [19]. In fact, ideas from translational approaches from DLs to disjunctive datalog [18] can also be integrated in tableau-based approaches. In the following section, we will evaluate how the optimization techniques introduced up to now provide a contribution to the data description scalability problem.

4   Evaluation

The significance of the optimization techniques introduced in this contribution is analyzed with the system RacerPro 1.9. RacerPro is freely available for research and educational purposes (http://www.racer-systems.com). The runtimes we present in this section are used to demonstrate the order of magnitude of time resources that are required for solving inference problems. They allow us to analyze the impact of proposed optimization techniques. We start with an evaluation of optimizations for (restricted) conjunctive queries with LUBM and turn to instance retrieval w.r.t. applications-specific knowledge bases afterwards.

An overview about the size of the LUBM benchmarks is given in Table 1. The runtimes for loading the data descriptions, transforming them into abstract syntax trees (preparation), and indexing are shown in Figure 1 (AMD 64bit processor, 4GB, Linux OS). It is important to note that these curves are roughly linear, thus, no reasoning is included in these phases. In Figure 1, also the runtimes for checking ABox consistency together with the computation of the precompletion are indicated (Consistency, black triangle). The "quadratic" shape reveals that this phase should be subject to further optimizations.

In Figure 2, average query-answering times for running all 14 LUBM queries on data descriptions for an increasing number of universities are presented (see Table 1 for an overview on the number of individuals, concept assertions, and role assertions). We use different modes (A, B, and C) to indicate the effects of optimization techniques. All modes are complete with respect to the Tbox and data descriptions (Abox) we used for the LUBM experiments in this paper.

Table 1. Linearly increasing number of individuals, concept assertions and role assertions for different numbers of universities

Univs	Inds	Concept Assertions	Role Assertions
1	17174	53738	49336
3	55664	181324	166682
5	102368	336256	309393
10	207426	685569	630753

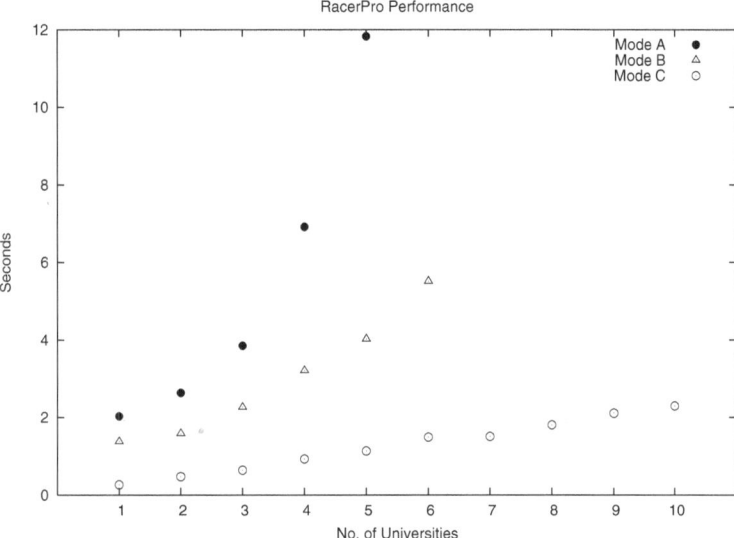

Fig. 2. Runtimes of 14 LUBM queries with different optimization settings (see text)

In mode A and B, concept definitions are not rewritten into conjunctive queries (see Section 3.6). In mode A, full constraint reasoning on OWL datatypes is provided. Thus, datatype properties are encoded in the obvious way as roles referring to individuals which, in turn, refer to values via concrete domain attributes. With concrete domains, arbitrary constraint systems can be specified in an Abox [12]. This means, (multiple) attribute values of *multiple* Abox individuals can constrained. In OWL one can only restrict (multiple) attributes of a *single* individual (nominal). For LUBM, however, only OWL datatypes are used, and no constraint reasoning is required because datatypes are used only to associate individuals with strings in the Abox. In order to answer queries, only "told values" must be retrieved. Therefore, in mode B, told value retrieval is performed only. As Figure 2 shows, this is much more efficient (but less powerful in the general case, of course). Mode C in Figure 2 presents the runtimes achieved when definitions of concept names are rewritten to conjunctive queries (and told value reasoning on datatypes only is employed, as in mode B). The results for mode C indicate that for deterministic knowledge bases such as LUBM, scalability for instance retrieval can be achieved with tableau-based retrieval engines.

5   Conclusion and Future Work

We take LUBM as a representative for largely deterministic data descriptions that can be found in practical applications. The investigations reveal that description logic systems can be optimized to also be able to deal with large bulks of deterministic descriptions quite effectively. Mode C indicates that performance scales well with an increasing number of data descriptions given the expressivity of the language used in the ontology meets certain requirements. Our work is based on a tableau calculus which has shown to be reliable if expressivity is increased (see the results in mode B). The linear shape of the curve in mode C suggests that the proposed technology ensures that performance scales if high expressivity is not required. LUBM is in a sense too simple but the benchmark allows us to study the data description scalability problem.

Note that we argue that the concept rewriting technique is advantageous not only for RacerPro but also for other tableau-based systems. Future work will investigate optimizations of large Aboxes and more expressive Tboxes. Our work is based on the thesis that for investigating optimization techniques for Abox retrieval w.r.t. more expressive Tboxes, we first have to ensure scalability for Aboxes and Tboxes such as those we discussed in this paper. We have shown that the results are encouraging.

References

1. F. Baader and U. Sattler. An overview of tableau algorithms for description logics. *Studia Logica*, 69:5–40, 2001.
2. Franz Baader, Diego Calvanese, Deborah McGuinness, Daniele Nardi, and Peter F. Patel-Schneider, editors. *The Description Logic Handbook: Theory, Implementation, and Applications*. Cambridge University Press, 2003.
3. S. Bechhofer, I. Horrocks, and D. Turi. The OWL instance store: System description. In *Proceedings CADE-20*, LNCS. Springer Verlag, 2005.
4. D. Calvanese, G. De Giacomo, D. Lembo, M. Lenzerini, and R. Rosati. Data complexity of query answering in description logics. In *Proc. of the 2005 Description Logic Workshop (DL 2005)*. CEUR Electronic Workshop Proceedings, http://ceur-ws.org/, 2005.
5. Diego Calvanese, Giuseppe De Giacomo, and Maurizio Lenzerini. On the decidability of query containment under constraints. In *Proc. of the 17th ACM SIGACT SIGMOD SIGART Symp. on Principles of Database Systems (PODS'98)*, pages 149–158, 1998.
6. H. Garcia-Molina, J.D. Ullman, and J. Widom. *Database Systems: The Complete Boook*. Prentice Hall, 2092.
7. Birte Glimm and Ian Horrocks. Handling cyclic conjunctive queries. In *Proc. of the 2005 Description Logic Workshop (DL 2005)*, volume 147. CEUR (http://ceur-ws.org/), 2005.
8. Y. Guo, J. Heflin, and Z. Pan. Benchmarking DAML+OIL repositories. In *Proc. of the Second Int. Semantic Web Conf. (ISWC 2003)*, number 2870 in LNCS, pages 613–627. Springer Verlag, 2003.
9. Y. Guo, Z. Pan, and J. Heflin. An evaluation of knowledge base systems for large OWL datasets. In *Proc. of the Third Int. Semantic Web Conf. (ISWC 2004)*, LNCS. Springer Verlag, 2004.
10. V. Haarslev and R. Möller. Optimization techniques for retrieving resources described in OWL/RDF documents: First results. In *Ninth International Conference on the Principles of Knowledge Representation and Reasoning, KR 2004, Whistler, BC, Canada, June 2-5*, pages 163–173, 2004.

11. Volker Haarslev, Ralf Möller, and Anni-Yasmin Turhan. Exploiting pseudo models for tbox and abox reasoning in expressive description logics. In *Proc. of the Int. Joint Conf. on Automated Reasoning (IJCAR 2001)*, volume 2083 of *Lecture Notes in Artificial Intelligence*. Springer-Verlag, 2001.

12. Volker Haarslev, Ralf Möller, and Michael Wessel. The description logic $\mathcal{ALCNH}_{R+}$ extended with concrete domains: A practically motivated approach. In *Proc. of the Int. Joint Conf. on Automated Reasoning (IJCAR 2001)*, pages 29–44, 2001.

13. I. Horrocks. *Optimising Tableaux Decision Procedures for Description Logics*. PhD thesis, University of Manchester, 1997.

14. I. Horrocks, O. Kutz, and U. Sattler. The even more irresistible $\mathcal{SROIQ}$. Technical report, University of Manchester, 2006.

15. I. Horrocks and S. Tobies. Reasoning with axioms: Theory and practice. In *Proc. of the 7th Int. Conf. on Principles of Knowledge Representation and Reasoning (KR 2000)*, pages 285–296, 2000.

16. Ian Horrocks, Ulrike Sattler, and Stephan Tobies. Reasoning with individuals for the description logic $\mathcal{SHIQ}$. In David McAllester, editor, *Proc. of the 17th Int. Conf. on Automated Deduction (CADE 2000)*, volume 1831 of *Lecture Notes in Computer Science*, pages 482–496. Springer-Verlag, 2000.

17. Ian Horrocks and Sergio Tessaris. A conjunctive query language for description logic ABoxes. In *Proc. of the 17th Nat. Conf. on Artificial Intelligence (AAAI 2000)*, pages 399–404, 2000.

18. B. Motik. *Reasoning in Description Logics using Resolution and Deductive Databases*. PhD thesis, Univ. Karlsruhe, 2006.

19. B. Motik, R. Volz, and A. Maedche. Optimizing query answering in description logics using disjunctive deductive databases. In *Proceedings of the 10th International Workshop on Knowledge Representation Meets Databases (KRDB-2003)*, pages 39–50, 2003.

20. M. Wessel and R. Möller. A high performance semantic web query answering engine. In *Proc. of the 2005 Description Logic Workshop (DL 2005)*. CEUR Electronic Workshop Proceedings, http://ceur-ws.org/, 2005.

21. Z. Zhang. Ontology query languages for the semantic web: A performance evaluation. Master's thesis, University of Georgia, 2005.

Relation Instantiation for Ontology Population Using the Web

Viktor de Boer, Maarten van Someren, and Bob J. Wielinga

Human-Computer Studies Laboratory, Informatics Institute,
Universiteit van Amsterdam
{vdeboer,maarten,wielinga}@science.uva.nl

Abstract. The Semantic Web requires automatic ontology population methods. We developed an approach, that given existing ontologies, extracts instances of ontology relations, a specific subtask of ontology population. We use generic, domain independent techniques to extract candidate relation instances from the Web and exploit the redundancy of information on the Web to compensate for loss of precision caused by the use of these generic methods. The candidate relation instances are then ranked based on co-occurrence with a seed set. In an experiment, we extracted instances of the relation between artists and art styles. The results were manually evaluated against selected art resources.

1 Introduction

The ongoing project of the Semantic Web [1] intends to add semantics to the World Wide Web through the use of ontologies. Following [2], we make a distinction between an ontology and a knowledge base. An ontology consists of the concepts (classes) and relations that make up a conceptualization of a domain, while a knowledge base contains the instances of the classes and of the relations in the ontology. The Semantic Web calls for a large number of both ontologies and knowledge base content. Since manual construction of these ontologies and knowledge bases proves to be costly, (semi-)automatic methods for the construction of ontologies (ontology learning and enrichment) and the construction of knowledge bases are needed. The latter task is called ontology population.

We decompose ontology population into the extraction of concept instances and the extraction of instances of relations. In this paper, we focus on this last sub-task of ontology population: the extraction of instances of a relation that is predefined in an ontology. We call this task *relation instantiation*.

In this paper, we describe a method that extracts these relation instances for existing ontologies. Our method extracts the information from heterogeneous sources on the Web and is not dependent on the type of structure of documents. We designed this general method to be also domain- and language-independent.

2 Relation Instantiation Task

We define an ontology as a set of labeled classes (the domain concepts) $C_1, ..., C_n$, hierarchically ordered by a subclass relation. Non-hierarchical relations between

C. Freksa, M. Kohlhase, and K. Schill (Eds.): KI 2006, LNAI 4314, pp. 202–213, 2007.

concepts are also defined $(R : C_i \times C_j)$. We speak of a (partly) populated ontology when, besides the ontology, a knowledge base with instances of both concepts and relations from the ontology is also present.

We define the task of relation instantiation from a corpus as follows:

> Given two classes C_i and C_j in a partly populated ontology, with sets of instances I_i and I_j and given a relation $R : C_i \times C_j$, identify for an instance $i \in I_i$ an instance $j \in I_j$ such that the relation $R(i,j)$ holds given the information in the corpus.

Furthermore, we make a number of additional assumptions:

- R is not a one-to-one relation. The instance i is related to multiple elements of I_j.
- We know all elements of I_j. With this method, we do not attempt to extract new instances of a class.
- We have a method available that recognizes these elements in the documents in our corpus. For a textual corpus such as the Web, this implies that the instances must have a textual label.
- In individual documents of the corpus, multiple instances of the relation are represented.
- We have a (small) example set of instances of C_i and C_j for which the relation R holds.

An example of such a relation instantiation task is the extraction of instances of the relation 'appears_in' between films (instances of class 'Film') and actors (instances of class 'Actor') in an ontology about movies. Another example is finding the relation 'has_artist' between instances of the class 'Art Style' and instances of the class 'Artist' in an ontology describing the Cultural Heritage domain. As a case study for our approach, we chose this latter example and we shall discuss this in Section 4.

3   Redundancy-Based Relation Instantiation

In Section 3.1, we present our general approach to this task, which we further specify in Section 3.2

3.1   Approach

Current approaches for Information Extraction or Question Answering tasks could also be used for ontology population. However, the methods in these domains assume a specific structure of the corpus documents. Wrapper-induction techniques such as [3] assume structured text. Other methods learn natural language patterns. These methods generally perform well on free text, but do not work as well for more structured data. We designed our method to be structure-independent.

Methods that use some form of supervised Machine Learning assume a large number of tagged example instances to be able to learn patterns for extracting new instances and this is a serious limitation for large scale use[4]. We designed our method to require only a small amount of examples that are used as a seed set.

A number of Information Extraction methods perform very well on the domain they were constructed for. Their performance drops however when they are applied in a new, unknown domain. Our method as presented in this section is domain-independent.

Our approach incorporates generic methods that do not rely on assumptions about the domain or the type of documents in the corpus. By using these general methods for the extraction, we will lose in precision since the general methods are not optimized for a specific corpus or domain. However, since we use more generic methods, we are able to extract information from a greater number of sources. The main assumption behind our method is that because of the redundancy of information on the Web and because we are able to combine information from heterogeneous sources, we can compensate for this loss of precision.

To extract instances of the relation $R : C_i \times C_j$, the method takes as input a single instance i of C_i and the set of instances of C_j. Further input is in the form of a (small) seed set of instances for which we already know that the given relation holds.

The method uses generic methods to identify instances of C_j in the individual documents from the Web Corpus and marks them as candidates for the right-hand side of a relation instance. The documents are then given a score that reflects how well the relation R is represented in those documents. For this we use the seed set. All candidates are then scored based on the Document Scores of the pages they appear on, resulting in a ranked list of right-hand side instances. From this ranked list, the top n candidates are added to the seed set and all scores are recalculated, thus ending up with an iterative method.

We further specify the method in the next section. We show the extraction methods used, as well as the formulas for scoring the documents and the candidates.

3.2   Method Specification

The method consists of three steps, shown in Figure 1. We first construct a 'working corpus' by feeding the label(s) of the instance i to a search engine (in our case, Google [1]). The size of this working corpus is a parameter of the method.

In step 2, we identify the instances of the concept C_j in the documents of the working corpus. Since we assume we already know all instances of C_j, this step consists of matching the instances to their representations in the documents. These representations are extracted from the document using a domain dependent extraction method as listed in our assumptions. Named Entity Recognizers can extract different types of entities such as dates, persons, locations, companies, etc. These extracted representations (strings) are then matched to the

---

[1] http://www.google.com

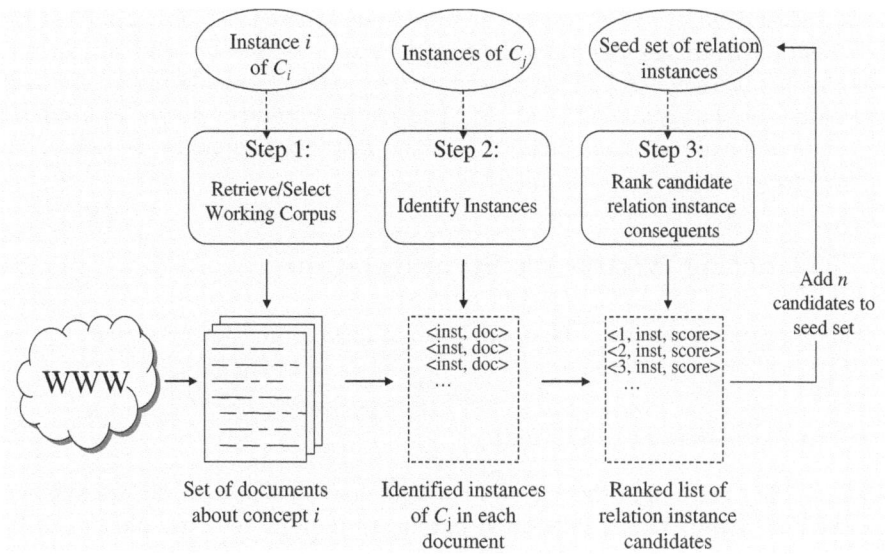

Fig. 1. Outline of the method

instances from the knowledge base. This matching process itself aims for a high precision and because of the large number of documents to extract from, the redundancy helps to raise the recall. The identified instances in the documents are the right-hand side instances of the candidate relation instances.

In step 3, the method combines the evidence from the different documents to produce a ranking for these candidates. We base this ranking on the assumption that on average in individual web pages, a target relation is either well represented (the web page contains a number of correct right-hand side instances) or not represented (it contains few or none of these instances). We therefore calculate a Document Score DS for each document. This is the probability that for all candidates in that document the relation R holds, according to the seed set. This is equal to the number of identified instances that are in the seed set divided by the total number of candidate instances in that document:

$$DS_{doc} = \frac{\mu_{doc}}{\nu_{doc}} \qquad (1)$$

where μ_{doc} is the number of instances of C_j identified in document doc for which the relation is already in our seed set and ν_{doc} is the total number of instances of C_j identified in document doc

We then combine all evidence for each of the candidate instances by taking the average of DS over all used documents N in the corpus resulting in an Instance Score IS:

$$IS_i = \frac{\sum^{doc} DS_{doc}}{N} \qquad (2)$$

where $i \in I_j, i \in doc$.

At the end of this step, we are left with an ordered list of candidates for new relation instances. We add the top n candidates to the seed set. In our experiments, we set $n = 1$. This procedure iterates by recalculating all DS and IS, based on the expanded seed set. The method iterates up to a threshold on the number of iterations or a drop in the Instance Scores. In Section 4, we explore the effects of these thresholds.

4   Extracting Artist-Art Style Relation

In this section, we describe the application of our method for an experiment in the Cultural Heritage domain.

4.1   Cultural Heritage Domain

We tested our method in the cultural heritage domain. We used two well-known art thesauri as our partly populated ontologies. One is the Art and Architecture Thesaurus[5] (AAT), a thesaurus defining more then 133.000 terms used to describe and classify art. The other is the Unified List of Artist names[6] (ULAN), a list of more then 255.000 names of artists. We also added a relation aua:has_artist [2] between the AAT concept aat:Styles_and_Periods and the top-level ULAN concept ulan:Artist. The aua:has_artist relation describes which artists represent a specific art style. This relation satisfies the requirement that it is not a one-to-one relation since a single art style is represented by a number of artists.

4.2   Experiment Setup

From the instances of aat:Styles_and_Periods, we chose nine modern European art styles to extract. We list their preferred labels from the AAT in Table 1. For each of these art styles, we applied the method.

Table 1. Art styles used

Art Deco	Dada	Neo-Impressionist
Art Nouveau	Expressionist	Neue Sachlichkeit
Cubist	Impressionist	Surrealist

We first populated the seed set with three well-known artists associated with that art style. Then in Step 1, 1000 pages were extracted as a working corpus by querying Google with a combination of the preferred and non-preferred labels from the AAT (for 'Dada' this resulted in the query 'Dada OR Dadaist OR Dadaism').

---

[2] aua denotes our namespace specifically created for these experiments.

Because the right-hand side instances in this task are persons, we first identified in Step 2 all person names in the documents. For this we used the Person Name Extractor from the TOKO toolkit [7]. The extracted names were then matched to the ULAN list of artists. This matching step is problematic as the number of artists in the ULAN is very large and so is the number of possible occurrences of person names in the texts. For example, 'Vincent van Gogh' can also appear as 'V. van Gogh', 'van Gogh, V.' or 'van Gogh'.

To tackle this matching problem, we performed tokenization on both the labels of all ULAN instances and the extracted Person Name strings. An ULAN instance is a possible match if all tokens in the extracted string are also tokens of that instance. If a string has exactly one possible match, we accept that match. If there still is ambiguity (the string 'van Gogh' matches three different artists), we reject the string and proceed to the next candidate string.

We expect that because of the redundancy of names from the corpus, a non-ambiguous name will eventually be extracted and correctly matched. However, as we found in earlier experiments, some names will still not be found as a result of this matching process. In addition, some names will not be extracted due to imperfections of the Person Name Extractor.

After the candidate instances have been extracted, we calculated the Document Scores and Instance Scores, resulting in an ordered list of candidates. We then added the top candidate to the seed set and re-iterated. For each of the art styles, we evaluated the results of 40 iterations.

4.3   Evaluation

As is often the case in ontology learning and population, evaluating the results is not trivial, in particular in a Web context. Since this task resembles Information Retrieval, we would like to evaluate the method using precision and recall. However, since we use an open domain and manually annotating the large number of relevant web pages is too time-consuming, we are unable to know the artists linked to an art style and therefore are unable to calculate the recall.

In our previous experiments, we solved this problem by constructing a small and very strict gold standard and calculated a form of recall with respect to that gold standard. However, even though this can be done for one art style, it is expensive to evaluate the method on multiple art styles. In the current experiment, we therefore opted to only calculate precision. We did this by having two annotators manually evaluate each of the 40 retrieved relation instances for each art style. For this, the annotators were allowed to consult a fixed set of sources: the articles on both the art style and the artist on the wikipedia web encyclopedia[3], the art style page on the artcyclopedia web site[4] and any encyclopedic web page that Google retrieved in the first ten results when queried with both the art style's label and the artist's name. If in any of these sources the artist was explicitly stated as a participant in the art style, the annotator was to mark the relation instance 'correct' and else mark it 'incorrect'.

---

[3] http://www.wikipedia.org
[4] http://www.artcyclopedia.com

After separately evaluating the relation instances in this way, inter-annotator agreement was calculated using Cohen's Kappa measure. Calculated over all nine ten art styles, this resulted in a value of 0.83. The annotators then reached agreement over the instances that initially differed. The consensus annotations are used to calculate precision.

4.4   Results for 'Neue Sachlichkeit'

We first illustrate the results for a single art style: 'Neue Sachlichkeit' ('New Objectivity'). The three artists we added to the seed set were 'George Grosz', 'Otto Dix' and 'Christian Schad'. Table 2 shows the top 16 results of the 40 artists iteratively extracted from the documents. For each of the artists, we also list the Instance Score with which they were extracted. The last column shows the evaluation (1='correct', 0='incorrect').

Table 2. Top ranked candidate artists for the has_artist relation for the art style 'Neue Sachlichkeit' for the first 16 iterations

iteration	AAT preferred label	Instance Score	correct
1	Beckmann, Max	0.0651	1
2	Schlichter, Rudolf	0.0291	1
3	Kanoldt, Alexander	0.0318	1
4	Schrimpf, Georg	0.0351	1
5	Gropius, Walter Adolf	0.0252	1
6	Griebel, Otto	0.0239	1
7	Chirico, Giorgio de	0.0260	1
8	Querner, Curt	0.0287	1
9	Grossberg, Carl	0.0299	1
10	Taut, Bruno	0.0300	1
11	Oelze, Richard	0.0312	1
12	Uzarski, Adolf	0.0291	1
13	Muthesius, Hermann	0.0303	1
14	Hubbuch, Karl	0.0191	1
15	Heckel, Erich	0.0131	0
16	Kollwitz, Kathe	0.0134	1
... ...		...	...

Figure 2 shows the Instance Scores for the top artists for all 40 iterations as well as the value for the precision (number of extracted candidates evaluated as correct divided by the total number of extracted candidates). The Instance Score represents the confidence at each iteration that for the top ranked artist a relation should be added to the knowledge base. As can be seen, this confidence for the first candidate instance is relatively high (0.0651), then drops to about 0.025 and stays relatively constant for a number of iterations. After 13 iterations, the Instance Score again drops to a new constant level of about 0.01.

After 13 iterations the method starts adding more and more false relation instances. For this art style, we achieve the best precision/number of extractions

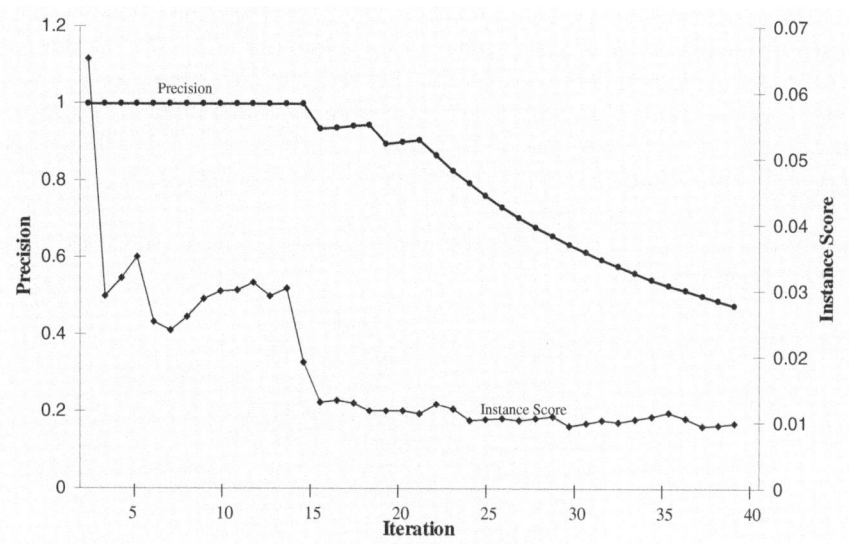

Fig. 2. Instance score versus rank number for 'Neue Sachlichkeit'

ratio if we set the maximum number of iterations somewhere between 13 and 21 iterations (after 21 iterations, only incorrect instances are added).

This maximum number of iterations depends on the specific art style: For popular art styles, with many associated artists, this drop in precision will occur after more iterations than for relatively small art styles such as 'Neue Sachlichkeit'. We also cannot cut off the iterations by setting an absolute threshold value for the Instance Score since it is highly variable for the different art styles.

As can be seen in the figure, the drop in precision co-occurs with a drop in the Instance Score. We choose the iteration threshold to be dependant on the relative drop in the Instance Score. We introduce a Drop Factor, (DF). The algorithm stops adding relation instances to the knowledge base if the Instance Score of the next candidate artist is less than DF multiplied by the maximum of the Instance Scores up till that iteration. We also stop adding instances after an absolute maximum number of iterations has been reached (Max).

For example, in the case of 'Neue Sachlichkeit', if we set DF to 0.2 and Max to 40, the algorithm stops adding new relation instances after iteration 16. This leads to a precision of 0.933, with 15 correct relations and one incorrect relation added to the knowledge base.

4.5 Results for the Nine Art Styles

In this section, we present the results for all nine art styles for which the relation instances were extracted.

In Table 3, we show the precision and the number of correct relation instances extracted for each of the nine art styles for an arbitrarily chosen value for the two threshold parameters (DF=0.3 and Max=20). For these values, the precision for each of the art styles is acceptable, with a minimum of 0.667. The number of correct extractions differs considerably between the art styles, for a 'small' art style such as 'Surrealist' only 5 correct new relation instances are extracted with a threshold at 7 iterations, resulting in a precision of 0.714. The average precision in this example is 0.84 with a standard deviation of 0.14.

Table 3. Precision and number of correct extractions (extr.) for the nine Art Styles for DF=0.3 and Max=20

	precision	extr.
Art Deco	0.900	18
Art Nouveau	1.000	16
Cubist	0.850	17
Dada	1.000	15
Expressionist	0.750	15
Impressionist	0.700	14
Neo-Impressionist	0.667	4
Neue Sachlichkeit	1.000	13
Surrealist	0.714	5

In Table 4, we list both the average precision and the total sum of the number of correct relation instances extracted for the nine art styles for 24 combinations of the two threshold parameters DF and Max. The lowest value for precision is 0.65. This occurs at DF=0 (the drop in the Instance Score is not used to set the threshold) and Max=40. In that case, for the nine art styles, all 360 (9×40) extractions are added to the knowledge base, of which 234 are evaluated correct.

Table 4. Average precision and total number of correct extractions (extr.) for the nine Art Styles

	Max							
	10		20		30		40	
DF	precision	extr.	precision	extr.	precision	extr.	precision	extr.
0	0.856	77	0.806	145	0.722	195	0.650	234
0.1	0.856	77	0.806	145	0.721	193	0.648	228
0.2	0.856	77	0.799	137	0.776	179	0.746	197
0.3	0.865	73	0.842	117	0.830	138	0.810	144
0.4	0.857	62	0.834	96	0.826	114	0.824	120
0.5	0.902	55	0.878	86	0.868	103	0.866	109
0.6	0.924	46	0.896	67	0.882	81	0.880	87

The highest average precision, 0.924 with a standard deviation of 0.11, is reached at DF=0.6 and Max=10, with only 46 correct relation instances added to the knowledge base. In this case, DF has a big effect. For some art styles (e.g. Expressionist, Impressionist) ten instances are extracted, while for other styles such as 'Neue Sachlichkeit', only one relation instance is extracted. Depending on further processing of these results, users might choose high precision, low number of correct extractions or vice versa by choosing the appropriate values for the two threshold parameters. The values for the standard deviation for each of these values of average precision ranged from 0.11 to 0.20.

We observe a tradeoff between precision and number of correct extractions comparable to that of the traditional precision/recall tradeoff.

4.6 Discussion

The results from the experiments show relatively good values for precision.

In some cases, the method yields false positives (relations that have been evaluated as 'incorrect'). One reason these occur is that in step 2, the Person Name Extraction module incorrectly extracts names and matches them to a single ULAN entity. For example, in extracting artists associated with 'Neo-Impressionist', the string "d'Orsay" (the name of a museum) is first misclassified by the Person Name Extraction module as a person name, then it is unambiguously matched to the ULAN entity "Comte d'Orsay". Other false positives are domain specific (Gustav Klimt is strictly speaking not an Art Deco artists, although he is frequently associated with that movement, especially in poster shops).

Also, not all artists that we would expect were found in the set of 40 candidate relation instances. As with precision, errors made by the Person Name Extraction module account for a part of these errors as some artist's person names were not recognized as such. Another cause for recall errors is the difficulty of the disambiguations of the artist names. From some extracted names, our strict matching procedure is not able to identify the correct ULAN entity. An example is the string 'Lyonel Feininger'. The ULAN has two different artists: one with the name 'Lyonel Feininger' and one with the name 'Andreas Bernard Lyonel Feininger'. The match is ambiguous and the string is discarded.

5   Related Work

The Armadillo system [8] is also designed to extract information from the World Wide Web. The Armadillo method starts out with a reliable seed set, extracted from highly structured and easily minable sources such as lists or databases and uses bootstrapping to train more complex modules to extract and combine information from different sources. Also, Armadillo does not require a complete list of instances as our method does. Armadillo's method, however requires the input of domain-dependant sources that are mined using wrappers. Our method requires no modification defined by the extraction task other than relevant instance extraction modules such as the Person Name Extraction module.

The KnowItAll system [9] aims to automatically extract the 'facts' (instances) from the web autonomously and domain-independently. The method, unlike our method, uses patterns to extract instances. It starts with universal extraction patterns and uses Machine Learning to learn more specific extraction patterns.

In combination with techniques that exploit list structures the method is able to extract information from heterogeneous sources.

The paper of Geleynse and Korst[10] also presents an automatic and domain-independent method for ontology population by querying Google. They also combine evidence from multiple sources (i.e. Google excerpts) and use a form of bootstrapping that enables the method to start with a small seed set. The method differs from our method in that it currently uses handcrafted rules to extract these instances.

6   Conclusions and Further Research

We presented a generic, domain-independent method for Relation Instantiation, a subtask of Ontology Population. Our method exploits the redundancy of information on the Web. As an example, we used the method in an experiment to extract instances of the Artist-Art Style relation. This was done using actual ontologies from the Cultural Heritage domain.

Results show a tradeoff of precision and the number of correct extractions analogous to the precision/recall tradeoff. Considering the method uses very generic methods and intuitive ranking scores, the results are encouraging but also suggest that further processing of the results could improve the relation instantiation.

Analysis of the documents from which information was extracted showed that the documents were highly heterogeneous in structure. Some documents were essays and consisted of free text while other documents such as art prints shops featured a list structure. Also, content was extracted from pages in a language different from English. How much this redundancy helped is a topic for further research.

Improvement in the Person Name Extraction module or combining different Person Name Extractors could improve the extraction. Using a different, less strict name-entity matching procedure is also a possible improvement. Also, other measures for the Document Score and Instance Score could be considered.

An obvious direction for further research is to test this method in other domains where relations that satisfy our assumptions are to be instantiated. Examples of these domains are geography (eg. which cities are located in a country) or sports (which players play for which teams).

Currently, we do not use any knowledge stored in the ontology in the extraction process other than the different labels of an instance. In the future we would like to develop general guidelines on how ontological knowledge derived from the class hierarchy or meta-properties can be used to aid the relation instantiation process.

Acknowledgements

This research was supported by the MultimediaN project (www.multimedian.nl) funded through the BSIK programme of the Dutch Government. We would like to thank Anjo Anjewierden and Jan Wielemaker for their extensive programming support.

References

1. Berners-Lee, T., Hendler, J., Lassila, O.: The semantic web. Scientific American (2001)
2. Maedche, A., Staab, S.: Ontology learning for the semantic web. IEEE Intelligent Systems **13** (2001) 993
3. Kushmerick, N., Weld, D., Doorenbos, R.: Wrapper induction for information extraction. In: in Proceedings of the Fifteenth International Joint Conference on Artificial Intelligence. (1997) 729737
4. Cimiano, P.: Ontology learning and population. Proceedings Dagstuhl Seminar Machine Learning for the Semantic Web (2005)
5. The Getty Foundation: Aat: Art and architecture thesaurus. http://www.getty.edu/research/tools/vocabulary/aat/ (2000)
6. The Getty Foundation: Ulan: Union list of artist names. http://www.getty.edu/research/tools/vocabulary/ulan/ (2000)
7. Anjewierden, A., Wielinga, B.J., de Hoog, R.: Task and domain ontologies for knowledge mapping in operational processes. Metis Deliverable 4.2/2003, University of Amsterdam. (2004)
8. Ciravegna, F., Chapman, S., Dingli, A., Wilks, Y.: Learning to harvest information for the semantic web. In: Proceedings of the 1st European Semantic Web Symposium. (2004)
9. Etzioni, O., Cafarella, M., Downey, D., Kok, S., Popescu, A., Shaked, T., Soderland, S., Weld, D.S., Yates, A.: Webscale information extraction in knowitall preliminary results. In: in Proceedings of WWW2004. (2004)
10. Geleijnse, G., Korst, J.: Automatic ontology population by googling. In: Proceedings of the Seventeenth Belgium-Netherlands Conference on Artificial Intelligence (BNAIC 2005), Brussels, Belgium (2005) 120 – 126

GeTS – A Specification Language for Geo-Temporal Notions

Hans Jürgen Ohlbach

Institut für Informatik, Universität München
ohlbach@lmu.de

Abstract. This paper contains a brief overview of the 'Geo-Temporal' specification language GeTS. The objects which can be described and manipulated with this language are time points, crisp and fuzzy time intervals and labeled partitionings of the time axis. The partitionings are used to represent periodic temporal notions like months, semesters etc. and also whole calendar systems. GeTS is essentially a typed functional language with a few imperative constructs and many built-ins. GeTS can be used to specify and compute with many different kinds of temporal notions, from simple arithmetic operations on time points up to complex fuzzy relations between fuzzy time intervals. A parser, a compiler and an abstract machine for GeTS is implemented.

1 Motivation and Introduction

The phenomenon of *time* has many different facets which are investigated by different communities. Physicists investigate the flow of time and its relation to physical objects and events. Temporal logicians develop abstract models of time where only the aspects of time are formalized which are sufficient to model the behaviour of computer programs and similar processes. Linguists develop models of time which can be used as semantics of temporal expressions in natural language. More and more information about facts and events in the real world is stored in computers, and many of them are annotated with temporal information. Therefore it became necessary to develop computer models of the use of time on our planet, which are sophisticated enough to allow the kind of computation and reasoning that humans can do. Examples are 'calendrical calculations' [7], i.e. formal encodings of calendar systems for mapping dates between different calendar systems. Other models of time have been developed in the temporal database community [5], mainly for dealing with temporal information in databases. This work is becoming more important now with the emergence of the Semantic Web [2]. Informal, semi-formal and formal temporal notions occur frequently in semistructured documents, and need to be 'understood' by query and transformation mechanisms.

The formalisms developed so far approximate the real use of time on our planet to a certain extent, but still ignore important aspects. In the CTTN–project ('Computational Treatment of Temporal Notions') [16] we aim at a very detailed modeling of the temporal notions which can occur in semi-structured data. The

C. Freksa, M. Kohlhase, and K. Schill (Eds.): KI 2006, LNAI 4314, pp. 214–228, 2007.

CTTN–system consists of a kernel and several modules around the kernel. The kernel itself consists of several layers. At the bottom layer there are a number of basic data types for elementary temporal notions. These are time points, crisp and fuzzy time intervals [20] and partitionings for representing periodical temporal notions like years, months, semesters etc. [22]. The partitionings can be specified algorithmically or algebraically. The algorithmic specifications allow one to encode phenomena like leap seconds, daylight savings time regulations, the Easter date, which depends on the moon cycle etc.

Partitionings can be arranged to form 'durations', e.g. '2 year + 1 month', but also '2 semester + 1 month', where *semester* is a user defined partitioning.

Sets of partitionings, together with certain procedures, form a *calendar*. The Gregorian calendar in particular can be formalized with the partitionings for years, months, weeks, days, hours, minutes and seconds.

A part of the second layer is presented in this paper. It uses the functions and relations of the first layer as building blocks in the specification language GeTS ('GeoTemporal Specifications'[1]) for specifying complex temporal notions. A very first version of this language has been presented in [17,18], but the new version has more than 20 times as many constructs and features than the old one. It is essentially a functional programming language with certain additional constructs for this application area. A flex/bison type parser and an abstract machine for GeTS has been implemented as part of the CTTN–system. GeTS is the first specification and programming language with such a rich variety of built-in data structures and functions for geo-temporal notions. The details of the language can be found in the technical report [21]. The third layer contains interfaces to GeTS and the other modules of the system. The standard interface is socket based. There are experimental RMI, SOAP and CORBA interfaces [1].

2   Basic Data Structures in CTTN and GeTS

2.1   Time Points and Time Intervals

The flow of time underlying most calendar systems corresponds to a time axis which is isomorphic to the real numbers $\mathbb{R}$. Therefore we take as time points just real numbers. Since the most precise clocks developed so far, atomic clocks, measure the time in discrete units, it is sufficient to restrict the representation of concrete time points to *integers*. In the standard setting these integers count the *seconds* from the Unix epoch, which is January 1st 1970. Nothing significant changes in GeTS, however, if the meaning of these integers is changed to count, for example, femtoseconds from the year 1.

The next important datatype is that of time intervals. Time intervals can be crisp or fuzzy [27,8]. With fuzzy intervals one can encode notions like 'around

---

[1] The prefix 'geo' in the word geo-temporal was chosen to distinguish it from temporal logics in the usual understanding. 'geo, i.e. 'earth', emphasizes that it is about temporal notions as used on our planet. There is a close analogy to the area of 'geo-spatial' in contrast to 'spatial' representation and reasoning.

noon' or 'late night' etc. This is more general and more flexible than crisp intervals. Therefore the CTTN–system uses fuzzy intervals as basic interval datatype.

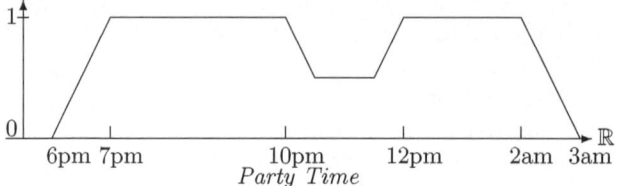

Party Time

This set may represent a particular party time, where the first guests arrive at 6 pm. At 7 pm all guests are there. Half of them disappear between 10 and 12 pm (because they go to the pub next door to watch an important soccer game). Between 12 pm and 2 am all of them are back. At 2 am the first ones go home, and finally at 3 am all are gone. The fuzzy value indicates in this case the number of people at the party.

The CTTN–system has an extensive implementation of fuzzy time intervals with a rich application programming interface [20].

2.2 Partitionings

The CTTN–system uses the concept of *partitionings* of the real numbers to model periodical temporal notions. In particular, the basic time units years, months etc. are realized as partitionings. Other periodical temporal notions, for example semesters, school holidays, sunsets and sunrises etc. can also be modeled as partitionings.

Partitionings of the time axis are infinite mathematical structures. Therefore they must be represented on a computer in a more indirect way. We distinguish three aspects of partitionings of the time axis:

1. the mathematical structure. It serves as the semantics for the more concrete descriptions of these objects;
2. the specification of concrete partitionings. There are different ways to specify them. Each type of specification comes with a mathematical structure that has also a serialized text form which can be stored in files;
3. the implementation. There is a common interface for all types of partitionings, such that the algorithms working with these partitionings are independent of the specification type. The methods of the partitioning application interface (API) are automatically compilable from the specification.

Different types of specifications for partitionings and a common API for all of them is implemented in the PartLib library [22]. The first type of partitionings are called 'algorithmic partitionings'. They are characterized by implementing an isomorphism to integers directly. All the standard periodic temporal notions, years, months, weeks etc., but also Easter time, sun rises, tides etc. are of this type. The implementation can in particular take into account all the nasty and irregular features of real calendar systems, leap years, leap seconds, daylight savings time, time zones etc.

Another type of specification are called 'duration partitionings'. They are specified by giving an anchor date and a list of durations. For example, one can specify semesters in this way. The anchor date could be first of October 2000. The durations could be '6 months' (for the winter semester) and '6 months' (for the summer semester). A further type are 'date partitionings', which are specified by concrete dates. An example could be the seasons. 2000/3/21 spring 2000/6/21 summer 2000/9/23 autumn 2000/12/21 winter 2001/3/21 specifies the seasons for one year. An extrapolation mechanism extrapolates them to the infinity. In principle, all partitionings are infinite, but there is a mechanism for constraining a 'validity region'. This way one can express, for example, 'I have this meeting every Monday 9:00-10:00 for the next 15 weeks'.

The next version of the PartLib library will contain 'tree partitionings' [23]. A bus timetable, for example, can be specified as a tree partitioning: '(in very winter, in every week, (in day 0-4, hour 5, minute 20, bus B1, hour 6, minute 20 bus B2 ...), (in day 5-6, hour 8, minute 20 bus B1, ...)), (in every spring ...)...

Partitions can be *labeled*. The labels are names for the partitions. They can be used for two purposes. The first purpose is to get access to the partitions via their names (labels). For example, the labels for the 'day' partitioning can be 'Monday', 'Tuesday' etc., and one can use these names in various GeTS functions. The second purpose is to use the labels to group partitions together to so called *granules*. The concept of 'working day', for example, can be modeled by taking an 'hour' partitioning, and attached labels 'working_hour' and 'gap' to the hour partitions. Groups of hour partitions labeled 'working_hour' yield a working day. The working days can be interrupted by 'gap' partitions, for example to take 'lunch time' out of a 'working day'. A group of partitions with the same label, possibly interrupted by 'gap'–partitions, is a *granule*.

Remark 1 (Calendar Systems). A *calendar* in the CTTN–system is a set of partitionings, for example the partitionings for seconds, minutes, hours, weeks, months and years, together with some extra data and methods. Calendars are not visible in the GeTS language because they are only special cases of sets of partitionings. Some GeTS constructs use partitionings which can not only be the partitionings of calendar systems, but any kind of partitioning. This is more general than sticking to particular calendar systems. ∎

2.3 Durations

The partitionings are the mathematical model of periodic time units, such as years, months etc. This offers the possibility to define *durations*. A duration may, for example, be '3 months + 2 weeks'. Months and weeks are represented as partitionings, and 3 and 2 denote the number of partitions in these partitionings. The numbers need not be integers, but they can be arbitrary real numbers.

A duration can be interpreted as the length of an interval. In this case the numbers should not be negative. A duration, however, can also be interpreted as a time shift. In this interpretation negative numbers make perfect sense. $d = -2$ *week* $+ 3$ *month*, for example, denotes a backward shift of 2 weeks followed by a forward shift of 3 months.

3 The GeTS Language

The design of the GeTS language was influenced by the following considerations:

- Although the GeTS language has many features of a functional programming language, it is not intended as a general purpose programming language. It is a specification language for temporal notions, however, with a concrete operational semantics.
- The parser, compiler, and in particular the underlying GeTS abstract machine are not standalone systems. They must be embedded into a host system which provides the data structures and algorithms for time intervals, partitionings etc., and which serves as the interface to the application. This excludes using an existing functional language like SML or Haskell.
- The language should be simple, intuitive, and easy to use. It should not be cluttered with too many features which are mainly necessary for general purpose programming languages.
- Developing GeTS from scratch has also the advantage that one can design the syntax of the language in a way which better reflects the semantics of the language constructs. As an example, the syntax for a time interval constructor is just $[expression_1, expression_2]$.

The GeTS language is a strongly typed functional language with a few imperative constructs. Let us get a flavor of the language, before the technical details are introduced.

Example 1 (tomorrow). The definition

$$\text{'tomorrow = partition(now(),day,1,1)'}$$

specifies 'tomorrow' as follows: now() yields the time point of the current point in time. day is the name of the day partitioning. Let i be the coordinate of the day-partition containing now(). partition(now(),day,1,1) computes the interval $[t_1, t_2[$ where t_1 is the start of the day-partition with coordinate $i + 1$ (i.e next midnight) and t_2 is the end of the day-partition with coordinate $i + 1$ (i.e. midnight tomorrow). ∎

Example 2 (Christmas). The definition

```
christmas(Time t) =
  dLet year = date(t,Gregorian_month) in
                  [time(year|12|25,Gregorian_month),
                   time(year|12|27,Gregorian_month)]
```

specifies Christmas for the year containing the time point t. ∎

date(t,Gregorian_month) computes a date representation for the time point t in the date format Gregorian_month (year/month/day/hour/minute/second). Only the year is needed. dLet year = ... therefore binds only the year to the integer variable year. If, for example, in addition the month is needed one can write dLet year|month = date(....

time(year|12|25,Gregorian_month) computes $t_1 =$ begin of the 25th of December of this year. time(year|12|27,Gregorian_month) computes $t_2 =$ begin of the 27th of December of this year. The expression [..., ...] denotes the interval between t_1 and t_2. The result is therefore the interval from the beginning of the 25th of December of this year until the end of the 26th of December of this year.

Example 3 (Point–Interval Before Relation). The function

```
PIRBefore(Time t, Interval I) =
    if (isEmpty(I) or isInfinite(I,left)) then false
    else (t < point(I,left,support))
```

specifies the standard crisp point–interval 'before' relation in a way which works also for fuzzy intervals. ∎

If the interval I is empty or infinite at the left side then PIRBefore(t,I) is false, otherwise t must be smaller than the left boundary of the support of I.

Now we define a parameterized fuzzy version of the interval–interval before relation.

Example 4 (Fuzzy Interval–Interval Before Relation). A fuzzy version of an interval–interval before relation could be

```
IIRFuzzyBefore(Interval I, Interval J, Interval->Interval B) =
case
  isEmpty(I) or isEmpty(J) or isInfinite(I,right) or isInfinite(J,left):0,
  (point(I,right,support) <= point(J,left,support))               :1,
  isInfinite(I,left) : integrateAsymmetric(intersection(I,J),B(J))
  else integrateAsymmetric(I,B(J))
```
∎

The input are the two intervals I and J and a function B which maps intervals to intervals. B is used to compute for the interval J an interval B(J), which represents the degree of 'beforeness' for the points before J.

The function first checks some trivial cases where I cannot be before J (first clause in the case statement), or where I definitely is before J (second clause in the case statement). If I is infinite at the left side then $\int (I \cap J)(x) \cdot B(J)(x) dx / |I \cap J|$ is computed to get a degree of 'beforeness', at least for the part where I and J intersect. If I is finite then $\int I(x) \cdot B(J)(x) dx / |I|$ is computed. This averages the degree of a point-interval 'beforeness', which is given by the product $I(x) \cdot B(J)(x)$, over the interval I.

The next example illustrates some procedural features of GeTS. The effect function takes two intervals and a function F, which maps the two intervals to a fuzzy value. F could for example be the relation IIRFuzzyBefore. The first interval I is now shifted step times by the given distance, and each time F(I,J) is computed. These values are inserted into a new interval, which is the result of the function. The 'effect' function turned out a useful test and debug tool for developing the fuzzy interval–interval relations [19,24].

Example 5 (effect).

```
effect(Interval I, Interval J, (Interval*Interval)->Float F,
       Time distance, Integer steps) =
     Let K = [] in
            while (steps >= 0) {
                   pushBack(K,point(I,right,kernel),F(I,J)),
                   I := shift(I,distance),
                   steps := steps - 1}
          K
```
■

'Let K = []' creates a new empty interval and binds it to the variable K. The while loop shifts the interval I steps times by the given distance (I := shift(I,distance)). Each time pushBack(K,point(I,right,kernel), F(I,J)) adds the pair (x, y) consisting of $x =$ right boundary of the kernel of the shifted I and $y = $ F(I,J) to the interval K.

The dashed line in the figure below shows the result of the effect function when applied to the two intervals I and J, and a suitable interval–interval 'before' relation as parameter F. The dotted figure shows the position of the shifted interval I when the F(I,J) drops down to 0.

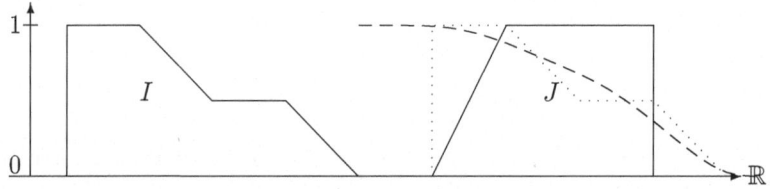

Effect of the effect *function*

3.1   Types in the GeTS Language

The GeTS language has a fixed number of basic types. They represent certain data structures and certain keywords. So far there is no mechanism for extending the basic types. The basic types can be combined to functional types $T_1 * \ldots * T_n \mapsto T$.

Definition 1 (Data Structure Types)

Integer	*standard integers*	Partitioning	*partitionings*
Time	*very long integers*	Label	*labels for partitions*
Float	*standard floating point numbers*	Duration	*durations*
String	*strings*	DateFormat	*date formats*
Interval	*fuzzy intervals*		

■

The data structure types abstract away from the concrete implementation. The Integer type, for example, is currently realized as 32 bit signed integer data, while the Time type is currently realized as 64 bit signed integer data.

Intervals are realized as *polygons* with integer coordinates. An interval is therefore a sequence of pairs $I = (x_0, y_0), \ldots, (x_n, y_n)$. The x_i are Time points and the y_i are fuzzy values. Internally the y_i are realized as short integers between 0 and 1000. From the GeTS point of view, however, the y_i are Float numbers between 0 and 1. The interval I is *negative infinite* if $y_0 \neq 0$. I is *positive infinite* if $y_n \neq 0$. The internal representation of Interval data, however, is completely invisible to the GeTS user. Details about the internal representation and the algorithms can be found in [20].

Partitionings are complex data structures. Fortunately, this is also not visible to the GeTS user. Partitionings are just parameters to some of the functions. They can be used without knowing anything about the internal details.

Durations are sequences of pairs $d_0\ P_0, \ldots, d_n\ P_n$ where the d_i are Float data and the P_i are *Partitionings*.

DateFormats are sequences $P_0/ \ldots /P_n$ of Partitionings.

A number of enumeration types is predefined in GeTS. They are used to control some of the algorithms. Their meaning therefore depends on the meaning of the built-in function where they occur as parameters.

3.2   Language Constructs for GeTS

The GeTS language has a number of general purpose functional and imperative language components. Additionally a number of language constructs are geared to manipulating time points, temporal intervals, partitionings, dates etc. As already mentioned, the language is strongly typed. This means, the type of each expression is determined by the top level function name together with the types of its arguments.

The language has an operational semantics. It is described in detail in [21] where all language constructs are introduced. Some aspects of the language depend on the context where it is used. For example, GeTS itself has no exception mechanisms. Nevertheless, exceptions are thrown and must be caught by the host programming system.

Definition 2 (Function Definitions). *A GeTS function definition has one of the forms*

$$(1) \qquad\qquad\qquad\qquad name = expression$$
$$(2) \qquad\qquad\qquad\qquad name() = expression$$
$$(3) \qquad name(type_1\ var_1, \ldots, type_n\ var_n) = expression$$
$$(4)\ type : name(type_1\ var_1, \ldots, type_n\ var_n) = expression$$
$$(5)\ type : name(type_1\ var_1, \ldots, type_n\ var_n)$$

∎

Version (1) and (2) are for constant expressions, i.e. the name at the left hand side is essentially an abbreviation for the expression at the right hand side. Version (3) is the standard function definition. The type of the function is $type_1*\ldots*type_n \mapsto T$ where T is the type of the *expression*. Version (4) declares the range type of the function explicitly. It can be used for recursive function definitions, where the name of the newly defined function occurs already in the body. In this case it is necessary to know the range type of the function, before the *expression* can be fully parsed. Finally, version (5) is a forward declaration. It must be used for mutually recursive functions.

Function definitions can be overloaded. They are distinguished by their argument types, not by the result type.

Standard Language Constructs. GeTS supports the same kind of arithmetic and Boolean expressions as many other programming languages. A small difference is the Time type, which is integrated in the arithmetics of GeTS. The obligatory 'if-then-else' construct is of course also available. In addition there is a case construct to avoid the need for nested if-then-elses. A 'while' loop is also available. Since GeTS is a functional language, the while construct needs a return value. Therefore in addition to the while loop body, it has a separate return expression. In the body, however, only imperative constructs (with return type Void) are allowed. The values of local variables can be changed with an assignment construct. The assignment operation returns no value. It can only occur in the body of the while statement.

A *function call* in GeTS is an expression $name(argument_1, \ldots, argument_n)$ where '*name*' is either the name of a built-in function, or the name of a previously defined function (or a function with forward declaration), or a variable with suitable functional type.

Since variables can have functional types, and GeTS allows overloading of function definitions, it needs a notation for functional arguments. A functional argument can either be just a variable with appropriate functional type, or a function name with argument type specifications, or a lambda expression. A function name with argument type specifications is necessary to choose among different overloaded functions.

A *functional argument* in GeTS is either

1. a variable with the appropriate functional type,
2. an expression $name[type_1 * \ldots * type_n]$ of a previously defined function with that name and with argument types $type_1 * \ldots * type_n$, (for distinguishing between overloaded functions), or
3. a lambda expression:
   $lambda(type_1\ variable_1, \ldots, type_n\ variable_n)\ expression$. '*expression*' can contain variables which are lexically bound outside the parameter list of *lambda*.

3.3   Built-ins for Time Intervals

Fuzzy time intervals (type Interval) are one of the built-in data structures in GeTS. It is possible to create new empty intervals and fill them up with coordinate points. The expression $[t_1, t_2]$ of type Time $*$ Time $\mapsto$ Interval, for example, constructs a new crisp interval with boundaries t_1 and t_2. pushBack($I, time, value$) of type Interval $*$ Time $*$ Float $\mapsto$ Void extends a fuzzy interval with a new point for the envelope polygon.

For crisp intervals there are the standard set operators: complement, intersection, union etc. These are uniquely defined. There is no choice. Unfortunately, or fortunately, because it gives you more flexibility, there are no such uniquely defined set operators for fuzzy intervals. Set operators are essentially transformations of the membership functions, and there are lots of different ones.

GeTS offers standard versions of the set operators, parameterized set operators of the Hamacher family, and finally set operators with transformation functions for the membership function as parameter. These allow one to customize the set operators in an arbitrary way.

Predicates like 'isCrisp', 'isEmpty', 'isConvex', 'isMonotone', 'isInfinite' can be used to check the corresponding properties of intervals. Basic relations between time points and intervals, or between key parts of two intervals can be checked with predicates like 'during' or 'subset' etc. The function member($time, I$) returns the fuzzy membership value for an interval. If an interval is non-convex, there a number of functions to count components, measure their size, extract them or map over them. n, m-*center points* are used to express temporal notions like 'the first half of the year', or 'the second quarter of the weekend'. This can be computed with the function centerPoint(I, n, m).

Intervals can be transformed in various ways: shifted, scaled, extended or shrinked, integrated or fuzzified with linear or Gaussian shapes. Parts can be cut out, it can be split into different parts. Different types of hulls can be calculated. Membership functions can be multiplied or exponentiated.

Example 6 (Birthday Party Time). Consider a database about, say, the institute's birthday parties. It may contain the entry that the birthday party for the director took place 'from around noon until early evening' of 20/7/2003. 'Around noon' is a fuzzy notion and 'early evening' is a fuzzy notion. Suppose, we have a formalization of 'around noon' and 'early evening' as the following fuzzy sets:

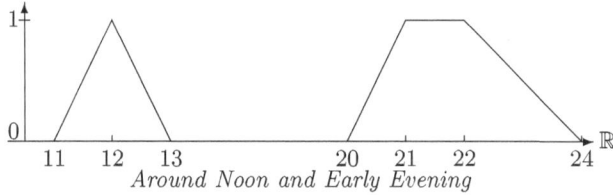

Around Noon and Early Evening

What is now the duration of the birthday party? It must obviously also be a fuzzy set. The fuzzy value of the birthday party duration at a time point t is 1 if the party definitely started before t and definitely ended after t. Therefore the fuzzy value at point t is computed by integrating over the membership functions of the start intervals and the end intervals. A natural definition would therefore be:

$$partyTime(\texttt{Interval I, Interval J})$$
$$= intersection(integrate(\texttt{I,positive}),integrate(\texttt{J,negative})) \qquad (1)$$

The resulting fuzzy set is:

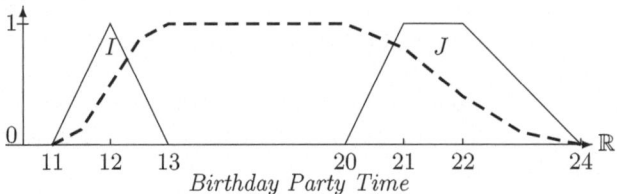

The dashed curve may, for example, represent the percentage of people at the party at a give time.   ∎

The next example illustrate a potential use of the fuzzify function. We want to realize a function beforeChristmas. It should accept a time point t and compute a fuzzy interval, whose membership function increases for a certain time period and then stays 1.0 until Christmas. The increase is determined by two parameters, offset and increase. offset = 50 means that the increase should start in the middle between t and Christmas. increase = 10 means that the duration of the actual linear increase should be 10% of the interval length.

If $t = 2004/7/1$ then beforeChristmas(t,50,10) yields an interval whose membership function rises from 2004/9/28 until 2004/10/6/19/12 (which is at 10% of the distance between 2004/9/28 and Christmas) and then stays at 1.0 until 2004/12/25.

Example 7 (Before Christmas).

```
1 beforeChristmas(Time t, Float offset, Float increase) =
2   dLet year = date(t,Gregorian_month) in
3     Let christmas = time(year|12|25,Gregorian_month) in
4       case (t < christmas) :
5         Let days = round(length(t,christmas,day,false),down) in
6             fuzzify([time(year|12|25-days+round((days*offset/100)),
7                           Gregorian_month),christmas],
8                     linear,left,increase,0),
9         (t < time(year|12|27,Gregorian_month)): []
10        else
11            Let christmas1 = time(year+1|12|25,Gregorian_month) in ...
```
    ∎

The beforeChristmas function considers the three cases, namely (1) that the time point t in the year y is before Christmas in this year, (2) that t is just on Christmas in this year, and (3) that t is after Christmas in this year. In case (1) the rounded number of days between t and Christmas is computed first (line 5). This number minus the offset is subtracted from christmas to get the left boundary of the interval to be fuzzified (line 6). The right boundary is christmas. The left part of the interval is fuzzified linearly with the given increase (line 6–8). If the time point t is just on Christmas (line 9) then the empty interval is returned. If t is after Christmas (case 3), then next year's Christmas is considered (line 11 and later).

Notions like 'in two weeks time' or 'three years from now' etc. denote time shifts. Time shifts are basic operations for many other temporal notions. Therefore GeTS provides a `shift` function which can shift single time points as well as whole intervals by a given duration. A time point or an interval can be shifted by a fraction of a partitioning (1.5 years, for example). Two different shift functions are available, a length oriented shift and a date oriented shift, which give slightly different results for fractional shifts.

Date and Time. In examples 2 and 7 we have already seen applications of functions which convert time points to dates and dates to time points. The dates are sequences of integers which correspond to date formats, and these are sequences of partitionings. An example for a date format is year/month/day/hour/minute/second in the Gregorian calendar. The sequence $2004|12|3|21|43|0$ in this date format is therefore the 3rd of December 2004, 9:43 pm.

The `time` function converts a date in a given date format to the corresponding time point. Examples are:

`time(2004,Gregorian_month) = 1072915231` (1st of January 2004)
`time(2004|1+1,Gregorian_month) = 1075593631` (1st of February 2004)
`time(2004|2|2,Gregorian_week) = 1073347231` (6th of January 2004)
`Gregorian_week` is the date format year/week/day/hour/minute/second. Therefore $2004|2|2$ is the second day in the second week in the year 2004 (The first week in 2004 started at Monday, 29th of December 2003).

The `dLet` construct is a kind of inverse to the `time` function. The expression `dLet` $year|month|...$ = `date`($time, dateFormat$) in $expression$ binds the variables $year, month, ...$ to the integers which correspond to the date denoted by '$time$', in the given date format. '$expression$' is then evaluated under this binding. **Example:** '`dLet` $y|m|d|h$ = `date`(0, `Gregorian_month`) in $y + m + d$' yields 1973 because the time point 0 corresponds to the first of January 1970. Therefore $y = 1970$, $m = 1$, $d = 1$ and $h = 0$.

Partitionings and Labels. GeTS has a number of functions for reckoning with time points, partitions and labels. The function `partition`($time, partitioning$) maps time points to intervals, which represent partitions.

The version `partition`($time, partitioning, n, m$) computes an interval $[t_1, t_2[$ as follows: If i is the coordinate of the partition containing $time$ then t_1 is the start of partition $i + n$ and t_2 is the end of the partition $i + m$.

If instead of the partition as interval, only the boundaries are needed, one can use the `partitionBoundary` function.

The next function is `which`($time, P, Q, inclusion, asGranule$). It can, for example, be used to compute which week in the year is now, or which day in the semester is now.

The further set of functions deals with labels of partitions. Labels are not just strings, but also special data structures. `label`($time, partitioning$) returns the label of the partition containing $time$. If there is no labeling defined, it returns a NULL label.

The function isLabel(*label*) checks whether the *label* is not the NULL label. isGap(*label*) checks whether the label is the gap label. LabelName(*name*) turns a string (without quotes) into a Label.

The function extractLabelled(*I, label, partitioning, inclusion, intersect*) can be used to extract from an interval all partitions with a given label, for example all Tuesdays of a labeled day partitioning. The extractLabelled function maps through all partitions of the given partitioning which are labeled with the given label, and which overlap with the interval $[a, b[$ where a is the left boundary of the interval and b is the right boundary of the interval. An error is thrown if a or b are the infinity. For each such partition p a condition is tested which depends on the parameter *inclusion*: *inclusion* = subset: p must be a subset of I's support. *inclusion* = overlaps: p must overlap with I's support. *inclusion* = bigger_part_inside: the bigger part of p must be a subset of I's support.

If the parameter *intersect* = false then all partitions p which meet the condition are joined into the resulting (crisp) interval. If the parameter *intersect* = true then the intersection of I with all partitions p which meet the condition are joined into the resulting interval. The result may now be a fuzzy interval.

The function nextGranule(*time, partitioning, label, n, withGaps*) is for constructing intervals which represent granules. The interval is constructed as follows: If *time* is inside a granule with the given *label* and if $n = 0$ then this granule is computed. Otherwise the nth next/previous (if $n < 0$) granule with this label is computed. If *time* lies outside a granule with the given *label* and $n = 0$ then the empty interval is returned. Otherwise the nth next/previous (if $n < 0$) granule with this label is computed.

4   Summary and Related Work

Most of the approaches for modeling every-day temporal notions are 'monolithic', i.e. there is one single formalism for specifying calendar systems. In particular there is all the work about the mathematical representation of periodic temporal notions as *time granularities*, or similar kind of mathematical objects. A good overview is given in the book of Bettini, Jajoda and Wang [5]. This work is particularly motivated by the need to represent time in temporal databases. A selection of papers about the abundant work in this area is [3,15,12,25,13,14,9,4,10,6,11,26]. In contrast to these approaches, the CTTN–system has various representation formalisms for the different aspects of temporal notions. One of the components is the GeTS language as a special purpose functional specification and programming language for temporal notions. It has a basic set of general purpose functional and imperative programming language features. In addition there are a number of built-in data structures and functions which are specific for this application. The most important ones are time points, fuzzy temporal intervals and labeled partitionings of the time line.

GeTS is not a stand alone programming language. It must be part of a host system which provides these data structures and which invokes the GeTS application programming interface.

The GeTS constructs were carefully chosen as a compromise between simplicity and easy usage. In a first application, various versions of fuzzy binary relations between fuzzy intervals have been defined [24]. Example 4 (fuzzy before) is one of them.

Acknowledgments. This research has been funded by the European Commission and by the Swiss Federal Office for Education and Science within the 6th Framework Programme project REWERSE number 506779 (cf. http://rewerse.net).

References

1. Julius Benkert. Integration of the CTTN system in Java. Master's thesis, Inst. for Computer Science, LMU Munich, 2006.
2. T. Berners-Lee, M. Fischetti, and M. Dertouzos. *Weaving the Web: The Original Design and Ultimate Destiny of the World Wide Web.* Harper, San Francisco, September 1999. ISBN: 0062515861.
3. C. Bettini and R.D.Sibi. Symbolic representation of user-defined time granularities. *Annals of Mathematics and Artificial Intelligence*, 30:53–92, 2000. Kluwer Academic Publishers.
4. Claudio Bettini, Curtis E. Dyreson, William S. Evans, Richard T. Snodgrass, and X. Sean Wang. *Temporal Databases, Rreseach and Practice*, volume 1399 of *LNCS*, chapter A Glossary of Time Granularity Concepts, pages 406–413. Springer Verlag, 1998.
5. Claudio Bettini, Sushil Jajodia, and Sean X. Wang. *Time Granularities in Databases, Data Mining and Temporal Reasoning.* Springer Verlag, 2000.
6. Claudio Bettini, Sergio Mascetti, and X. Sean Wang. Mapping calendar expressions into periodical granularities. In C. Combi and G. Ligozat, editors, *Proc. of the 11th International Symposium on Temporal Representation and Reasoning*, pages 87–95, Los Alamitos, California, 2004. IEEE.
7. Nachum Dershowitz and Edward M. Reingold. *Calendrical Calculations.* Cambridge University Press, 1997.
8. Didier Dubois and Henri Prade, editors. *Fundamentals of Fuzzy Sets.* Kluwer Academic Publisher, 2000.
9. Curtis E. Dyreson, Wikkima S. Evans, Hing Lin, and Richard T. Snodgrass. Efficiently supporting temporal granularities. *IEEE Transactions on Knowledge and Data Engineering*, 12(4):568–587, 2000.
10. Lavinia Egidi and Paolo Terenziani. A lattice of classes of user-defined symbolic periodicities. In C. Combi and G. Ligozat, editors, *Proc. of the 11th International Symposium on Temporal Representation and Reasoning*, pages 13–20, Los Alamitos, California, 2004. IEEE.
11. I.A. Goralwalla, Y. Leontiev, M.T. Ozsu, D. Szafron, and C. Combi. Temporal granularity: Completing the picture. *Journal of Intelligent Information Systems*, 16(1):41–63, 2001.
12. Nick Kline, Jie Li, and Richard Snodgrass. Specifying multiple calendars, calendric systems and field tables and functions in timeadt. Technical Report TR-41, Time Center Report, May 1999.

13. B. Leban, D. Mcdonald, and D.Foster. A representation for collections of temporal intervals. In *Proc. of the American National Conference on Artificial Intelligence (AAAI)*, pages 367–371. Morgan Kaufmann, Los Altos, CA, 1986.

14. M. Niezette and J. Stevenne. An efficient symbolic representation of periodic time. In *Proc. of the first International Conference on Information and Knowledge Management*, volume 752 of *Lecture Notes in Computer Science*, pages 161–169. Springer Verlag, 1993.

15. Peng Ning, X. Sean Wang, and Sushil Jajodia. An algebraic representation of calendars. *Annals of Mathematics and Artificial Intelligenc*, 36(1-2):5–38, September 2002. Kluwer Academic Publishers.

16. Hans Jüergen Ohlbach. Computational treatement of temporal notions – the CTTN system. In François Fages, editor, *Proceedings of PPSWR 2005*, Lecture Notes in Computer Science, pages 137–150, 2005. see also URL: http://www.pms.ifi.lmu.de/publikationen/#PMS-FB-2005-30.

17. Hans Jürgen Ohlbach. About real time, calendar systems and temporal notions. In H. Barringer and D. Gabbay, editors, *Advances in Temporal Logic*, pages 319–338. Kluwer Academic Publishers, 2000.

18. Hans Jürgen Ohlbach. Calendar logic. In I. Hodkinson D.M. Gabbay and M. Reynolds, editors, *Temporal Logic: Mathematical Foundations and Computational Aspec ts*, pages 489–586. Oxford University Press, 2000.

19. Hans Jürgen Ohlbach. Relations between fuzzy time intervals. In *Proceedings of 11th International Symposium on Temporal Representation and Reasoning, Tatihoui, Normandie, France (1st–3rd July 2004)*, pages 44–51. IEEE Computer Society, 2004. See also http://www.pms.ifi.lmu.de/publikationen/#PMS-FB-2004-33.

20. Hans Jürgen Ohlbach. Fuzzy time intervals – the FuTI-library. Research Report PMS-FB-2005-26, Inst. für Informatik, LFE PMS, University of Munich, June 2005. URL: http://www.pms.ifi.lmu.de/publikationen/#PMS-FB-2005-26.

21. Hans Jürgen Ohlbach. GeTS – a specification language for geo-temporal notions. Research Report PMS-FB-2005-29, Inst. für Informatik, LFE PMS, University of Munich, June 2005. URL: http://www.pms.ifi.lmu.de/publikationen/#PMS-FB-2005-29.

22. Hans Jürgen Ohlbach. Modelling periodic temporal notions by labelled partitionings – the PartLib library. In S. Artemov, H. Barringer, A. d'Avila Garces, L. C. Lamb, and J. Woods, editors, *Essays in Honour of Dov Gabbay*, volume 2, pages 453–498. College Publications, King's College, London, 2005. ISBN 1-904987-12-5. See also http://www.pms.ifi.lmu.de/publikationen/#PMS-FB-2005-28.

23. Hans Jürgen Ohlbach. Periodic temporal notions as 'tree partitionings', March 2006. Submitted to PPSWR06.

24. Hans Jürgen Ohlbach. Relations between fuzzy time intervals. Research Report PMS-FB-2006-26, Inst. für Informatik, LFE PMS, University of Munich, June 2006. URL: http://www.pms.ifi.lmu.de/publikationen/#PMS-FB-2006-26.

25. Michael D. Soo and Richard T. Snodgrass. Mixed calendar query language support for temporal constants. Technical Report TR 92-07, Dept. of Computer Science, Univ. of Arizona, February 1992.

26. Stephanie Spranger. *Calendars as Types – Data Modeling, Constraint Reasoning, and Type Checking with Calendars*. Dissertation/Ph.D. thesis, Institute of Computer Science, LMU, Munich, Munich, 2005. PhD Thesis, Institute for Informatics, University of Munich, 2005.

27. L. A. Zadeh. Fuzzy sets. *Information & Control*, 8:338–353, 1965.

Active Monte Carlo Recognition

Felix v. Hundelshausen[1] and Manuela Veloso[2]

[1] Computer Science Department, Freie Universität Berlin, 14195 Berlin, Germany
hundelsh@googlemail.com
[2] Computer Science Department, Carnegie Mellon University
Pittsburgh, PA 15213, USA
veloso@cs.cmu.edu

Abstract. In this paper we introduce *Active Monte Carlo Recognition (AMCR)*, a new approach for object recognition. The method is based on seeding and propagating "relational" particles that represent hypothetical relations between low-level perception and high-level object knowledge. AMCR acts as a filter with each individual step verifying fragments of different objects, and with the sequence of resulting steps producing the overall recognition. In addition to the object label, AMCR also yields the point correspondences between the input object and the stored object. AMCR does not assume a given segmentation of the input object. It effectively handles object transformations in scale, translation, rotation, affine and non-affine distortion. We describe the general AMCR in detail, introduce a particular implementation, and present illustrative empirical results.

1 Introduction

As Computer Vision researchers, we are interested in processing and extracting information from a sequence of images. Since image sequences are subject to continuity constraints, *iterative methods* should be applied, rather than treating the images as being independent of one another.

In the field of state estimation by vision, e.g. recursively estimating the position and motion of a vehicle on a highway [4][3], the concept of iterative processing has been fully adopted. Until now, it has not been adopted for the task of object recognition. Almost all existing approaches for object recognition treat the images as being independent from one another, processing each in a pipeline of steps, not considering the results of earlier processing.

In this paper, we propose a new, iterative approach for object recognition. The framework deals with a sequence of input images of an object, and iteratively finds the best match in a given set of prototype objects. Besides recognizing the object, this approach finds the mapping that transforms the input object to the object stored in memory. It can handle differences in scale, translation, rotation, affine and even non-affine distortions.

The most important aspect of our approach is that recognition does not take place in a one-way pipeline: It is recursive and exploits feedback information.

C. Freksa, M. Kohlhase, and K. Schill (Eds.): KI 2006, LNAI 4314, pp. 229–243, 2007.

While the recognition runs, it can guide low-level operations according to the current recognition state. For instance, when the system cannot decide whether the input object is a camel or a dromedary, our method allows to focus attention on the back of the animal, thus finding out whether there are one or two humps.

2   Related Work

Our method is different and contrasts with two common approaches for object recognition: *Independent Feature Extraction* and *Directed Processing*. The first means running low-level image operations in each image of a video stream from scratch. In [6], for instance, building the scale space has to be done for each image anew. On a Pentium 4 3GHz processor just building the scale space for a 640×480 image takes approximately one second[1]. In object recognition from video rather than from photographs, this processing time is too long. Other examples for *Independent Feature Extraction* include Shape Contexts [2], Maximally Stable Extremal Regions [8], and affine interest point detectors [9].

Directed Processing means that current approaches treat object recognition as being solvable in a one-way sequence of processing steps, hopefully ending with the recognition of the object (for example, [6] builds the scale space, extracts keypoints, builds feature descriptors, matches these descriptors with descriptors in a database).

In contrast to *Independent Feature Extraction*, our approach allows to speedup processing by making use of earlier results. When a feature was detected in frame I_k the same feature is likely to occur at a close position in I_{k+1}.

In contrast to *Directed Processing*, our approach is iterative and does allow feedback loops. The current state of recognition can guide attention to parts and features that help to discriminate between objects.

3   Active Monte Carlo Recognition (AMCR)

In this section we introduce our new approach. We call it *Active Monte Carlo Recognition (AMCR)*, because it is based on sequential Monte Carlo filtering [11]. The word *active* stresses the fact that the approach allows the integration of information feedback. It can guide low-level feature extraction based on the current recognition state, as well as focus attention on important image locations. In this sense, the approach integrates a *visual behavior*, that is, a policy for guiding attention and feature extraction to parts of the object which allow its recognition.

Sequential Monte Carlo methods, or particle filters, have extensively been applied for robust object tracking. The best-known approach is the Condensation algorithm. For example, in [5] it is shown that a leaf can robustly be tracked in the presence of background clutter. Also, particle filtering has become the standard approach for mobile-robot localization [10]. Here, a probability distribution

---

[1] Using the C++ implementation of [6].

for the robot's position is approximated and propagated by a set of particles, each representing a hypothesis for the robot's position. Starting with a uniform distribution for example, the particles start to build clusters at highly likely points in a map, while the robot moves and observes its environment [10].

3.1 Analogy Between Object Recognition and Mobile Robot Localization

One inspiration for our approach came from realizing that object recognition and mobile robot-localization are essentially the same problem: When a robot finds its position in one of a set of maps (e.g. of different buildings), one could say that it recognized that its current environment matches one of the maps. When considering the environment as an object, finding the correct map actually means object recognition. Besides identifying the correct map, the correct position within this map is found, too. Hence, the correspondences between environment and map are also found. In the following section we will adapt and modify the Monte Carlo Localization approach to perform object recognition. Here, the main additional complication is, that we want to be not only invariant with respect to translation and rotation, but also to scale and affine distortion.

3.2 An Example

The initial task is a follows. Given are:

- $\mathcal{I} = I_1, ..., I_l$: a sequence of l input images
- $P = \{M_1, M_2, ..., M_r\}$: a set of r prototype images

The goal is to find the image M_k that corresponds to the image sequence. We assume that $\mathcal{I}$ shows an object of one of the prototype images M_i but that it can be arbitrarily scaled, translated, rotated or even sheared, in short, we want to allow any affine transformation. Furthermore, we do not assume that the object in I_k is already segmented from the background, i.e. we allow background clutter. However, we do assume that the prototype images are perfectly segmented. Besides identifying M_k we also want to find the affine transformation for a good match. Figure 1 shows an example of this setup.

3.3 Overview

We deal with two types of particles, *V-particles* and *M-particles*. The V-particles refer to positions in the input image, while the M-particles refer to positions in the prototype images. One V-particle is linked to several M-particles as shown in figure 2. One important property of the algorithm is that the particles move. While the V-particles move in the input image, the M-particles form moving clusters in the prototype images, at positions that correspond to the shape surrounding the V-particles.

To get an initial idea of how the algorithm works, Figure 3 illustrates the algorithm using only one V-particle. Initially, the M-particles are randomly distributed in the prototype images. While the V-particle moves, the M-particles

Fig. 1. The goal is to identify the bird shown in a sequence of input images I_k in the set of prototype images M_1, M_2, M_3 and to determine the affine transformation that maps the input object to its corresponding prototype object. Background clutter in the input image can be present. The shape shown in the image sequence can move. We assume that the movement is restricted to affine transformations.

move accordingly, subject to their estimate of the affine transformation. Step by step the M-Particles start to build clusters at probable locations and finally only one cluster remains in the prototype that corresponds to the input image. The resulting particles hold the correct affine transformation.

3.4   Definitions

To give a mathematically sound description of our algorithm we have to define several terms:

Definition 1. *A **V-particle** $\mathbf{v}$ is defined as $\mathbf{v} := (\mathbf{p}, \mathcal{E}_a, \mathcal{E}_f, \mathcal{O}, \mathcal{Q}, \mathcal{F})$ where $\mathbf{p} = (x, y, \phi)^T$ describes a position and orientation in the input image I, $\mathcal{E}_a$ is an affine estimator, $\mathcal{E}_f$ is a feature extractor, $\mathcal{O}$ is an observation model, $\mathcal{Q}$ is a motion model and $\mathcal{F}$ is a feedback strategy. (The latter five terms will be described with an example in section 4.)*

Definition 2. *A **prototype pose** $\tilde{\mathbf{x}} = (x, y, \phi, i)$, $x, y \in \mathbb{R}$, $\phi \in [0, ..., 2\pi]$, $i \in \{1, ..., r\}$, specifies a position (x, y) with orientation ϕ in the i_{th} of all r prototype images.*

Definition 3. *An **M-particle** $\mathbf{m}$ is defined as $\mathbf{m} := (\tilde{\mathbf{x}}, \mathbf{A}, \pi)$ where $\tilde{\mathbf{x}}$ specifies a pose in one of the prototype images. The 3×3 matrix $\mathbf{A}$ defines a 2D affine mapping in homogeneous coordinates. The value $\pi \in [0, ..1]$ is a probability.*

Definition 4. *A **particle configuration** $\mathcal{C}$ is a triple $\mathcal{C} = (\mathcal{V}, \mathcal{M}, \mathcal{R})$, where $\mathcal{V} := \{\mathbf{v}_i\}, i = 1, ..., m$ is a set of m V-particles, $\mathcal{M} := \{\mathbf{m}_i\}, i = 1, ..., n$ is a set of n M-particles and $R : \mathcal{M} \longrightarrow \mathcal{V}$ is a mapping relating each M-particle to a V-particle. The mapping is surjective but not injective, that is, different M-particles can be mapped to the same V-particle.*

The mapping R induces an equivalence relation in the set $\mathcal{M}$, dividing it in subsets whose elements are mapped to the same V-particle. The subset belonging to $\mathbf{v} \in \mathcal{V}$ is

$$R^{-1}(\mathbf{v}) = \{\mathbf{m} \in \mathcal{M} : R(\mathbf{m}) = \mathbf{v}\} \tag{1}$$

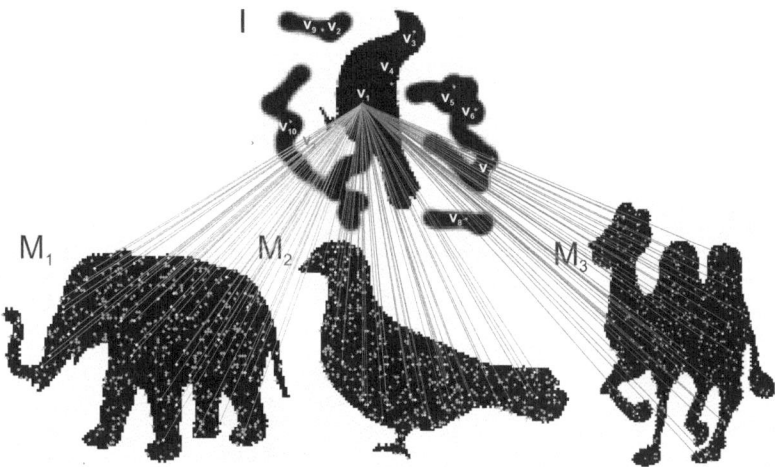

Fig. 2. Each of the $m = 10$ V-particles in the input image I is linked to 50 M-particles in each prototype image M_1, M_2, M_3. All particles are drawn in gray, except the $\mathbf{v_1}$ and the M-particles that are linked to it. The links are drawn by straight thin lines. Note, that only the connections of $\mathbf{v_1}$ are drawn, but that each V-particle has the same number of connections (to different M-particles).

Thus, each V-particle is linked to a whole set of M-particles and the sets of M-particles are disjoint. Relating our approach to mobile robot localization, each V-particle can be thought of a "simulated robot" discovering the input environment (the image I) and the related M-particles represent a probability distribution for the robot's position within a set of maps (the prototype images M_i). An example is shown in figure 2 where each of $m = 10$ V-particles is linked to 50 M-particles in each prototype image resulting in a number of $n = 10 \times 3 \times 50 = 1500$ M-particles. The set $R^{-1}(\mathbf{v})$ contains M-particles that are distributed over all prototype images. Often, we are interested in only the M-particles of a V-particle $\mathbf{v}$ that are in the same prototype image M_k. We will denote this set by

$$R_k^{-1}(\mathbf{v}) := \{\mathbf{m} = (i, \mathbf{p}, \mathbf{A}) \in \mathcal{M} : R(\mathbf{v}) = \mathbf{m} \wedge i = k\} \tag{2}$$

This set $R_k^{-1}(\mathbf{v})$ approximates the probability distribution for the corresponding position $\mathbf{v}$ in the prototype image M_k.

Often, we will consider an M-particle $\mathbf{m}$ in conjunction with its linked V-particle $\mathbf{v_m} := \mathcal{R}(\mathbf{m})$. Such a pair $(\mathbf{v_m}, \mathbf{m})$ represents a hypothetical relation between the input image and a prototype image. Thus we call the entity $(\mathbf{v_m}, \mathbf{m})$ a *relational particle*. The affine mapping $\mathbf{A}$, stored in $\mathbf{m}$, defines how the local neighborhood around $\mathbf{v_m}$ has to be transformed in order to match the local neighborhood around $\mathbf{m}$.

Definition 5. *A relational particle* $\mathbf{r}$ *is a tuple* $(\mathbf{v_m}, \mathbf{m})$, *with* $\mathbf{m} \in \mathbf{M}$ *and* $\mathbf{v_m} := \mathcal{R}(\mathbf{m})$.

Input   Memory (Prototypes)

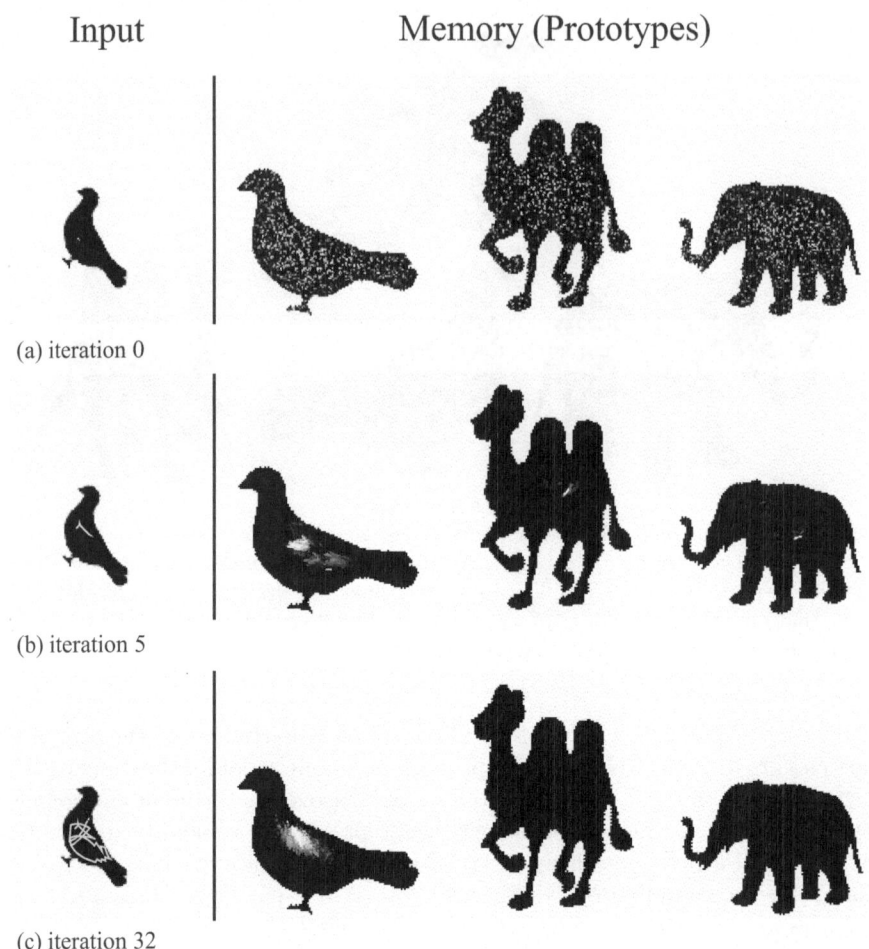

(a) iteration 0

(b) iteration 5

(c) iteration 32

Fig. 3. This figure illustrates the algorithm using only one V-particle connected to 500 M-particles in each prototype image. (a) Initially, the M-particles are distributed randomly. (b)While the V-particle moves, the M-particles start to build clusters at probable locations. (c)Finally, only one cluster remains in the prototype image that corresponds to the input image. The trajectory of the V-particle corresponds to the thin curve in the input images. While the V-particles move, the M-particles move accordingly, subject to their current estimate of how the input and prototype images are related in terms of an affine transformation.

3.5 Probabilistic Formulation

For each V-particle $\mathbf{v}$ we reformulate the problem of object recognition as the problem of localizing $\mathbf{v}$. Here, localizing $\mathbf{v}$ means finding its corresponding prototype pose $\tilde{\mathbf{x}} = (x, y, \phi, i)$ (see definition 2), that is identifying the correct prototype image M_i and finding the corresponding pose (x, y, ϕ) within M_i. Since we do not assume that the object shown in the input images is segmented, the algorithm will try to recognize whatever is at the position of **v**. Since several V-particles exist, different objects can be recognized simultaneously.

With each new image I_k, each V-particle $\mathbf{v} = (\mathbf{p}, \mathcal{E}_a, \mathcal{E}_f, \mathcal{O}, \mathcal{Q}, \mathcal{F})$ will perform a movement and a measurement $\mathbf{z}_k$. The movement is controlled by the feedback policy $\mathcal{F}$ that returns a control command $\mathbf{u}_k$ in each iteration. It consists of a rotation and a translation. The measurement $\mathbf{z}_k$ is a feature descriptor returned as a result of applying feature extractor $\mathcal{E}_f$. We want to estimate the prototype pose $\tilde{\mathbf{x}}$ based on all measurements $Z^k = \{\mathbf{z}_k\}, i = 1, ..., k$, up to the current time, and knowledge about the movements and the initial state of $\tilde{\mathbf{x}}$. The initial knowledge about the state $\tilde{\mathbf{x}}_0$ is given by the apriori probability distribution $p(\tilde{\mathbf{x}}_0)$ over the space of $\tilde{\mathbf{x}}$. For instance, it could be that certain prototype images are known to be more likely, in a given context. However, we will often assume an initial uniform distribution. Thus, for each V-Particle, we are interested in constructing the posterior density $p(\tilde{\mathbf{x}}_\mathbf{k}|Z^k)$ of the current prototype pose $\tilde{\mathbf{x}}_\mathbf{k}$ conditioned on all measurements.

In analogy to mobile robot localization [10], to localize $\tilde{\mathbf{x}}_\mathbf{k}$ the corresponding position of the V-particle in time step k, we need to recursively compute the density $p(\tilde{\mathbf{x}}_\mathbf{k}|Z^k)$ at each time step, which is done in two phases, the prediction phase and the update phase:

- **Prediction Phase.** We use the *motion model* $\mathcal{Q}$ of the V-particle to predict $\tilde{\mathbf{x}}$ in the form of a predicted probability density function (PDF) $p(\tilde{\mathbf{x}}_k|Z^{k-1})$ taking only motion into account. In contrast to real mobile robot localization we know the effect of the control input precisely, without noise. However it still makes sense to include a noise model, since the overall approach will then be able to handle even non-affine distortions (to some limited degree). We assume that the current state $\tilde{\mathbf{x}}_\mathbf{k}$ is only dependent on the previous state $\tilde{\mathbf{x}}_{k-1}$ (the Markov assumption) and the known control input $\mathbf{u}_{k-1}$. The motion model is specified as a conditional density $p(\tilde{\mathbf{x}}_k|\mathbf{x}_{k-1}, \mathbf{u}_{k-1})$ and the predictive density over $\tilde{\mathbf{x}}_k$ is then calculated by integration:

$$p(\tilde{\mathbf{x}}_k|\mathbf{Z}^{k-1}) = \int p(\tilde{\mathbf{x}}_k|\mathbf{x}_{k-1}, \mathbf{u}_{k-1})p(\tilde{\mathbf{x}}_{k-1}|Z^{k-1})d\tilde{\mathbf{x}}_{k-1} \qquad (3)$$

- **Update Phase.** In this phase we take a measurement $\mathbf{z}_k$ by applying feature extractor $\mathcal{F}_e$ around the V-Particle and use its *measurement model* $\mathcal{O}$ to obtain the posterior $p(\tilde{\mathbf{x}}_k|Z^k)$. We assume that the measurement $\mathbf{z}_k$ is conditionally independent from earlier measurements Z^{k-1} given $\tilde{\mathbf{x}}_k$ and the measurement model is given in terms of a likelihood $p(\mathbf{z}_k|\tilde{\mathbf{x}}_k)$. This term expresses the likelihood that the V-particle is at a location corresponding to $\tilde{\mathbf{x}}_k$, given that $\mathbf{z}_k$ was observed. The posterior density over $\tilde{\mathbf{x}}_k$ is obtained using Bayes theorem:

$$p(\tilde{\mathbf{x}}_k|Z^k) = \frac{p(\mathbf{z}_k|\tilde{\mathbf{x}}_k)p(\tilde{\mathbf{x}}_k|Z^{k-1})}{p(\mathbf{z}_k|Z^{k-1})} \qquad (4)$$

3.6 The AMCR-Algorithm

For each V-particle $\mathbf{v} = (\mathbf{p}, \mathcal{E}_a, \mathcal{E}_f, \mathcal{O}, \mathcal{Q}, \mathcal{F})$ the density $p(\tilde{\mathbf{x}}_k | Z_k)$ is approximated by its set $R^{-1}(\mathbf{v})$ of connected M-Particles. The overall algorithm is then summarized by the following procedure:

Algorithm 3.1. AMCR()

$$
\text{for each } I_k
$$

do $\left\{ \begin{array}{l} \textbf{for each } \text{V-particle } \mathbf{v} = (\mathbf{p}, \mathcal{E}_a, \mathcal{E}_f, \mathcal{O}, \mathcal{Q}, \mathcal{F}) \\[4pt] \quad \text{do} \left\{ \begin{array}{l} \text{Get control command } \mathbf{u}_k = \mathcal{F}(\mathbf{v}) \\ \text{Move } \mathbf{v} \text{ according to } \mathbf{u}_k \\ \textbf{for each } \mathbf{m} \in R^{-1}(\mathbf{v}) \\ \quad \text{do} \left\{ \begin{array}{l} \text{PREDICT}(\mathbf{m}, \mathcal{Q}) \\ \text{UPDATE}(\mathbf{v}, \mathbf{m}, \mathcal{E}_a, \mathcal{E}_f, \mathcal{O}) \end{array} \right. \\ \text{RESAMPLE}(R^{-1}(\mathbf{v})) \end{array} \right. \\[4pt] \textbf{every } w_{th} \text{ frame: } \left\{ \begin{array}{l} \text{UPDATE}(V) \\ \text{RESAMPLE}(V) \end{array} \right. \\ \text{Run other algorithms like tracking in parallel} \end{array} \right.$

Here, updating and resampling $\mathcal{V}$ is done only every wth frame ($w = 10$ in our implementation). To update $\mathcal{V}$, each $\mathbf{v} \in \mathcal{V}$ is assigned a weight proportional to the number of M-particles connected to $\mathbf{v}$. When resampling $\mathcal{V}$, particles that are created in several instances receive an exact copy of the M-Particles connected to the original V-particle. By resampling V the focus of attention is directed to parts of the input image that are interpretable in terms of the prototype images. The entire procedure of propagating the particles by sampling, reweighting, and resampling (Sampling/Importance Resampling, or SIR) is discussed in full detail in [12].

One important property of AMCR is that the input images are only accessed at the position of the V-particles. In each iteration algorithms such as determining the optical-flow can be processed in parallel and the V-particles can be moved according to the optical flow. In this way recognition can be distributed over several frames, even though the input object moves.

4 Radial-AMCR: AMCR for Shape Recognition

In this section, we describe a particular application of the AMCR-algorithm, allowing it to perform shape recognition. Here, we assume that all V-particles use the same affine estimator, feature extractor, motion and measurement model. We call this particular instantiation *Radial-AMCR* because it is based on radial edge scans.

4.1 The Affine Estimator and the Measurement Model

For Radial-AMCR the affine estimator $\mathcal{E}_a$ estimates an affine mapping for each M-particle by considering two triangles, one in the input image I_k and one in

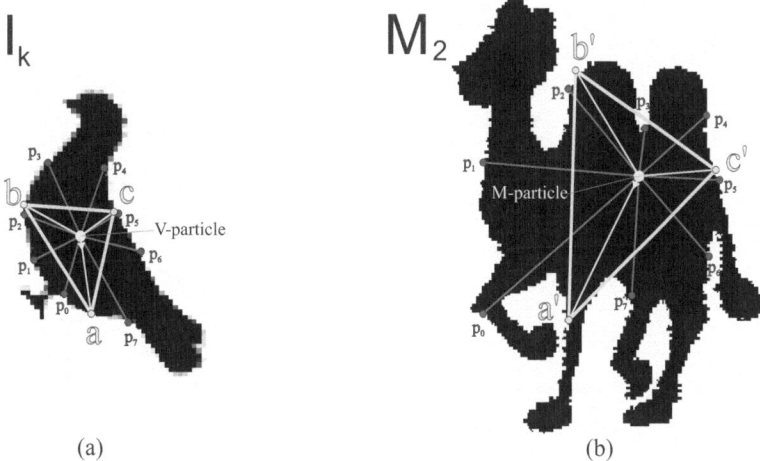

(a)   (b)

Fig. 4. Each pair $(\mathbf{v}, \mathbf{m})$ of particles creates a hypothesis for an affine transformation by constructing two triangles. The corner points of the triangles are found by three edge scans per particle. To compare the local shape around the particles 8 further points are determined by edge scans.

the prototype image of the M-particle. The three points of each triangle are found by applying an edge-detector along three scan lines, radially emerging from the V- and M-particle's position. The orientation of the V- and M-particle specifies the direction of the first scan line. The remaining angles are at 120 and 240 degrees relative to the first direction. In doing so, three points $\mathbf{a}, \mathbf{b}$ and $\mathbf{c}$ are determined for the V-particle $\mathbf{v}$ in I_k and three points $\mathbf{a}', \mathbf{b}'$ and $\mathbf{c}'$ are determined for the M-particle $\mathbf{m}$ in its prototype image. The pair $(\mathbf{v}, \mathbf{m})$, a relational particle, will now represent the hypothesis that the shape in the input image has to be transformed in such a way, that both triangles match. That is, we compute the affine transformation $\mathbf{A}$ that maps $\mathbf{a}$ to $\mathbf{a}', \mathbf{b}$ to $\mathbf{b}'$ and $\mathbf{c}$ to $\mathbf{c}'$. Depending on the position of the V- and M-particles different hypotheses arise. Although, the affine mapping $\mathbf{A}$ is stored in the M-particle, it is an attribute of the whole relational particle $(\mathbf{v}, \mathbf{m})$. It is possible to store it in $\mathbf{m}$, because each M-particle can only be linked to one V-particle.

Based on the estimate of the affine transformation, we are now able to specify a measurement model. Similar to how the triangles are found, the feature extractor $\mathcal{E}_f$ performs radial edge scans along 8 rays. For each relational particle $(\mathbf{v}, \mathbf{m})$, two sets of 8 points are found, $\mathbf{z}_k = (\mathbf{p}_0, ... \mathbf{p}_7)$ in input image I_k (which constitutes the measurement of the V-particle) and $\mathbf{p}'_0, ... \mathbf{p}'_7$ in the prototype image of $\mathbf{m}$. Based on these points and the affine transformation we specify the measurement model $\mathcal{O}$, that is the likelihood $p(\mathbf{z}_k | \tilde{\mathbf{x}})$. The underlying consideration is that if the current M-particle was in the correct prototype image (the one that corresponds to the input images) and if its position and orientation exactly corresponded to the position of the V-particle in the current input image, then, when transforming the points $\mathbf{p}_0, ..., \mathbf{p}_7$ using $\mathbf{A}$, they would exactly match the points $\mathbf{p}'_0, ..., \mathbf{p}'_7$. We then calculate the deviations $\mathbf{w}_j = \mathbf{p}'_j - \mathbf{A}\mathbf{p}_j, j = 0, ..., 7$

and assume that each deviation is either an outlier or subject to a Gaussian distribution,

$$p(||\mathbf{w}_j||) = p_{outlier} + (1 - p_{outlier})\frac{1}{\sigma_w\sqrt{2\pi}}e^{-\frac{||w_j||^2}{2\sigma_w^2}}. \tag{5}$$

Here, $p_{outlier}$ is the probability of an outlier, in case of which we assume a uniform distribution over all ranges of $||\mathbf{w}_j||$ and σ_w is a constant ($p_{outlier} = 0.1$ and $\sigma_w = 10$ pixel in our implementation). Assuming that the individual scan line measurements are independent the measurement model then is

$$\pi = p(\mathbf{z}_k|\tilde{\mathbf{x}}_k^i) := \Pi_{j\in I_{valid}}\, p(||\mathbf{w}_j||)\Pi_{j\in I_{invalid}}\, p_{invalid}. \tag{6}$$

Here, I_{valid} is the subset of indices $\{i = 1..m\}$ for which both v_i and v_i' indicate the validity of the ith edge measurement, and $I_{invalid}$ is its complementary set. The constant $p_{invalid}$ is the modeled probability of an invalid range measurement ($p_{invalid} = 0.1$ in our implementation).

4.2   The Motion Model

The motion model $\mathcal{Q}$ defines how the M-particles move, when a V-particle moves as a response to its control command $\mathbf{u}$. It is specified in terms of the conditional density function $p(\tilde{\mathbf{x}}_k|\tilde{\mathbf{x}}_{k-1}, \mathbf{u}_{k-1})$. Rather than explicitly specifying this function we specify how to sample from it. Given $\mathbf{u}$ specifying the translation and rotation of the V-particle we transform the movement by the M-particle's current estimate of the affine transformation and add zero-mean Gaussian noise to both translation and rotation. The M-particle will then perform this transformed movement. When the prototype image is not an exact affine transformation of the input image, non-affine distortions can be compensated by the noise in the M-particle's movement.

4.3   Feedback Loops

There are two different feedback loops: *Feature Feedback* and *Attention Feedback*. Feature feedback means that different V-particles can have different feature extractors and that the extractors are selected depending on the recognition state. For instance, when recognition is ambiguous at a given iteration, feature extractors could change from edges to texture, if texture would better discriminate among the hypotheses. This aspect is not dealt with in more detail in this paper. Attention feedback has two mechanisms: One automatically occurs during V-resampling. V-particles with many connected M-particles will be reproduced more likely, which lets the V-particles concentrate on interpretable parts of the input shapes. V-particles which cover non-interpretable background clutter, will vanish automatically. The second mechanism involves the motion guidance of the V-particles.

Consider the case where a V-particle determines its movement solely based on the input image. For instance, a V-particle could always move forward and be

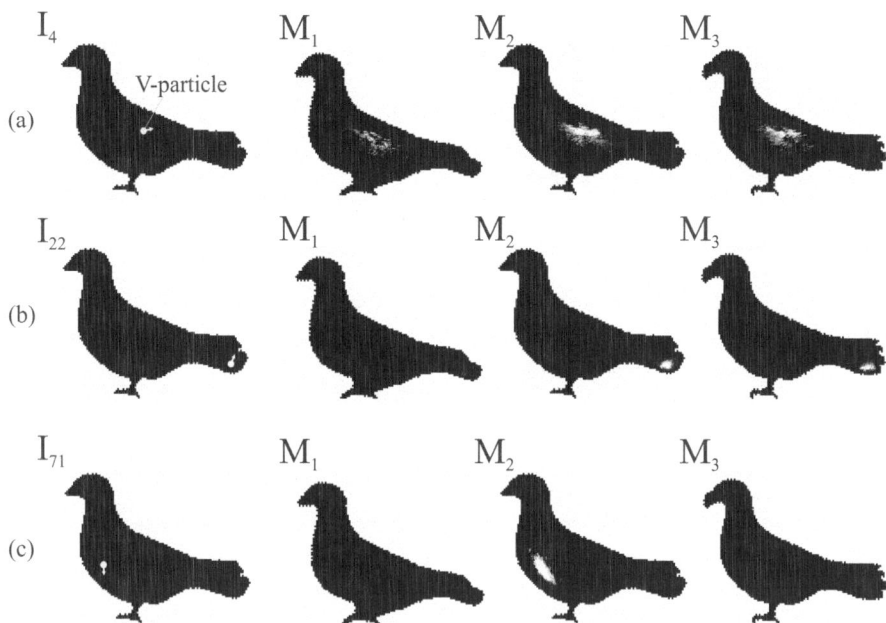

Fig. 5. Even if the prototype images are very similar the method converges. The figures shown here, are snapshots of a real experiment. However, convergence is slow in the case of high similarity. To speed it up, a feedback policy is needed that guides the V-particles to parts of the figures that let discriminate them (e.g. the feed, tail and head).

reflected at an edge in the image. This simple behavior would let the V-particle explore its input shape. Although the approach works, it is not optimal. To see why, consider the case where the prototype images are very similar, i.e. as in the scenario shown in figure 5. Since the birds are very similar, it takes 71 iterations till the method converges to only one remaining cluster in the correct prototype image. The reason is that the M-particles often scan the shapes at parts that are very similar. A strategy of guiding the V- and M-particles to parts that help to discriminate the shapes is required to increase the performance. In the situation shown in figure 5b) the recognition process is not sure, whether the input image is M_2 or M_3 and a feedback control that would move the V-particle to a part that best discriminates M_2 and M_3 (i.e the head of the birds) is desirable in this situation. One difficulty in implementing such an approach is to automatically compare all pairs of prototype images and to find the discriminative locations. This issue will be subject of a separate paper.

4.4   Lookup Tables

In each iteration, each M-particle has to apply its feature extractor in its prototype image. But since, the prototype images (the memory) is fixed, we can pre-compute lookup tables to speed up the extraction process. For instance, for Radial-AMCR each pixel in each prototype images holds a pre-computed radial scan, and the edge points can simply be looked up. Thus, while the V-particles actually have to access the input image, the M-particles will just operate on lookup tables. In this sense, the prototype images M_i are rather feature lookup tables, instead of the actual prototype images.

4.5   The Focus of Attention

In our approach, the positions of the V-particles constitute the focus of attention. Since we do not assume that the input image is segmented, the V-particles will try to recognize whatever is at their position. Consider i.e an image showing different shapes the same time. Starting with 100 V-particles, initially randomly distributed in the image, some of the particles will lay within the figures, some between the figures, and some at image parts that contain background clutter. V-particles that are in figures, that are represented in the prototype images, will produce clusters after several iterations, the other V-particles will remain unsuccessful, that is they will not yield a focussed cluster of M-particles in a prototype image. This implies, that different shapes can be recognized simultaneously. Context based recognition can be achieved through the initial distribution of M-particles in the prototype images. If a prototype has no M-particles no processing power is consumed in trying to recognize it. That is, depending on the situation in which recognition takes place, certain prototypes can be biased.

5   Experimental Results

It is difficult to compare our algorithm against other methods for shape or object recognition, because they typically are not iterative. But the iterativeness is one of the most important properties of AMCR, because it allows to distribute the computational load over several frames. Even though recognition might take a second, the video input can be processed at full frame rate during this second. Thus other algorithms like tracking can run in parallel. This is the important philosophy behind our approach.

Despite this difficulty, we performed a comparison against shape contexts [1] by repeating the iterations over static images. Some of the input images that were correctly classifed are shown in figure 6. These had to be identified within

Fig. 6. Some examples of input images that were correctly classified

a database of 17 prototype shapes. Because of the severe background clutter our method performed clearly better than shape contexts. However, this is not the point, because shape contexts themselves could be integrated in AMCR by defining an appropriate observation model. Thus, our approach is more a framework that allows to incorporate existing approaches, rather than being opposed to them.

With the following example we try to show all properties of AMCR. We work with a prototype memory of only three shapes. We printed two of them on a sheet of paper, together with outlier shapes, such that the figures of interest cannot easily be segmented from the background clutter (see figure 7a). Then we took a video, while moving the sheet around. Running Radial-AMCR with 30 V-particles and 60 M-particle per prototype and V-particle, our method iteratively recognizes the given shapes. Using the lookup tables, one iteration takes approximately 61 milliseconds on a Pentium IV, 3 GHz processor. Splitting the update phases over two frames, the processing time per frame can be reduced to 30 milliseconds, which allows to process the input video at a rate of 30 frames per second. While recognition runs, we simultaneously determine the optical flow (using Lukas Kanade [7]) in the input sequence, and move the V-particles according to the flow, such that the relative position between the input shapes

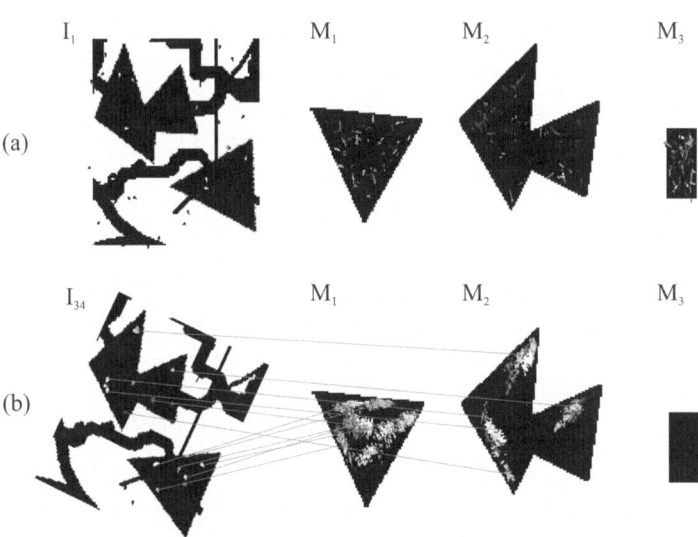

Fig. 7. At the beginning (a) the V and M-particles have random positions in the input image and the prototype images. After 34 iterations only V-particles remain, that are interpretable in terms of the prototype images. The M-particles form clusters at corresponding positions. The thin straight lines show these correspondences. While recognition runs, the input image is rotated and translated randomly and the V-particles are moved according to the optical flow. Thus, while tracking, the process of recognition is distributed over the sequence of images.

and the V-particles remains the same. This example is illustrated in figure 7. It shows how recognition and tracking can be performed simultaneously, how the process of recognition can be distributed over several frames and how several objects can be recognized simultaneously.

6   Conclusions

In this paper we have introduced *Active Monte Carlo Recognition (AMCR)* as a new framework for object recognition and a result of our realization of the similarity between mobile robot localization and object recognition. In relation to existing approaches, AMCR is based on an image sequence rather than on single images, and it includes a feedback loop integrated in the recognition process to guide attention to discriminative parts. At the core of AMCR are *relational particles* that represent hypotheses for relations between an input image and a set of prototype images. Low-level image access is hence performed with a relationship to high-level memory. We have shown the potential of our approach by implementing a particular instantiation of the algorithm for shape recognition. In summary, our approach has several contributions, including:

- **Iterativeness.** The process of recognition is distributed over a sequence of images showing the same object. Tracking can be performed simultaneously with recognition.
- **Local image access.** The input images are only accessed at the position of the V-particles. By moving the V-particles recognition can be combined with tracking. For instance, the V-particles can be moved according to the optical flow such that the relative position between V-particles and the object remains the same.
- **Multi-Modality.** AMCR maintains several hypothetical interpretations during its iterations. Also, several objects can be recognized simultaneously.
- **Simultaneous Segmentation and Recognition.** Our method does not require the object to be segmented from the background. Rather, segmentation and recognition are performed simultaneously.
- **Integration of Feedback Loops.** During iterations, an object might not be uniquely classified. Our approach allows to guide attention to parts and features that help discriminate the different hypotheses.

Future work will concentrate on the simultaneous application of different feature-extractors, a hierarchical organization of the prototype objects and the learning of feedback strategies.

References

1. S. Belongie, J. Malik, and J. Puzicha. Shape matching and object recognition using shape contexts, 2001.
2. S. Belongie, J. Malik, and J. Puzicha. Shape matching and object recognition using shape contexts. *IEEE Trans. Pattern Anal. Mach. Intell.*, 24(4):509–522, 2002.

3. E. D. Dickmanns. The 4d-approach to dynamic machine vision. In *Proceedings of the 33rd IEEE Conference on Decision and Control*, volume 4, pages 3770–3775, 1994.

4. E. D. Dickmanns and B. D. Mysliwetz. Recursive 3-d road and relative ego-state recognition. *IEEE Transaction on Pattern Analysis and Machine Intelligence*, 14:199–213, February 1992.

5. M. Isard and A. Blake. Condensation: conditional density propagation for visual tracking. *International Journal of Computer Vision*, 1998.

6. D. G. Lowe. Distinctive image features from scale-invariant keypoints. *International Journal of Computer Vision*, 60(2):91–110, 2004.

7. B. Lucas and T. Kanade. An iterative image registration technique with an application to stereo vision. In *IJCAI81*, pages 674–679, 1981.

8. J. Matas, O. Chum, M. Urban, and T. Pajdla. Robust wide baseline stereo from maximally stable extremal regions. In *BMVC*, pages 384–393, 2002.

9. K. Mikolajczyk and C. Schmid. Scale and affine invariant interest point detectors. *International Journal of Computer Vision*, 60(1):63–86, 2004.

10. W. B. S. Thrun, D. Fox and F. Dellaert. Robust Monte Carlo Localization for Mobile Robots. *Artificial Intelligence*, 2001.

11. A. Smith, A. Doucet, and N. de Freitas. *Sequential Monte Carlo Methods in Practice*. Springer, New-York, 2001. ISBN 0-387-95146-6.

12. A. F. M. Smith and A. E. Gelfand. Bayesian statistics without tears: A sampling-resampling perspective. *American Statistician*, 46(2):84–88, 1992.

Cross System Personalization and Collaborative Filtering by Learning Manifold Alignments

Bhaskar Mehta[1] and Thomas Hofmann[2]

[1]Fraunhofer IPSI, Damstadt 64293, Germany
[2]Darmstadt University of Technology, Darmstadt 64289, Germany
mehta@ipsi.fhg.de, hofmann@int.tu-darmstadt.de

Abstract. Today, personalization in digital libraries and other information systems occurs separately within each system that one interacts with. However, there are several potential improvements w.r.t. such isolated approaches. Investments of users in personalizing a system, either through explicit provision of information, or through long and regular use are not transferable to other systems. Moreover, users have little or no control over the information that defines their profile, since user profiles are deeply buried in personalization engines. Cross-system personalization, i.e. personalization that shares personalization information across different systems in a usercentric way, overcomes the aforementioned problems. Information about users, which is originally scattered across multiple systems, is combined to obtain maximum leverage. The key idea is that when a large number of users cross over from one system to another, carrying their user profiles with them, a mapping between the user profiles of the two systems can be discovered. In this paper, we discuss the use of manifold learning for the purpose of computing recommendations for a new user crossing over from one system to another.

1   Introduction

The World Wide Web provides access to a wealth of information and services to a huge and heterogeneous user population on a global scale. One important and successful design mechanism in dealing with this diversity of users is to *personalize* Web sites and services, i.e. to customize system contents, characteristics, or appearance with respect to a specific user. The ultimate goal is to optimize access to relevant information or products by tailoring search results, displays, etc. to a user's presumed interests and preferences. More specifically, this optimization may aim at, for example, increasing the efficiency of system usage or improving the quality and relevance of results. Given the huge and rapidly growing amount of data available online as well as an ever growing user population that uses the World Wide Web, the relevance of personalized access is likely to further increase in the future.

While most users will interact with different systems and sites on the Web, personalization most often occurs separately within each system. Each system independently builds up user profiles, for instance, by locally storing information

C. Freksa, M. Kohlhase, and K. Schill (Eds.): KI 2006, LNAI 4314, pp. 244–259, 2007.
© Springer-Verlag Berlin Heidelberg 2007

about a user's likes and dislikes, interests, and further characteristics, and may then use this information to personalize the system's content and service offering. Such isolated approaches have two major drawbacks: Firstly, investments of users in personalizing a system either through explicit provision of information or through long and regular use are not transferable to other systems. Secondly, users have little or no control over the information that defines their profile, since user data are deeply buried in personalization engines running on the server side.

Cross system personalization [14] allows for sharing information across different information systems in a user-centric way and can overcome the aforementioned problems. Information about users, which is originally scattered across multiple systems, is combined to obtain maximum leverage and reuse of information. Previous approaches to cross system personalization [15] rely on each user having a *unified profile* which different systems can understand. The unified profile will contain facets modeling aspects of a multidimensional user. The basis of *understanding* in this approach is of a semantic nature, i.e. the semantics of the facets and dimensions of the unified profile are known, so that the latter can be aligned with the profiles maintained internally at a specific site. The main challenge in this approach is to establish some common and globally accepted vocabulary and to create a standard every system will comply with. Without such a convention, the exact mapping between the unified user profile and the system's internal user profile would not be known.

Machine learning techniques provide a promising alternative to enable cross system personalization without the need to rely on accepted semantic standards or ontologies. The key idea is that one can try to learn dependencies between profiles maintained within one system and profiles maintained within a second system based on data provided by users who use both systems and who are willing to share their profiles across systems – which we assume is in the interest of the user. Here, instead of requiring a common semantic framework, it is only required that a sufficient number of users cross between systems and that there is enough regularity among users that one can learn within a user population, a fact that is commonly exploited in social or *collaborative filtering* [20].

2   Automatic Cross System Personalization

For simplicity, we consider a two system scenario in which there are only two sites or systems denoted by A and B that perform some sort of personalization and maintain separate profiles of their users; generalization to an arbitrary number of systems is relatively straightforward and is discussed later. We assume that there is a certain number of c common users that are known to both systems. For simplification, we assume that the user profiles for a user u_i are stored as vectors $\mathbf{x}_i \in \mathcal{X} \subseteq \mathbb{R}^n$ and $\mathbf{y}_i \in \mathcal{Y} \subseteq \mathbb{R}^m$ for systems A and B, respectively. Given the profile $\mathbf{x}_i$ of a user in system A, the objective is to find the profile $\mathbf{y}_i$ of the same user in system B, so formally we are looking to find a mapping

$$F_{AB} : \mathbb{R}^n \to \mathbb{R}^m, \qquad \text{s.t.} \quad F_{AB}(\mathbf{x}_i) \approx \mathbf{y}_i \tag{1}$$

for users u_i. Notice that if users exist for which profiles in both system are known, i.e. a training set $\{(\mathbf{x}_i, \mathbf{y}_i) : i = 1, \ldots, l\}$, then this amounts to a standard supervised learning problem. However, regression problems typically only involve a single (real-valued) response variable, whereas here the function F_{AB} that needs to be learned is *vector-valued*. In fact, if profiles store say rating information about products or items at a site, then the dimensionality of the output can be significant (e.g. in the tens of thousands). Moreover, notice that we expect the outputs to be highly correlated in such a case, a crucial fact that is exploited by recommender systems. For computational reasons it is inefficient and often impractical to learn independent regression functions for each profile component. Moreover, ignoring inter-dependencies can seriously deteriorate the prediction accucracy that is possible when taking such correlations into account. Lastly, one also has to expect that a large fraction of users are only known to one system (either A or B). This brings up the question of how to exploit data without known correspondence in a principled manner, a problem generally referred to as *semi-supervised learning*. Notice that the situation is symmetric and that unlabeled data may be available for both systems, i.e. sets of vectors $\mathbf{x}_i$ without corresponding $\mathbf{y}_i$ and vice versa. In summary, we have three conceptual requirements from a machine learning method:

- Perform vector-valued regression *en bloc* and not independently
- Exploit correlations between different output dimensions (or response variables)
- Utilize data without known correspondences

In addition, the nature of the envisioned application requires:

- Scalability of the method to large user populations and many systems/sites
- Capability to deal with missing and incomplete data

There are some recent learning methods that can be utilized for vector-valued regression problem, but some of them do not fulfill the above requirements. Kernel dependency estimation [21] (KDE) is a technique that performs kernel PCA [19] on the output side and then learns independent regression functions from inputs to the PCA-space. However, KDE can only deal with unlabeled data on the output side and requires to solve computationally demanding pre-image problems for prediction [1]. Another option is Gaussian process regression with coupled outputs [12]. Here it is again difficult to take unlabeled data into account while preserving the computational efficiency of the procedure. The same is true for more traditional approaches like Multi-Layer-Perceptrons with multiple outputs. Instead of using regression methods, we thus propose the use of *manifold learning* in this context. Manifold learning methods generalize linear dimension reduction techniques that have already been used successfully in various ways for collaborative filtering. Moreover, they are usually motivated in an unsupervised setting that can typically be extended to semi-supervised learning in a rather straightforward manner. More specifically, we propose to use the *Laplacian Eigenmaps*[3] and *Locally Linear Embedding* (LLE)[18] approaches as our core method. LLE constructs a low-dimensional data representation for a given set of data points by embedding the points in a way that preserves the local (affine) geometry. Compared to other manifold learning and non-linear dimension reduction algorithms, such as Sammon's MDS [16] or Isomap [6], the LLE approach is computationally attractive and highly scalable, since it only relies on distances within local neighborhoods. Moreover, as presented in [9], constrained LLE (CLLE) can be utilized to learn mappings between two vector spaces by semi-supervised alignment of manifolds. The former work also provides empirical evidence that CLLE can outperfom standard regression methods. The key idea is to embed user profiles from different systems in low-dimensional manifolds such that profiles known to be in correspondence (i.e. profiles of the same user) are mapped to the same point. This means the manifolds will be aligned at correspondence points. A more general version of CLLE has been derived in [10], which takes the Laplacian Eigenmap approach [3] as the starting point. In the next section, we will provide more detail on these methods.

3   Non Linear Dimensionality Reduction and Manifold Alignment

3.1   Laplacian Eigenmaps

Suppose we are given l data points in $\mathcal{S} = \{\mathbf{x}_i \in \mathbb{R}^n \colon i = 1, \ldots, l\}$. When the data lie approximately on a low-dimensional manifold embedded in the n-dimesional Euclidean space, manifold learning methods such as Laplacian Eigenmaps [3], Hessian Eigenmaps [7], Isomap [6] or locally linear embeddings [18] can be used to recover the manifold from a sample $\mathcal{S}$. We pursue the Laplacian Eigenmap approach, which has been used sucessfully in semi-supervised learning [10] and for which rigorous convergence results exists in the large sample limit [11].

The starting point in Laplacian Eigenmaps is the construction of a weighted graph whose nodes are the sample points and whose edges connect the nearest neighbors of each node. Neighborhoods may consist of the k-nearest neighbors of a sample point or the set of all points that are within an ϵ-ball. We write $i \sim j$ as a shorthand for sample points $\mathbf{x}_i$ and $\mathbf{x}_j$ that are neighbors. The weights W_{ij} between neighbors are usually assumed to be non-negative and symmetric, $W_{ij} = W_{ji} \geq 0$ and are summarized in an affinity matrix $\mathbf{W}$. There are several alternatives on how to define these weights when starting from a vector-valued representation over $\mathbb{R}^n$, one popular choice being the Gaussian kernel,

$$W_{ij} \equiv \exp\left[-\beta \|\mathbf{x}_i - \mathbf{x}_j\|^2\right], \tag{2}$$

where $\beta > 0$ is a suitably chosen bandwidth parameter. Another choice is to compute weights based on a local affine approximation over neighbors, as discussed in the following subsection on LLE.

The heart of the Laplacian Eigenmap approach is the generalized graph Laplacian $\mathbf{L}$ defined as,

$$\mathbf{L} = (L_{ij})_{i,j=1}^n, \quad L_{ij} = \begin{cases} \sum_{j \sim i} W_{ij}, & \text{if } i = j \\ -W_{ij}, & \text{if } i \sim j \\ 0, & \text{otherwise}. \end{cases} \tag{3}$$

An Laplacian Eigenmap is a function $f : \mathcal{S} \to \mathbb{R}$ for which $\mathbf{L}f = \lambda f$ and $\|f\|^2 = 1$, where we think of f as a vector of function values for convenience. Moreover, in order to remove the trivial solution with $\lambda = 0$ one can add the constraints $(1, \ldots, 1)f = \sum_{i=1}^l f_i = 0$. It can be shown that the eigenmap corresponding to the smallest eigenvalue $\lambda > 0$ minimizes the criterion

$$f^T \mathbf{L} f = \sum_{i,j} W_{ij}(f_i - f_j)^2. \tag{4}$$

The eigenmaps corresponding to the d smallest eigenvalues span a d-dimensional coordinate system on the low-dimensional data manifold.

In the case of semi-supervised learning one may utilize $f^T \mathbf{L} f$ as a regularizer and combine it with supervised information about target values t_i that may be available at some subset $\mathcal{S}' \subseteq \mathcal{S}$ of the nodes of the graph to define the regularized solution (cf. [2])

$$f^* = \arg \min_f \sum_{\mathbf{x}_i \in \mathcal{S}'} (f_i - t_i)^2 + \lambda f^T \mathbf{L} f. \tag{5}$$

3.2   Aligned Manifold Learning

Consider now the case where two sets of points are given $\mathcal{S}_x \equiv \{\mathbf{x}_i \in \mathbb{R}^n : i = 1, \ldots, l_x\}$ and $\mathcal{S}_y \equiv \{\mathbf{y}_j \in \mathbb{R}^m : i = 1, \ldots, l_y\}$ where we assume without loss of generality that the first $l \leq \min\{l_x, l_y\}$ points are in correspondence. In the case of cross system personalization, $\mathbf{x}_i$ will denote a user profile in system A, $\mathbf{y}_j$ will denote a user profile in system B and $\mathbf{x}_i \leftrightarrow \mathbf{y}_i$ for users u_i, $i = 1, \ldots, l$, who are known in both systems. We will separately construct graphs $\mathcal{G}_x$ on $\mathcal{S}_x$ and $\mathcal{G}_y$ on $\mathcal{S}_y$ in order to find low-dimensional embeddings of the points in $\mathcal{S}_x$ and $\mathcal{S}_y$, respectively. In addition, we will follow the approach in [10] and utilize the correspondence information to enforce that embeddings of user profiles for the same user are close to one another. To that extend we compute a simultaneous embedding f of $\mathcal{S}_x$ and g of $\mathcal{S}_y$ by minimizing the objective

$$C(f, g) = \sum_{i=1}^l (f_i - g_i)^2 + \lambda \left(f^T \mathbf{L}_x f + g^T \mathbf{L}_y g. \right) \tag{6}$$

More specifically, in order to deal with simultaneous re-scaling of f and g, one minimizes the Rayleigh quotient

$$\tilde{C}(f, g) = \frac{C(f, g)}{f^T f + g^T g}. \tag{7}$$

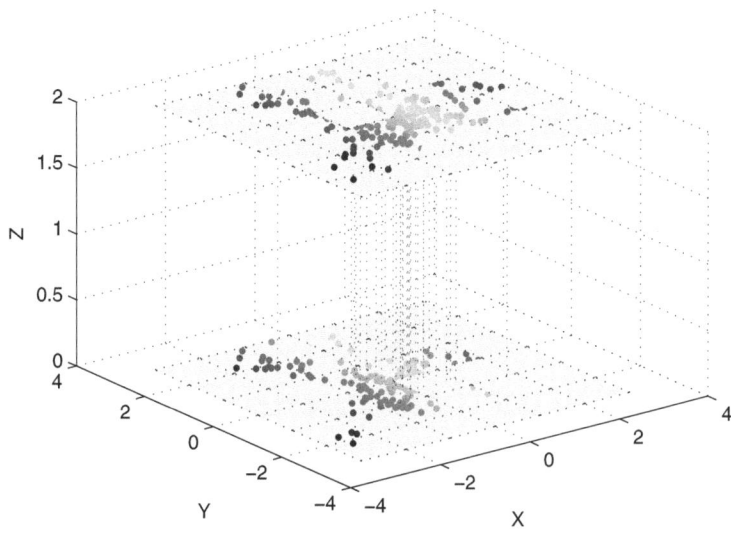

Fig. 1. Aligned 2D manifolds for two subsets of MovieLens dataset. The vertical lines show the points which are in correspondance.

By defining the combined graph $\mathcal{G} \equiv \mathcal{G}_x \cup \mathcal{G}_y$ with Laplacian $\mathbf{L}$ and combined functions $h = (f^T, g^T)^T$ the above objective can be rewritten as

$$\tilde{C}(h) = \frac{h^T \mathbf{H} h}{h^T h}, \quad \text{where } \mathbf{H} \equiv \lambda \mathbf{L} + \begin{pmatrix} \mathbf{U}^{nn} & \mathbf{U}^{nm} \\ \mathbf{U}^{mn} & \mathbf{U}^{mm} \end{pmatrix} \tag{8}$$

and $\mathbf{U}^{nm} \in \mathbb{R}^{n \times m}$ is diagonal with $U_{ii}^{nm} = 1$ for $1 \leq i \leq l$ and 0 otherwise. Again, a solution is obtained as before by finding the eigenvectors of the matrix $\mathbf{L}$.

One can also enforce the embeddings of points in correspondence to be the same on both manifolds [10]. In this case, one identifies the first l points in $\mathcal{S}_x$ and $\mathcal{S}_y$, resulting in a combined graph $\mathcal{G}$ with $l_x + l_y - l$ nodes with a combined weight matrix. Notice that weights between pairs of nodes with indices $1 \leq i, j \leq l$ are simply given by the sum of the weights from $\mathcal{G}_x$ and $\mathcal{G}_y$. Introducing functions h one then minimizes

$$\tilde{C}(h) = \frac{h^T \mathbf{L} h}{h^T h}, \quad \text{s.t.} \sum_i h_i = 0. \tag{9}$$

3.3 Locally Linear Embedding

One way to define the weights W_{ij} for neighboring nodes in the graph is to compute them based on a local affine approximation. This idea has originally presented in the context of the Locally Linear Embedding (LLE) method [18]. Its use as a preprocessing step in conjunction with Laplacian Eigenmaps has been proposed in [10]. For a sample of l data points $\mathcal{S} = \{\mathbf{x}_i \in \mathbb{R}^n : i = 1, \ldots, l\}$, LLE proceeds as follows:

- For each data point $\mathbf{x}_i$, compute the k nearest neighbors in $\mathcal{S}$ which are closest to $\mathbf{x}_i$ in Euclidean distance.
- Compute for each $\mathbf{x}_i$ the optimal approximation weights for an affine local regression over the neighbors. This is equivalent to approximating the nonlinear manifold at $\mathbf{x}_i$ by the linear hyperplane that passes through the neighboring points. This step of the algorithm amounts to solving a quadratic optimization problem:

$$W_{ij}^* = \arg\min_W |\mathbf{x}_i - \sum_{j \sim i} W_{ij}\mathbf{x}_j|^2 \,, \text{s.t.} \sum_j W_{ij} = 1 \,, \tag{10}$$

where $j \sim i$ indicates that $\mathbf{x}_j$ is a neighbor of $\mathbf{x}_i$ (notice that the relation is in general not symmetric).

- Finally, a low-dimensional representation $\hat{\mathbf{x}}_i$ is computed by solving the minimization problem

$$\hat{\mathbf{X}}^* = \arg\min_{\hat{\mathbf{X}}} \sum_i \|\hat{\mathbf{x}}_i - \sum_{j \sim i} W_{ij}\hat{\mathbf{x}}_j\|^2 \tag{11}$$

This can be shown to be equivalent to an eigenvector decomposition problem involving the matrix

$$M = (I - W^*)^T (I - W^*) \tag{12}$$

where I is the $l \times l$ identity matrix. The bottom $d + 1$ eigenvectors of M (excluding the smallest, which is 1) form a co-ordinate system for the low dimensional data manifold.

While the LLE algorithm can be used in its own right for manifold learning, we have employed it here to compute the affinity matrix for the Laplacian Eigenmap method. Note that the matrix $\mathbf{L}^* = I - W^*$ corresponds to the graph Laplacian $\mathbf{L}$ (defined in eq. 3) for a graph with $\sum_j W_{ij} = 1$ for all graph nodes. Also note that the graph Laplacian thus formed is not symmetric and the weights can be negative. Multiplying $\mathbf{L}^*$ with its transpose gives a symmetric matrix M. [3] explains that under some conditions, the matrix M is approximately the same as $\mathbf{L}^2$, which has the same eigenvectors as $\mathbf{L}$. It has been shown in [10] that the matrix M can be substituted for the graph Laplacian $\mathbf{L}$ in the aligned manifold method.

3.4   Reconstructing Points from Alignments

The remaining problem we would like to discuss is how to map a point on the low-dimensional manifold back into the original data space. This is particularly relevant in the context of manifold alignment, where one ultimately may want to realize a mapping from $\mathbb{R}^n \rightarrow \mathbb{R}^m$. After mapping a point $\mathbf{x} \in \mathbb{R}^n$ to a k-dimensional representation $\hat{\mathbf{x}}$, we would thus like to compute an approximation $\mathbf{y} \in \mathbb{R}^m$ by finding a pre-image to $\hat{\mathbf{y}}$ and identifying $\hat{\mathbf{y}} = F(\hat{\mathbf{x}})$. The next section explains the method used to compute the pre-images.

4   The Manifold Alignment Collaborative Filtering Algorithm

Manifold alignment using non-linear dimensionality reduction has the promise of a fast and effective supervised learning technique for the correspondence problem. It has been reported that dimensionality reduction techniques are effective for k-NN algorithms used typically for collaborative filtering[17]. *Non Linear dimensionality Reduction* (NLDR) techniques in turn have performed better than linear dimensionality reduction techniques like Factor Analysis and PCA[9]. Therefore we expect manifold alignment for the purposes of cross system personalization to be an effective approach. The algorithm essentially works as a k-NN algorithm as well. After projecting the user profile vectors from two (or more) systems on a low dimension manifold, we are able to find nearest neighbors based on distance measures like Euclidean distance. The additional constraint of aligning profiles belonging to the same user aligns the two submanifolds and helps in finding more effective neighborhoods. Our algorithm has 4 steps, the first 3 of which form the manifold projection phase (see *Algorithm 1*), and the last step does the pre image computation and is outlined in *Algorithm 2*. Our algorithm assumes two datasets $\mathcal{X}$ and $\mathcal{Y}$ of sizes $n_{\mathcal{X}}$ and $n_{\mathcal{Y}}$ with c common users and is as follows:

1. **Neighborhood identification:** For each point $\mathbf{x}_i \in \mathcal{X}$, we find the k-nearest neighbors. NLDR techniques usually use Euclidean distance to identify the nearest points. In our setting, data is sparse, therefore Euclidean distance on pure data is not neccesarily effective unless missing data is imputed. Options here include mean imputation (with item mean), measuring distance only on commonly-voted items, or using a distance based on a similarity measure like Pearson's correlation coefficient. This procedure also has to be repeated for $\mathbf{y}_i \in \mathcal{Y}$. Note that choosing exactly k-nearest neighbors for every node may result in a graph Laplacian thats not symmetric. Using LLE, one selects *exactly* k neighbors, while for the Laplacian one does not impose this constraint. As a result, the neighborhood of some points in the Laplacian Eigenmaps method can be much larger than k. This usually shows the importance of a node and is similar to the notion of *authority* nodes in the HITS algorithm[13].

2. **Calculate Affinity Matrix:** After the k nearest neighbors have been identified for every point, an affinity weight with every neighbor has to be computed. Options here include an affine decomposition (like in LLE), an exponential weight (aka the heat kernel used in Laplacian Eigenmaps) based either on euclidean distance or on a similarity measure like Pearson's correlation.

$$W_{ij} = \exp\left[-\beta \|\mathbf{x}_i - \mathbf{x}_j\|^2\right] \ or \tag{13}$$

$$W_{ij} = \exp\left[-\beta \|1 - correlation(\mathbf{x}_i, \mathbf{x}_j)\|\right] \ or \tag{14}$$

$$W_{ij} = 1 \ or \tag{15}$$

$$W_{ij} = \arg\min_W |\mathbf{x}_i - \sum_{j \sim i} W_{ij}^* \mathbf{x}_j|^2 \ , \text{s.t.} \ \sum_j W_{ij}^* = 1 \tag{16}$$

Algorithm 1. ComputeManifold-NLDR $(\mathcal{X}, \mathcal{Y}, c, k, d)$

Input: Matrices $\mathcal{X}, \mathcal{Y}$ with the first c columns aligned. k is the number of neighbors, d is the dimensionality of the manifold.

1: Impute missing values with mean item votes respectively for $\mathcal{X}$ and $\mathcal{Y}$ to get $\mathcal{X}_{norm}, \mathcal{Y}_{norm}$.
2: Calculate adjacency matrices $A_{\mathcal{X}}, A_{\mathcal{Y}}$ for graphs representing $\mathcal{X}_{norm}, \mathcal{Y}_{norm}$ by choosing k-nearest neighbors for every $\mathbf{x}_i \in \mathcal{X}, \mathbf{y}_i \in \mathcal{Y}$.

$$A_{\mathcal{X}}(i,j) = \begin{cases} 1, & \text{if } i \sim j \\ 0, & \text{otherwise}. \end{cases}$$

3: Compute reconstruction weights $W_{\mathcal{X}}, W_{\mathcal{Y}}$.
4: Compute the graph Laplacians $\mathbf{L}_{\mathcal{X}}, \mathbf{L}_{\mathcal{Y}}$ from the Weight Matrices as defined in equation 3. For constrained LLE, use $\mathbf{L}_{\mathcal{X}}^* = (I - W_{\mathcal{X}})^T (I - W_{\mathcal{X}})$, etc.
5: Compute $\mathbf{L}_{\mathcal{X}\mathcal{Y}} = \begin{bmatrix} \mathbf{L}_{\mathcal{X}}^{cc} + \mathbf{L}_{\mathcal{Y}}^{cc} & \mathbf{L}_{\mathcal{X}}^{cs} & \mathbf{L}_{\mathcal{Y}}^{cs} \\ \mathbf{L}_{\mathcal{X}}^{sc} & \mathbf{L}_{\mathcal{X}}^{ss} & 0 \\ \mathbf{L}_{\mathcal{Y}}^{sc} & 0 & \mathbf{L}_{\mathcal{Y}}^{ss} \end{bmatrix}$
c represents the points in alignment, while s represents the *single* points.
6: Find the low dimensional manifold $\mathbf{H}$ for the matrix $\mathbf{L}_{\mathcal{X}\mathcal{Y}}$. $\mathbf{H}$ has a dimensionality of $(n_{\mathcal{X}} + n_{\mathcal{Y}} - c) \times d$.

Output: Low dimensional manifold $\mathbf{H}$

In our experiments, we use the similarity measures defined in eq. 14. Finally, the Laplacians $\mathbf{L}_{\mathcal{X}}, \mathbf{L}_{\mathcal{Y}}$ of the graphs characterized by affinity matrices for $\mathcal{X}$ and $\mathcal{Y}$ are computed.

3. **Compute points on manifold:** This is usually done by solving an eigenvalue problem, and finding the eigenvectors of the Laplacian $\mathbf{L}$ (or $\mathbf{L}^*$ in case of LLE). For points in alignment, a modified eigenvalue problem has to be solved: A joint graph of the two datasets is formed and the eigenvectors of this Laplacian matrix $\mathbf{L}_{\mathcal{X}\mathcal{Y}}$ are computed (see eq. 8). The only parameter here is the dimensionality of the manifold (the number of eigenvectors that are chosen).

4. **Compute preimages for points not in correspondence:** In this step, neighborhoods for points not in correspondence are formed in a manner similar to the first step. The normal method to follow here is to do find the nearest neighbors (based on Euclidean distance) and compute a weight distribution over this neighborhood. We do this in the following manner: For a point $\mathbf{x}_i \in \mathcal{X}$ with $i > c$ and manifold coordinates $\hat{\mathbf{x}}_i$ we first identify a set of k nearest neighbors $\hat{\mathbf{y}}_r$ on the manifold among the points that are images of points in $\mathcal{Y}$, resulting in some set of image/pre-image pairs $\{(\mathbf{y}_r, \hat{\mathbf{y}}_r)\}$. We then compute the optimal affine combination weights w_r that optimally reconstruct $\hat{\mathbf{x}}_i \approx \sum_r w_r \hat{\mathbf{y}}_r$. Then the pre-image prediction is given by $F(\mathbf{x}_i) = \sum_r w_r \mathbf{y}_r$. Similarly, we can compute an inverse map by exchanging the role of the $\mathbf{x}_i$ and $\mathbf{y}_j$. Notice that one can also generalize this for arbitrary new samples $\mathbf{x} \in \mathbb{R}^n$ by generalizing the manifold mapping $\mathbf{x} \mapsto \hat{\mathbf{x}}$ to new points, which can be done along the lines presented in [4].

Algorithm 2. ComputePreimage $(\mathbf{H}, c, n_{\mathcal{X}}, n_{\mathcal{Y}}, k, \mathcal{X}_{norm}, \mathcal{Y}_{norm}, n_{\mathcal{X}}, n_{\mathcal{Y}})$

Input: Matrix $\mathbf{H}$ of length $(n_{\mathcal{X}} + n_{\mathcal{Y}} - c)$ representing the aligned manifold with c points overlapping between manifolds of $\mathcal{X}$ and $\mathcal{Y}$. k is the number of neighbors. $\mathbf{H}_M$ has the first c points representing the overlapping users. The next $n_{\mathcal{X}} - c$ points represents single points of $\mathcal{X}$ and the last $n_{\mathcal{Y}} - c$ points represent the single points of $\mathcal{Y}$. $\mathbf{H}(i)$ denotes the ith d-dimensional point on the manifold.

1: Extract submanifold $\mathbf{H}_{\mathcal{Y}}$ by combining the first c and the last $n_{\mathcal{Y}} - c$ points of $\mathbf{H}$.
2: **for** $i = (c + 1)$ to $(n_{\mathcal{X}})$ **do**
3:   $\hat{\mathbf{x}}_{\mathbf{i}} \leftarrow \mathbf{H}(i)$
4:   Compute the k nearest neighbors $\hat{\mathbf{y}}_r$ of $\hat{\mathbf{x}}_{\mathbf{i}}$ on the sub-manifold $\mathbf{H}_{\mathcal{Y}}$. Let $\mathbf{y}_r$ represent the preimage of $\hat{\mathbf{y}}_r$ in $\mathcal{Y}_{norm}$.
5:   Compute affine weights $\mathbf{W}^* = (\mathrm{w}_r)_{r=1}^k$ for the neighborhood.
6:   Compute Preimage prediction $F(\hat{\mathbf{x}}_{\mathbf{i}}) = \sum_r \mathrm{w}_r \mathbf{y}_r$.
7:   $\hat{\mathcal{X}}_{\mathbf{s}}(i - c) = F(\hat{\mathbf{x}}_{\mathbf{i}})$
8: **end for**
9: Repeat above procedure by exchanging $\mathcal{X}$ and $\mathcal{Y}$ to compute preimages for single points of $\mathcal{Y}$.

Output: Preimages $\hat{\mathcal{X}}_{\mathbf{s}}, \hat{\mathcal{Y}}_{\mathbf{s}}$

5 Evaluation

The manifold algorithm seeks to predict the ratings of a user who has not rated even a single item on the current system so far. In this scenario, truly nothing is known about the active user. The best rating prediction that a system can provide is the popularity vote based on the mean votes of every item. The items with the highest mean votes are then recommended to the user. Our algorithm seeks to do better than this non-personalized recommendation. On the other extreme, given all the data for both systems, the best possible rate prediction could be calculated if all the data was known to one system. In this scenario, a SVD or Pearson's correlation based algorithm could compute the predictions (which serve as the gold standard for us). In order to be useful, our algorithm should perform better than predictions from the popularity vote and perform as close as possible to the gold standard.

5.1 Dataset and Evaluation Scheme

We chose the Movielens[1] data with 100,000 votes for the purposes of our evaluation. This data set consists of rates in 1682 items by 944 different users. This data is quite sparse ($\sim 6\%$) as is typically for user ratings. We split the data into two subsets X and Y by spiliting the movie ratings for all users (e.g. two matrices 840×944 and 842×944). In principle the overlap between datasets can be varied anging from no overlap to all items overlapping. While in the earlier case, the movie ratings are effectivly spilt into half, the complete data is

---

[1] http://www.grouplens.org/

available to both systems in the second case. However, in real world scenarios, item overlaps are very small. Therefore we chose a random 5% from the item-set as an overlap. The other free parameter is the number of users set to be in correspondence, which we vary from 0 to 800. The last 144 users form the test set for our evaluations. We randomly choose the test set and the item set for every run of the algorithm. Individual NLDR methods(LLE and Laplacian) have other parameters which need to be varied in order to judge their effect. There parameters are (a) the dimensionality of the manifold, (b) the size of the neigh-borhood for the adjacency matrix, and (c) the size of the neighborhood for the user profile reconstruction. Additionally, the Laplacian Eigenmap method has a free parameter β which can take any real value. In our experiments, we have varied the parameter and present the results for the optimized values. Further increases in neighborhood sizes offers some advantage, but at a much increased computational cost. We have chosen these values: $k = 36$, $d = 6$, and size of neighborhood on the manifold $k_1 = 55$. In addition, we choose different values of β, namely $0, 0.4$ and 4.

Evaluation Metrics

1. Mean Average Error $= \frac{1}{m}|p_v - a_v|$, where p_v is the predicted vote and a_v is the actual vote. The average is taken only over known values (assume the active user has provided m votes).
2. Ranking score of top-20 items. $R_{score} = 100 * (\sum R / \sum R_{max})$. This metric gives a value between 0 and 100 and was proposed in [5]. Higher values indicate a better ranking with top items as the most highly rated ones. We measure this metric only over known ratings. One big advantage of this metric is that it gives the same score for permutations of items with the same score. Thus if a user has rated 6 items with the maximum score 5, then the R_{score} is the same for any permutation of the ranking. This removes the problem of breaking ties.

6   Discussion

The results of the evaluation are encouraging. A simple NLDR to a manifold even with any explicit alignment of user profiles performs better than popular voting. Expectedly, the predicted votes become more accurate as more users cross over and their profiles are aligned. While the predictions are not as good as the gold standard even in the case of complete overlap according to the MAE, the algorithm provides a 4-5% improvement over the baseline after $\sim$ 35% user profiles have been aligned. For collaborative filtering, this is not an insignifi-cant improvement: the gold standard is only 12.6% better than the baseline. Experimental results also show that the top-N recommendation using manifold alignment is a significantly higher quality than the baseline. In case of complete overlap, Laplacian Eigenmap based manifold aligment can provide a $top - 20$ ranked list which is more relavant than the gold standard. The results presented here are obtained after a $10-$fold validation; in some cases, the algorithm was

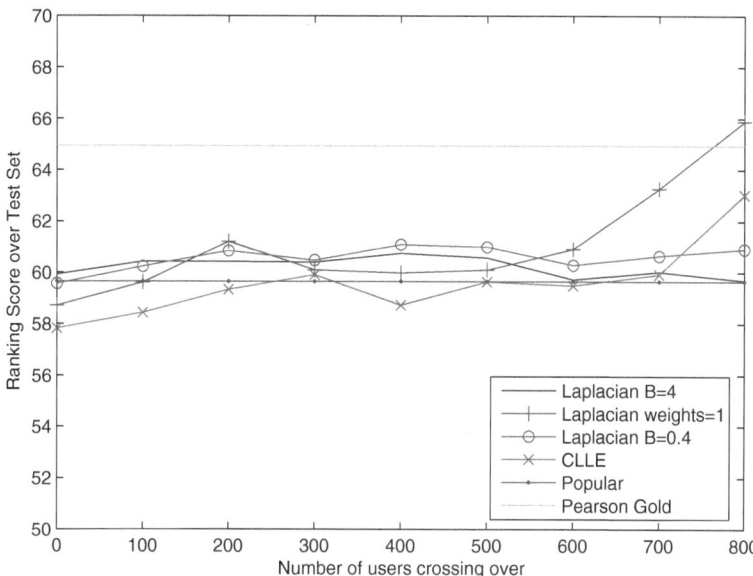

Fig. 2. Precision for the different NLDR methods within the manifold alignment framework. Numbers plotted are after 10-fold validation and averaging.

able to outperform the gold standard for MAE as well. One possible reason for the lower performance is the small size of data which is very sparse. With more training data, we expect to find more neighbors for every user which have votes for many items. Due to the sparsity of data, the majority of the normalized user database consists of mean. Therefore, the reconstructed values are heavily weighted towards the mean votes, especially for items that are note frequently rated. Previous research [8] has shown that learning from incomplete data offers significant advantage over strategies like mean imputation. Given that our approach works better than popularity votes even with a heavy bias towards mean values, algorithmic enhancements which offer a probablistic interpretation to manifold alignment are likely to be more accurate and form our future work.

6.1   Implementation and Performance

The manifold algorithm outlined in this paper has been implemented using Matlab R14 on a Pentium 4 based Desktop PC. Standard Matlab routines have been used and sparse matrices are used wherever possible. For the smaller MovieLens data with 100,000 rates, the algorithm uses around 100 MB of RAM. It performs reasonably w.r.t. to time as well. Each run of the *Algorithm 1* followed by *Algorithm 2* runs in approximately 5 seconds using Laplacian Eigenmaps. The LLE algorithm runs slower (70 seconds) since a quadratic program has to be solved for every point. The memory requirements of the LLE algorithm are also higher.

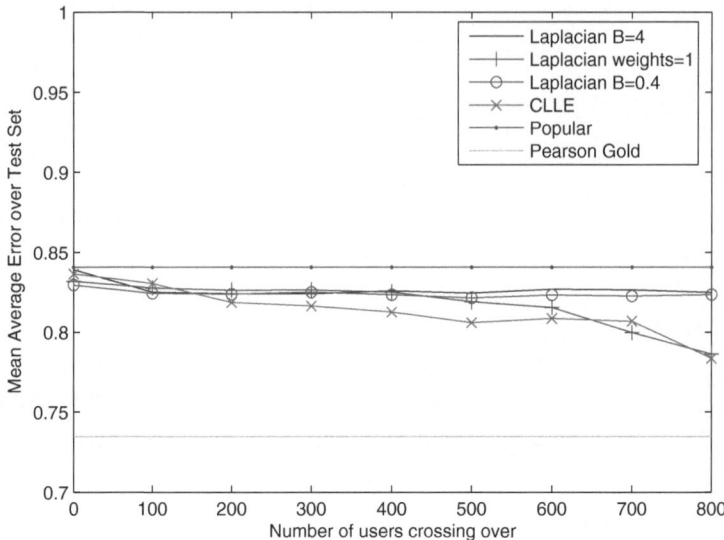

Fig. 3. Precision for the different NLDR methods within the manifold alignment framework. Numbers plotted are after 10-fold validation and averaging.

6.2 Computation Complexity

The Laplacian Eigenmap method clearly offers computational advantages over the LLE method. The LLE method has 3 basic steps: a) find nearest neighbors, b) compute reconstruction weights, and c) find eigenvalues and eigenvectors. For two datasets of sizes $m_\mathcal{X} \times n_\mathcal{X}$ and $m_\mathcal{X} \times n_\mathcal{X}$ with c common points, the size of the common graph is $n_\mathcal{X} + n_\mathcal{Y} - c$ nodes. The complexity of the LLE method for a matrix with n points each of dimensionality m is thus $O(mn) + O(dnk^3) + O(dn^2) \equiv O(dn(n + k^3))$. The Laplacian Eigenmap method essentially skips the second step, and hence has a complexity of $O(dn^2)$. Therefore the overall complexity of *Algorithm 1* (without the reconstruction of user profiles) is $O(dn(n + k^3))$ where $n = n_\mathcal{X} + n_\mathcal{Y} - c$. For our experiments, k typically had a value between $24 - 48$, while n was around 1000. In this range, k^3 was 1-2 orders of magnitude higher than n, thus explaining the difference between the running times of LLE and LapE based NLDR. Note however that this entire alignment computation can be performed off line. For a new user, out of sample extensions for LLE and Laplacian Eigenmaps[4] can be used. These typically have a computational complexity of $O(m) + O(dk^3)$. Importantly, the neighborhood formation step can be reused in the second part of the algorithm where user profiles have to be reconstructed.

The reconstruction of a user profile(*Algorithm 2*) involves (a) neighborhood formation (b) finding reconstruction weights, and (c) combining neighbor votes. The complexity reconstructing the profile for one user therefore is $O(dn) + O(dk^3) + O(mk)$. The significant term here depends on the values of the parameters: for a large neighborhood, the second term dominates. However if the number of items is very large (say a million), then the last term is the most significant one.

6.3 Usefulness in Practical Scenarios

With a variety of systems using personalization engines, there is a lot of data being collected about users as they go about their day to day pursuits. Combining this data from various sources in a secure and transparent way can significantly improve the level of personalization that electronic systems currently provide. In this scenario, creating an approach which makes very few assumptions about systems and users is of paramount importance. While our algorithm has been demonstrated in a collaborative filtering setting, there is no binding to use only rating data. The profiles of a content based system can be just as easily plugged in, as can be a profile from a hybrid system. Importantly, we also hypothesize that user profiles should be stored on the user's side in *Context Passport* which can leverage data about the user available with multiple systems. We envision that even data from operating systems and email clients can be plugged into the *Context Passport*. Our approach makes all of this possible in principle. However, the absence of relevant data, where user profiles of the *same users* at multiple sites are available, makes it difficult to evaluate the effectiveness of our algorithm in a real life setting. Our attempts are on to collect such a dataset in a scientific conference setting.

While our currently implementation performs all calculations in an online fashion, it is possible to implement the learning phase and the actual user profile reconstruction as separate phases. In a practical environment, the alignment learning would be performed off line and a new user will approach a new system with dimensionally reduced profile. This profile will then be projected on to the manifold using out-of-sample extensions, and reconstruction of the user profile can be done performed in real time. In case the item space is huge, the reconstruction phase can be performed online only on a sample set of item extracted from the entire item space. The entire profile can then be reconstructed offline.

6.4 Privacy

One important aspect of cross system personalization is privacy. People and companies alike are likely to have reservations against sharing their data with our systems. Users fear the loss of their anonymity, while companies fear a loss of their competitive edge. With our method, the important thing is to discover the underlying social similarity between people and not their exact buying/rating patterns. A less accurate, but more secure (w.r.t privacy) approach could start with a dimensionally reduced user database from say 1 million items to 1000 dimensions. Also the complete user database does not need to be known: a random selection of a sufficient number of users might be sufficient to learn the mapping from one system to another.

6.5 Scaling to a $n-$System Scenario

The manifold alignment algorithm needs only a minor modification in case some users are common to all n systems. This modification in in the step where a joint graph $\mathcal{G}$ is formed. The low dimensional embedding of this graph will have all the submanifolds aligned. More fined tuned modifications are required in case the set of overlapping users is different between different users. Manifold alignment in n-system scenario is successful only if a small fraction of users need to cross from one system to another. In order to test this scenario, larger datasets are needed.

7 Conclusions and Future Work

This paper outlines a novel approach to leverage user data distributed across various electronic systems to provide a better personalization experience. One major benefit of this approach is dealing with the *new user* problem: A new user of a collaborative filtering system can usually be provided only the non-personalized recommendation based on popular items. Our approach allows system to make a better prediction using the user's profile in other systems. The contribution of this paper is in describing an algorithm which offers a satisfactory improvement over *status quo* for a potentially important application scenario. Future work includes developing a practical framework around the manifold alignment algorithm. Also, there is a potential for improvement in performance both from an algorithmic and methodological point of view.

References

1. G. Bakir, J. Weston, and B. Schlkopf. Learning to find pre-images. *Advances in Neural Information Processing Systems*, 16:449–456, 2004.
2. M. Belkin, I. Matveeva, and P. Niyogi. Regularization and semi-supervised learning on large graphs. In J. Shawe-Taylor and Y. Singer, editors, *COLT*, volume 3120 of *Lecture Notes in Computer Science*, pages 624–638. Springer, 2004.
3. M. Belkin and P. Niyogi. Laplacian eigenmaps for dimensionality reduction and data representation. *Neural Computation*, 15(6):1373–1396, 2003.
4. Y. Bengio, J.-F. Paiement, P. Vincent, O. Delalleau, N. L. Roux, and M. Ouimet. Out-of-sample extensions for lle, isomap, mds, eigenmaps, and spectral clustering. In *NIPS*, 2003.
5. J. S. Breese, D. Heckerman, and C. M. Kadie. Empirical analysis of predictive algorithms for collaborative filtering. In *UAI*, pages 43–52, 1998.
6. V. de Silva and J. B. Tenenbaum. Global versus local methods in nonlinear dimensionality reduction. In *NIPS*, pages 705–712, 2002.
7. D. L. Donoho and C. E. Grimes. Hessian eigenmaps: locally linear embedding techniques for highdimensional data. *Proceedings of the National Academy of Arts and Sciences*, 100(10):5591–5596, 2003.
8. Z. Ghahramani and M. I. Jordan. Learning from incomplete data. Technical Report AIM-1509, MIT, 1994.
9. J. Ham, D. Lee, and L. Saul. Learning high dimensional correspondence from low dimensional manifolds. In *ICML Workshop on The Continuum from Labeled to Unlabeled Data in Machine Learning and Data Mining*, pages 34–41, 2003.
10. J. Ham, D. Lee, and L. Saul. Semisupervised alignment of manifolds. In R. G. Cowell and Z. Ghahramani, editors, *AISTATS 2005*, pages 120–127. Society for Artificial Intelligence and Statistics, 2005.

11. M. Hein, J.-Y. Audibert, and U. von Luxburg. From graphs to manifolds - weak and strong pointwise consistency of graph laplacians. In *COLT*, pages 470–485, 2005.
12. S. Keerthi and W. Chu. A matching pursuit approach to sparse gaussian process regression. In Y. Weiss, B. Schlkopf, and J. Platt, editors, *Advances in Neural Information Processing Systems 18*. MIT Press, Cambridge, MA, 2006.
13. J. M. Kleinberg. Authoritative sources in a hyperlinked environment. *J. ACM*, 46(5):604–632, 1999.
14. B. Mehta, C. Niederee, and A. Stewart. Towards cross-system personalization. In *UAHCI*, 2005.
15. B. Mehta, C. Niederée, A. Stewart, M. Degemmis, P. Lops, and G. Semeraro. Ontologically-enriched unified user modeling for cross-system personalization. In *User Modeling*, pages 119–123, 2005.
16. J. W. Sammon. A non-linear mapping for data structure analysis. In *IEEE Transactions on Computing*, volume C18 (5), pages 401–409, May 1969.
17. B. Sarwar, G. Karypis, J. Konstan, and J. Riedl. Application of dimensionality reduction in recommender systems–a case study. In *ACM WebKDD 2000 Web Mining for E-Commerce Workshop*, 2000.
18. L. K. Saul and S. T. Roweis. Think globally, fit locally: Unsupervised learning of low dimensional manifold. *Journal of Machine Learning Research*, 4:119–155, 2003.
19. B. Schölkopf, A. J. Smola, and K.-R. Müller. Nonlinear component analysis as a kernel eigenvalue problem. *Neural Computation*, 10(5):1299–1319, 1998.
20. U. Shardanand and P. Maes. Social information filtering: algorithms for automating word of mouth. In *CHI '95: Proceedings of the SIGCHI conference on Human factors in computing systems*, pages 210–217, New York, NY, USA, 1995. ACM Press/Addison-Wesley Publishing Co.
21. J. Weston, O. Chapelle, A. Elisseeff, B. Schölkopf, and V. Vapnik. Kernel dependency estimation. In *NIPS*, pages 873–880, 2002.

A Partitioning Method for Mixed Feature-Type Symbolic Data Using a Squared Euclidean Distance

Renata M.C.R. de Souza, Francisco de A.T. de Carvalho, and Daniel F. Pizzato

Centro de Informatica - CIn / UFPE, Av. Prof. Luiz Freire, s/n - Cidade Universitaria, CEP: 50740-540 - Recife - PE - Brasil
{rmcrs,fatc,dfp}@cin.ufpe.br

Abstract. A partitioning cluster method for mixed feature-type symbolic data is presented. This method needs a previous pre-processing step to transform Boolean symbolic data into modal symbolic data. The presented dynamic clustering algorithm has then as input a set of vectors of modal symbolic data (weight distributions) and furnishes a partition and a prototype to each class by optimizing an adequacy criterion based on a suitable squared Euclidean distance. To show the usefulness of this method, examples with synthetic symbolic data sets and applications with real symbolic data sets are considered.

1 Introduction

Cluster analysis has been widely used in numerous fields including pattern recognition, data mining and image processing. Their aim is to summarize data sets in homogeneous clusters that may be organized according to different structures ([12], [13], [16]): hierarchical methods yield complete hierarchy, i.e., a nested sequence of partitions of the input data, whereas partitioning methods seek to obtain a single partition of the input data into a fixed number of clusters by, usually, optimizing a criterion function.

The partitioning dynamic cluster algorithms [10] are iterative two steps relocation algorithms involving at each iteration the construction of the clusters and the identification of a suitable representative or prototype (means, factorial axes, probability laws, groups of elements, etc.) of each cluster by locally optimizing an adequacy criterion between the clusters and their corresponding prototypes. This optimization process begins from a set of prototypes or an initial partition and interactively applies an *allocation step* (the prototypes are fixed) in order to assign the patterns to the clusters according to their proximity to the prototypes. This is followed by a *representation step* (the partition is fixed) where the prototypes are updated according to the assignment of the patterns in the allocation step, until achieving the convergence of the algorithm, when the adequacy criterion reaches a stationary value.

In classical data analysis, the items to be grouped are usually represented as a vector of quantitative or qualitative measurements where each column represents

C. Freksa, M. Kohlhase, and K. Schill (Eds.): KI 2006, LNAI 4314, pp. 260–273, 2007.

a variable. In particular, each individual takes just one single value for each variable. In practice, however, this model is too restrictive to represent complex data since to take into account variability and/or uncertainty inherent to the data, variables must assume sets of categories or intervals, possibly even with frequencies or weights.

The aim of *Symbolic Data Analysis* (SDA) is to extend classical data analysis techniques (clustering, factorial techniques, decision trees, etc.) to these kinds of data (sets of categories, intervals, or weight (probability) distributions) called symbolic data [2]. SDA is a domain in the area of knowledge discovery and data management related to multivariate analysis, pattern recognition and artificial intelligence.

SDA has provided partitioning methods in which different types of symbolic data are considered. Ralambondrany [17] extended the classical k-means clustering method in order to deal with data characterized by numerical and categorical variables. El-Sonbaty and Ismail [11] have presented a fuzzy k-means algorithm to cluster data on the basis of different types of symbolic variables. Bock [1] has proposed several clustering algorithms for symbolic data described by interval variables, based on a clustering criterion and thereby generalized similar approaches in classical data analysis. Chavent and Lechevallier [4] proposed a dynamic clustering algorithm for interval data where the class representatives are defined based on a modified Hausdorff distance. Souza and De Carvalho [18] have proposed partitioning clustering methods for interval data based on city-block distances. More recently, De Carvalho et al [8] proposed an algorithm using an adequacy criterion based on adaptive Hausdorff distances.

In this paper, we introduce a partitioning method for mixed feature-type symbolic data using the dynamic clustering methodology. To be able to manage ordered and non-ordered mixed feature-type symbolic data, this method assumes a previous pre-processing step the aim of which is to obtain a suitable homogenization of mixed symbolic data into modal symbolic data. In order to show the usefulness of this method, synthetic interval data sets ranging from different degree of difficulty to be clustered and applications with real data sets were considered. The evaluation of the clustering results is based on an external validity index.

This paper is organized as follow. Section 2 presents the data homogenization pre-processing step. Section 3 presents the dynamic clustering algorithm for mixed feature-type symbolic data. In Section 4 it is presented the evaluation of this method. The accuracy of the results furnished by this clustering method is assessed by the corrected Rand index [15] considering synthetic interval data sets in the framework of a Monte Carlo experience and applications with real data sets. Finally, Section 5 gives the conclusions and final remarks.

2   Data Homogenization Pre-processing Step

Usual data allow exactly one value for each variable. However, this data is not able to describe complex information, which must take into account variability and/or uncertainty. It is why symbolic variables have been introduced: multi-valued variables, interval variables and modal variables [2].

Let $\Omega = \{1, \ldots, n\}$ be a set of n items indexed by i described by p symbolic variables $X_1, \ldots, X_p$. A symbolic variable X_j is categorical multi-valued if, given an item i, $X_j(i) = x_i^j \subseteq A_j$ where $A_j = \{t_1^j, \ldots, t_{H_j}^j\}$ is a set of categories. A symbolic variable X_j is an interval variable when, given un item i, $X_j(i) = x_i^j = [a_i^j, b_i^j] \subseteq A_j$ where $A_j = [a, b]$ is an interval. Finally, a symbolic variable X_j is a modal variable if, given un item i, $X_j(i) = (S(i), \mathbf{q}(i))$ where $\mathbf{q}(i)$ is a vector of weights defined in $S(i)$ such that a weight $w(m)$ corresponds to each category $m \in S(i)$. $S(i)$ is the support of the measure $\mathbf{q}(i)$.

Each object i ($i = 1, \ldots, n$) is represented as a vector of mixed feature-type symbolic data $\mathbf{x}_i = (x_i^1, \ldots, x_i^p)$. This means that $x_i^j = X_j(i)$ can be a (ordered or non ordered) set of categories, an interval or a weight distribution according to the type of the corresponding symbolic variable.

Concerning the methods described in Chavent et al. [5], there is one of them which is a dynamic clustering algorithm based on a suitable squared Euclidean distance to cluster interval data. This method assumes a pre-processing step which transform interval data into modal data. However, the approach considered to accomplish this data transformation is not able to take into consideration the ordered nature inherent to interval data.

In this paper we consider a new data transformation pre-processing approach, the aim of which is to obtain a suitable homogenization of mixed symbolic data into modal symbolic data, which is able to manage ordered and non-ordered mixed feature-type symbolic data in the framework of a dynamic clustering algorithm. In this way, the presented dynamic cluster algorithm has as input data only vectors of weight distributions.

The data homogenization is accomplished according to type of symbolic variable: categorical non-ordered or ordered multi-valued variables, interval variables.

2.1   Categorical Multi-valued Variables

If X_j is a categorical non-ordered multi-valued variable, its transformation into a modal symbolic variable $\widetilde{X}_j$ is accomplished in the following way: $\widetilde{X}_j(i) = \widetilde{x}_i^j = (A_j, \mathbf{q}^j(i))$, where $\mathbf{q}^j(i) = (q_1^j(i), \ldots, q_{H_j}^j(i))$ is a vector of weights $q_h^j(i)$ ($h = 1, \ldots, H_j$), a weight being defined as [6]:

$$q_h^j(i) = \frac{c(\{t_h^j\} \cap x_i^j)}{c(x_i^j)} \qquad (1)$$

$c(A)$ being the cardinality of a finite set A.

If X_j is a categorical ordered multi-valued variable, its transformation into a modal symbolic variable $\widetilde{X}_j$ is accomplished in the following way: $\widetilde{X}_j(i) = \widetilde{x}_i^j = (A_j, \mathbf{Q}^j(i))$, where $\mathbf{Q}^j(i) = (Q_1^j(i), \ldots, Q_{H_j}^j(i))$ is a vector of cumulative weights $Q_h^j(i)$ ($h = 1, \ldots, H_j$), a cumulative weight being defined as:

$$Q_h^j(i) = \sum_{r=1}^{h} q_r^j(i), \quad \text{where } q_r^j(i) = \frac{c(\{t_r^j\} \cap x_i^j)}{c(x_i^j)} \tag{2}$$

It can be shown [6] that $0 \leq q_h^j(i) \leq 1$ $(h = 1, \ldots, H_j)$ and $\sum_{h=1}^{H_j} q_h^j(i) = 1$. Moreover, $q_1^j(i) = Q_1^j(i)$ and $q_h^j(i) = Q_h^j(i) - Q_{h-1}^j(i)$ $(h = 2, \ldots, H_j)$.

2.2   Interval Variables

In this case, the variable X_j is transformed into a modal symbolic variable $\widetilde{X}_j$ in the following way ([5], [6], [7]): $\widetilde{X}_j(i) = \widetilde{x}_i^j = (\widetilde{A}_j, \mathbf{Q}^j(i))$, where $\widetilde{A}_j = \{I_1^j, \ldots, I_{H_j}^j\}$ is a set of elementary intervals, $\mathbf{Q}^j(i) = (Q_1^j(i), \ldots, Q_{H_j}^j(i))$ and $Q_h^j(i)$ $(h = 1, \ldots, H_j)$ is defined as:

$$Q_h^j(i) = \sum_{r=1}^{h} q_r^j(i), \quad \text{where } q_r^j(i) = \frac{l(I_r^j \cap x_i^j)}{l(x_i^j)} \tag{3}$$

$l(I)$ being the length of a closed interval I.

The bounds of these elementary intervals I_h^j $(h = 1, \ldots, H_j)$ are obtained from the ordered bounds of the $n+1$ intervals $\{x_1^j, \ldots, x_n^j, [a, b]\}$. They have the following properties:

1. $\bigcup_{h=1}^{H_j} I_h^j = [a, b]$
2. $I_h^j \cap I_{h'}^j = \emptyset$ if $h \neq h'$
3. $\forall h \; \exists i \in \Omega$ such that $I_h^j \cap x_i^j \neq \emptyset$
4. $\forall i \; \exists S_i^j \subset \{1, \ldots, H_j\} : \bigcup_{h \in S_i^j} I_h^j = x_i^j$

It can be shown [6] that also in this case $0 \leq q_h^j(i) \leq 1$ $(h = 1, \ldots, H_j)$ and $\sum_{h=1}^{H_j} q_h^j(i) = 1$. Moreover, again $q_1^j(i) = Q_1^j(i)$ and $q_h^j(i) = Q_h^j(i) - Q_{h-1}^j(i)$ $(h = 2, \ldots, H_j)$.

2.3   Example

In order to illustrate this data homogenization pre-processing step, we considere here a symbolic data table which shows four countries (items), each country being described by a symbolic interval variable X_1 and a symbolic categorical non-ordered multi-valued variable X_2. In Table 1, symbolic variable X_1 is the minimum and maximum of the gross national product (in millions) whereas symbolic variable X_2 indicates the main industries from the list $A_2 = \{A = $agriculture, $C = $chemistry, $Co = $commerce, $E = $engineering, $En = $energy, $I = $informatic$\}$.

Concerning the symbolic variable X_1, the set of elementary intervals $\widetilde{A}_1 = \{I_1^1, \ldots, I_{H_1}^1\}$ are obtained as follows: at first, we consider the set of values formed by every bound (lower and upper) of all the intervals associated to the items. Then, such set of bounds is sorted in a growing way. Therefore,

Table 1. Countries described by symbolic variables

Country	X_1	X_2
1	[10,30]	$\{A, Co\}$
2	[25,35]	$\{C, Co, E\}$
3	[90,130]	$\{A, C, E\}$
4	[125,140]	$\{A, C, Co, E\}$

the set of elementary intervals is $\widetilde{A}_1 = \{I_1^1, I_2^1, I_3^1, I_4^1, I_5^1, I_6^1, I_7^1\}$, where $I_1^1 = [10, 25[, I_2^1 = [25, 30[, I_3^1 = [30, 35[, I_4^1 = [35, 90[, I_5^1 = [90, 125[, I_6^1 = [125, 130[$ and $I_7^1 = [130, 140]$.

Concerning symbolic variable X_2, $\widetilde{A}_2 = A_2 = \{A =$agriculture, $C =$chemistry, $Co =$commerce, $E =$engineering, $En =$energy, $I =$informatic$\}$.

In this way, each item (country) i $(i = 1, \ldots, 4)$ is represented as a vector of modal symbolic data $\widetilde{\mathbf{x}}_i = (\widetilde{x}_i^1, \widetilde{x}_i^2)$, where $\widetilde{x}_i^1 = (A_1, \mathbf{Q}^1(i))$ and $\widetilde{x}_i^2 = (A_2, \mathbf{q}^2(i))$, $\mathbf{q}^2(i)$ and $\mathbf{Q}^1(i)$ being obtained, respectively, as described in sections 2.1 and 2.2.

Finally, Table 2 shows the new symbolic data table obtained after the application of the data homogenization pre-processing step to the original symbolic data table:

Table 2. Countries described by two modal symbolic variables

Country	$\widetilde{X}_1$	$\widetilde{X}_2$
1	$(A_1, \mathbf{Q}^1(1) = (0.75, 1, 1, 1, 1, 1))$	$(A_2, \mathbf{Q}^2(1) = (0.5, 0.5, 1, 1, 1))$
2	$(A_1, \mathbf{Q}^1(2) = (0, 0.5, 1, 1, 1, 1))$	$(A_2, \mathbf{Q}^2(2) = (0, 0.33, 0.67, 1, 1, 1))$
3	$(A_1, \mathbf{Q}^1(3) = (0, 0, 0, 0, 0.88, 1, 1))$	$(A_2, \mathbf{Q}^2(3) = (0.33, 0.67, 0.67, 1, 1, 1))$
4	$(A_1, \mathbf{Q}^1(4) = (0, 0, 0, 0, 0, 0.33, 1))$	$(A_2, \mathbf{Q}^2(4) = (0.25, 0.50, 0.75, 1, 1, 1))$

3   A Dynamic Clustering Algorithm for Mixed Feature-Type Symbolic Data

This section presents a dynamic clustering method which allows to cluster mixed feature-type symbolic data. The aim of this method is to determine a partition $P = \{C_1, \ldots, C_K\}$ of Ω into K classes such that the resulting partition P is (locally) optimum with respect to a given clustering criteria.

Let $\Omega = \{1, \ldots, n\}$ be a set of n items. After the pre-processing step, each object i $(i = 1, \ldots, n)$ is represented by a vector of modal symbolic data $\widetilde{\mathbf{x}}_i = (\widetilde{x}_i^1, \ldots, \widetilde{x}_i^p)$, $\widetilde{x}_i^j = (\mathcal{D}_j, \mathbf{u}^j(i))$, where $\mathcal{D}_j$ is a (ordered or non-ordered) set of categories if $\widetilde{X}_j$ is a modal variable, $\mathcal{D}_j$ is a non-ordered set of categories if $\widetilde{X}_j$ is a categorical non-ordered multi-valued variable, $\mathcal{D}_j$ is an ordered set of categories if $\widetilde{X}_j$ is a categorical ordered multi-valued variable and $\mathcal{D}_j$ is a set of elementary intervals if $\widetilde{X}_j$ is an interval variable. Moreover, $\mathbf{u}^j(i) = (u_1^j(i), \ldots, u_{H_j}^j(i))$ is a vector of weights if $\mathcal{D}_j$ is a non-ordered set of categories and $\mathbf{u}^j(i)$ is a vector of cumulative weights if $\mathcal{D}_j$ is an ordered set of categories or a set of elementary intervals.

As in the standard dynamical clustering algorithm [10], this clustering method for symbolic data aims to provide a partition of Ω in a fixed number K of clusters $P = \{C_1, \ldots, C_K\}$ and a corresponding set of prototypes $L = \{L_1, \ldots, L_K\}$ by locally minimizing a criterion W that evaluates the fit between the clusters and their representatives.

Here, each prototype L_k of C_k $(k = 1, \ldots, K)$ is also represented as a vector of modal symbolic data $\mathbf{g}_k = (g_k^1, \ldots, g_k^p)$, $g_k^j = (\mathcal{D}_j, \mathbf{v}^j(k))$ $(j = 1, \ldots, p)$, where $\mathbf{v}^j(k) = (v_1^j(k), \ldots, v_{H_j}^j(k))$ is a vector of weights if $\mathcal{D}_j$ is a non-ordered set of categories and $\mathbf{v}^j(k)$ is a vector of cumulative weights if $\mathcal{D}_j$ is an ordered set of categories or a set of elementary intervals. Notice that for each variable the modal symbolic data presents the same support $\mathcal{D}_j$ for all individuals and prototypes. The criterion W is then defined as:

$$W(P, L) = \sum_{k=1}^{K} \sum_{i \in C_k} \phi(\widetilde{\mathbf{x}}_i, \mathbf{g}_k) \tag{4}$$

where

$$\phi(\widetilde{\mathbf{x}}_i, \mathbf{g}_k) = \sum_{j=1}^{p} d^2(\mathbf{u}^j(i), \mathbf{v}^j(k)) \tag{5}$$

The comparison between the two vectors of (non-cumulative or cumulative) weights $\mathbf{u}^j(i)$ and $\mathbf{v}^j(k)$ for the variable j is accomplished by a suitable squared Euclidean distance:

$$d^2(\mathbf{u}^j(i), \mathbf{v}^j(k)) = \sum_{h=1}^{H_j} (u_h^j(i) - v_h^j(k))^2 \tag{6}$$

The cumulative weights obtained in the pre-processing step will allow the dynamic clustering algorithm to take into account the order inherent to the categorical multi-valued or interval symbolic data.

As in the standard dynamical clustering algorithm [10], this algorithm starts from an initial partition and alternates a *representation step* and an *allocation step* until convergence when the criterion W reaches a stationary value representing a local minimum.

3.1 Representation Step: Definition of the Best Prototypes

In the representation step, each cluster C_k is fixed and the algorithm looks for the prototype $\mathbf{g}_k = (g_k^1, \ldots, g_k^p)$ of class C_k $(k = 1, \ldots, K)$ which minimizes the clustering criterion W in equation (4).

As the criterion W is additive, the optimization problem becomes to find for $k = 1, \ldots, K$, $j = 1, \ldots, p$ and $h = 1, \ldots, H_j$, the weight $v_h^j(k)$ minimizing

$$W(C_k, L_k) = \sum_{i \in C_k} (u_h^j(i) - v_h^j(k))^2 \qquad (7)$$

The solution for $v_h^j(k)$ is :

$$\hat{v}_h^j(k) = \frac{1}{n_k} \sum_{i \in C_k} u_h^j(i) \qquad (8)$$

where n_k is the cardinality of the class C_k. The prototype of class C_k is then $\hat{\mathbf{g}}_k = (\hat{g}_k^1, \ldots, \hat{g}_k^p)$, where $\hat{g}_k^j = (\mathcal{D}_j, \hat{v}_h^j(k))$.

3.2 Allocation Step: Definition of the Best Partition

In this step, the vector of prototypes $L = (L_1, \ldots, L_K)$ is fixed. The clusters C_k ($k = 1, \ldots, K$), which minimize the criterion W, are updated according to the following allocation rule:

$$C_k = \{i \in \Omega : \phi(\tilde{\mathbf{x}}_i, \mathbf{g}_k) \le \phi(\tilde{\mathbf{x}}_i, \mathbf{g}_m), \forall m \ne k \, (m = 1, \ldots, K)\} \qquad (9)$$

3.3 The Algorithm

The algorithm has the following steps:

SCHEMA OF DYNAMIC CLUSTERING ALGORITHM FOR MIXED FEATURE-TYPE SYMBOLIC DATA

1. **Initialization.**
   Randomly choose a partition $\{C_1 \ldots, C_K\}$ of Ω or randomly choose K distinct objects $L_1, \ldots, L_K$ belonging to Ω and assign each objects i to the closest prototype L_{k*}, where $k* = \arg \min_{k=1,\ldots,K} \phi(\tilde{\mathbf{x}}_i, \mathbf{g}_k)$.
2. **Representation step: definition of the best prototypes.**
   (the partition P is fixed)
   For $k = 1, \ldots, K$, compute the vector of modal symbolic data $\mathbf{g}_k = (g_k^1, \ldots, g_k^p)$, $g_k^j = (\mathcal{D}_j, \mathbf{v}^j(k))$ ($j = 1, \ldots, p$), representing the prototype L_k, where $\mathbf{v}^j(k) = (v_1^j(k), \ldots, v_{H_j}^j(k))$ and $v_h^j(k)$ ($h = 1, \ldots, H_j$) is given by equation (8).
3. **Allocation step: definition of the best partition.**
   (the set of prototypes L is fixed)
   $test \leftarrow 0$
   for $i = 1$ to n do
       define the cluster C_{k*} such that
           $k* = arg \min_{k=1,\ldots,K} \phi(\tilde{\mathbf{x}}_i, \mathbf{g}_k)$
       if $i \in C_k$ and $k* \ne k$
           $test \leftarrow 1$
           $C_{k*} \leftarrow C_{k*} \cup \{i\}$
           $C_k \leftarrow C_k \setminus \{i\}$
4. **Stopping criterion.**
   If $test = 0$ then STOP, else go to (2).

4   Experimental Evaluation

To show the usefulness of this method, experiments with synthetic symbolic interval data sets of different degrees of clustering difficulty (clusters of different shapes and sizes, linearly non-separable clusters, etc) and applications with real data sets are considered. For synthetic symbolic interval data sets, the evaluation was performed in the framework of a Monte Carlo experience: 100 replications are considered for each interval data set.

To evaluate the clustering results furnished by this dynamic clustering method an external index, the adjusted Rand index (CR), will be considered [15]. The CR index measures the similarity between an a priori partition and a partition furnished by the clustering algorithm.

The definition of the index CR is as follows.

Let $U = \{u_1, \ldots, u_i, \ldots, u_R\}$ and $V = \{v_1, \ldots, v_j, \ldots, v_C\}$ be two partitions of the same data set having respectively R and C clusters. The corrected Rand index is:

$$CR = \frac{\sum_{i=1}^{R} \sum_{j=1}^{C} \binom{n_{ij}}{2} - \binom{n}{2}^{-1} \sum_{i=1}^{R} \binom{n_{i.}}{2} \sum_{j=1}^{C} \binom{n_{.j}}{2}}{\frac{1}{2}[\sum_{i=1}^{R} \binom{n_{i.}}{2} + \sum_{j=1}^{C} \binom{n_{.j}}{2}] - \binom{n}{2}^{-1} \sum_{i=1}^{R} \binom{n_{i.}}{2} \sum_{j=1}^{C} \binom{n_{.j}}{2}} \tag{10}$$

where $\binom{n}{2} = \frac{n(n-1)}{2}$ and n_{ij} represents the number of objects that are in clusters u_i and v_i; $n_{i.}$ indicates the number of objects in cluster u_i; $n_{.j}$ indicates the number of objects in cluster v_j; and n is the total number of objects in the data set. CR takes its values from the interval [-1,1], where the value 1 indicates perfect agreement between partitions, whereas values near 0 (or negatives) correspond to cluster agreement found by chance.

Our aim is to compare the approach presented in [5], which transforms interval symbolic data on modal symbolic data represented by non-cumulative weight distributions, with the approach presented in this paper, which transforms interval symbolic data on modal symbolic data represented by cumulative weight distributions.

4.1   The Monte Carlo Experiences

In each experiment, we considered two standard quantitative data sets in $\Re^2$. Each data set has 300 points scattered among three classes of unequal sizes 150, 100 and 50, respectively, and elliptical shapes. Each class in these quantitative data sets were drawn according to a bi-variate normal distribution with vector $\boldsymbol{\mu}$ and covariance matrix $\boldsymbol{\Sigma}$ represented by:

$$\boldsymbol{\mu} = \begin{bmatrix} \mu_1 \\ \mu_2 \end{bmatrix} \text{ and } \boldsymbol{\Sigma} = \begin{bmatrix} \sigma_1^2 & \rho\sigma_1\sigma_2 \\ \rho\sigma_1\sigma_2 & \sigma_2^2 \end{bmatrix}$$

We will consider two different configurations for the standard quantitative data sets: 1) data drawn according to a bi-variate normal distribution with well separated classes and 2) data drawn according to a bi-variate normal distribution with overlapping classes.

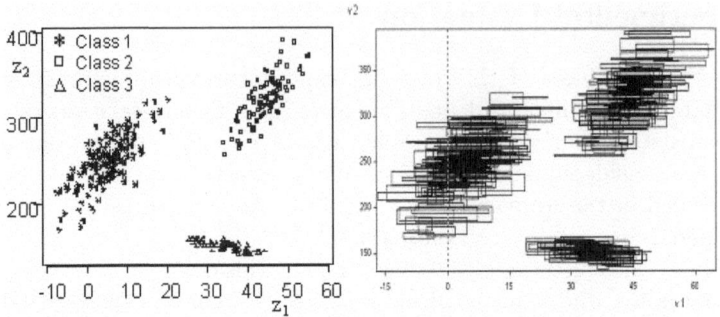

Fig. 1. Symbolic interval (right side) and seed (left side) data sets 1

Each data point (z_1, z_2) of each one of these synthetic quantitative data sets is a seed of a vector of intervals (rectangle): $([z_1 - \gamma_1/2, z_1 + \gamma_1/2], [z_2 - \gamma_2/2, z_2 + \gamma_2/2])$. These parameters γ_1, γ_2 are randomly selected from the same predefined interval. The intervals considered in this paper are: $[1, 10], [1, 20], [1, 30]$ and $[1, 40]$.

Symbolic interval data set 1 (Figure 1, right side), showing classes well separated, was constructed from the standard data set 1 (Figure 1, left side) drawn according to the following parameters:

a) Class 1: $\mu_1 = 5$, $\mu_2 = 250$, $\sigma_1^2 = 25$, $\sigma_2^2 = 90$ and $\rho_{12} = 0.7$;
b) Class 2: $\mu_1 = 45$, $\mu_2 = 320$, $\sigma_1^2 = 25$, $\sigma_2^2 = 90$ and $\rho_{12} = 0.8$;
c) Class 3: $\mu_1 = 35$, $\mu_2 = 150$, $\sigma_1^2 = 25$, $\sigma_2^2 = 25$ and $\rho_{12} = -0.7$;

Symbolic interval data set 2 (Figure 2, right side), showing overlapping classes, was constructed from standard data set 2 (Figure 2, left side) drawn according to the following parameters:

a) Class 1: $\mu_1 = 45$, $\mu_2 = 22$, $\sigma_1^2 = 25$, $\sigma_2^2 = 90$ and $\rho_{12} = 0.7$;
b) Class 2: $\mu_1 = 60$, $\mu_2 = 30$, $\sigma_1^2 = 25$, $\sigma_2^2 = 90$ and $\rho_{12} = 0.8$;
c) Class 3: $\mu_1 = 52$, $\mu_2 = 38$, $\sigma_1^2 = 25$, $\sigma_2^2 = 25$ and $\rho_{12} = -0.7$;

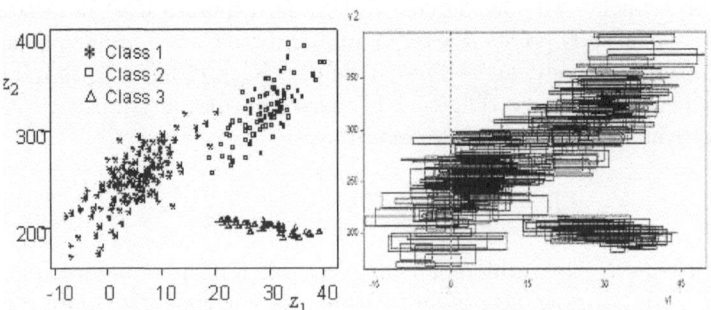

Fig. 2. Symbolic interval (right side) and seed (left side) data sets 2

In each replication of a Monte Carlo experience a clustering method is run (until the convergence to a stationary value of the adequacy criterion) 50 times and the best result, according to the corresponding criterion, is selected. The average of the corrected Rand (CR) index [15] among these 100 replications is calculated.

Tables 1 and 2 show the values of the average and standard deviation of the CR index according to different methods and interval data sets 1 and 2, respectively.

Table 3. Comparison between clustering methods for interval data set 1

Range of values	Cumulative Weight Data Homogenization Method		Non-Cumulative Weight Data Homogenization Method	
of γ_i $i = 1, 2$	Average	Standard Deviation	Average	Standard Deviation
$\gamma_i \in [1, 10]$	0.9940	0.0000	0.7264	0.0091
$\gamma_i \in [1, 20]$	0.9925	0.0000	0.8167	0.0049
$\gamma_i \in [1, 30]$	0.9866	0.0001	0.8355	0.0086
$\gamma_i \in [1, 40]$	0.9799	0.0002	0.8520	0.0045

In conclusion, for these data configurations (showing well separated and overlapping classes), the clustering cumulative weight approach, as it takes into consideration the ordered nature inherent to symbolic interval data, outperforms the clustering non-cumulative weight one.

Table 4. Comparison between clustering methods for interval data set 2

Range of values	Cumulative Weight Data Homogenization Method		Non-Cumulative Weight Data Non-Cumulative Weight Method	
of γ_i $i = 1, 2$	Average	Standard Deviation	Average	Standard Deviation
$\gamma_i \in [1, 10]$	0.3515	0.0016	0.1680	0.0014
$\gamma_i \in [1, 20]$	0.3264	0.0035	0.1329	0.0006
$\gamma_i \in [1, 30]$	0.2548	0.0117	0.0956	0.0006
$\gamma_i \in [1, 40]$	0.1148	0.0100	0.0851	0.0007

4.2 Applications with Real Data Sets

We apply the dynamic clustering algorithm using the non-cumulative and cumulative weight data homogenization approaches to two real symbolic interval data sets.

Car data set. The car data set consists of a set of 33 car models described by 8 interval, 3 multi-valued variables. In this application, the 8 interval variables - *Price, Engine Capacity, Top Speed, Acceleration, Step, Length, Width* and *Height*, 2 categorical non-ordered multi-valued variables - *Alimentation* and *Traction* - have been considered for clustering purposes and 1 nominal variable

Table 5. Car data set with 8 interval, two non-ordered multi-valued and one nominal variables

Model	Price	Engine Capacity	Alimentation	Traction	...	Height	Category
Alfa 145	[27806, 33596]	[1370, 1910]	(1,2)	(1)	...	[143, 143]	Utilitarian
Alfa 156	[41593, 62291]	[1598, 2492]	(1)	(1)	...	[142, 142]	Berlina
Alfa 166 L	[64499, 88760]	[1970, 2959]	(1)	(1)	...	[142, 142]	Luxury
Aston Martin S	[260500, 460000]	[5935, 5935]	(1)	(2)	...	[124, 132]	Sport
Audi A3 U	[40230, 68838]	[1595, 1781]	(1)	(1,3)	...	[142, 142]	Utilitarian
Audi A6 B	[68216, 140265]	[1781, 4172]	(1)	(1,3)	...	[145, 145]	Berlina
Audi A8 L	[123849, 171417]	[2771, 4172]	(1)	(3)	...	[144, 144]	Luxury
Bmw serie 3 B	[45407, 76392]	[1796, 2979]	(1)	(2,3)	...	[142, 142]	Berlina
...	...	...	...	...	...	...	...
Rover 25 U	[21492, 33042]	[1119, 1994]	(1,2)	(1)	...	[142, 142]	Utilitarian
Rover 75 B	[50490, 65399]	[1796, 2497]	(1)	(1)	...	[143, 143]	Berlina
Skoda Fabia U	[19519, 32686]	[1397, 1896]	(1,2)	(1)	...	[145, 145]	Utilitarian
Skoda Octavia B	[27419, 48679]	[1585, 1896]	(1,2)	(1)	...	[143, 143]	Berlina
Passat L	[39676, 63455]	[1595, 2496]	(1,2)	(1,3)	...	[146, 146]	Luxury

Car Category has been used as a *a priori* classification. Table 5 shows part of this car data set.

Table 6 shows for each cluster their individuals and respective *a priori* class labels given by the clustering algorithms using non-cumulative and cumulative weight data homogenization methods. Concerning the result in this table, the CR indices taken with respect to Car category were 0.558 and 0.2035, respectively, for the cumulative and non-cumulative weight data homogenization methods. This indicates that, for this data set, the cumulative version outperforms the non-cumulative.

Table 6. Clustering Results for the Car data set

Method	Cluster 1	Cluster 2	Cluster 3	Cluster 4
Cumulative weights	12/U 13/U 17/U 24/U 25/U 28/U 29/U 31/U	6/B 7/L 9/L 10/L 22/L 23/L	4/S 11/S 15/S 16/S 19/S 20/S 27/S	1/U 2/B 3/L 5/U 8/B 14/B 18/L 21/B 26/B 30/B 32/B 33/L
Non-Cumulative weights	4/S 9/L 11/S 15/S 19/U 20/S 21/B 22/L 23/L	6/B 12/U 13/U 17/U 18/L 24/U 25/U 29/U 31/U	1/U 14/B 26/B 30/B 32/B 30/B 32/B	2/B 3/L 5/U 7/L 8/B 16/B 27/S 28/U 33/L

Ecotoxicology data set. Several studies realized in French Guyana indicated abnormal levels of mercury contamination in some Amerindian populations. This contamination was connected to their high consumption of contaminated freshwater fish [3]. In order to get a better knowledge of this phenomenon, a data set has been collected by researchers from the LEESA (Laboratoire d'Ecophysiologie et d'Ecotoxicologie des Systèmes Aquatiques) laboratory.

This data set concerns 12 species of freshwater fish, each species being described by 13 interval variables. These species are grouped into four a priori clusters of unequal sizes according to diet: two clusters (Carnivorous and Detritivorous) of size 4 and two clusters of size 2 (Omnivorous and Herbivorous). Table 7 shows the freshwater fish data set.

Table 7. Freshwater Fish Data Set described by 13 interval symbolic variables

Individuals/labels	Interval Variables				
	Length	Weight	...	Intestin/ Muscle	Stomach/ Muscle
Ageneiosusbrevifili: 1	[1.8 : 7.1]	[2.1 : 7.2]	...	[7.8 : 17.9]	[4.3 : 11.8]
Cynodongibbus: 1	[19 : 32]	[77 : 359]	...	[0 : 0.5]	[0.2 : 1.24]
Hopliasaïmara: 1	[25.5 : 63]	[340 : 5500]	...	[0.11 : 0.49]	[0.09 : 0.4]
Potamotrygonhy: 1	[20.5 : 45]	[400 : 6250]	...	[0 : 1.25]	[0 : 0.5]
Leporinusfasciatus: 3	[18.8 : 25]	[125 : 273]	...	[0 : 0]	[0.12 : 0.17]
Leporinusfrederici: 3	[23 : 24.5]	[290 : 350]	...	[0.18 : 0.24]	[0.13 : 0.58]
Dorasmicropoeus: 2	[19.2 : 31]	[128 : 505]	...	[0 : 1.48]	[0 : 0.79]
Platydorascostatus: 2	13.7 : 25]	[60 : 413]	...	[0.3 : 1.45]	[0 : 0.61]
Pseudoancistrus: 2	[13 : 20.5]	[55 : 210]	...	[0 : 2.31]	[0.49 : 1.36]
Semaprochilodusvari: 2	[22 : 28]	[330 : 700]	...	[0.4 : 1.68]	[0 : 1.25]
Acnodonoligacanthus: 4	[10 : 16.2]	[34.9 : 154.7]	...	[0 : 2.16]	[0.23 : 5.97]
Myleusrubripinis: 4	[2.7 : 8.4]	[2.7 : 8.7]	...	[8.2 : 20]	[5.1 : 13.3]

Table 8. Clustering Results for the Freshwater fish data set

Method	Cluster 1	Cluster 2	Cluster 3	Cluster 4
Non-cumulative weights	4/1 7/2 8/2 10/2	9/2 11/4 12/4	1/1 2/1 3/1	5/3 6/3
Cumulative weights	4/1 9/2	6/3 11/4	1/1 2/1 3/1 7/2 8/2 10/2	5/3 12/4

Table 8 shows for each cluster their individuals and respective *a priori* class labels given by the clustering algorithms using cumulative and non-cumulative weight data homogenization methods. The CR indices obtained from the results displayed in Table 8 are: 0.488 and 0.179 for the clustering cumulative and non-cumulative weight approaches, respectively.

In conclusion, for these real data sets, the performance of the clustering cumulative weight approach measured by the CR index is superior to the clustering non-cumulative weight one.

5   Concluding Remarks

A partitioning clustering method for mixed feature-type symbolic data using a dynamic cluster algorithm based on the squared Euclidean distance was presented in this paper. To be able to manage ordered and non-ordered mixed feature-type symbolic data, it was introduced a preceding pre-processing step the aim of which is to obtain a suitable homogenization of mixed symbolic data into modal symbolic data represented by weight distributions. These weights are non-cumulative, if the symbolic data are non-ordered sets of categories and cumulative, if the symbolic data are ordered sets of categories or a set of elementary intervals. The algorithm locally optimizes an adequacy criterion that measures the fitting between the classes and their representatives (prototypes).

To compare classes and prototypes, a suitable squared Euclidean distance for modal data is introduced.

The dynamic clustering algorithm starts from an initial partition and alternates a representation step and an allocation step until convergence when the adequacy criterion reaches a stationary value representing a local minimum. In the representation step, the solution for the best prototype of each class, presented in this paper, is a vector of (cumulative or non-cumulative) weight distributions whose weights, for a given variable, are the average of the (cumulative or non cumulative) weights computed for the objects belonging to this class.

An experimental evaluation in order to compare the results of the dynamic clustering algorithm using non-cumulative and cumulative weight vectors to represent interval data has been carried out. The accuracy of the results furnished by these clustering methods were assessed by the corrected Rand index considering synthetic interval data sets in the framework of a Monte Carlo experience and applications with real data sets. These results clearly show that the accuracy of the clustering method using cumulative weight vectors to represent interval data is superior to that which uses non-cumulative weight vectors.

Acknowledgments. The authors would like to thank CNPq (Brazilian Agency) for its financial support.

References

1. Bock, H. H.: Clustering algorithms and kohonen maps for symbolic data. Proc. ICNCB, Osaka, 203-215. J. Jpn. Soc. Comp. Statistic, **15**, (2002) 1–13
2. Bock, H. H. and Diday, E.: *Analysis of Symbolic Data, Exploratory methods for extracting statistical information from complex data.* Springer, Heidelberg, (2000)
3. Bobou, A. and Ribeyre, F.: Mercury in the food web: accumulation and transfer mechanisms, in Sigrel A. and Sigrel H. Eds., Metal Ions in Biological Systems. M. Dekker, New York, (1998) 289-319.
4. Chavent, M. and Lechevallier, Y.: Dynamical Clustering Algorithm of Interval Data: Optimization of an Adequacy Criterion Based on Hausdorff Distance. In: A. Sokolowski and H.-H. Bock (Eds.): *Classification, Clustering and Data Analysis.* Springer, Heidelberg, (2002) 53–59
5. Chavent, M., De Carvalho, F. A. T., Lechevallier, Y. and Verde, R.: Trois nouvelles mthodes de classification automatique de donnes symboliques de type intervalle. *Revue de Statistique Applique*, v. LI, n. 4, p. (2003) 5–29
6. De Carvalho, F. A. T.: Histograms In Symbolic Data Analysis. *Annals of Operations Research*, v. 55, p. (1995) 229–322
7. De Carvalho, F. A. T.; Verde, R.; Lechevallier, Y.: A dynamical clustering of symbolic objcts based on a context dependent proximity measure. In : *Proceedings of the IX International Symposium on Applied Stochastic Models and Data analysis. Lisboa*, Universidade de Lisboa, p. (1999) 237–242
8. De Carvalho, F.A.T, Souza, R.M.C.R., Chavent, M. and Lechevallier, Y.: Adaptive Hausdorff distances and dynamic clustering of symbolic data. Pattern Recognition Letters, **27** (3) (2006) 167–179
9. Diday, E. and Brito, P.: Symbolic Cluster Analysis. In: O. Opitz (Ed.): *Conceptual and Numerical Analysis of Data.* Springer, Heidelberg, (1989) 45-84

10. Diday, E. and Simon, J. J.: Clustering Analysis. In: Fu, K. S. (Eds): *Digital Pattern Recognition.* Springer-Verlag, Heidelberg, (1976) 47-94
11. El-Sonbaty, Y. and Ismail, M. A.: Fuzzy Clustering for Symbolic Data. *IEEE Transactions on Fuzzy Systems* **6**, (1998) 195-204
12. Everitt, B.: *Cluster Analysis.* Halsted, New York (2001)
13. Gordon, A. D.: *Classification.* Chapman and Hall/CRC, Boca Raton, Florida (1999)
14. Gordon, A. D.: An Iteractive Relocation Algorithm for Classifying Symbolic Data. In: W. Gaul *et al* (Eds.): *Data Analysis: Scientific Modeling and Practical Application.* Springer-Verlag, Berlin, (2000) 17-23
15. Hubert, L. and Arabie. P.: Comparing Partitions. *Journal of Classification,* **2** (1985) 193-218
16. Jain, A.K., Murty, M.N. and Flynn, P.J.: Data Clustering: A Review. ACM Computing Surveys, **31**, (3) (1999) 264-323
17. Ralambondrainy, H.: A conceptual version of the *k*-means algorithm. *Pattern Recognition Letters* **16**, (1995) 1147-1157
18. Souza, R. M. C. R. and De Carvalho, F. A. T.: Clustering of interval data based on city-block distances. *Pattern Recognition Letters,* **25** (3), (2004) 353-365

On Generalizing Orientation Information in $\mathcal{OPRA}_m$

Frank Dylla and Jan Oliver Wallgrün

SFB/TR 8 Spatial Cognition
Universität Bremen
Bibliothekstr. 1, 28359 Bremen, Germany
{dylla,wallgruen}@sfbtr8.uni-bremen.de

Abstract. Research on qualitative spatial reasoning has produced a variety of calculi for reasoning about orientation or direction relations between point objects or line segments. Altough it is obvious that some calculi are more general than others, the exact relationships between the calculi have not been investigated thoroughly. We show that many well-known orientation calculi can be expressed in the more general $\mathcal{OPRA}_m$ calculus which allows to translate information from one calculus into another. In addition, we demonstrate that the mapping can be exploited to automate typically complex tasks like determining or verifying composition tables.

1 Introduction

Qualitative representation of space abstracts from the physical world and qualitative spatial reasoning (QSR) enables computers to make predictions about spatial relations, even when precise quantitative information is not available [2]. The two main research directions in QSR are topological reasoning about regions [16,5,18] and positional reasoning about configurations of point objects [8,6,11,10,14,13,17] or line segments [15,19,4]. Calculi dealing with such information have been well investigated over the recent years and provide general and sound reasoning mechanisms. An overview is given in [3].

In this text we focus on calculi for reasoning about orientation or direction relations. A variety of such calculi have been proposed, among them the FlipFlop Calculus [10] and its $\mathcal{LR}$ [20] refinement, the Dipole Relation Algebra [15,19] with its fine grained variants $\mathcal{DRA}_f$ and $\mathcal{DRA}_{fp}$ [4], the Single Cross Calculus [8] and Double Cross Calculus [8], the Qualitative Trajectory Calculus [22], as well as the Oriented Point Relation Algebra ($\mathcal{OPRA}_m$) with adjustable granularity [13,12]. While some relationships between these calculi are rather obvious – e.g. the Double Cross Calculus is a refinement of the Single Cross Calculus – their relationships have not been investigated in general.

On the other hand, we will argue that knowing the exact relationship between two calculi can be exploited to build tools that reduce the typically complex tasks involved in designing and implementing a spatial calculus. For instance, for every calculus the composition operation has to be specified which is usually done by creating the composition table by hand. Determining the correct composition of two relations is a time-consuming, difficult and error-prone process for most of the calculi listed above.

C. Freksa, M. Kohlhase, and K. Schill (Eds.): KI 2006, LNAI 4314, pp. 274–288, 2007.

However, if the calculus can be expressed in another already completely specified calculus this can facilitate the automatic generation or verification of the composition table. Another problematic task which could be supported in a similar way is to determine the neighborhood structure of a spatial calculus, if we want to employ it for neighborhood-based reasoning [7,4].

We take a step towards understanding the relationships between the above-mentioned calculi by showing that and how they all can be mapped to the $\mathcal{OPRA}_m$ calculus which turns out to be the most general one among those we consider here. Mapping other calculi into the algebraically well-defined framework of $\mathcal{OPRA}_m$ facilitates reasoning over relations from different calculi. We will apply it to translate information from one calculus to another which can be valuable for applications that employ multiple calculi. In addition, we will consider the problem of automatically determining or verifying the composition operation for a given calculus.

The remainder of the text is structured as follows: In Section 2 we give a brief overview on the orientation calculi relevant for the rest of the paper. We then proceed by showing how each of these calculi can be mapped into the $\mathcal{OPRA}_m$ calculus (Section 3). In Section 4, we demonstrate the merits of these mappings by employing it to translate information from the Double Cross Calculus to the FlipFlop calculus and by considering the task of determining the composition table for the Double Cross Calculus.

2   Qualitative Orientation Calculi

Qualitative calculi are based on a certain type of entity, time intervals in the case of Allen's Interval Calculus [1], or points, line segments or regions in typical spatial calculi. A calculus consists of a finite set of base relations forming a partition of all possible relations. Generally speaking, a qualitative relation describes a property between an n-tuple of entities abstracting from numerical values. In the following, we will investigate a selected set of binary and ternary calculi proposed for reasoning about orientation information. We will begin by briefly describing each of these calculi.

FlipFlop Calculus (FFC) and the $\mathcal{LR}$ Refinement. The FlipFlop calculus proposed in [10] describes the position of a point C (the referent) in the plane with respect to two other points A (the origin) and B (the relatum) as illustrated in Figure 1. It can for instance be used to describe the spatial relation of C to B as seen from A or as the spatial representation perceived after moving from A to B. For configurations with $A \neq B$ the following base relations are distinguished: C can be to the **left** or to the **right** of the oriented line going through A and B, or C can be placed on the line resulting in one of the five relations **inside**, **front**, **back**, **start** ($C = A$) or **end** ($C = B$) (cp. Figure 1). Relations for the case where A and B coincide were not included in Ligozat's original definition [10]. This was done with the $\mathcal{LR}$ refinement [20] that introduces the relations **dou** ($A = B \neq C$) and **tri** ($A = B = C$) as additional relations, resulting in 9 base relations overall. A $\mathcal{LR}$ relation $rel_{\mathcal{LR}}$ is written as $A, B\ rel_{\mathcal{LR}}\ C$, e.g. $A, B\ \mathbf{r}\ C$ as depicted in Figure 1.

276   F. Dylla and J.O. Wallgrün

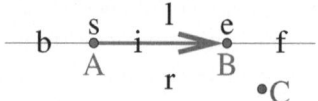

Fig. 1. The reference frame for the $\mathcal{LR}$ calculus, an enhanced version of the FlipFlop Calculus

Single Cross Calculus (SCC). The single cross calculus is a ternary calculus that describes the direction of a point C (the referent) with respect to a point B (the relatum) as seen from a third point A (the origin). It has originally been proposed in [8]. The plane is partitioned into regions by the line going through A and B and the perpendicular at B. This results in eight possible directions for C as illustrated in Figure 2(a). We denote these base relations by numbers from 0 to 7 instead of using linguistic prepositions, e.g. *left* instead of 2, as originally done in [8]. Relations 0,2,4,6 are linear ones, while relations 1,3,5,7 are planar. In addition, three special relations exist for the cases $A \neq B = C$ (**bc**), $A = B \neq C$ (**dou**), and $A = B = C$ (**tri**). A single cross relation rel_{SCC} is written as $A, B \; rel_{SCC} \; C$, e.g. $A, B \; \mathbf{4} \; C$ or $A, B \; \mathbf{dou} \; C$. The relation depicted in Figure 2(a) is the relation $A, B \; \mathbf{5} \; C$.

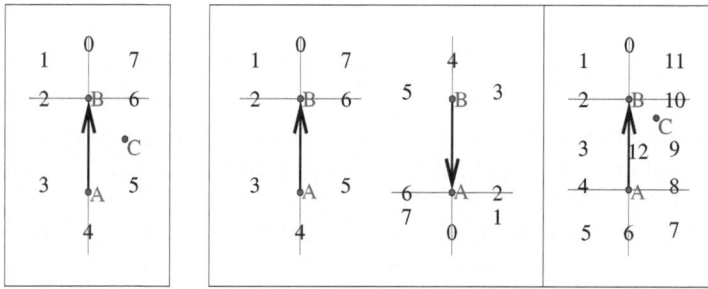

(a) Single Cross Calculus reference frame.

(b) The two Single Cross reference frames resulting in the overall Double Cross Calculus reference frame.

Fig. 2. The Single and Double Cross Reference System

Double Cross Calculus (DCC). The double cross calculus [8] can be seen as an extension of the single cross calculus adding another perpendicular, this time at A (see Figure 2(b) (right)). It can also be interpreted as the combination of two single cross relations, the first describing the position of C with respect to B as seen from A and the second with respect to A as seen from B (cp. Figure 2(b) (left)). The resulting partition distinguishes 13 relations (7 linear and 6 planar) denoted by numbers from 0 to 12 and four special cases, $A = C \neq B$ (**a**), $A \neq B = C$ (**b**), $A = B \neq C$ (**dou**), and $A = B = C$ (**tri**), resulting in 17 base relations overall. In Figure 2(b) the relation $A, B \; \mathbf{9} \; C$ is depicted.

Dipole Relation Algebra: $\mathcal{DRA}_f$ and $\mathcal{DRA}_{fp}$. A dipole is an oriented line segment as e.g. determined by a start and an end point. We will write d_{AB} for a dipole defined by start point A and end point B. The idea was first introduced in [19] by Schlieder and extended in [15]. The fine-grained dipole calculus ($\mathcal{DRA}_f$) [4] describes the orientation relation between two dipoles d_{AB} and d_{CD}. Each base relation is a 4-tuple (r_1, r_2, r_3, r_4) of FlipFlop relations. r_1 describes the relation of C with respect to the dipole d_{AB}, r_2 of D with respect to d_{AB}, r_3 of A with respect to d_{CD}, and r_4 of b with respect to d_{CD}. The relations are usually written without the commas, e.g. $(rrll)$. Thus, the example in Figure 3 shows the relation d_{AB} $(rlll)$ d_{CD}. $\mathcal{DRA}_f$ has 72 base relations.

An extended version called $\mathcal{DRA}_{fp}$ [4] further classifies the angle s that would result from translating d_{CD} so that both start points coincide. Four cases are distinguished: **Parallel** ($s = 0°$), **Anti-Parallel** ($s = 180°$), **+** ($s \in \,]0°..180°[$), and **-** ($s \in \,]180°..360°[$). This results in 80 base relations and the example from Figure 3 depicts the $\mathcal{DRA}_{fp}$ relation $(rlll+)$.

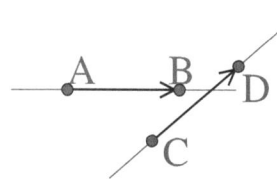

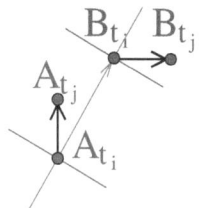

Fig. 3. A dipole configuration: d_{AB} $(rlll)$ d_{CD} in the fine-grained dipole relation algebra ($\mathcal{DRA}_f$) or d_{AB} $(rlll+)$ d_{CD} in the extended version $\mathcal{DRA}_{fp}$

Fig. 4. A spatial motion configuration resulting in A $(-+--)$ B in QTC

Qualitative Trajectory Calculus (QTC). The Qualitative Trajectory Calculus [21,22] was developed recently handling changes between two moving objects explicitly with respect to a 'double cross' reference system build up by the observation of two objects' positions A and B at a certain time point t_i, denoted by A_{t_i} and B_{t_i} in the following. The relative position changes between t_i and t_j (with $j > i$) are expressed within this reference system. The relations are denoted by sequences of four symbols taken from +, -, or 0. The first symbol describes the change in distance of A wrt. the perpendicular to the line $A_{t_i} B_{t_i}$ at point B_{t_i} that occurred during t_i and t_j. In Figure 4, A_{t_j} is on the same side of the perpendicular at A_{t_i} as B_{t_i}, and thus it is closer and the symbol takes the value -. The second symbol expresses the change in distance of B_{t_j} compared to the perpendicular at A_{t_i}. B_{t_j} is further away from it, therefore the value is +. The third and fourth symbol abstract the relative motion to the side regarding the directed line from A_{t_i} to B_{t_i}, respectively from B_{t_i} to A_{t_i}. In our example A moves to the left (-) of $A_{t_i} B_{t_i}$ and B to the left (-) of $B_{t_i} A_{t_i}$. This results in the relation A $(-+--)$ B.

Oriented Point Relation Algebra $\mathcal{OPRA}_m$: A calculus based on two oriented points. The $\mathcal{OPRA}_m$ calculus [13,12] relates two oriented points A and B (points in the plane with an additional direction parameter) and describes their relative orientation towards each other. $\mathcal{OPRA}_m$ is well suited for dealing with objects that have an intrinsic front or a move in a particular direction and can be abstracted as points. The granularity factor $m \in \mathbb{N}_+$ determines the number of distinguished relations. For each of the points, m lines are used to partition the plane into $2m$ planar and $2m$ linear regions. Figure 5 shows the partitions for the cases $m = 2$ (5(a)) and $m = 4$ (5(b)). The orientation of the two points is depicted by the arrows starting at A and B, respectively. The regions are numbered from 0 to $(4m-1)$, region 0 always coincides with the orientation of the point. An $\mathcal{OPRA}_m$ relation $rel_{\mathcal{OPRA}_m}$ consist of pairs (reg_i, reg_j) where reg_i is the region of A in which B falls into, while reg_j is the region of B that contains A. They are usually written as A $_m\angle_i^j$ B with $i, j \in \mathcal{Z}_{4m}$ [1]. Thus, the examples in Figure 5 depict the relations A $_2\angle_7^1$ B and A $_4\angle_{13}^3$ B. Additional relations describe situations in which both oriented points coincide. In these cases, the relation is determined by the region reg of A the orientation arrow of B falls into as illustrated in Figure 5(c). These relations are written as A $_2\angle reg$ B (A $_2\angle 1$ B in the example).

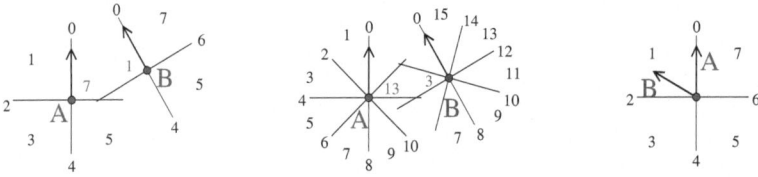

(a) with granularity $m = 2$: (b) with granularity $m = 4$: (c) case where A and B coin-
A $_2\angle_7^1$ B A $_4\angle_{13}^3$ B cide: A $_2\angle 1$ B

Fig. 5. Two oriented points related at different granularities

3   Mapping Orientation Calculi into $\mathcal{OPRA}_m$

The idea of this text is to encode relations from different orientation calculi within a single framework such that they can be combined in a single reasoning system and information can be translated from one calculi into the other. As we will see, the $\mathcal{OPRA}_m$ calculus is a suitable candidate for this framework as it is expressive enough to encode the relations from the other considered calculi. In addition, it offers a clear algebraic definition for transforming relations between different granularities especially if they are multiples of each other [13,12]. In the following, we will give transformations for representing the base relations of the other calculi in the $\mathcal{OPRA}_m$ framework.

3.1   Preliminaries

In this section, we will on the one hand deal with *normal points* P being defined by their position in the plane ($P = (x_P, y_P) \in \mathbb{R}^2$). On the other hand we will deal with

---

[1] $\mathcal{Z}_{4m}$ defines a cyclic group with $4m$ elements.

oriented points as required for the $\mathcal{OPRA}_m$ calculus. An oriented point, written as $\boldsymbol{O}$, is an ordered pair of a point O represented by its Cartesian coordinates x_O and y_O, with $x_O, y_O \in \mathbb{R}$ and a direction $\phi_{\boldsymbol{O}}$ ($\boldsymbol{O} = ((x_O, y_O), \phi_{\boldsymbol{O}})$). $\phi_{\boldsymbol{O}} \in [0, 2\pi]$ denotes the angle between the oriented point orientation and an absolute reference direction. The following notations will be used assuming A, B and C are normal points: The direction ϕ^{BC} is defined as the direction from B towards C. We write $\boldsymbol{A}^{BC}$ for the oriented point $((x_A, y_A), \phi^{BC})$ that has the same position as the normal point A and the direction ϕ^{BC}. We just write $\boldsymbol{A}$, if the direction is unknown or unspecified, e.g. if we want to define an oriented point that coincides with A but can have an arbitrary direction. Note that $\boldsymbol{A}$, $\boldsymbol{A}^{AC}$, and $\boldsymbol{A}^{BC}$ are all different oriented points, coinciding in position, but possibly differing in orientation. Additionally, we want to emphazise that oriented point names like $\boldsymbol{A}^{AC}$ are only identifiers we use for making their role intuitively comprehensible. The knowledge that one oriented point coincides with or points at another has to be explicitly represented by respective relations.

We use the abbreviation $\boldsymbol{A}\ _m\angle_{\{i-j\}}^{\{k-l\}}\ \boldsymbol{B}$ with $i, j, k, l \in \mathcal{Z}_{4m}$ for the disjunction

$$\bigvee_{a=i}^{j} \bigvee_{b=k}^{l} \boldsymbol{A}\ _m\angle_a^b\ \boldsymbol{B}.$$

$*$ abbreviates all members 0 to $(4m - 1)$ of $\mathcal{Z}_{4m}$ and $\{i, j\}$ a disjunction of i and j such that for example $\boldsymbol{A}\ _m\angle_{\{i,j\}}^*\ \boldsymbol{B}$ denotes

$$\left(\bigvee_{b=0}^{4m-1} \boldsymbol{A}\ _m\angle_i^b\ \boldsymbol{B} \right) \vee \left(\bigvee_{b=0}^{4m-1} \boldsymbol{A}\ _m\angle_j^b\ \boldsymbol{B} \right).$$

3.2   Encoding FlipFlop Calculus and $\mathcal{LR}$ in $\mathcal{OPRA}_m$

As stated in Section 2, $\mathcal{LR}$ is an enhanced variant of the FlipFlop Calculus. In the remainder of this text, we will not distinguish between those two variants, but simply talk about the FlipFlop calculus (FFC), always meaning the enhanced version. To encode the nine ternary FlipFlop relations $A, B\ rel_{FFC}\ C$ within $\mathcal{OPRA}_1$, we utilize three oriented points: $\boldsymbol{A}^{AB}$, $\boldsymbol{B}^{AB}$ and $\boldsymbol{C}^{BC}$. For cases with $A \neq B$, the FlipFlop relations right, left, front, inside, back, start and end are distinguished.

To formulate the condition $A \neq B$ in $\mathcal{OPRA}_m$ and define the reference frame, we introduce the following *reference frame constraint 1* (rfc_1^{FFC}), that has to hold for all these relations:

$$rfc_1^{FFC} = \boldsymbol{A}^{AB}\ _1\angle_0^2\ \boldsymbol{B}^{AB}\ .$$

This describes first, that $\boldsymbol{A}^{AB}$ and $\boldsymbol{B}^{AB}$ have different positions but the same orientation, and second, that $\boldsymbol{B}^{AB}$ is in front of $\boldsymbol{A}^{AB}$.

We can now describe the individual FlipFlop relations as configurations of the three oriented points $\boldsymbol{A}^{AB}$, $\boldsymbol{B}^{AB}$ and $\boldsymbol{C}^{BC}$. We always provide a complete description of the configuration, though it is often sufficient to relate the referent to only one point of the reference frame. For instance for the front case the following $\mathcal{OPRA}_1$ relations have to hold in addition to rfc_1^{FFC}: $\boldsymbol{A}^{AB}\ _1\angle_0^2\ \boldsymbol{C}^{BC}$ and $\boldsymbol{B}^{AB}\ _1\angle_0^2\ \boldsymbol{C}^{BC}$.

For inside, we get the following relations: $A^{AB} {}_1\angle_0^0 C^{BC}$ and $B^{AB} {}_1\angle_2^2 C$. And for left the resulting relations are $A^{AB} {}_1\angle_1^1 C^{BC}$ and $B^{AB} {}_1\angle_1^2 C^{BC}$.

We now turn to the special cases **dou** and **tri**. We need to formalize that A and B have the same position in our second reference frame constraint that has to hold only for **dou** and **tri**:

$$rfc_2^{FFC} = A {}_1\angle_* B .$$

The orientation of the oriented points for A and B is unimportant in this case and we thus leave it unspecified.

In the **dou** case, we have to make sure that C is different from A and B. We achieve this by the following relations: $A {}_1\angle_*^2 C^{BC}$ and $B {}_1\angle_*^2 C^{BC}$. For **tri**, C has to be the same as A and B: $A {}_1\angle_* C$ and $B {}_1\angle_* C$.

The complete listing of formalizations for all FlipFlop relations is given in Table 1. It describes a jointly exhaustive and pairwise disjoint set of configurations of three oriented points that can be used to translate between FlipFlop and $\mathcal{OPRA}_m$.

Table 1. A mapping of FlipFlop relations to $\mathcal{OPRA}_1$ relations

$A, B \, rel_{FFC} \, C$	$\mathcal{OPRA}_1$ representation
front	$rfc_1^{FFC} \wedge A^{AB} {}_1\angle_0^2 C^{BC} \wedge B^{AB} {}_1\angle_0^2 C^{BC}$
end	$rfc_1^{FFC} \wedge A^{AB} {}_1\angle_0^* C \wedge B^{AB} {}_1\angle_* C$
inside	$rfc_1^{FFC} \wedge A^{AB} {}_1\angle_0^0 C^{BC} \wedge B^{AB} {}_1\angle_2^2 C^{BC}$
start	$rfc_1^{FFC} \wedge A^{AB} {}_1\angle_2 C^{BC} \wedge B^{AB} {}_1\angle_2^2 C^{BC}$
back	$rfc_1^{FFC} \wedge A^{AB} {}_1\angle_2^2 C^{BC} \wedge B^{AB} {}_1\angle_2^2 C^{BC}$
left	$rfc_1^{FFC} \wedge A^{AB} {}_1\angle_1^1 C^{BC} \wedge B^{AB} {}_1\angle_1^2 C^{BC}$
right	$rfc_1^{FFC} \wedge A^{AB} {}_1\angle_3^3 C^{BC} \wedge B^{AB} {}_1\angle_3^2 C^{BC}$
dou	$rfc_2^{FFC} \wedge A {}_1\angle_*^2 C^{BC} \wedge B {}_1\angle_*^2 C^{BC}$
tri	$rfc_2^{FFC} \wedge A {}_1\angle_* C \wedge B {}_1\angle_* C$

3.3 Encoding DCC in $\mathcal{OPRA}_m$

As stated above, the Double Cross Calculus (DCC) is a combination of two SCC[2] reference frames with not all combinations possible. We adopt the original idea to conjunct the two reference frames used in [7], although we are not using the inverse orientation for the second reference frame for reasons of simplicity, i.e. we also use orientation AB for the second reference frame instead of orientation BA (compare Fig. 2(b)). We will now give the mapping of the 17 base relations based on the three oriented points C^{BC} to A^{AB} and B^{AB} and summarize it in Table 2.

For building the reference frame for the cases with $A \neq B$ we need to define the *reference frame constraint 1* as for the FlipFlop calculus, but this time on the basis of granularity $m = 2$, because the perpendiculars to A^{AB} and B^{AB} are needed for classification:

$$rfc_1^{DCC} = A^{AB} {}_2\angle_0^4 B^{AB}.$$

---

[2] Because an $\mathcal{OPRA}_m$ specification for SCC can be easily derived from the DCC representation, we omit it here.

The second reference frame constraint required for **dou** and **tri** is:

$$rfc_2^{DCC} = \boldsymbol{A} \; {}_1\angle* \; \boldsymbol{B} \; .$$

We need to relate referent $\boldsymbol{C}^{BC}$ to $\boldsymbol{A}^{AB}$ and $\boldsymbol{B}^{AB}$ for representing DCC relations. Given A, B **1** C in DCC C is left front of $\boldsymbol{A}^{AB}$ ($\mathcal{OPRA}_2$ region 1) as well as left front of $\boldsymbol{B}^{AB}$. $\boldsymbol{B}^{AB}$ is straight back of $\boldsymbol{C}^{BC}$ (region 4) and $\boldsymbol{A}^{AB}$ can only be positioned in region 3 of $\boldsymbol{C}^{BC}$. Therefore, it follows that A, B **1** C has to be encoded as $rfc_1^{DCC} \wedge \boldsymbol{A}^{AB} \; {}_2\angle_1^3 \; \boldsymbol{C}^{BC} \wedge \boldsymbol{B}^{AB} \; {}_2\angle_1^4 \; \boldsymbol{C}^{BC}$.

Cases with C being positioned between $\boldsymbol{A}^{AB}$ and $\boldsymbol{B}^{AB}$ ($rel_{DCC} \in \{3, 9, 12\}$) are a little more complex. For instance, if relation A, B **3** C is given, $\boldsymbol{C}^{BC}$ is left front of $\boldsymbol{A}^{AB}$ (region 1) and left back of $\boldsymbol{B}^{AB}$ (region 3). $\boldsymbol{B}^{AB}$ is straight back of $\boldsymbol{C}^{BC}$ (region 4). Now imagine $\boldsymbol{C}^{BC}$ being positioned quite close to the line segment AB. Then $\boldsymbol{A}^{AB}$ is positioned in region 1 of $\boldsymbol{C}^{BC}$. But $\boldsymbol{C}^{BC}$ could be also positioned very far away from this line, then $\boldsymbol{A}^{AB}$ is positioned in region 3. The case with $\boldsymbol{A}^{AB}$ being straight left is also valid. Therefore $rfc_1^{DCC} \wedge \boldsymbol{A}^{AB} \; {}_2\angle_1^{\{1-3\}} \; \boldsymbol{C}^{BC} \wedge \boldsymbol{B}^{AB} \; {}_2\angle_3^4 \; \boldsymbol{C}^{BC}$ follows as the correct encoding.

The special cases with $A \neq B = C$ (**b**) $A = C \neq B$ (**a**) as well as **dou** and **tri** can be handled just like the according cases in FFC. **Dou** results in $rfc_2^{DCC} \wedge \boldsymbol{A} \; {}_2\angle_*^4 \; \boldsymbol{C}^{BC} \wedge \boldsymbol{B} \; {}_2\angle_*^4 \; \boldsymbol{C}^{BC}$ and **tri** in $rfc_2^{DCC} \wedge \boldsymbol{A} \; {}_2\angle* \; \boldsymbol{C} \wedge \boldsymbol{B} \; {}_2\angle* \; \boldsymbol{C}$. A complete list is given in Table 2.

Table 2. A mapping of DCC relations to $\mathcal{OPRA}_m$ relations

$A, B \; rel_{DCC} \; C$	$\mathcal{OPRA}_m$ representation
0	$rfc_1^{DCC} \wedge \boldsymbol{A}^{AB} \; {}_2\angle_0^4 \; \boldsymbol{C}^{BC} \wedge \boldsymbol{B}^{AB} \; {}_2\angle_0^4 \; \boldsymbol{C}^{BC}$
1	$rfc_1^{DCC} \wedge \boldsymbol{A}^{AB} \; {}_2\angle_1^3 \; \boldsymbol{C}^{BC} \wedge \boldsymbol{B}^{AB} \; {}_2\angle_1^4 \; \boldsymbol{C}^{BC}$
2	$rfc_1^{DCC} \wedge \boldsymbol{A}^{AB} \; {}_2\angle_1^3 \; \boldsymbol{C}^{BC} \wedge \boldsymbol{B}^{AB} \; {}_2\angle_2^4 \; \boldsymbol{C}^{BC}$
3	$rfc_1^{DCC} \wedge \boldsymbol{A}^{AB} \; {}_2\angle_1^{\{1-3\}} \; \boldsymbol{C}^{BC} \wedge \boldsymbol{B}^{AB} \; {}_2\angle_3^4 \; \boldsymbol{C}^{BC}$
4	$rfc_1^{DCC} \wedge \boldsymbol{A}^{AB} \; {}_2\angle_2^3 \; \boldsymbol{C}^{BC} \wedge \boldsymbol{B}^{AB} \; {}_2\angle_3^4 \; \boldsymbol{C}^{BC}$
5	$rfc_1^{DCC} \wedge \boldsymbol{A}^{AB} \; {}_2\angle_3^3 \; \boldsymbol{C}^{BC} \wedge \boldsymbol{B}^{AB} \; {}_2\angle_3^4 \; \boldsymbol{C}^{BC}$
6	$rfc_1^{DCC} \wedge \boldsymbol{A}^{AB} \; {}_2\angle_4^4 \; \boldsymbol{C}^{BC} \wedge \boldsymbol{B}^{AB} \; {}_2\angle_4^4 \; \boldsymbol{C}^{BC}$
7	$rfc_1^{DCC} \wedge \boldsymbol{A}^{AB} \; {}_2\angle_5^5 \; \boldsymbol{C}^{BC} \wedge \boldsymbol{B}^{AB} \; {}_2\angle_5^4 \; \boldsymbol{C}^{BC}$
8	$rfc_1^{DCC} \wedge \boldsymbol{A}^{AB} \; {}_2\angle_6^5 \; \boldsymbol{C}^{BC} \wedge \boldsymbol{B}^{AB} \; {}_2\angle_5^4 \; \boldsymbol{C}^{BC}$
9	$rfc_1^{DCC} \wedge \boldsymbol{A}^{AB} \; {}_2\angle_7^{\{5-7\}} \; \boldsymbol{C}^{BC} \wedge \boldsymbol{B}^{AB} \; {}_2\angle_5^4 \; \boldsymbol{C}^{BC}$
10	$rfc_1^{DCC} \wedge \boldsymbol{A}^{AB} \; {}_2\angle_7^5 \; \boldsymbol{C}^{BC} \wedge \boldsymbol{B}^{AB} \; {}_2\angle_6^4 \; \boldsymbol{C}^{BC}$
11	$rfc_1^{DCC} \wedge \boldsymbol{A}^{AB} \; {}_2\angle_7^5 \; \boldsymbol{C}^{BC} \wedge \boldsymbol{B}^{AB} \; {}_2\angle_7^4 \; \boldsymbol{C}^{BC}$
12	$rfc_1^{DCC} \wedge \boldsymbol{A}^{AB} \; {}_2\angle_0^0 \; \boldsymbol{C}^{BC} \wedge \boldsymbol{B}^{AB} \; {}_2\angle_4^4 \; \boldsymbol{C}^{BC}$
a	$rfc_1^{DCC} \wedge \boldsymbol{A}^{AB} \; {}_2\angle_4 \; \boldsymbol{C}^{BC} \wedge \boldsymbol{B}^{AB} \; {}_2\angle_4^4 \; \boldsymbol{C}^{BC}$
b	$rfc_1^{DCC} \wedge \boldsymbol{A}^{AB} \; {}_2\angle_0^* \; \boldsymbol{C} \wedge \boldsymbol{B}^{AB} \; {}_2\angle* \; \boldsymbol{C}$
dou	$rfc_2^{DCC} \wedge \boldsymbol{A} \; {}_2\angle_*^4 \; \boldsymbol{C}^{BC} \wedge \boldsymbol{B} \; {}_2\angle_*^4 \; \boldsymbol{C}^{BC}$
tri	$rfc_2^{DCC} \wedge \boldsymbol{A} \; {}_2\angle* \; \boldsymbol{C} \wedge \boldsymbol{B} \; {}_2\angle* \; \boldsymbol{C}$

3.4 Encoding $\mathcal{DRA}_f$ in $\mathcal{OPRA}_m$

We will now give a mapping of the 72 base relations of the fine grained Dipole Relation Algebra ($\mathcal{DRA}_f$). Introducing the parallelity distinction will cause some problems, therefore we will investigate the additional feature of $\mathcal{DRA}_{fp}$ in Section 3.5 separately.

We have to define the reference frame constraint such that $A \neq B \wedge C \neq D$ and line AB as well as CD form a reference line:

$$rfc_1^{\mathcal{DRA}_f} = \boldsymbol{A}^{AB} {}_1\angle_0^2 \boldsymbol{B}^{AB} \wedge \boldsymbol{C}^{CD} {}_1\angle_0^2 \boldsymbol{D}^{CD}.$$

A $\mathcal{DRA}_f$ relation consists of four letters each of the case **l**, **r**, **f**, **b**, **i**, **s**, or **e**. Since each letter is derived in the same way, just for different triples of points, and each describes a FlipFlop relation, we begin by providing the FlipFlop encodings from Table 1, but with the concrete oriented points replaced by variables X, Y, and Z in Table 3. **dou** and **tri** are not listed because of the preliminaries $A \neq B$ and $C \neq B$. Table 4 lists the instantiations that have to be chosen for X, Y, and Z for each of the four letters. The first letter describes the FlipFlop relation between A, B and C, the second for A, B and D, the third for C, D and A, and the fourth C, D and B. We give two examples on how complete $\mathcal{DRA}_f$ relations are mapped:

Example (1): $d_{AB} (rfll) d_{CD} \equiv rfc_1^{\mathcal{DRA}_f}$
$$\wedge \boldsymbol{A}^{AB} {}_1\angle_3^3 \boldsymbol{C}^{BC} \wedge \boldsymbol{B}^{AB} {}_1\angle_3^2 \boldsymbol{C}^{BC}$$
$$\wedge \boldsymbol{A}^{AB} {}_1\angle_0^2 \boldsymbol{D}^{BD} \wedge \boldsymbol{B}^{AB} {}_1\angle_0^2 \boldsymbol{D}^{BD}$$
$$\wedge \boldsymbol{C}^{CD} {}_1\angle_1^1 \boldsymbol{A}^{DA} \wedge \boldsymbol{D}^{CD} {}_1\angle_1^2 \boldsymbol{A}^{DA}$$
$$\wedge \boldsymbol{C}^{CD} {}_1\angle_1^1 \boldsymbol{B}^{DB} \wedge \boldsymbol{D}^{CD} {}_1\angle_1^2 \boldsymbol{B}^{DB}$$

Example (2): $d_{AB} (ebis) d_{CD} \equiv rfc_1^{\mathcal{DRA}_f}$
$$\wedge \boldsymbol{A}^{AB} {}_1\angle_0^2 \boldsymbol{C}^{AB} \wedge \boldsymbol{B}^{AB} {}_1\angle 0 \, \boldsymbol{C}^{AB}$$
$$\wedge \boldsymbol{A}^{AB} {}_1\angle_2^2 \boldsymbol{D}^{BD} \wedge \boldsymbol{B}^{AB} {}_1\angle_2^2 \boldsymbol{D}^{BD}$$
$$\wedge \boldsymbol{C}^{CD} {}_1\angle_0^0 \boldsymbol{A}^{DA} \wedge \boldsymbol{D}^{CD} {}_1\angle_2^2 \boldsymbol{A}^{DA}$$
$$\wedge \boldsymbol{C}^{CD} {}_1\angle 2 \, \boldsymbol{B}^{DB} \wedge \boldsymbol{D}^{CD} {}_1\angle_2^2 \boldsymbol{B}^{DB}$$

Table 3. Describing the local $\mathcal{DRA}_f$ point configurations (cp. Table 1)

	$X^{XY} {}_1\angle_i^j Z$	$\wedge\, Y^{XY} {}_1\angle_k^l Z$
front	$X^{XY} {}_1\angle_0^2 Z^{YZ}$	$\wedge\, Y^{XY} {}_1\angle_2^2 Z^{YZ}$
end	$X^{XY} {}_1\angle_0^* Z$	$\wedge\, Y^{XY} {}_1\angle * Z$
inside	$X^{XY} {}_1\angle_0^0 Z^{YZ}$	$\wedge\, Y^{XY} {}_1\angle_2^2 Z^{YZ}$
start	$X^{XY} {}_1\angle 2 \, Z^{YZ}$	$\wedge\, Y^{XY} {}_1\angle_2^2 Z^{YZ}$
back	$X^{XY} {}_1\angle_2^2 Z^{YZ}$	$\wedge\, Y^{XY} {}_1\angle_2^2 Z^{YZ}$
left	$X^{XY} {}_1\angle_1^1 Z^{YZ}$	$\wedge\, Y^{XY} {}_1\angle_1^2 Z^{YZ}$
right	$X^{XY} {}_1\angle_3^3 Z^{YZ}$	$\wedge\, Y^{XY} {}_1\angle_3^2 Z^{YZ}$

Table 4. Instantiation mapping of X, Y, Z in Table 3 according to the position in the $\mathcal{DRA}_f$ relation tuple

position	X Y Z
1st position (r_1)	A B C
2nd position (r_2)	A B D
3rd position (r_3)	C D A
4th position (r_4)	C D B

3.5   The $\mathcal{DRA}_{fp}$ Enhancement in $\mathcal{OPRA}_m$

The $\mathcal{DRA}_{fp}$ is an enhancement of $\mathcal{DRA}_f$ by four additional distinctions about relative orientation of the two dipoles indicated by a fifth letter at the end of the relation string. The classes are **P**arallel, **A**nti-Parallel, B oriented mathematically positive regarding A (+), and B negative towards A (-). Only four $\mathcal{DRA}_f$ relations ($rrrr$, $rrll$, $llrr$, $llll$)

have to be split into three new ones, for all other relations this parameter is uniquely defined by the original configuration in $\mathcal{DRA}_f$. For details we refer to [4]. As seen in Section 3.4, a mapping for $\mathcal{DRA}_f$ to $\mathcal{OPRA}_m$ exists. The problem now is to formalize the additional relative orientation of $\boldsymbol{d}_{AB}$ and $\boldsymbol{d}_{CD}$ because the concept of parallelity is not contained in $\mathcal{OPRA}_m$.

The keypoint is shifting point C onto A while keeping the relative orientation defined by points C and D. Table 5 shows the formalization of the four additional distinctions. The problem with this definition is that the point position is not part of orientation determination. Unfortunately, in perception based scenarios with an observer looking from current position towards an object such a definition will cause problems. Nevertheless, the given definition is valid in the sense of defining $\mathcal{DRA}_{fp}$ correctly.

Table 5. The definitions for the fifth part of $\mathcal{DRA}_{fp}$ relations

Parallel	$\boldsymbol{A}^{AB}\ _1\angle 0\ \boldsymbol{A}^{CD}$
Anti-Parallel	$\boldsymbol{A}^{AB}\ _1\angle 2\ \boldsymbol{A}^{CD}$
+	$\boldsymbol{A}^{AB}\ _1\angle 1\ \boldsymbol{A}^{CD}$
-	$\boldsymbol{A}^{AB}\ _1\angle 3\ \boldsymbol{A}^{CD}$

3.6 Encoding QTC in $\mathcal{OPRA}_m$

QTC relations are based on a double cross reference frame spanned by the objects' original positions A_{t_i} and B_{t_i} at time point t_i. The relations are defined on the differentiated motion direction given by the positions of A and B at t_j with $i < j$. We abbreviate A_{t_i} with A_i and A_{t_j} with A_j. We define the reference frame in DCC manner based on A_i and B_i.

$$rfc^{QTC} = \boldsymbol{A}_i^{A_iB_i}\ _2\angle_0^4\ \boldsymbol{B}_i^{A_iB_i}.$$

Table 6 shows the according mapping of the four single $-, 0, +$ literals of the QTC relation string. The first two literals are defined by the front/back dichotomy regarding the perpendiculars relative to the reference orientation from A_i to B_i, the last two by left/right regarding the reference orientation. For the example in Figure 4 (A $(- +$ $--)$ B) it follows $\boldsymbol{A}_i^{A_iB_i}\ _2\angle_{\{7-1\}}^4\ \boldsymbol{A}_j^{A_iA_j} \wedge \boldsymbol{B}_i^{A_iB_i}\ _2\angle_{\{7-1\}}^4\ \boldsymbol{B}_j^{B_iB_j}$ for the front/back part and $\boldsymbol{A}_i^{A_iB_i}\ _1\angle_1^2\ \boldsymbol{A}_j^{A_iA_j} \wedge \boldsymbol{B}_i^{A_iB_i}\ _1\angle_3^2\ \boldsymbol{B}_j^{B_iB_j}$ for the left/right part.

Table 6. An $\mathcal{OPRA}_m$ representation schema for QTC relations

	-	0	+
1st parameter	$\boldsymbol{A}_i^{A_iB_i}\ _2\angle_{\{7-1\}}^4\ \boldsymbol{A}_j^{A_iA_j}$	$\boldsymbol{A}_i^{A_iB_i}\ _2\angle_{\{2,6\}}^4\ \boldsymbol{A}_j^{A_iA_j}$	$\boldsymbol{A}_i^{A_iB_i}\ _2\angle_{\{3-5\}}^4\ \boldsymbol{A}_j^{A_iA_j}$
2nd parameter	$\boldsymbol{B}_i^{A_iB_i}\ _2\angle_{\{3-5\}}^4\ \boldsymbol{B}_j^{B_iB_j}$	$\boldsymbol{B}_i^{A_iB_i}\ _2\angle_{\{2,6\}}^4\ \boldsymbol{B}_j^{B_iB_j}$	$\boldsymbol{B}_i^{A_iB_i}\ _2\angle_{\{7-1\}}^4\ \boldsymbol{B}_j^{B_iB_j}$
3rd parameter	$\boldsymbol{A}_i^{A_iB_i}\ _1\angle_1^2\ \boldsymbol{A}_j^{A_iA_j}$	$\boldsymbol{A}_i^{A_iB_i}\ _1\angle_{\{0,2\}}^2\ \boldsymbol{A}_j^{A_iA_j}$	$\boldsymbol{A}_i^{A_iB_i}\ _1\angle_3^2\ \boldsymbol{A}_j^{A_iA_j}$
4th parameter	$\boldsymbol{B}_i^{A_iB_i}\ _1\angle_3^2\ \boldsymbol{B}_j^{B_iB_j}$	$\boldsymbol{B}_i^{A_iB_i}\ _1\angle_{\{0,2\}}^2\ \boldsymbol{B}_j^{B_iB_j}$	$\boldsymbol{B}_i^{A_iB_i}\ _1\angle_1^2\ \boldsymbol{B}_j^{B_iB_j}$

4   Applications of $\mathcal{OPRA}_m$ Mappings

In this section, we demonstrate the merits of expressing other orientation calculi within $\mathcal{OPRA}_m$ by dealing with the problem of deriving FlipFlop relations from DCC relations and the problem of automatically determining the composition for the DCC.

4.1   From DCC Relations to FlipFlop Relations

So far it was not possible to translate relations between different calculi without an explicit description of such a mapping, or to reason with relations represented in different calculi. By expressing arbitrary orientation calculi in $\mathcal{OPRA}_m$ we provide the facilities to do so. We give a simple example how a FFC relation can be derived from a DCC relation on the basis of the $\mathcal{OPRA}_m$ representation. The idea can be generalized for more complex transformations as well.

Given $rel_{DCC} = A, B$ **3** C we get $rfc_1^{DCC} \wedge A^{AB} \,_2\angle_1^{\{1-3\}} C^{BC} \wedge B^{AB} \,_2\angle_3^4 C^{BC}$ as $\mathcal{OPRA}_m$ representation from Table 2. By changing the granularity from $m = 2$ to $m = 1$ we get the FFC relation between A, B and C. How to map relations between different granularities is explained in [13].

From $rfc_1^{DCC} = A^{AB} \,_2\angle_0^4 B^{AB}$ follows $A^{AB} \,_1\angle_0^2 B^{AB} = rfc_1^{FFC}$. From $A^{AB} \,_2\angle_1^{\{1-3\}} C^{BC}$ follows $A^{AB} \,_1\angle_1^1 C^{BC}$ and from $B^{AB} \,_2\angle_3^4 C^{BC}$ follows $B^{AB} \,_1\angle_1^2 C^{BC}$.

The combined result $rfc_1^{FFC} \wedge A^{AB} \,_1\angle_1^1 C^{BC} \wedge B^{AB} \,_1\angle_1^2 C^{BC}$ is the $\mathcal{OPRA}_m$ formalization for the FFC relation A, B **l** C (cf. Table 1).

4.2   DCC Composition with $\mathcal{OPRA}_m$

We want to derive the composition table for DCC based on the completely specified composition of $\mathcal{OPRA}_m$. Thus, we have four points A, B, C, and D and want to infer the relation $A, B\ rel_{DCC}\ D$ from the given relations $A, B\ rel_{DCC}\ C$ and $B, C\ rel_{DCC}\ D$ (cp. Figure 6(a)). The configuration modeled in $\mathcal{OPRA}_m$ is shown in Figure 6(b).

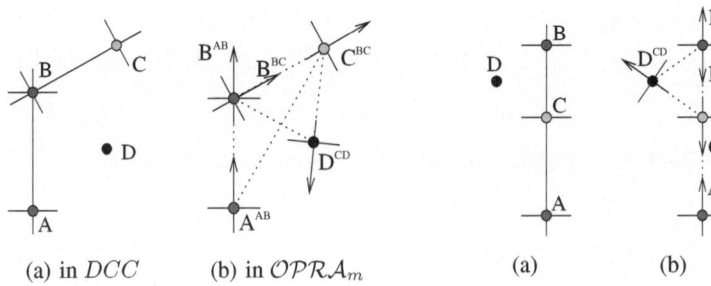

(a) in DCC (b) in $\mathcal{OPRA}_m$ (a)   (b)

Fig. 6. A general spatial configuration for composition

Fig. 7. A, B **12** C and B, C **9** D for composition in DCC and $\mathcal{OPRA}_m$

To demonstrate the process of deriving the composition, we consider the example of composing the DCC relations **12** and **9** (cf. Figure 7(a)). We will employ several additional theorems that we give here without proof due to space restrictions:

$$X \; {}_m\angle_i^j \; Y \Rightarrow \qquad Y \; {}_m\angle_j^i \; X \qquad \text{(converse)} \tag{1}$$

$$X \; {}_m\angle_i^j \; Y \Rightarrow \qquad X \; {}_m\angle_i^0 \; Y^{YX} \qquad \text{(orientation shift)} \tag{2}$$

$$X \; {}_m\angle_i^0 \; Y^{YX} \Rightarrow X \; {}_m\angle(2m+i) \; X^{YX} \quad \text{(projection type 1)} \tag{3}$$

$$X \; {}_m\angle_i^{2m} \; Y^{XY} \Rightarrow \qquad X \; {}_m\angle i \; X^{XY} \qquad \text{(projection type 2)} \tag{4}$$

We start by turning the given DCC relations into $\mathcal{OPRA}_m$ relations as described in Section 3.3 (cf. Table 2). To do this, five oriented points have to be introduced (see Figure 7(b)). A, B **12** C and B, C **9** D yield the following relations:

$$A^{AB} \; {}_2\angle_0^0 \; C^{BC} \tag{5}$$

$$B^{AB} \; {}_2\angle_4^4 \; C^{BC} \tag{6}$$

$$A^{AB} \; {}_2\angle_0^4 \; B^{AB} \quad (rfc_1^{DCC}) \tag{7}$$

$$B^{BC} \; {}_2\angle_7^{\{5-7\}} \; D^{CD} \tag{8}$$

$$C^{BC} \; {}_2\angle_5^4 \; D^{CD} \tag{9}$$

$$B^{BC} \; {}_2\angle_0^4 \; C^{BC} \quad (rfc_1^{DCC}) \tag{10}$$

We will later need the relation between D^{CD} and D^{BD}. This can be derived from (8) by employing first the converse theorem (1), then the orientation shift theorem (2), and finally the projection type 1 theorem (3):

$$B^{BC} \; {}_2\angle_7^{\{5-7\}} \; D^{CD} \overset{(1)}{\Longrightarrow} D^{CD} \; {}_2\angle_{\{5-7\}}^7 \; B^{BC} \overset{(2)}{\Longrightarrow}$$

$$D^{CD} \; {}_2\angle_{\{5-7\}}^0 \; B^{BD} \overset{(3)}{\Longrightarrow} D^{CD} \; {}_2\angle\{1-3\} \; D^{BD} \tag{11}$$

From applying the projection type 2 theorem to (6) follows:

$$B^{AB} \; {}_2\angle_4^4 \; C^{BC} \Rightarrow B^{AB} \; {}_2\angle 4 \; B^{BC} \tag{12}$$

There are two paths that allow to derive information about the position of B^{AB} and D^{CD}. First, by composing (12) with (10) and then with (9). And second, by composing (12) with (8). Since both can constrain the possible position, we need to compute both and take the intersection of the resulting disjunctions of base relations. Overall, we thus compute $(12) \circ (((10) \circ (9)) \cap (8))$. Composing (10) and (9) yields:

$$B^{BC} \; {}_2\angle_0^4 \; C^{BC} \circ C^{BC} \; {}_2\angle_5^4 \; D^{CD} = B^{BC} \; \left({}_2\angle_{\{5-7\}}^5 \vee {}_2\angle_7^{\{6-7\}} \right) \; D^{CD} \tag{13}$$

Taking the intersection with (8) we get:

$$B^{BC} \; {}_2\angle_7^{\{5-7\}} \; D^{CD} \tag{14}$$

And composing this with (12) results in the following disjunction for the relation between B^{AB} and D^{CD}:

$$B^{AB} \; {}_2\angle 4 \; B^{BC} \circ B^{BC} \; {}_2\angle_7^{\{5-7\}} \; D^{CD} = B^{AB} \; {}_2\angle_3^{\{5-7\}} \; D^{CD} \tag{15}$$

We can now apply the orientation shift theorem to (15) to derive the following relation between $\boldsymbol{B}^{AB}$ and $\boldsymbol{D}^{BD}$:

$$\boldsymbol{B}^{AB} \; {}_2\angle_3^{\{5-7\}} \; \boldsymbol{D}^{CD} \Rightarrow \boldsymbol{B}^{AB} \; {}_2\angle_3^{4} \; \boldsymbol{D}^{BD} \tag{16}$$

To compute the relation between $\boldsymbol{A}^{AB}$ and $\boldsymbol{D}^{BD}$ there are again two paths to consider. First, composing (7) with (16) and second, composing (5) and (9) and (11). This results in the overall computation of $((7) \circ (16)) \cap ((5) \circ (9) \circ (11))$. Composing (7) and (16) yields:

$$\boldsymbol{A}^{AB} \; {}_2\angle_0^{4} \; \boldsymbol{B}^{AB} \circ \boldsymbol{B}^{AB} \; {}_2\angle_3^{4} \; \boldsymbol{D}^{BD} = \boldsymbol{A}^{AB} \left({}_2\angle_1^{\{1-3\}} \vee {}_2\angle_2^{3} \vee {}_2\angle_3^{3} \right) \boldsymbol{D}^{BD} \tag{17}$$

Composing (5) and (9) and (11) yields:

$$\boldsymbol{A}^{AB} \; {}_2\angle_0^{0} \; \boldsymbol{C}^{BC} \circ \boldsymbol{C}^{BC} \; {}_2\angle_5^{4} \; \boldsymbol{D}^{CD} \circ \boldsymbol{D}^{CD} \; {}_2\angle\{1-3\} \; \boldsymbol{D}^{BD} = \boldsymbol{A}^{AB} \; {}_2\angle_1^{\{7-3\}} \; \boldsymbol{D}^{BD} \tag{18}$$

Taking the intersection of (17) and (18) we get:

$$\boldsymbol{A}^{AB} \; {}_2\angle_1^{\{1-3\}} \; \boldsymbol{D}^{BD} \tag{19}$$

The resulting $\mathcal{OPRA}_m$ relations (16) and (19) together with rfc_1^{DCC} can be looked up in Table 2. The result is that they describe the DCC relation **3**. Thus we have correctly derived that **3** is the composition of **12** and **9**.

5   Conclusion

As we have shown in this paper, $\mathcal{OPRA}_m$ is a very expressive spatial calculus for reasoning about orientation and direction information and it is possible to map other point or line-based orientation calculi to $\mathcal{OPRA}_m$. For a selected set of well-known calculi these mappings have been presented. This does in *no* way imply that these calculi are obsolete. In scenarios for which they are sufficiently expressive they can be the better choice due to computational advantages (e.g. when employed for constraint-based reasoning). The merits of analyzing the relationships between different calculi have been exemplarily demonstrated by showing that FFC relations can be extracted very easily from a DCC relation and additionally, by showing the composition of DCC relations can be derived from $\mathcal{OPRA}_m$ composition. Other forms of application still need to be investigated. One such application is the automatic construction of conceptual neighborhood graphs from the neighborhood structure of $\mathcal{OPRA}_m$ relations. However, to do this, we first need to extend the notion of conceptual neighborhoods to configurations of more than two objects on the $\mathcal{OPRA}_m$ side. Moreover, we plan to provide $\mathcal{OPRA}_m$ formalizations for additional orientation calculi.

Acknowledgments. The authors like to thank Nico van de Weghe, Lutz Frommberger, Diedrich Wolter, Reinhard Moratz, and Christian Freksa for fruitful discussions and impulses. We would also like to thank the anonymous reviewers for their valuable comments. Our work was supported by the DFG Transregional Collaborative Research Center SFB/TR 8 Spatial Cognition.

References

1. J. F. Allen. Maintaining knowledge about temporal intervals. *Communications of the ACM*, pages 832–843, Nov. 1983.
2. A. G. Cohn. Qualitative spatial representation and reasoning techniques. In G. Brewka, C. Habel, and B. Nebel, editors, *KI-97: Advances in Artificial Intelligence, 21st Annual German Conference on Artificial Intelligence, Freiburg, Germany, September 9-12, 1997, Proceedings*, volume 1303 of *Lecture Notes in Computer Science*, pages 1–30, Berlin, 1997. Springer.
3. A. G. Cohn and S. M. Hazarika. Qualitative spatial representation and reasoning: An overview. *Fundamenta Informaticae*, 46(1-2):1–29, 2001.
4. F. Dylla and R. Moratz. Exploiting qualitative spatial neighborhoods in the situation calculus. In Freksa et al. [9], pages 304–322.
5. M. J. Egenhofer. A formal definition of binary topological relationships. In *3rd International Conference on Foundations of Data Organization and Algorithms*, pages 457–472, New York, NY, USA, 1989. Springer.
6. A. Frank. Qualitative spatial reasoning about cardinal directions. In *Proceedings of the American Congress on Surveying and Mapping (ACSM-ASPRS)*, pages 148–167, Baltimore, Maryland, USA, 1991.
7. C. Freksa. Conceptual neighborhood and its role in temporal and spatial reasoning. In M. Singh and L. Travé-Massuyès, editors, *Decision Support Systems and Qualitative Reasoning*, pages 181 – 187. North-Holland, Amsterdam, 1991.
8. C. Freksa. Using orientation information for qualitative spatial reasoning. In A. U. Frank, I. Campari, and U. Formentini, editors, *Theories and methods of spatio-temporal reasoning in geographic space*, pages 162–178. Springer, Berlin, 1992.
9. C. Freksa, M. Knauff, B. Krieg-Brückner, B. Nebel, and T. Barkowsky, editors. *Spatial Cognition IV. Reasoning, Action, Interaction: International Conference Spatial Cognition 2004*, volume 3343 of *Lecture Notes in Artificial Intelligence*. Springer, Berlin, Heidelberg, 2005.
10. G. Ligozat. Qualitative triangulation for spatial reasoning. In A. U. Frank and I. Campari, editors, *Spatial Information Theory: A Theoretical Basis for GIS, (COSIT'93), Marciana Marina, Elba Island, Italy*, volume 716 of *Lecture Notes in Computer Science*, pages 54–68. Springer, 1993.
11. G. Ligozat. Reasoning about cardinal directions. *Journal of Visual Languages and Computing*, 9:23–44, 1998.
12. R. Moratz. Representing relative direction as a binray relation of oriented points. In *ECAI 2006 Proceedings of the 17th European Conference on Artificial Intelligence*, 2006. to appear.
13. R. Moratz, F. Dylla, and L. Frommberger. A relative orientation algebra with adjustable granularity. In *Proceedings of the Workshop on Agents in Real-Time and Dynamic Environments (IJCAI 05)*, 2005.
14. R. Moratz, B. Nebel, and C. Freksa. Qualitative spatial reasoning about relative position: The tradeoff between strong formal properties and successful reasoning about route graphs. In C. Freksa, W. Brauer, C. Habel, and K. F. Wender, editors, *Spatial Cognition III*, volume 2685 of *Lecture Notes in Artificial Intelligence*, pages 385–400. Springer, Berlin, Heidelberg, 2003.
15. R. Moratz, J. Renz, and D. Wolter. Qualitative spatial reasoning about line segments. In W. Horn, editor, *Proceedings of the 14th European Conference on Artificial Intelligence (ECAI)*, Berlin, Germany, 2000. IOS Press.

16. D. A. Randell, Z. Cui, and A. Cohn. A spatial logic based on regions and connection. In B. Nebel, C. Rich, and W. Swartout, editors, *Principles of Knowledge Representation and Reasoning: Proceedings of the Third International Conference (KR'92)*, pages 165–176. Morgan Kaufmann, San Mateo, California, 1992.

17. J. Renz and D. Mitra. Qualitative direction calculi with arbitrary granularity. In C. Zhang, H. W. Guesgen, and W.-K. Yeap, editors, *PRICAI 2004: Trends in Artificial Intelligence, 8th Pacific RimInternational Conference on Artificial Intelligence, Auckland, New Zealand, Proceedings*, volume 3157 of *Lecture Notes in Computer Science*, pages 65–74. Springer, 2004.

18. J. Renz and B. Nebel. On the complexity of qualitative spatial reasoning: A maximal tractable fragment of the region connection calculus. *Artificial Intelligence*, 108(1-2):69–123, 1999.

19. C. Schlieder. Reasoning about ordering. In *Spatial Information Theory: A Theoretical Basis for GIS (COSIT'95)*, volume 988 of *Lecture Notes in Computer Science*, pages 341–349. Springer, Berlin, Heidelberg, 1995.

20. A. Scivos and B. Nebel. The finest of its class: The practical natural point-based ternary calculus $\mathcal{LR}$ for qualitative spatial reasoning. In Freksa et al. [9], pages 283–303.

21. N. van de Weghe. *Representing and Reasoning about Moving Objects: A Qualitative Approach*. PhD thesis, Ghent University, 2004.

22. N. van de Weghe, B. Kuijpers, P. Bogaert, and P. Maeyer. A qualitative trajectory calculus and the composition of its relations. In *First International Conference on GeoSpatial Semantics (GeoS 2005)*, volume 3799, pages 181–211. Springer, 2005.

Towards the Visualisation of Shape Features
The Scope Histogram

A. Schuldt, B. Gottfried, and O. Herzog

Centre for Computing Technologies (TZI)
University of Bremen, Am Fallturm 1, D-28359 Bremen

Abstract. Classifying objects in computer vision, we are faced with a great many features one can use. This paper argues that diagrammatic representations help to comprehend properties of features. This is important for the purpose of deciding which features should be used for a given classification task. We introduce such a diagrammatic representation for a shape feature and show how it enables one to decide whether this feature helps to distinguish some categories given. Additionally, it shows that the proposed feature keeps up with other features falling into the same complexity class.

1 Introduction

In computer vision several features have been devised in order to describe objects. While some of them are more intuitive than others, even the intuitive ones raise the question whether they are appropriate given a number of categories which are to be distinguished. For instance, a feature like the compactness obviously allows two object categories to get distinguished which are clearly different regarding their roundness. However, determining this feature for a number of categories it is quite difficult to comprehend its meaning. What does a difference, lets say, of 0.1 tell us about the shapes of two object categories? This question is especially to be asked from the point of view of the expert, who possesses particular skills about his domain but not about some methods he needs to apply to this domain.

It is frequently difficult or even impossible to determine which object properties, and as a consequence, which categories can be distinguished by specific features without thoroughly analysing them regarding that feature. Furthermore, it is even more difficult to comprehend how a number of different features combine. The problem we are faced with is more general and concerns the difficulty to choose among a number of methods. Normally, the appropriateness of the chosen method is not revealed before having used it. Here, we argue in favour of visualisation techniques which relate to diagrammatic representations. Such techniques enable us to assess the meaning of some method or feature by visualising some of its properties. Thereby, its inherent structure is made explicit, allowing attention to be guided in a specific way.

From the point of view of diagrammatic reasoning, [8] point out that attention mechanisms are relevant when interpreting diagrams. In particular, the ease of recognition is strongly affected by what information is explicit in a representation. The idea is to make use of these findings by devising diagrams which relate

C. Freksa, M. Kohlhase, and K. Schill (Eds.): KI 2006, LNAI 4314, pp. 289–301, 2007.

a feature's semantics to its range of values. In this paper we shall demonstrate how this works for a representation which comprises a structure that enables a specific feature defined on it to be visualised, so that the relationship between semantics and structure is made visible. As a consequence, it can intuitively be decided how well a number of categories can be distinguished by this feature.

In Sect. 2 previous work on a qualitative shape description is reviewed. At the same time, a diagrammatic representation for a shape feature is introduced which is based on this shape description. Applying this feature to shapes of the MPEG test dataset, Sect. 3 shows that this feature is in fact more expressive than others which pertain to the same class of features in terms of the complexity required for comparisons. But more importantly, it also shows that the feature's meaning can be comprehended very well through its diagrammatic representation. Section 4 discusses for a number of pairwise similar as well as pairwise different categories how their diagrammatic representation enables one to decide in favour or against the application of our feature. Finally, Sect. 5 summarises our conclusions.

2   Scope Histograms

For the purpose of dealing with shape data we choose polygons to be the underlying representation for our approach. Polygons can easily be extracted from binary raster images containing the silhouette of an object. Additionally, they allow the creation of a more compact representation through the application of polygonal simplification algorithms — even with only little influence on the perception of shape [2]. We apply especially the method proposed in [11], choosing an approximation error of one percent of a polygon's perimeter, in this way being invariant regarding the scale.

Consider the 13 qualitative relations between one-dimensional time intervals [1]. Their generalisation to two dimensions define the notion of positional-contrast [7] which allows intervals in the two-dimensional plane to be characterised. Using the orientation grid of [12], [6] introduces 23 $\mathcal{BA}_{23}$ relations that can occur between two line segments (Fig. 1 left and centre). This description is not only restricted to single lines. Instead, using an ordered sequence of $\mathcal{BA}_{23}$ relations, it is also possible to describe the course of a polygon w. r. t. one of its segments. This description has linear space complexity, since each of the polygon's segments is described by one $\mathcal{BA}_{23}$ relation. Indeed, an even more compact description is possible. This can be accomplished by characterising the polygon by its position as a whole, instead of describing the positions of its segments individually.

2.1   Representing Polygons by Their Scope

While $\mathcal{BA}_{23}$ relations are suitable in order to describe arrangements between line segments, they are not devised for the purpose of describing polygons as a whole. This is due to the fact that, in contrast to single line segments, polygons are not restricted to being straight. On this account, there exist polygons that cannot be characterised by single $\mathcal{BA}_{23}$ relations, namely those with a course circulating one or more times completely around the reference segment. $\mathcal{BA}_{12}$,

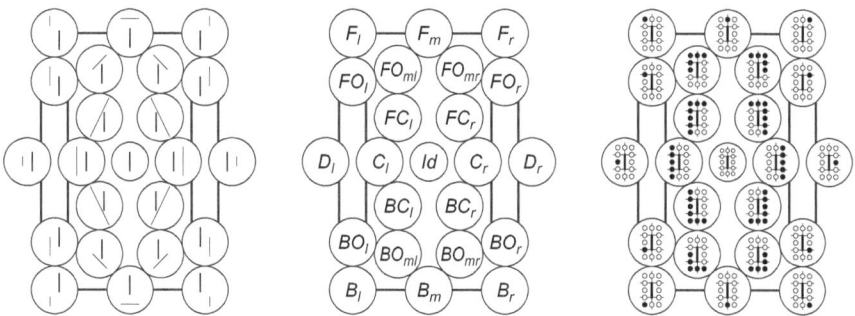

Fig. 1. The 23 $\mathcal{BA}_{23}$ relations that can be distinguished between two line segments in the two-dimensional plane. Left: Example arrangements. Centre: Mnemonic labels. Right: Iconic representation of the respective scopes.

which is a subset of $\mathcal{BA}_{23}$, can be used instead. It contains only the relations B_l, BO_l, D_l, FO_l, F_l, F_m, F_r, FO_r, D_r, BO_r, B_r, and B_m. These relations are referred to as atomic $\mathcal{BA}$ relations. They have the property of being located either in only one of the orientation grid's sectors or of passing exactly one of the orientation grid's singularities connecting two adjacent sectors.

Applying the notion of a *scope* [7], which is nothing other than a set of connected $\mathcal{BA}_{12}$ relations, it is possible to describe the location of any other segment or even the whole course w. r. t. the reference segment. As depicted on the right hand side of Fig. 1 the scopes of the atomic relations contain only the respective relation itself. In contrast, the scopes of the non-atomic $\mathcal{BA}_{23}$ relations are obtained by simply computing the union of the atomic relations comprising their positions. Starting from this, the description of a course is straightforward by simply aggregating $\mathcal{BA}_{12}$ relations which together make up this course.

2.2   Conceptual Neighbourhoods of Scopes

Applying the scope representation described above, the characterisation of a polygon's course w. r. t. one of its segments can be accomplished with constant space complexity. This is due to the fact that we are always dealing with a set of at most twelve atomic relations. In theory, $2^{12} = 4096$ different scopes can be distinguished. When dealing with simple, closed polygons, we can confine ourselves to those scopes, which do not have gaps between their atomic relations. The application of this definition restricts the number of realisable scopes to 133.

After having defined the set of realisable scopes, the question arises as to how they relate concerning their similarity. As an example, the scope of relation F_l may be considered to be more similar to F_m than it is to B_r. As suggested in [4], it is useful to connect qualitative relations explicitly by defining a conceptual neighbourhood structure. This does not only allow coarse knowledge to be dealt with, as described in [4], but also defines a distance between scopes. Starting from the scope's representation, we define two scopes to be conceptual neighbours if they can directly be transformed into one another by shortening or elongating them with exactly one atomic relation. Since the underlying reference system is circular, the visualisation of the conceptual neighbourhood structure is circular, too. The scopes of the $\mathcal{BA}_{12}$ relations are located at the outmost positions. The more atomic relations a scope contains the shorter is its distance to the centre, where the universal scope is located which contains all atomic relations. The top left quadrant of this neighbourhood structure is depicted in Fig. 2. For instance, in accordance to the neighbourhood structure the distance of the scopes of F_l and F_m is two. In contrast, the scopes of F_l and B_r are located at converse locations within the neighbourhood structure, which corresponds to a distance of twelve edges within this graph.

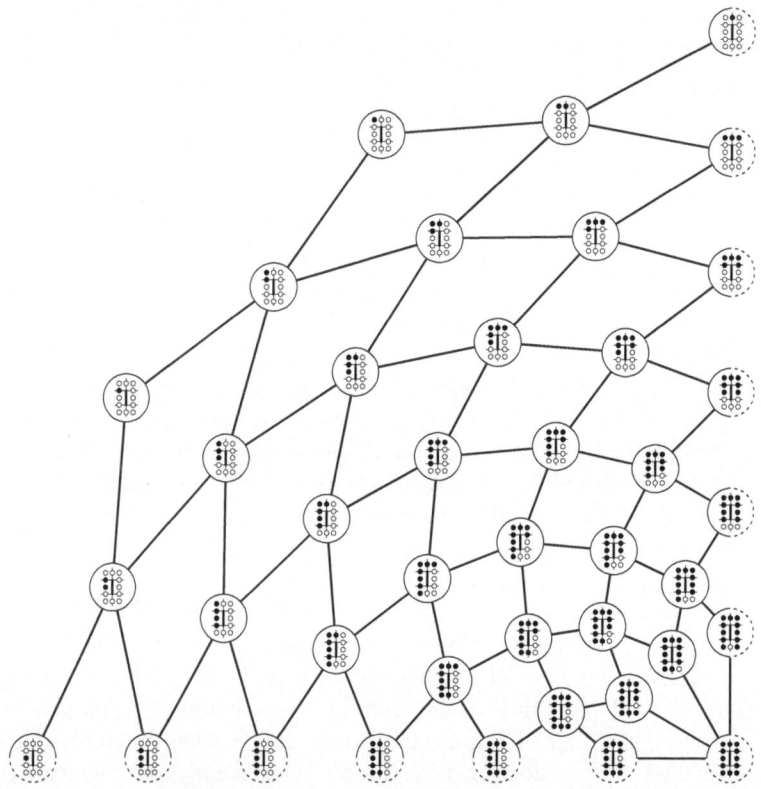

Fig. 2. The top left quadrant of the scope's conceptual neighbourhood structure. The missing quadrants can be created by reflecting the image along the bottom axis and the right axis. Two adjacent scopes can be transformed into one another by shortening or elongating one scope by exactly one atomic relation.

2.3   Computing Scope Histograms

The scope representation introduced in Sect. 2.1 has the advantage of allowing a polygon to be described w. r. t. one of its segments with constant space complexity. That is to say, we always deal with a set of at most twelve atomic relations regardless of how many line segments are required to represent a specific polygon. Since this is a very coarse description, it is necessary to characterise a polygon not only w.r.t. one of its segments, but w.r.t. to all of its segments. In doing so, we gain a description with linear space complexity, $O(n)$. This approach has the disadvantage that its time complexity for the comparison of two polygons is accordingly higher. This is due to the ordered sequences of scopes which are circularly permutable and due to the fact that different polygons do not have the same number of scopes, from what follows, that time complexity for the comparison of two scope sequences is $O(mn^3)$. Since efficient retrieval algorithms are generally desired, it is worth analysing how performance can be improved. Abandoning the scopes' order, time complexity can be reduced to $O(1)$ by computing a histogram of the occurring scopes. This also leads to constant space complexity for the description, since only the frequencies of occurring scopes have to be described.

It is worth mentioning that when dealing with simple, closed polygons not all of the 133 scopes introduced in Sect. 2.2 actually occur. By confining ourselves to polygons with a mathematically positive order only 86 out of the 133 distinguishable scopes can be realised (Fig. 3 left). That is, only the frequencies of this reduced set of scopes have to be determined.

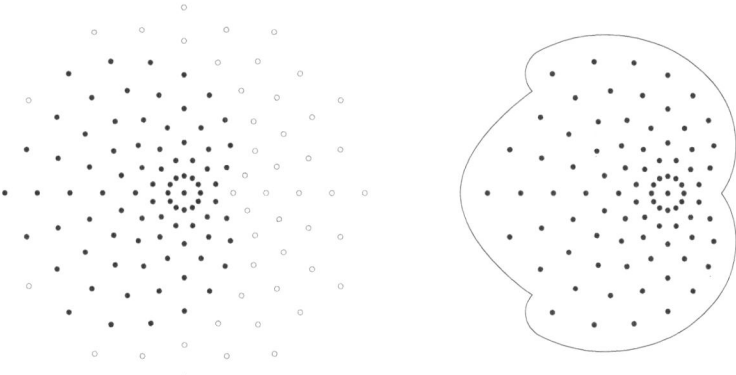

Fig. 3. Left: Each of the 133 distinguishable scopes is depicted by a point which is located accordingly to its position in the conceptual neighbourhood structure. The 86 scopes that are realisable by simple, closed polygons are depicted opaque, the others transparent. Right: Scope histograms can be visualised by only taking into account realisable scopes.

2.4   Visualising Scope Histograms

As mentioned before, an important property of scope histograms is that they allow two polygons to be described and to be compared with constant complexity. Another important property is the fact that a simple visualisation for scope histograms exists. For this visualisation we arrange the scopes accordingly to the conceptual neighbourhood structure (Fig. 2). As depicted on the right hand side of Fig. 3 only the position of the 86 realisable scopes have to be taken into consideration. The size of each entry in the histogram's visualisation depends on how often the scope appears for a given polygon. Fig. 4 shows two examples.

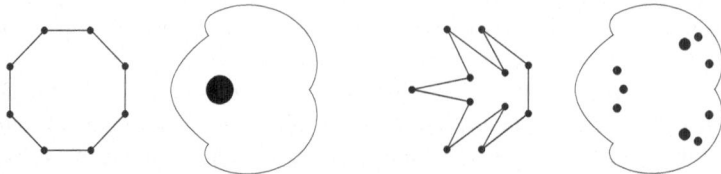

Fig. 4. Two polygons and their respective scope histograms. The size of an entry's depiction correlates with its frequency. Left: Since the polygon has the same position w. r. t. each of its segments, there exists only one (particularly large) entry in the histogram. Right: This polygon has different positions w. r. t. its segments. As a consequence, its scope histogram also contains several different entries.

Since our visualisation arranges the histogram's entries in accordance to their position in the conceptual neighbourhood structure, even the non expert is able to estimate how similar two entries are. This similarity can directly be derived from their distance in the visualisation. Extending this observation to histograms as a whole, it is even possible to judge the similarity of two polygons solely on the basis of their histograms' visualisations. Another property of this visualisation is that it can easily be determined whether two scopes are mirror images of each other. This is the case if their entries in the histogram can be mapped onto each other by reflecting them along the middle axis. As depicted in Fig. 4, polygons that have an axis of symmetry have a histogram which is symmetric w. r. t. its middle axis. In contrast, not every scope histogram having this property describes a reflection-symmetric polygon. With some experience, it is to some extent possible to identify scopes only by their position in the histogram. It is at least possible to get a coarse impression of the properties of a scope, relating to the correspondence of structure (e.g. large entry for the universal scope) and semantics (parts of the polygon circulating around other parts).

3   Categorising Objects

After having introduced the scope histogram in Sect. 2 we will now concentrate on the evaluation of its performance before we discuss its visualisation in the following Sect. 4.

A well-known method for the comparison of shape classification approaches is the core experiment CE-Shape-1 [10] for the MPEG-7 standard. This experiment compares approaches only by their retrieval results, instead of comparing the underlying methods directly. We apply especially part B which tests the capability of similarity-based retrieval techniques with a database of 1400 images. These are semantically grouped into 70 shape classes, each one consisting of 20 object instances. During the test all 1400 images are used as a query one after another. For each query all images in the database are ordered w. r. t. their similarity by the approach under consideration. Subsequently, the images belonging to the same class as the query are counted within the first 40 results. Since each class contains 20 instances, for each query up to 20 correct matches can be found. For all 1400 queries the total number of correct matches is 28000. The result of the test is the ratio of the number of found objects and the total number of correct matches.

In order to measure the performance of the scope histogram, we apply features that pertain to the same class of complexity required for comparisons, i. e. that they also allow two shapes to be compared with constant time complexity. Especially we use the compactness [3], which corresponds to the ratio $\frac{4\pi A}{P^2}$ of area and perimeter, the radius ratio $\frac{R_{min}}{R_{max}}$ of the minimum enclosing circle and the maximal contained circle [5], and the aspect ratio $\frac{H_r}{W_r}$ of the minimal enclosing rectangle [3]. Each of these features characterises the shape of an object by a single number and therefore compares two objects with constant time complexity. The performance of these features as well as that of the scope histogram is listed in Table 1. The results show that the numeric features, namely compactness, radius ratio, and aspect ratio gain results between about 16% and 24%. In contrast, our scope histogram achieves results of about 46% and therefore outperforms the other approaches. Better results can be achieved by combining features. The results show that a combination of the three numeric features already gains about 52% correct matches. Combining these features with the scope histogram almost 64% can be achieved. Eventually, it is worth mentioning that this retrieval result is only about twelve percentage points less than the 76.46% achieved by the correspondence of visual parts of [9]. However, this latter approach has a significantly higher time complexity of $O(mn^3)$ for the comparison of two objects, while the 64% are achieved with constant time complexity.

Table 1. Classification results of compactness (CO), radius ratio (RR), aspect ratio (AR), and scope histogram (SH) for CE-Shape-1 Part B. Furthermore, the classification has been evaluated for the combination of all numeric features (NF) as well as their combination with the scope histogram (NS).

CO	RR	AR	SH	NF	NS
21.86	16.82	24.12	45.52	51.58	**63.75**

There is yet another advantage of scope histograms, namely that they allow prototypes of categories to be defined. This is useful since a common categorisation technique is based on the definition of clusters which define classes by training examples. In order to analyse whether the scope histogram qualifies itself as such a clustering method, we shall define clusters upon the MPEG dataset and rerun our evaluation on this basis. For this purpose the average of the values of all features is taken for each class. For the scope histogram we determine the average of the corresponding entries. In proceeding this way we define a number of 70 prototypes, one for each class. Using these prototypes we achieve the results listed in Table 2. It shows that the classification results of the three numeric features do not significantly change when they are solely applied in a clustering scenario. By contrast, the scope histogram's results can be improved by 20 percentage points to 66%. The scope histogram now even outperforms the combination of the three numeric features, which achieve together about 63%. A combination of the numeric features and the scope histogram leads to almost 83% correct matches.

Table 2. The classification results of Table 1 can be improved if a prototype is computed for each class of the MPEG test dataset

CO	RR	AR	SH	NF	NS
22.14	15.43	24.43	65.57	62.57	**82.93**

Summarising these results, the scope histogram combined with other features offering constant time complexity achieves about 64% in the MPEG test. By computing a prototype for each class the retrieval results for the MPEG test dataset can even be improved to almost 83%. As elaborated above, this result reflects the scope histogram's qualification as a clustering method.

4   Discussion

In this section we will take a closer look at some selected scope histograms. On this basis we will discuss how the visualisation introduced in Sect. 2.4 supports decisions on whether the application of our approach is useful or not, given a number of classes which are to be distinguished. For this purpose we analyse the classes from the MPEG test dataset and especially their prototypes. The prototypes' scope histograms as well as one example instance from each class are depicted in Figs. 5 to 7.

Comparing the polygons it shows that the prototype of the "Spring" class can be distinguished from all other prototypes by the universal scope which frequently occurs in this class; this scope consists of all twelve atomic $\mathcal{BA}_{12}$ relations, and it represents those line segments which are completely circulated by their polygon's course. In the case of the spring class, the frequency of the universal scope can be attributed to those segments which are located within its ends. This is a particular salient feature of this class in that the universal scope appears in no other class so frequently. Analysing all classes the universal scope is found rather seldom. Further remarkable occurrences can be determined within the body of the "Sea Snake", in the inner half of the "Horseshoe", at the inner side of the tail of the "Lizzard", and within the handle of the "Cup" class.

Another notable scope is that one of the $\mathcal{BA}_{23}$ relation C_l (Fig. 1), i.e. the relation containing all atomic relations located on the left side of the orientation grid. It does not only appear if line segments are in relation C_l but whenever the polygon is located completely left w. r. t. the reference segment. Since this scope describes a reference segment that is convex w. r. t. its polygon (i. e. the segment is part of the polygon's convex hull), it is not surprising that it appears much more frequently than the universal scope. As an example, it has remarkable occurrences in the "Bottle", "Cellular Phone", "Face", and the "Pencil" classes. It also appears quite often in the scopes of the "Apple" and the "Pocket" class. As objects pertaining to these classes show, apples and pockets have in common that they have a big round body with a significantly smaller part on the top. This is reflected in their scope histograms, which show that there are five entries in the case of the apple and similar frequencies at the very same entries in the scope histogram of the pocket. It is worth mentioning, that due to the fact, that we do not consider the order in which the scopes appear, there are yet other classes which have similar scope histograms. This holds in particular for the "Heart" class and the "Jar" class.

Apparently, the prototypes of some classes are described by very simple scope histograms, i. e. they are made up of only a few different scopes. This can be explained by the fact, that all instances in these classes have very similar scope histograms. On the one hand this brings in the advantage that distances between two instances of this class are small. On the other hand this poses a problem if the dominant scopes are very common and if their combination appears in other classes, too. Examples for classes that do not contain dominant scopes and which have similar scope histograms are the "Guitar", the "Key", as well as the "Spoon" class.

Another problem arises from classes containing instances with completely different shapes. This is due to the fact that the classes for the MPEG test dataset have been compiled semantically, so that it is impossible to classify shapes correctly using solely shape information. Examples for classes of this kind are the "Elephant" class as well as the various "Device" classes.

Coming to a conclusion, two classes of shapes can be distinguished very well by our method if their scope histograms have frequencies which are both dominant and different. It has been shown that the visualisations of the scope histograms enable us to directly judge how promising their application is. Defining those visualisations only on the basis of the size of the spots depicting frequencies in the histograms, such diagrams are not difficult to interpret. Moreover, in contrast to other histograms which are used to show, for example, the colour distribution of images, the entries of the scope histogram have been arranged in a specific way. By this means relations between semantics and structure are made explicit in that the position of a spot in the diagram reflects the complexity of the polygon w. r. t. specific reference segments as well as to specific shape properties such as the convexity for specific parts.

5   Summary

We introduced a new approach for the description and comparison of objects by their shape. Due to its intuitive diagrammatic visualisation, it can be easily estimated whether the application of our approach is promising or not. Besides, our approach offers a constant time complexity for the comparison of two objects. It shows that by combining it with other features offering the same time complexity retrieval results of about 64% can be achieved in the MPEG test. Furthermore, retrieval performance for the MPEG test dataset can even be improved to almost 83% by the application of prototype shapes.

References

1. J. F. Allen. Maintaining Knowledge about Temporal Intervals. *Communications of the ACM*, 26(11):832–843, 1983.
2. F. Attneave. Some Informational Aspects of Visual Perception. *Psychological Review*, 61:183–193, 1954.
3. R. O. Duda and P. E. Hart. *Pattern Classification and Scene Analysis*. John Wiley and Sons, Inc., 1973.
4. C. Freksa. Temporal Reasoning Based on Semi-Intervals. *Artificial Intelligence*, 54(1):199–227, 1992.
5. G. D. Garson and R. S. Biggs. *Analytic Mapping and Geographic Databases*. Sage Publications, Newbury Park, CA, 1992.
6. B. Gottfried. Reasoning about Intervals in Two Dimensions. In W. Thissen, P. Wieringa, M. Pantic, and M. Ludema, editors, *IEEE International Conference on Systems, Man and Cybernetics*, pages 5324–5332, The Hague, 2004. IEEE Press.
7. B. Gottfried. *Shape from Positional-Contrast — Characterising Sketches with Qualitative Line Arrangements*. Deutscher Universitäts-Verlag, 2007.
8. J. H. Larkin and H. A. Simon. Why a Diagram is (Sometimes) Worth Ten Thousand Words. *Cognitive Science*, 11:65–99, 1987.
9. L. J. Latecki and R. Lakämper. Shape Similarity Measure Based on Correspondence of Visual Parts. *IEEE Transactions on Pattern Analysis and Machine Intelligence*, 22(10):1185–1190, 2000.
10. L. J. Latecki, R. Lakämper, and U. Eckhardt. Shape Descriptors for Non-rigid Shapes with a Single Closed Contour. In *IEEE International Conference on Computer Vision and Pattern Recognition*, pages 424–429, 2000.
11. D. A. Mitzias and B. G. Mertzios. Shape Recognition with a Neural Classifier Based on a Fast Polygon Approximation Technique. *Pattern Recognition*, 27:627–636, 1994.
12. K. Zimmermann and C. Freksa. Qualitative Spatial Reasoning Using Orientation, Distance, and Path Knowledge. *Applied Intelligence*, 6:49–58, 1996.

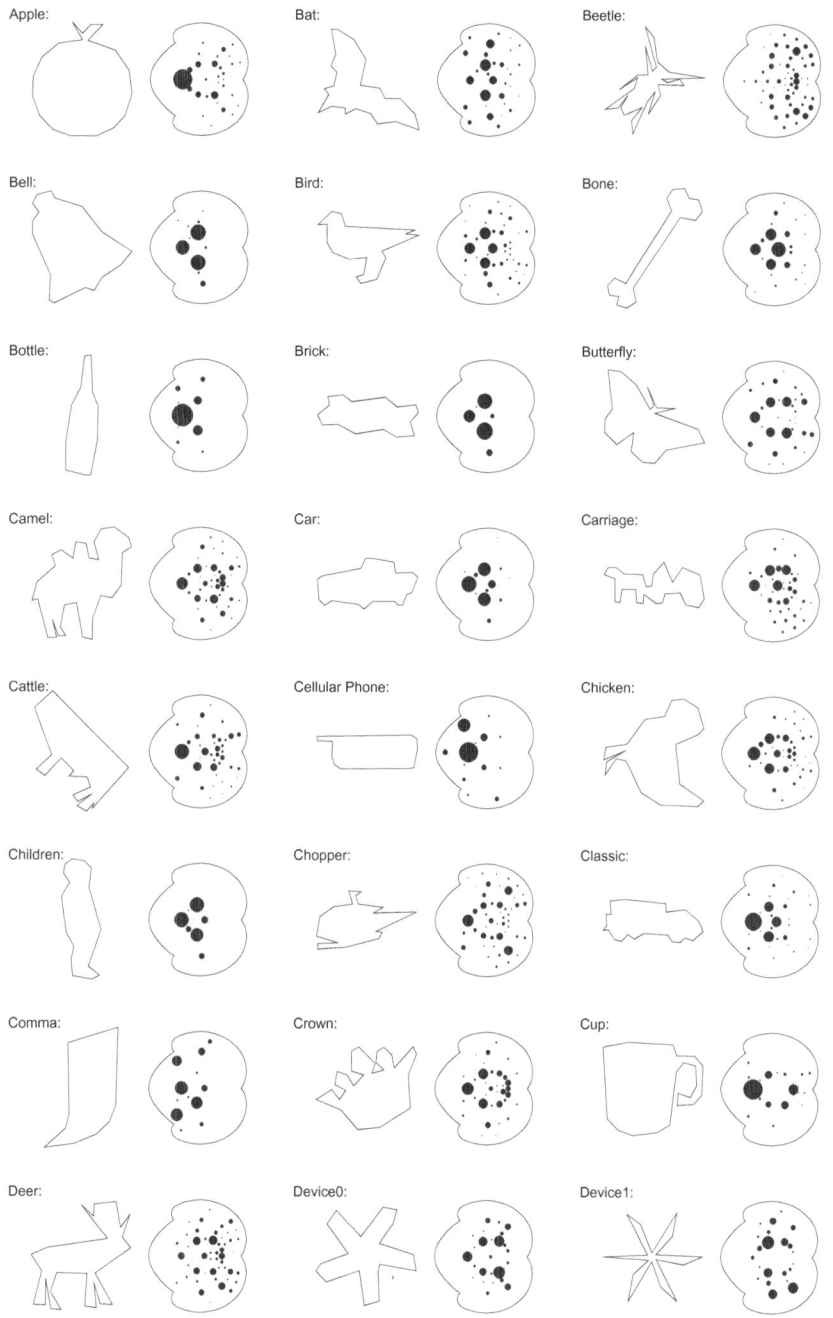

Fig. 5. Histogram prototype and example instance for each MPEG class (Part I)

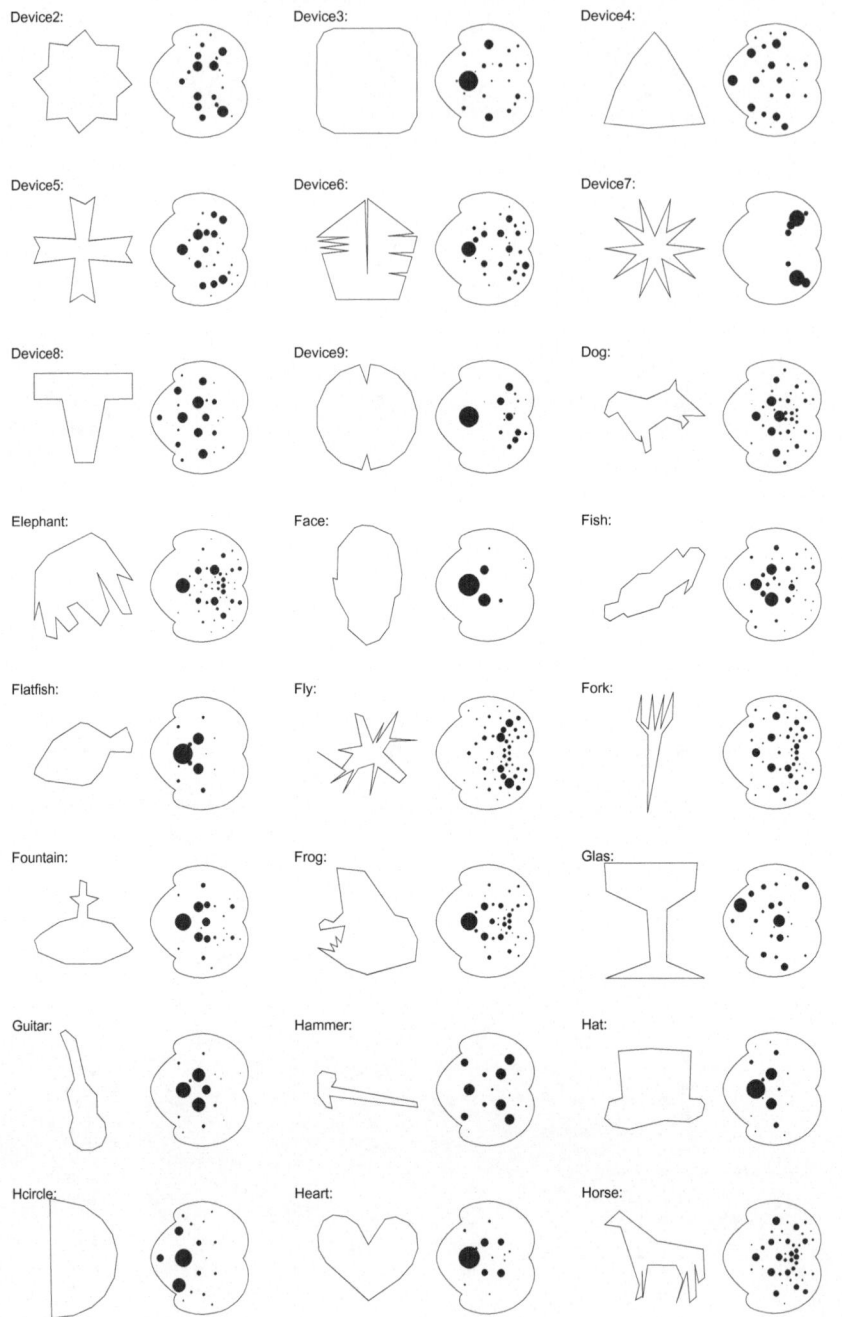

Fig. 6. Histogram prototype and example instance for each MPEG class (Part II)

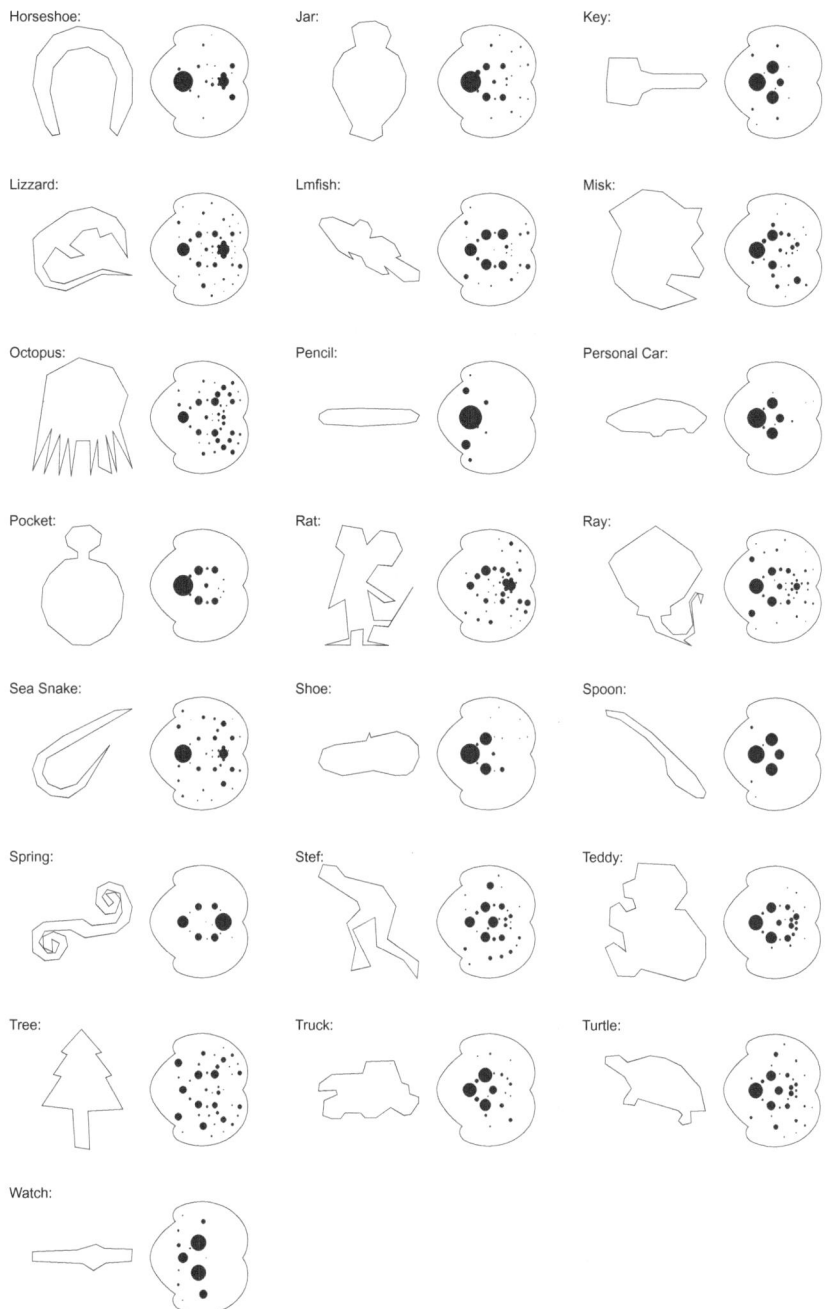

Fig. 7. Histogram prototype and example instance for each MPEG class (Part III)

A Robot Learns to Know People—First Contacts of a Robot

Hartwig Holzapfel[1], Thomas Schaaf[2], Hazım Kemal Ekenel[1],
Christoph Schaa[1], and Alex Waibel[1,2]

[1] InterACT Research, Interactive Systems Labs,
Universität Karlsruhe, Germany
{hartwig,ekenel}@ira.uka.de
[2] InterACT Research, Interactive Systems Labs,
Carnegie Mellon University Pittsburgh, USA
{tschaaf,waibel}@cs.cmu.edu

Abstract. Acquiring knowledge about persons is a key functionality for humanoid robots. In a natural environment, the robot not only interacts with different people who he recognizes and who he knows. He will also have to interact with unknown persons, and by acquiring information about them, the robot can memorize these persons and provide extended personalized services. Today, researchers build systems to recognize a person's face, voice and other features. Most of them depend on pre-collected data. We think that with the given technology it is about time to build a system that collects data autonomously and thus gets to know and learns to recognize persons completely on its own.

This paper describes the integration of different perceptual and dialog components and their individual functionality to build a robot that can contact persons, learns their names, and learns to recognize them in future encounters.

1   Introduction

Recognizing and memorizing other people is an important part of human-human communication. A humanoid robot, if equipped with such functionality, can offer more natural ways of communication and provides a basis to provide personalized services. Today, systems exist that can recognize a person's face, voice or identify persons using other biometric features. In addition, speech recognition and dialog management supply robots with a natural way of communication. We think that with the given technology it is about time to build a system that collects data autonomously and thus gets to know and learns to recognize persons completely on its own. The task of the robot is to meet persons that walk by the system, to establish contact and to find out some information about the person. Using speech recognition, the robot can understand a person's name and learns previously unknown names. The system stores information about these encounters and recognizes the person the next time with a face-ID recognizer. To our knowledge, such a system has not been presented before.

C. Freksa, M. Kohlhase, and K. Schill (Eds.): KI 2006, LNAI 4314, pp. 302–316, 2007.

In this paper we present such a system by describing its architecture, its major components, and first experiments with parts of the system. Techniques that are integrated in the system are speech recognition, unknown word detection, phoneme recognition, spelling, extensible grammars, multimodal dialog management, visual person detection, face detection, face-ID and recognition. We describe the system in two main parts. First, we describe experiments to obtain attention from a person and to maximize success rates in initiating a dialog. Learning mechanisms were applied to classify the intention of the user and to choose actions by the system to obtain attention. We present results of this experiment in the next section. Second, we describe the components and their integration to conduct a dialog with the person during which the system learns the person's name and face-id. The main components that are employed are speech recognition with unknown-word-detection (OOV), face recognition (face-ID) and multimodal dialog management. The components are described and evaluated separately.

The long-term vision of the system is a robot that can patrol the entrance area of a building or a corridor and get in touch with previously unknown people completely on his own. The field of related work is broad due to its interdisciplinary nature. Recently, several robotic systems and humanoid robots that interact with humans have been developed. Our system, as a robot, is broadly related to the field of social robotics [1,2], especially given its style of interaction and the scenario of its employment. An interactive and self controlled robot for example is Lewis [3], a robot with the task to take pictures of groups of persons, visually triggered by certain regions of interest. In our task, the system tracks (moving) persons with whom the robot wants to interact, which relates to visual tracking e.g. [4], but also multimodal tracking e.g. [5] is helps finding the right conversation partner. The following section, where we describe experiments to obtain attention and to initiate a dialog with persons, also relates to studies on engagement [6] in Human-Robot interaction.

Besides controlling a robot that interacts with people, the second great challenge is to conduct a dialog with the person to learn the person's name. Some work exists that describe language learning and learning of words on a multimodal basis. Dusan and Flanagan describe a system for learning words and their meaning in a multimodal setting [7,8]. [9] describes understanding and grounding of new words in situated dialog. Other work focuses on understanding new words, which is a speech recognition task. Chung et al. [10] combine phoneme recognition of spoken input to obtain phonetic representation of names with telephone keypad input to obtain textual representation of names. Even if the name is known but a very large vocabulary list is used, special attention is required. Chung et al. [11] describe a system with a dynamic vocabulary that can be updated according to the given context. The approach from Scharenborg and Seneff [12] runs multiple recognition passes on speech input, with a phone-based OOV word-model in the first step, which is used to constrain the vocabulary in the second step that best matches the resulting phone graph. More related work on unknown words is described in section 3.3.

The paper is organized as follows. Section 2 describes experiments with proactive behavior to obtain the user's attention and initiate a dialog. Section 3 describes the system architecture, the dialog manager, speech recognition and vision. Section 4 concludes the paper and gives an outlook to future work.

2   Obtaining Attention and Initiating Dialogs

A pretest with the system was conducted to evaluate if and how well the robot can obtain the attention of persons passing by, and then initiate a dialog with that person. We furthermore wanted to see which actions of the robot are most important to obtain attention. For the experiment we first recorded and labeled a series of persons passing by the robot, and a series of persons that were told to walk towards the robot. Second, we trained a classifier to classify the person's 'interest' in the system. Finally, we evaluated the success rate of the system to initiate a dialog with interested persons.

For this purpose, we placed the robot in the corridor of our research institute and observed the behavior of people passing by. Each person was then interviewed about the robots behavior and his/her own interest in the system and how much attention was spent to the system when walking by. During each iteration, the robot chose a single action or a combination of the following actions: head movement (turn the head towards the person), play sounds, spoken output. Spoken output was "Hello! Please come closer!" and "Please use the Headset to say Hello!". The success of initiating a dialog was later measured by how many people took the headset to talk to the robot. Figure 1 shows a picture of the robot waiting for persons.

The baseline for attention was estimated by interviewing six persons that have never been in the building before. They were sent to an office at the other end of the corridor and had to pass by the robot, which didn't show any behavior. When they arrived at the office they were interviewed how they had perceived the robot. It was interesting to hear that only one person had even noticed that there was a robot. Afterward, ten more people passed by the robot, in this case the robot reacted by playing a moderate honking sound. Four persons didn't notice the sound, six persons noticed the sound, three of them also interpreted the sound as a reaction of the robot. The baseline can be used to compare the following set of artificial experiments, where persons walked intentionally past the robot to judge its actions with this set of "uninformed" persons. It also shows that the robot itself is not eye-catching at all, and thus the robot's actions become more important to establish a dialog.

In a second experiment, we instructed eleven persons to walk by the robot and judge its actions. Each user had to do this five times, each iteration with different combination of actions. Table 1 shows the results of the experiment. The evaluated categories are 'eye-catching' (does the action influence my attention?) and 'suitable' (do you feel you should start a dialog with the system?). The values for eye-catching are scaled to 0 (no influence), 1 (medium), 2 (annoying). The values for 'suitable' are scaled to -1 (not suitable), 0 (a little bit), 1 (yes). The

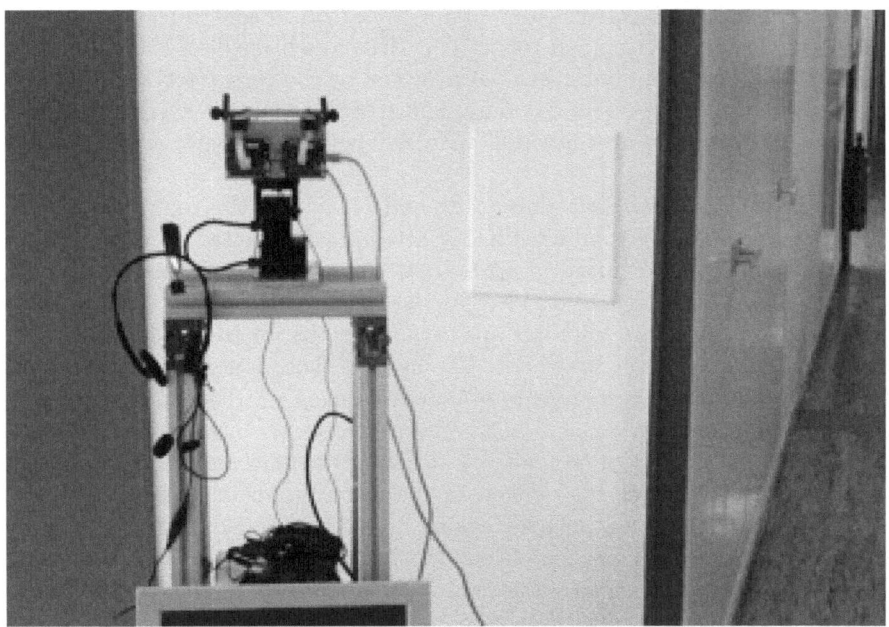

Fig. 1. A picture of the robot, waiting for persons to interact with

Table 1. Evaluation of different system actions

action	eye-catching	suitable
play sound	0.9	-0.3
turn head	0.9	-0.3
say 'hello'	0.9	0.8
play sound then say 'hello'	0.9	0.3
turn head then say 'hello'	1.0	0.9

experiment shows that playing a simple sound is already judged as eye-catching. It should be noted that turning the head also makes some noise induced by the pan-tilt unit. Turning the head offers no increase in attention over playing a simple sound, but leads to the highest attention rate and suitability for initiating a dialog when combined with speech.

The final experiment combines actions to obtain the user's attention and actions to initiate a dialog with an interest classifier. The classifier was trained only on 3D tracking data from the robot's stereo camera. This differs from other work that use a laser scanner to determine the position of persons, and vision-based face tracking doesn't seem to achieve the same reliability. Interest classification thus had to be made robust against recognition and tracking errors. We trained a multi-layer perceptron (MLP) with two hidden layers to classify '0' (not interested) or '1' (interested). The input features that led to the best results are (i) distance between person and robot (ii) angle between straight robot view (straight ahead) and person (iii) walking speed of person (iv) angle between person's walking direction and robot. The MLP was trained on 1000 samples of the (partly) noisy input data that was provided by the person tracker and hand-labeled interested/not-interested tags. The error rate was on average 14.6% on unseen data, averaged over five runs with different clustering of training, cross-evaluation and test-sets.

The final evaluation then aims to evaluate how well the robot was able to initiate a dialog. It was conducted in 100 attempts distributed over five persons. The experiment was artificial in a way that the persons redid the same experiment a couple of times and decided for themselves if they would interact with the robot. The absolute numbers are thus subjective to the willingness of the persons to interact with the robot. The system chose among four modes. All four modes use different actions to obtain the person's attention, but share the same spoken output once the person is recognized to be 'interested'. In the first and second mode, the system first plays a sound when detecting the person and moves the head towards the person. In the second mode the state transitions don't rely on a single continuous tracking in contrary to the first mode. In the third mode, the system first plays a sound when detecting the person, moves the head towards the person and then follows the person with the head, and also doesn't rely on a single continuous tracking. In the fourth mode, the system only plays a sound when detecting the person, and requires a continuous track again. In all modes, the system says "hello, please come closer!" when the user is interested and then "use the headset to say hello" when the user remains interested to start a dialog.

During some of the iterations the users could not be tracked correctly, e.g. due to changing light conditions. When the system failed to track the user no interaction could be initiated. Table 2 (left columns) shows for each user first the tracker-recognition rate and second the success rate to start a dialog. The table shows a dependency of recognition rate to success rate, but also different behavior by different users. Evaluated on a per system-action modes base 2 (right columns), the results show a smoother distribution of the success rates. It also shows that the modes that were more forgiving regarding the person tracks received higher success rates. Figure 2 shows a series of pictures taken by the robot camera during a single interaction.

Table 2. Evaluation of the success rates per user (left three columns), and evaluation of the success rates per mode (right three columns)

person no.	recognition rate	success rate	mode no.	recognition rate	success rate
1	80%	35%	1	76%	42%
2	100%	85%	2	80%	55%
3	60%	30%	3	76%	58%
4	70%	15%	4	96%	45%
5	100%	50%			

Fig. 2. Series of pictures taken by the robot camera during a single interaction

Tracking of persons was realized with a state-of-the art multi-person tracker developed within the CHIL-project[1]. It uses single 2D-images to detect face-candidates with a haar-cascade based face-detector. 3D-information is obtained for each face-candidate by using disparity information from matching the left and right camera images, including a size estimation for head candidates. More details on the implementation can be found in [13].

3   Recognizing Persons and Names

After initiating a dialog, it is the task of the system to identify the person using snapshots of the person's face and speech input to understand the person's name. To facilitate this task we use a face tracker and keep the person to communicate with in the field of view. A face-ID recognizer computes hypotheses of all snapshots and recognizes either a known or an unknown person. The spoken dialog part is then to ask for the unknown person's name or confirm the name of a known person.

Information about persons, i.e. video snapshots, pictures, voice input, their ID, their names as string and phonetic representation, and information about their latest interactions are stored in a MYSQL database. This database is filled during interaction and is used by the face-ID recognizer to build a model of known persons. It is also used by the speech recognizer and the dialog manager, which read names and their phonetic representations to create grammars for speech recognition, understanding and spoken output.

[1] *http://chil.server.de*

3.1   Face Recognition

Face recognition first tries to determine whether the person in question is known to the system or not. If the person hasn't been stored in the database yet, the robot asks his/her name to enroll the person's face images to the face database labeled with his/her person name.

The face recognition system uses a face sequence, which is provided by a face tracker module, as input. It processes the frames to locate the eyes and then aligns the face images according to the eye center coordinates. The features that will be used for classification purposes are extracted from each face image by using a local appearance-based face recognition approach [14]. In this feature extraction approach, the input face image is divided into 8x8 pixel blocks, and on each block discrete cosine transform (DCT) is performed. The most relevant DCT features are extracted using the zig-zag scan and the obtained features are fused either at the feature level or at the decision level for face recognition [14]. The approach is extensively tested on the publicly available face databases and compared with the other well known face recognition approaches. The experimental results showed that the proposed local appearance based approach performs significantly better than the traditional face recognition approaches. Moreover, this approach is tested on face recognition grand challenge (FRGC) version 1 data set for face verification [15], and a recent version of it is tested on FRGC version 2 data set for face recognition [16]. In both tasks the system provided better and more stable results than the baseline face recognition system. For example, in the conducted experiments on the FRGC version 2 data set, 96.8% correct recognition rate is obtained under controlled conditions and 80.5% correct recognition rate is obtained under uncontrolled conditions. In these experiments, there are 120 individuals in the database and each individual has ten training and testing images. There is a time gap of approximately +six months between the capturing time of the training and the test set images. The approach is also tested under video-based face recognition evaluations and again provided better results [17,18]. For details please see [14,15,16,17,18].

After extracting the feature vectors from each face image in the sequence, they are compared with the ones in the database using a nearest neighborhood classifier. Each frame's distance scores are normalized with Min-Max normalization method [19], and then these scores are fused over the sequence using the sum rule [20]. The obtained highest match score is compared with a threshold value to determine whether the person is known or unknown. If the similarity score is above the threshold, the identity of the person is assigned with that of the closest match. If the similarity match score is below the threshold, then the person is classified as unknown. In this case, the robot asks for person's name and saves his/her face images to the database.

3.2   System Integration and Dialog Components

The dialog implementation is based on the Tapas dialog manager [21,22]. It uses a language and domain independent dialogue engine with discourse representation and goal-based dialogue strategies. The dialogue engine follows the Ariadne dialogue manager [23].

In our communication centered scenario, the dialogue manager is the main component to decide which actions to take. Its dialog strategy defines which information to request or which actions the system should take. The dialogue manager also comprises interpretation of multimodal input and is responsible for storing information in the database. All components communicate over a message-based architecture. Figure 3 shows a diagram visualizing the integration of the recognition and understanding components into the dialogue system. The speech recognizer sends an n-best list of parse trees to the NLU-component that converts the parse trees to semantics, formulated as typed feature structures (TFS). The input TFS is converted and interpreted in the dialogue context and finally updates the discourse. Typed feature structures (TFS) also represent referenced database objects and user model data.

The dialogue engine's strategy matches discourse states to dialogue goals, each layer in the discourse corresponds to an unfinished dialog goal.

All semantic concepts that are used to represent user input and discourse representations are defined in an ontology which provides inheritance information and relations between concepts. The dialogue engine is language and domain independent. Language specific parts are semantic grammars for natural language understanding and generation templates for spoken output. Grammar resources for input understanding are an extension to JSGF grammars that are extended with inheritance rules and semantic construction rules. The grammars are shared by the dialogue manager's NLU parts and the speech recognizer.

For speech recognition we use the Janus speech recognizer (JRTK)[24] with the Ibis single-pass decoder[25]. Ibis allows to decode with context-free grammars (CFGs) instead of statistical n-gram language models, and offers a tighter integration of the dialogue manager and Janus by being able to weight grammar rules depending on the dialogue context [26].

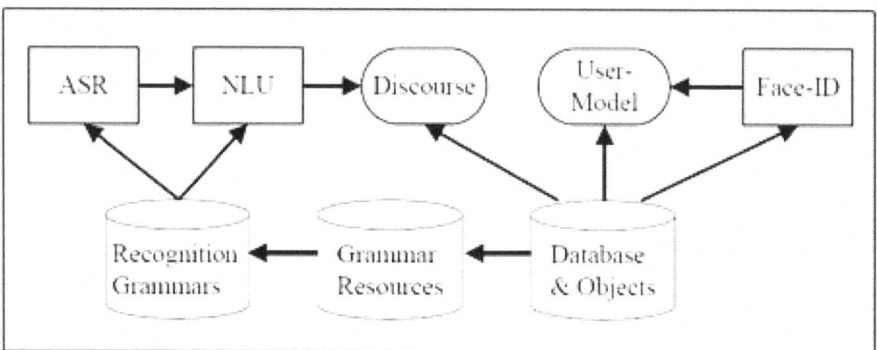

Fig. 3. Flow diagram visualizing the integration of recognition and understanding components, as well as face-ID into the dialogue system

To be able to operate in dynamic environments, i.e. especially to be able to update and understand new names, the system ontology and grammars need to be updated during runtime. In our system, we store person information such as names in an external database. The database information is used by dynamic (grammar) nodes that generate grammar parts from database information, and are update during runtime [27]. These nodes implement a caching mechanism and can be updated from the database when necessary, e.g. when new user names are added.

The following example shows a spoken interaction between the system and a user (table 3). The dialog shows a simple interaction where a person is asked for his name. In case of an unknown name, the system asks a second time to get a better phoneme hypothesis and then asks for the spelling. In the scenario shown here, the dialog manager doesn't try to confirm the spelling. In contrast, the strategy doesn't need to rely on the correct spelling (letters) but uses in the first place a phonetic description, respectively a joint hypothesis as described in section 3.3 that can be used to recognize the name again the next time.

Table 3. A typical dialog with the system

System	Hi, please tell me your name
User	My name is Stephan
System	Can you please repeat this?
User	I Said, my name is Stephan
System	I haven't heard this name before, please spell it.
User	S T E P H A N
System	Thank you. I will recognize you next time.

3.3 OOV Recognition

The above section describes how information about a person and especially the name, can be updated in the system to be recognized with the next input. However, to recognize a new name for the first time (i.e. previously unknown name of a person) a different approach has to be taken. Firstly, the speech recognizer needs to detect that the user has spoken a word which is not covered by the grammar at this point and also not in the vocabulary. This is called unknown-word or out-of-vocabulary (OOV) detection [28,29].

The Hidden Markov Model framework is currently the state of the art in speech recognition. In this framework, the possible output is constrained to the search vocabulary of the recognizer. The search vocabulary usually consists of words with some additional words to cover also acoustic events like hesitations (aem), lip-smack, silence and general noise events. For English we use the simple definition of a word as everything that is written between two blanks. Therefore if a word is spoken that is not in the search vocabulary the recognizer cannot hypothesize it. Actually, the recognizer usual comes up with one or more words that are the closest match to the spoken word. Because the recognizer combines an acoustic model and a language model, in general the words of the closest match do not always sound similar to what was said, or otherwise sounds similar but doesn't fit in the context, or even a combination of both. Hetherington [30] found that in English on average the number of errors a speech recognizer makes per OOV word is larger than one (1.5 - 1.8).

The standard approach to address the Out Of Vocabulary (OOV) problem is to increase the vocabulary size until the average number of OOV words becomes small and the effect on the word error rate becomes small. However, the word error rate does not measure the importance of words and it is obvious that names are very important. If they are missing, it is very hard to understand what is going on. If we would know all names we need in advance this would be nice, but this is in general not the case. Unfortunately, adding all names we know to the search vocabulary can harm the recognition also by increasing the acoustic confusability or because of the low probability of the names, they are still misrecognized. Nevertheless, most important, we just do not know all names that can come up, especially if we think about names that do not exist yet. Therefore we need a dynamic way to extend the vocabulary on demand.

With our approach, the model is extended with special 'words' that represent any unknown sets of words. In a dialog, missing information is acquired and the search vocabulary and language model - here a CFG-grammar - is updated.

Similar to the pioneer work by Asadi [31] who was the first to address the OOV problem in speech recognition, the possible location for OOV words is constrained by a grammar. He investigated two extensions of the recognition model to cover the acoustics of an unknown ship-name, a simple (flat-) acoustic model that is trained on multiple phonemes at the same time and duplicated this model to achieve a minimum length constrained. His findings inspired many researchers to find good acoustic models, and is still the prototype of the current state of the art.

The main goal is to extend the acoustic recognition model with generic words, which give a lower probability compared to the correct word model if the word is known, but are preferred if the word is unknown. In our approach we use Head-Tail-Models (HT-Models) [29] which are a combination of the same precise acoustic models that is used to model the context dependent phonemes of the known words and a generic phoneme model that is trained with multiple phonemes similar to the flat model of Asadi. In [28] this approach was advanced to define the head part of the HT-Models based on the search vocabulary of the recognizer. This has lead to vocabulary optimized HT-Models that also fit very well in a phonetic prefix tree or other search graph structure.

The basic idea of the HT-Models is that it is not possible to decide already after the first phonemes if a word is out of vocabulary because usually the prefix is shared by many known words. Therefore, the head part of the model has the task to compete with the known words until the OOV word starts to diverge from all known words, in which the average likelihood of the exact models gets worse.

If the hypothesis of the speech recognizer indicates the presence of an unknown word, this triggers a dialog with the user to verify and learn the new word.

To add a new word, it has to be included in the search vocabulary and the language model. To extend the search vocabulary, we need a phonetic description of the new word that represents the pronunciation. For the robot domain, this pronunciation usually serves two purposes. One is to allow the pronunciation of the word, e.g. if it is a name, to address a person and therefore it requires a high quality for speech synthesis. The other purpose is to describe the word close enough, that it is not confused with other known words and can be recognized later again. However, for this purpose the chosen phoneme sequence can be less than perfect. There are two sources from which a phoneme sequence can be derived, from a spoken version of the word and from the written form using grapheme to phoneme rules. To get the written form of the name to learn the robot asks the user to spell it. The recognition accuracy for this task is about 90% using a statistical N-Gram model allowing only letters and some human noises. On average, 60% of spelled words have no error, which is comparable to the performance from Hild without vocabulary constrained [32]. Therefore, it is useful to use the top N hypotheses to generate pronunciations. However, generating phoneme sequences for names from graphemes is difficult, especially if the names also come from different languages. In addition, each written form can generate more than one possible pronunciation. So it is necessary to select a small set of pronunciations that is closes to the spoken form. This can be done by using the original utterance to cross-validate. If a reasonable pronunciation was generated by the speech recognizer it replaces the OOV-symbol after the hypothesis is redecoded.

We use the same approach to find a good phoneme sequence, by asking the user to say the word again and do phoneme recognition. The phoneme recognition accuracy is about 65%. Therefore the phoneme recognizer also creates an n-best list of phoneme sequences which are merged with the list generated by the spelling recognizer. Actually, the cross-validation is performed on the merged set. The pronunciation that wins is used to extend the speech recognizer. If the winning pronunciation was generated by grapheme to phoneme, then the attached grapheme sequence is used, otherwise the first best grapheme sequence. If none of the learned pronunciations appear in the redecoded utterance, this indicates that the found phoneme sequences do not generalize well and therefore are not suitable for recognition. This allows to ask the user for more help until the name is learned or to give up learning the name.

3.4   The Dialog Manager

The dialog manager is based on the Tapas dialog framework [21,22] which is described briefly above. Here, we want to describe the extensions that were made to the framework to handle special requirements of the given task. We use most of its existing functionality, such as language understanding, discourse modeling, and knowledge representation, but implemented new dialog strategies that lead in the direction of better social communication. Also the abilities to handle multimodal information [33,27] was extended to handle new types of visual perception with communicative meaning.

The dialog manager uses a short-term memory for 'active' information and a database that represents long-term memory, similar to Kawamura et al. [34]. In addition, we maintain a standard discourse model to represent communicative information. The short term memory is organized in chunks that contain semantic information, which is, like all other semantic data structures in our system, represented by typed feature structures (TFS). One part of the short-term memory is a user (focus) model that represents information about currently available communication partner(s). The user model collects information from different modalities about the user. The collected information is compared against database entries where the information either matches a single person, a set of persons, or allows creating a new entry.

The dialog manager applies a method that we call behavior selection. Each behavior defines a specific operation mode, and defines how the dialog manager interprets incoming events, how information is interpreted in the discourse, and which actions are taken by the strategy. In the given task we have found the following four states: idle, obtain attention, initiate conversation, conduct dialog. The full state-transition model is shown in figure 4. The state model contains a number of 'hard-coded' transitions. These are the transitions to idle if the system recognizes that the user leaves the conversation or after a timeout where no communication happens. The other transitions are defined from any state (i.e. idle, obtain attention and initiate conversation) to 'dialog', if the user has picked up conversation with the system. The remaining transitions are defined by the classifier and selection mechanism described in section 2. The behavior model also provides a basis for further extending the transitions influenced by motivations and drives.

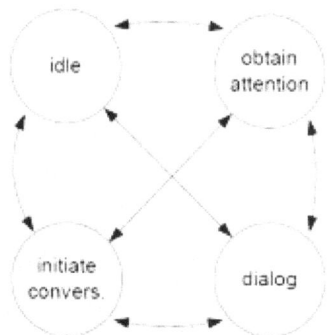

Fig. 4. State model with full transitions

4   Conclusions and Future Work

4.1   Conclusions

We have presented a first approach for a robot that learns to known persons completely on its own. The main challenges addressed are obtaining attention from the person, initiating a dialog, understanding new names input, building a face-ID database and recognizing persons, integration and decision-taking in dialog. The first experiments conducted show that such a system is possible with current technology, and motivate further development.

We have presented experiments on proactive behavior to obtain the user's attention and to initiate a dialog, so the robot doesn't need to wait for a person to initiate the dialog. During these experiments the robot has learned a behavior to obtain the attention and interest of persons walking by, without molesting other people, and a successive series of actions to initiate a dialog with interested persons.

We have further presented our speech recognizer with OOV recognition capabilities, as well as phoneme recognition and word spelling, and the integration of the face-ID recognizer as the main component for recognizing known and unknown persons prior to initiating a dialog. The dialog manager combines and coordinates the different components within the system. The different components have been evaluated individually, future experiments will show how the system operates in a real-world environment.

4.2 Future Works

In the future we plan to conduct further experiments with the fully integrated system, and evaluate the system in different environments and on different robots (especially our SFB588-robot Armar III) in out-of-the-lab scenarios. Further improvement is necessary for robust processing of the persons IDs - both vision and speech recognition can produce errors, so the database might contain different IDs for same person, or a mixture of different persons with the same ID. During the previous experiments, the system did not distinguish which persons to talk to. Selective dialogs will help in the future to remove this fortuity and to solve these errors by explicitly talking to specific persons. Furthermore, robustness issues and more elaborate methods to obtain better spelling - letters or phonemes - will be considered. The approach further motivates to obtain other personal information from the communication partners that exceed the name-only task.

Acknowledgments

This work was supported in part by the German Research Foundation (DFG) as part of the Collaborative Research Center 588 "Humanoid Robots - Learning and Cooperating Multimodal Robots" and by the interACT research scholarship.

References

1. Fong, T., Nourbakhsh, I., Dautenhahn, K.: A survey of socially interactive robots. Robotics and Autonomous Systems 42 (2003) 143–166
2. Breazeal, C.: Social Interactions in HRI: The Robot View. IEEE Transactions on Man Cybernetics and Systems Vol. 34, Issue 2, (2004) 181–186

3. Byers, Z., Dixon, M., Smart, W.D., Grimm, C.M.: Say Cheese!: Experiences with a Robot Photographer. AI Magazine 25, Nr. 3 (2004) 37–46
4. Schulz, D., Burgard, W., Fox, D., Cremers, A.B.: Tracking Multiple Moving Targets with a Mobile Robot using Particle Filters and Statistical Data Association. Int. Conf. on Robotics and Automation (ICRA), IEEE Press, New Orleans (2001)
5. Lang, S., Kleinehagenbrock, M., Hohenner, S., Fritsch, J., Fink, G.A., Sagerer, G.: Providing the basis for human-robot-interaction: A multi-modal attention system for a mobile robot. Proceedings of the Int. Conf. on Multimodal Interfaces, Vancouver, Canada, (2003) 28–35
6. Sidner, C., Kidd, C., Lee, C., Lesh, N.: Where to look: A study of human-robot engagement. Proceedings of the International Conference on Intelligent User Interfaces (IUI), ACM, (2004) 78–84
7. Dusan, S., Flanagan, J.: Adaptive Dialog Based upon Multimodal Language Acquisition. Proceedings of the Fourth Int. Conf. on Multimodal Interfaces, Pittsburgh, PA, USA, (2002)
8. Dusan, S., Flanagan J.: A System for Multimodal Dialogue and Language Acquisition. Invited, The 2nd Romanian Academy Conference on Speech Technology and Human-Computer Dialogue, Romanian Academy, Bucharest, Romania (2003)
9. Gorniak, P., Roy, D.: Probabilistic Grounding of Situated Speech using Plan Recognition and Reference Resolution. Proceedings of the Seventh International Conference on Multimodal Interfaces (2005)
10. Chung, G., Seneff, S.: Integrating Speech with Keypad Input for Automatic Entry of Spelling and Pronunciation of New Words. Proceedings of ICSLP'02, Denver, CO, USA (2002) 2061–2064
11. Chung, G., Seneff, S., Wang, C., Hetherington I.: A Dynamic Vocabulary Spoken Dialogue Interface. Proceedings of ICSLP'04, Jeju Island, Korea (2004)
12. Scharenborg, O., Seneff, S.: Two-pass strategy for handling OOVs in a large vocabulary recognition task. Proceedings of Interspeech'05 (2005) 1669–1672
13. Schaa, C.: Proaktive Initiierung von Dialogen für humanoide Roboter. Diploma Thesis, Universität Karlsruhe (2005)
14. Ekenel, H.K., Stiefelhagen, R.: Local Appearance based Face Recognition Using Discrete Cosine Transform. Proceedings of the 13th European Signal Processing Conference (EUSIPCO), Antalya, Turkey, (2005)
15. Ekenel, H.K., Stiefelhagen, R.: A Generic Face Representation Approach for Local Appearance based Face Verification. Proceedings of the CVPR IEEE Workshop on FRGC Experiments, San Diego, CA, USA (2005)
16. Ekenel, H.K., Stiefelhagen, R.: Analysis of Local Appearance-based Face Recognition on FRGC 2.0 Database. Face Recognition Grand Challenge Workshop (FRGC), Arlington, VA, USA (2006)
17. Ekenel, H.K., Pnevmatikakis, A.: Video-Based Face Recognition Evaluation in the CHIL Project - Run 1. Proceedings of the 7th Intl. Conf. on Automatic Face and Gesture Recognition (FG 2006), Southampton, UK (2006)
18. Ekenel, H.K., Jin, Q.: ISL Person Identification System in the CLEAR Evaluations. Proceedings of the CLEAR Evaluation Workshop, Southampton, UK (2006)
19. R. Snelick, M. Indovina: Large Scale Evaluation of Multimodal Biometric Authentication Using State-of-the-Art Systems. IEEE Trans. Pattern Anal. Mach. Intell., 27(3) (2005) 450–455
20. J. Kittler, M. Hatef, R. Duin, J. Matas: On Combining Classifiers. IEEE Trans. Pattern Anal. Mach. Intell., 20(3) (1998)
21. H. Holzapfel: Building multilingual spoken dialogue systems. Special issue of Archives of Control Sciences, G.eds. Z. Vetulani, 4 (2005)

22. H. Holzapfel and P. Gieselmann: A Way Out of Dead End Situations in Dialogue Systems for Human-Robot Interaction. Humanoids (2004)
23. M. Denecke: Rapid Prototyping for Spoken Dialogue Systems. Proc. of the 19th Int. Conf. on Computational Linguistics (COLING), Taiwan (2002)
24. Finke, M., P. Geutner, H. Hild, T. Kemp, K. Ries, and M. Westphal: The karlsruhe-verbmobil speech recognition engine. Proc. of ICASSP'97, Germany (1997)
25. Soltau, H., F. Metze, C. Fuegen, and A. Waibel: A one pass- decoder based on polymorphic linguistic context assignment. Proceedings of ASRU'01, Madonna di Campiglio, Trento, Italy (2001)
26. Fügen, C., Holzapfel, H., Waibel, A.: Tight coupling of speech recognition and dialog management - dialog-context grammar weighting for speech recognition. Proceedings of ICSLP'04, Jeju Island, Korea (2004)
27. Gieselmann, P., Holzapfel, H.: Multimodal Context Management within Intelligent Rooms. Proceedings of the 10th International Conference on Speech and Computer (SPECOM'05), Patras, Greece (2005)
28. Schaaf, T.: Erkennen und Lernen neuer Wörter. PhD Thesis, Universität Karlsruhe (2004)
29. Schaaf, T.: Detection Of OOV Words Using Generalized Word Models And A Semantic Class Language Model. Proceedings of Eurospeech (2001)
30. Hetherington, L.: A Characterization of the problem of new, out-of-vocabulary words in continuous speech recognition and understanding. Ph.D.-Thesis, MIT (1995)
31. Asadi, A., Schwartz, R., Makhoul, J.: Automatic detection of new words in a large-vocabulary continuous speech recognition system. Proceedings of ICASSP'90, IEEE Signal Processing Society, Albuquerque, New Mexico, USA (1990)
32. Hild, H.: Buchstabiererkennung mit neuronalen Netzen in Auskunftssystemen. Ph.D.-Thesis, Universität Karlsruhe, Fakultät für Informatik, Shaker Verlag, Aachen, ISBN 3-8265-3155-8, Karlsruhe, Germany (1997)
33. Holzapfel, H., Nickel, K., Stiefelhagen, R.: Implementation and Evaluation of a Constraint-Based Multimodal Fusion System for Speech and 3D Pointing Gestures. Proc. of the Int. Conf. on Multimodal Interfaces, State College, PA, USA (2004)
34. Kawamura, K.: Cognitive Approach to a Human Adaptive Robot Development. Proceedings of the Int. Workshop on Robot and Human Interactive Communication (RO-MAN), Nashville, TN, USA (2005)

Recombinant Rule Selection in Evolutionary Algorithm for Fuzzy Path Planner of Robot Soccer

Jong-Hwan Park[1], Daniel Stonier[2], Jong-Hwan Kim[2], Byung-Ha Ahn[1], and Moon-Gu Jeon[1]

[1] Dept. of Mechatronics, Gwangju Institute of Science and Technology, South Korea
{jhpark, bayhay, mgjeon}@gist.ac.kr
[2] Dept. of Electrical Engineering and Computer Science, Korea Advanced Institute of Science and Technology, South Korea
{stonierd, johkim}@vivaldi.kaist.ac.kr

Abstract. A rule selection scheme of evolutionary algorithm is proposed to design fuzzy path planner for shooting ability in robot soccer. The fuzzy logic is good for the system that works with ambiguous information. Evolutionary algorithm is employed to deal with difficulty and tediousness in deriving fuzzy control rules. Generic evolutionary algorithm, however, evaluate and select chromosomes which may include inferior genes, and generate solutions with uncertainty. To ameliorate this problem, we propose a recombinant rule selection method for gene level selection, which grades genes at the same position in the chromosomes and recombine new parent for next generation. The method was evaluated with application of designing the fuzzy path planner, where each fuzzy rule was encoded as a gene. Simulation and experimental results showed the effectiveness and the applicability of the proposed method.

1 Introduction

To control fast mobile robots, a simple controller is required, which satisfies the mechanical properties such as limitations of wheel speed or translational speed of the robot center. Efficiency of trajectories and short navigation time are also to be ensured. When we consider dribbling and kicking action in robot soccer, robot posture (position and orientation) is of utmost importance. This the paper aims to address the specific problem of robot posture at the target position with emphasis on optimizing the navigational path of the robot.

In the early stages of robot soccer, traditional navigation methods were popular, where the option was to use the simple shortest paths, Dubins path [1] or the composition of rotation, circular motion and straight motion for the path planning step [2,3]. Recently research interest is being focused on the application of fuzzy logic, evolutionary computation, reinforcement learning, unified navigation method, and so on [4,5,6,7]. The use of fuzzy control and behavior-based architectures has been intensively researched in the field of robot navigation,

C. Freksa, M. Kohlhase, and K. Schill (Eds.): KI 2006, LNAI 4314, pp. 317–330, 2007.
© Springer-Verlag Berlin Heidelberg 2007

because fuzzy logic is a mathematical formulation that copes with uncertainty in information [4].

However a commonly encountered problem is that the derivation of fuzzy control rules is often time consuming, difficult and relies to a great extent on process experts. An automated way to design fuzzy systems is preferable. In this regard, numerous researches have been dedicated to exploring the use of evolutionary algorithms (EAs) to automate the knowledge acquisition base and construct appropriate rules for a given task [8,9,10].

The evolutionary algorithms employed for this purpose use individuals with a single chromosome whose component genes are characterized as rules for the fuzzy control system. During the evolutionary process, a chromosome's performance is tested and the best chromosomes (individuals) are selected for reproduction. A common drawback to this approach is that during testing, some genes (rules) may be seldomly used. These genes contribute negligibly to a global fitness function and consequently the evolutionary process is to a high degree, insensitive to them.

This may lead to a stunted development in the evolution of certain genes as chromosomes are typically selected for reproduction with little dependance on these seldomly used genes. In the worst case scenario it may completely halt development as random processes begin to dominate very slowly evolving genes. In these situations, chromosomes can evolve that have high fitness and perform well for most objectives, but are below par for the few that require the seldomly used genes which have evolved poorly. This causes a variance in the consistency of a chromosome's performance and brings forth uncertainty in the solutions for the real-world problem. These issues are highlighted with an illustrative path planning example in Section 2.

In this paper, a scheme is formulated that automatically determines the sensitivity of various genes in contributing to the fittest solution. A rule-scoring mechanism ranks genes with the same role (allele) according to their performance under scenarios in which the gene is actually utilized. The parents for the next generation are then formed on the basis of this ranking. The method is then finally trialled on a fuzzy path planner for developing shooting strategies for mobile soccer robots.

This paper is organized as follows. In Section 2 we highlight the difficulties discussed with an illustrative example. Section 3 we present our proposed mechanisms for the evolutionary algorithm. The proposed scheme is analyzed in both simulations and experiments for a robot soccer system in Section 4. Finally, concluding remarks follow in Section 5.

2   Illustrative Example - Path Planning

In this example, the path planning problem for mobile robot navigation is investigated in detail. For this task a path must be planned from a large number of starting points contained in a discretised map of its environment to a specified target. The rule set to be learned is simply a table of desired heading values for each point on the discretised map. This rule set is expected to provide a good so-lution regardless of where the mobile robot is initially placed. This now becomes a multi-objective optimization problem, where each objective is characterized uniquely by the initial conditions, or posture, of the robot relative to the target.

Figure 1 illustrates three sample paths generated for a conventionally evolved rule set that has been developed for 48 different initial conditions (starting pos-tures). Here the path planner uses a map of its environment that has been discretised by a fuzzy segmentation of the robot's relative distance and angle from the target. The fitness function used in the evolutionary algorithm is a linear combination of sub-fitness functions for each starting posture and these sub-fitness functions are designed so that the mobile robot approaches its target from the left and in the shortest time possible. For path generation, we also assume the robot uses a controller that adequately tracks the desired heading angle determined by the evolved rule set - this allows us to focus on the path planner.

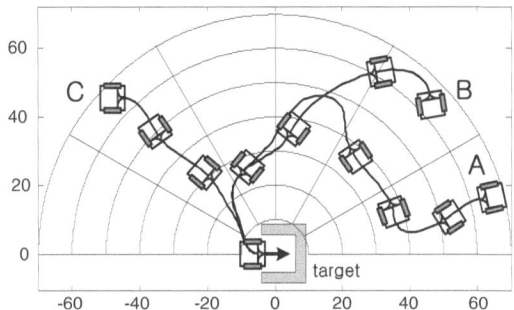

Fig. 1. Comparison of a chromosome's performance for multiple objectives

For this particular scenario, rules that are located in the lower right corner of the figure are utilized by fewer starting postures than rules elsewhere on the map, particularly those immediately to the left of the target. For a uniformly dispersed set of starting postures and a fitness function that weights the performance of each starting posture equally, evolution of these rules may be poor as discussed earlier. This is evident in figure 1 which exhibits reasonably optimal paths for B and C but a less than optimal path for A.

These characteristics are highlighted in figure 2 by analyzing the strength and the frequency with which the fuzzy rules are triggered for A, B and C (note that the graph displays the normalized contribution of each fuzzy rule for a particular trajectory - the equations used to generate these firing ratios are presented in Section 3).

For paths B and C fewer rules are triggered - there are many redundancies in the rule space for a particular objective. The rules that initially affect movement from A, rules r_1 to r_{12} (shaded grey in figure 2), have evolved poorly since very few starting postures require them. Subsequently these contribute negligibly to the global fitness function. They do not provide a very good solution for A, but

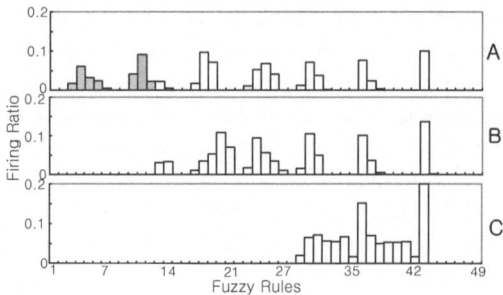

Fig. 2. Firing ratios of fuzzy rules for A, B and C

they are kept since the fuzzy rule set (chromosome) in general performs very well. Note that the evolved rule set here provides very good directions in the immediate vicinity to the left of the target - this is to be expected as almost every starting posture must utilize these rules for a successful approach. As a result the global fitness function will be highly sensitive to these rules and they evolve both rapidly and accurately.

In order to ensure improved evolution for these underdeveloped rules, a means for determining their sensitivity in the evolutionary process is needed as well as an alternate method ranking the performance of the chromosomes.

3   Proposed Evolutionary Technique

The conventional evolutionary algorithm used for the illustrative example utilized a selection scheme based on a q-tournament. A (μ, λ)-evolution strategy [11] was used and the elite chromosome saved by the elitist strategy [12]. Here, various modifications are proposed for the parent selection process that assist in the evolution of optimal solutions for multi-objective path planning problems.

3.1   Parent Selection Process

For a highly redundant problem, a gene (rule) may only be a necessary parameter for the optimization of a small portion of the objectives. Consequently evolving a gene for a more globally optimal solution by considering the performance of objectives that do not utilize it is redundant. It introduces complexity and in the course of an evolutionary algorithm, this will slow or halt its evolution.

The usual process of parent selection in an evolutionary algorithm is to find the best chromosome for all genes when tested against all objectives simultaneously. Our approach is to develop a unique scoring method that finds the highest ranked chromosomes for each gene when tested against only the objectives that utilize that gene. We then form the parents for the next generation from this pool of chromosomes. Some important aspects of the process are as follows:

- The method reverts to a conventional parent selection process if there is no redundancy.
- The final solution will converge more rapidly for systems with redundancy.
- Where evolution of particular genes comes to a halt, increasing the rate of gene evolution with this approach may provide an improvement in the final solution.

3.2   Implementation

Implementing this method in an evolutionary algorithm can be broken down into the following steps:

- Determining and prioritizing an objective's dependence on a particular gene.
- Scoring method to find a gene's 'best' chromosome.
- Forming the parents for the next generation.

Gene-Dependence. Given a chromosome and an objective, an objective's dependence on a particular gene can be found by prioritizing the frequency and strength with which the fuzzy rule is triggered as a solution is generated for the chromosome-objective pair. This is a process which is also gradually learned by the evolutionary algorithm as chromosomes evolve. Figure 2 illustrated the concept in the introductory example.

For the navigation problem, path generation is broken down into path planning and path following operations that are performed at discrete time intervals. At each step, the chromosome is used to determine the genes that are triggered and generate a desired path to follow. The path following controller is used to track the desired path until the next time interval at which point the process is repeated. By combining these discrete steps, a path is generated.

The mechanism for determining the strength and frequency with which rules are triggered along these paths is presented as follows. Discrete time steps are defined by t_i, $i = 0 \ldots n$ where $t_0 = 0$ and t_n is the elapsed time taken for the path to terminate. At the i-th step on the path generated for the j-th objective (starting posture) using the k-th chromosome, we collect the **normalized firing strength** of the l-th gene, the weight of each rule for center average defuzzifier [13], is defined by

$$NFS_{i,j,k,l} = \frac{w_{j,k,l}}{\sum_m w_{j,k,m}} \bigg|_{t_i}, \tag{1}$$

where $w_{j,k,l}$ is the strength with which the l-th gene (fuzzy rule) is triggered. The **total firing strength** of the l-th gene over the whole path is then

$$FS_{j,k,l} = \sum_{i=1}^{n} NFS_{i,j,k,l}. \tag{2}$$

The **firing ratio** for the l-th gene on this path is simply the firing strength normalized for the firing strength of all genes triggered on this path. This is defined by

$$FR_{j,k,l} = \frac{FS_{j,k,l}}{\sum_m FS_{j,k,m}}. \tag{3}$$

It is the firing ratio of each gene for a particular chromosome and starting posture that can be seen graphically represented on the bar graph in figure 2.

Scoring Method. The next step is to rank chromosomes for each gene in a prioritized order to find a gene's 'best' chromosome. For this we define the **Rule Score** of the k-th chromosome for the l-th gene with

$$RS_{k,l} = \sum_j (FR_{j,k,l} \times CW_{j,k}), \tag{4}$$

where $CW_{j,k}$ is the count of wins for the k-th chromosome by q-tournament for the j-th objective (starting posture). Note that if an objective does not have any dependence on a gene, it will not contribute to the rule score for that gene since its firing ratio will be zero. It also works more effectively than a simple test for each chromosome as it weights results according to the objective's dependence on the gene.

Forming the Parents. The final step is to assemble the parents (μ parents) for the next generation from the existing group of highly ranked individuals (λ chromosomes). The first parent is selected by the elitist strategy as the fittest individual for the global solution. The remainder are formed by recombining chromosomes at the gene level. The genes for a chromosome are ranked on the basis of their score amongst others in the same column (allele) of chromosomes. The genes of the first rank in each position, h, are collected into the first chromosome, R'_1. The genes of the second rank form a second chromosome, R'_2 and so forth. A single instance of this process is illustrated in figure 3. From these, an extended family of parent chromosomes is formed that disperses itself through the search space in a manner that allows solutions for seldomly triggered genes (rules) to be more readily found.

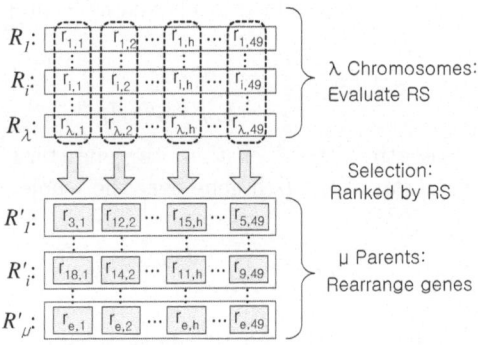

Fig. 3. Reproduction with recombinant rule selection

4   Experiments

4.1   Robot Soccer System

To demonstrate the effectiveness and applicability of the proposed method, a fuzzy evolutionary system was developed for enhancing the performance of mobile robot behaviours in the MIROSOT soccer system [14,15,16]. Behaviours for the Mirosot system that can be enhanced with the approach presented in this paper include passing, defending and shooting. In this experiment the path planning approach is applied for improving the speed, reliability and consistency of the shooting behaviour. Localisation of the robots in the MIROSOT system is achieved via an overhead vision system connected to an external PC that develops strategies and controls which are then transmitted to the robots on the playing field.

4.2   Fuzzy System

The fuzzy system that was developed is comprised of two modules connected in hierarchical fashion, a *fuzzy path planner* and a *fuzzy path-following controller* [17,18]. This is illustrated in figure 4. The fuzzy path planner is responsible for generating the desired paths from the initial posture to the ball position that are optimized for various shooting performance criteria and also satisfy any necessary constraints. It is again assumed the fuzzy controller can adequately track the desired heading angle so that the focus remains on the fuzzy path planner.

The fuzzy planner is designed to accept fuzzified information describing the mobile robot's relative position with respect to the ball (ρ, φ) as inputs for a set of fuzzy rules that determine the appropriate desired heading angle θ_D corresponding to each input. These fuzzy inputs are used to discretise the map

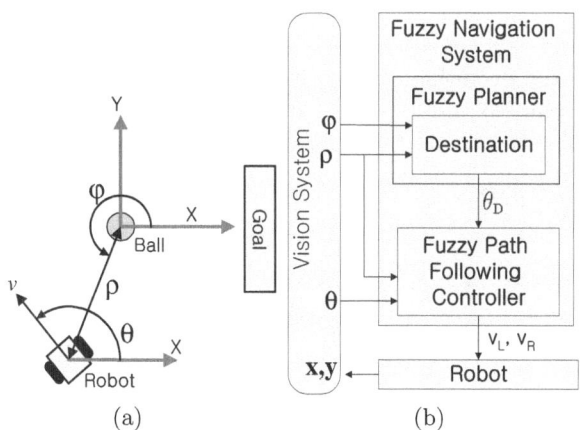

Fig. 4. Overall structure of fuzzy navigation system (a) Localisation variables (b) Fuzzy navigation system

of the robot's environment and the rules are used to generate paths consisting of singleton values for the direction at sampled positions for a trajectory to the ball, like a univector field [19,20].

4.3 The Evolutionary Algorithm

An evolutionary algorithm with the modifications presented in Section 3 is used to learn the fuzzy rules. The algorithm uses a (μ, λ)-strategy where the number of parents (μ) and offspring (λ) is set to 10 and 20, respectively. The q-tournament selects 10 competitors in each round.

Each chromosome in the algorithm represents an entire fuzzy rule set and each gene represents an individual fuzzy rule mapping the inputs (ρ, φ) to the output θ_D. Inputs are constrained to $0\text{cm} \leq \rho \leq 80\text{cm}$ and $0° \leq \varphi \leq 180°$ (there exists a geometrical symmetry for the simple case containing no obstacles). A typical chromosome is illustrated in Table 1 where each input variable is spanned by seven membership functions (here the fuzzified ρ values range from very near to very far and the angle φ from very small to very large). Subsequently, there are in total 49 genes within the chromosome for this experiment.

Table 1. Rules for desired heading angle

θ_D φ	VN	AN	SN	MD	SF	AF	VF
VS	11	144	119	125	76	74	206
LS	36	121	151	189	213	200	170
SS	315	169	231	234	203	239	242
MD	335	199	212	255	238	254	237
SL	153	268	269	267	301.7	276	295
AL	296	305	315	334	346.1	276	350
VL	6	344	351	349	332.3	335	331

Solutions to the evolutionary algorithm are required to be successful whilst minimizing for elapsed time t_l and vertical drift/orientation errors, y_e, θ_e. The x-axis represents the desired heading direction at the target. These variables are considered to be the necessary performance criteria for shooting and are illustrated in figure 5.

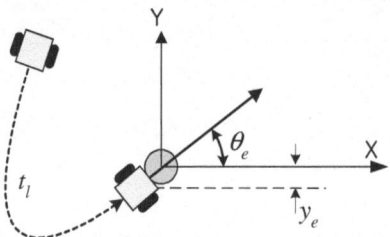

Fig. 5. Performance factors for shooting ability

To evaluate the fitness of the i-th chromosome for the j-th objective (initial posture), the performance index (PI) is defined as

$$PI_{i,j} = K_t \cdot t_l + K_p \cdot |\theta_e| + K_d \cdot y_e^2, \tag{5}$$

where K_t, K_p and K_d are positive constance. The performance index of the i-th chromosome for a group, Σ of objectives is simply defined as the cumulative sum of its performance indexes for each objective:

$$PI_i = \sum_{j \in \Sigma} PI_{i,j} \tag{6}$$

This is used in both the q-tournament and as a tool to evaluate the fitness of chromosomes once the algorithm has completed. The coefficients were manually tuned and finalized at $K_t = 10, K_p = 1$ and $K_d = 3$. These provided almost equal weightings for each criteria in the performance index - that is $K_t \cdot t_l \approx K_p \cdot |\theta_e| \approx K_d \cdot y_e^2$ for nearby solutions in this experiment.

Forty-eight points (initial postures) were selected for an exercise in which all the genes were used at least once. The evolutionary algorithm was evolved for 3,000 generations in both the conventional and proposed rule-scoring methods. To compare results, the evolutionary algorithm was applied using both conventional and proposed scoring methodologies 62 times each. Performance index values were recorded during evolution and utilized to calculate statistical data. For the statistical analysis, consistency of the results was also defined as an important measure of the reliability of the evolved chromosome as a solution. For this, the coefficient of variation (CV) was used since it provides a degree of invariance for comparing solutions with different initial postures and travel lengths. The coefficient of variation for a variable x is defined as

$$CV(x) = \frac{Std(x)}{Mean(x)} \times 100 \quad (\%). \tag{7}$$

Table 2 shows the simulation results for each method based on the overall performance index of the evolved rule set. Clearly the proposed method performs better with a smaller mean value and has a standard deviation less than the one-third that of the conventional method. It also shows a decrease in the co-efficient of variation of the PI for individual starting postures (figure 6) and exhibits faster convergence (figure 7) than the conventional method. Moreover, the coefficient of variation was reduced by about 50% on average, and the coefficient of variation of the performance index for each objective (training point or

Table 2. Converged global performance index

Algorithm	Mean	Std
Conventional selection	434.39	45.88
Recombinant selection	382.51	15.22

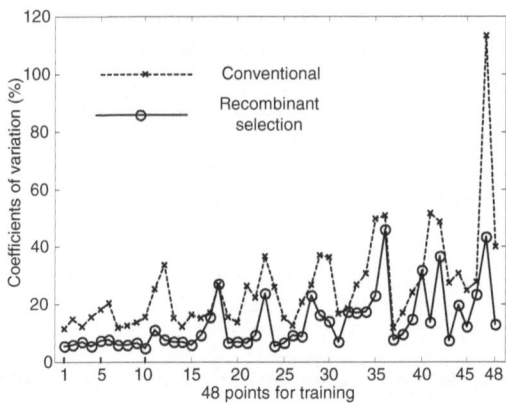

Fig. 6. Coefficient of variation of PI for individual training points

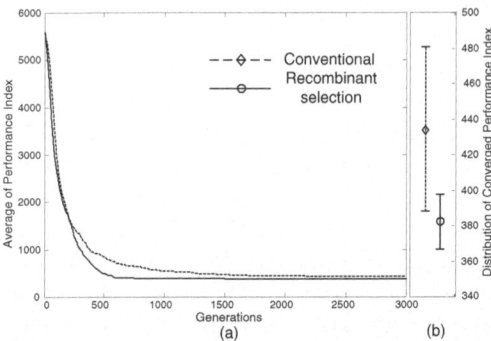

Fig. 7. Comparison of global performance index (a) Convergence (b) Variance

initial condition) also improved evenly (figure 6). These results imply a tendency for the proposed algorithm to consistently find more optimal solutions than the conventional method.

4.4 Experiment Results

The applicability of the proposed method was physically tested on a mobile robot where an improvement in shooting ability was desired. From the 62 fuzzy rule sets generated by both conventional and proposed methods, 27 were randomly selected for testing of the robot's shooting ability. The robot was initialized from five distinct postures with various pre-specified facings. To compensate for variation from the noise caused by physical disturbances and errors, the robot was tested five times for each combination of starting posture and applied rule set.

The elapsed traveling time and the direction of the ball in motion were used to evaluate *path effectiveness*. The direction of the ball (shooting angle error) was used to represent the effects of both vertical drift and heading angle errors as these were individually difficult to measure reliably. A statistical analysis is shown in Tables 3 and 4. The results indicate the proposed rule-scoring method consistently generates paths that have shorter elapsed times with significantly reduced variation (CV). These satisfied the original goals (speed and improved reliability of shooting) whilst maintaining a heading angle error reasonable for the dimensions of the field. Improved heading angle error can be achieved by a suitable tuning of the weightings in the Performance Index function.

Table 3. Elapsed time for shooting in experiments (sec)

Start point	Conventional selection			Recombinant selection		
	Mean	Std	CV(%)	Mean	Std	CV(%)
A	2.60	0.33	12.61	2.33	0.14	6.00
B	2.23	0.28	12.71	2.00	0.11	5.60
C	1.93	0.28	14.46	1.71	0.10	5.60
D	1.73	0.23	13.63	1.60	0.06	3.61
E	1.77	0.22	12.45	1.59	0.07	4.09
Average	2.05	0.27	13.14	1.85	0.09	4.98

Table 4. Shooting Angle error for shooting in experiments (deg)

Start point	Conventional selection			Recombinant selection		
	Mean	Std	CV(%)	Mean	Std	CV(%)
A	5.11	3.72	72.78	6.46	4.24	65.75
B	4.67	3.45	73.98	6.26	3.01	48.19
C	4.41	3.19	72.19	6.20	3.46	55.78
D	4.68	3.13	66.74	6.11	3.23	52.87
E	3.95	2.62	66.24	5.28	3.66	69.21
Average	4.36	3.22	70.20	6.06	3.52	58.8

Figure 8 illustrates several shooting solutions generated by both conventional and proposed methods. Figure 8(a) shows several trajectories generated by an applied rule set derived using the conventional method. As discussed earlier, chromosomes that performed well (relatively low PI) were selected to evolve the population. These often had genes (rules) that were triggered for a select few paths on which they performed poorly. Consequently evolution of these genes did not occur and the final rules needed for these poorly evolved points remained inferior. This is clearly seen in the figure where paths generated for D and E provide successful solutions, however the remaining paths for A, B and C deviate undesirably.

The proposed rule-scoring method discriminates among genes using the strength of the rules. This eliminates inferior rules and allows for uniform evolution of rules across the entire input space. This is highlighted in figure 8(b).

It is worth noting that this procedure can be used to derive the fuzzy path following controller (refer to figure 4) by determining wheel velocities given a relative heading angle error and the radial distance from the target as the inputs.

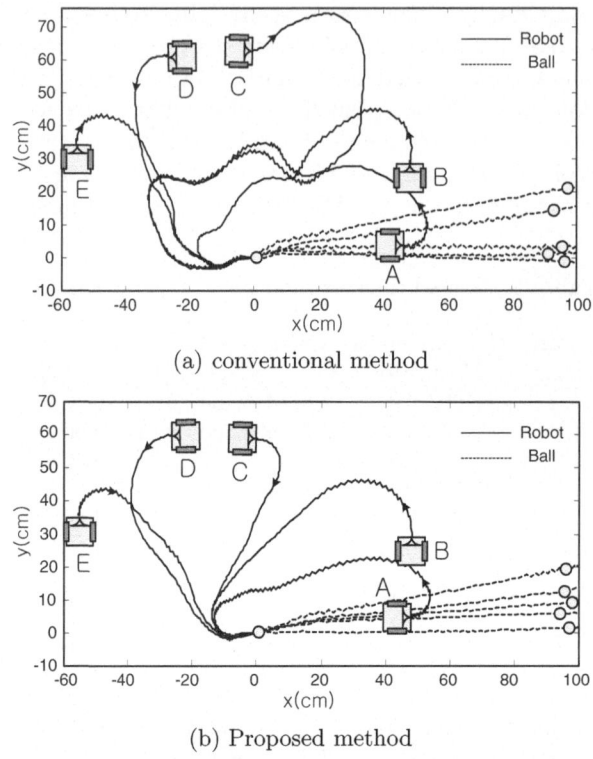

(a) conventional method

(b) Proposed method

Fig. 8. Experiment of robot shooting

This problem exhibits the same properties as the path planner and the proposed algorithm can assist in learning of the seldomly triggered rules.

4.5   Experimental Conclusions

One of the most important features of the path planning problem demonstrated here is that there exists an optimal solution that is shared by all objectives (consider two points in line with the target - the rule set for the rearmost point is equally valid for the closer point). This property is typical of such path planning problems in general. In these situations, the rule-scoring method helps identify chromosomes that perform well for seldomly triggered rules and the gene recombination used to form parents for the next generation more aggressively directs the solution to progress towards the common goal.

For more complex nonlinear optimization problems with opposing objectives, a pareto optimal solution must be found and an alternative process may be needed for forming the non-elite parents of the next generation. Stochastic contributions may be necessary to ensure gene recombination do not consistently and adversely affect each other as they strive toward differing goals. This avenue is currently being explored by the authors and is open to further research.

5 Conclusion

Conventional evolutionary algorithms for multi-objective problems evolve genes by ranking chromosomes for the reproduction process based on their performance on all objectives simultaneously. This often results in poorly evolved rules for those rules which are needed by only a few objectives. In this paper, new methods are proposed for the mobile robot path planning problem which prioritize objectives that utilize a particular gene and rank chromosomes against these objectives on a gene by gene basis. The reproduction process then selects the information from the best of each of these groups and rearranges the rules for reproducing the parents of the next generation. This enables the seldomly triggered rules to evolve at a much higher rate. Experimental results on a path planning problem in robot soccer verifies these results. Solutions evolved with the proposed method were faster and most importantly they exhibited a higher consistency of performance than was possible using a conventional method.

Acknowledgement

This research is supported by the Ubiquitous Computing and Network (UCN) Project, the Ministry of Information and Communication (MIC) 21st Century Frontier R&D Program in Korea.

References

1. L. E. Dubins: On curves of minimal length with a constraint on average curvature, and with prescribed initial and terminal positions and tangents. Amer. J. Math., **79**, (1957) 497-516
2. J.-H. Kim, H.-S. Shim, H.-S. Kim, M.-J. Jung, I.-H. Choi, and J.-O. Kim: A cooperative multi-agent system and its real time application to robot soccer. IEEE Proc. Int. Conf. Robot. Automat. Albuquerque, NM, (1997)638-643
3. W.-M. Shen et al: Building integrated mobile robots for soccer competition. IEEE Proc. Int. Conf. Robot. Automat. vol. 3, Leuven, Belgium, May, (1998) 2613-2618
4. M.-J. Jung H. S. Kim, H. S. Shim, and J. H. Kim: Fuzzy rule extraction for shooting action controller of soccer robot. IEEE Proc. Int. Conf. Fuzzy Syst. **1**, (1999) 556-561
5. D.-H. Kim et al.: Vector field based path planning and Petri-net based role selection mechanism with Q-learning for the soccer robot system. Intell. Automat. Soft Comput. In: J.-H. Kim et al. (eds)., **6**, (2000) 75-87
6. W.-G. Han et al.: GA based on-line path planning of mobile robots playing soccer games. Proc. IEEE 40th Midwest Symp. Circuit Syst. Sacramento, CA, Sept.(1998) 522-525
7. J.-H. Kim et al.: Path planning and role selection mechanism for soccer robots. Proc. IEEE Int. Conf. Robot. Automat. Vol. 4. Leuven, Belgium, (1998) 3216-3221
8. F. Hoffmann: Evolutionary Algorithms for Fuzzy Control System Design. Proceedings of the IEEE, **89** (9) (2001) 1318–1333.

9. S.-J. Kang, C.-H. Woo, H.-S. Hwang, K.B. Woo: Evolutionary design of fuzzy rule base for nonlinear system modeling and control. IEEE Transactions on Fuzzy Systems. **8** (1) (2000), 37 – 45.

10. D.-H. Park, A. Kandel: Genetic-Based New Fuzzy Resoning Models with Application to Fuzzy Control. IEEE Transactions on Sys., Man, and Cybernetics. **24**(1) (1994), 39–47

11. H.-P. Schwefel: Numerical Optimization of Computer Models. UK:John Wiley, Chichester, (1981)

12. D. E. Goldberg: Genetic Alogrithms in Search, Optimization, and Machine Learning. Addison-Wesley (1989)

13. Li-Xin Wang: A Course in Fuzzy Systems and Control. Prentice Hall(Int. ed.). (1997) 110

14. J.-H. Kim, (ed): Robotics and Autonomous Systems. Special Issue: First Micro-Robot World Cup Soccer Tournament, MiroSot, **21**(2)(1997).

15. J.-M. Yang and J-H. Kim: Sliding Mode Control for Trajectory Tracking of Non-holonomic Wheeled Mobile Robots. IEEE Transactions on Robotics and Automation. **15**(3) (1999), 578-587.

16. J.-H. Kim, D.-H. Kim, Y.-J. Kim, K.-T. Seow, Soccer Robotics (Springer Tracts in Advanced Robotics), Springer Verlag, 3540218599, 326, 2004.

17. M.-J. Jung, H.-S. Shim, H.-S. Kim, J.-H. Kim: Fuzzy rule extraction for shooting action controller of soccer robot. Fuzzy Systems Conference Proceedings. **1** (1999), 556 – 561.

18. M.-S. Lee, M.-J. Jung, J.-H. Kim: Evolutionary programming-based fuzzy logic path planner and follower for mobile robots. Proceedings of the 2000 Congress on Evolutionary Computation. **1** (2000), 139 – 144

19. Y.-J. Kim, J.-H. Kim and D.-S. Kim: Evolutionary Programming-Based Uni-vector Field Navigation Method for Fast Mobile Robots. IEEE Trans. on Systems Man and Cybernetics - Part B - Cybernetics. **31**(3) (2001) 450–458

20. M. Mizumoto: Fuzzy controls by fuzzy sington-type resoning method. Proc. of the Fifth IFSA world congress. Seoul, Korea, (1993) 945-948

A Framework for Quasi-exact Optimization Using Relaxed Best-First Search

Rüdiger Ebendt* and Rolf Drechsler

Institute of Computer Science, University of Bremen
28359 Bremen, Germany
{ebendt,drechsle}@informatik.uni-bremen.de

Abstract. In this paper, a framework for previous and new quasi-exact extensions of the A^*-algorithm is presented. In contrast to previous approaches, the new methods guarantee to expand every state at most once if guided by a so-called monotone heuristic. By that, they account more effectively for aspects of run time while still guaranteeing that the cost of the solution will not exceed the optimal cost by a certain factor. First a general upper bound for this factor is derived. This bound is $(1 + \epsilon)^{\lfloor \frac{N}{2} \rfloor}$ where N is (an upper bound on) the maximum depth of the search. Next, we look at specific instances of the algorithm class described by our framework. For one of the new methods a linear, i.e. much tighter upper bound is obtained: the cost of the solution will not exceed the optimal cost by a factor greater than $1 + \epsilon$. The parameter $\epsilon \geq 0$ can be chosen by the user. Within a range of reasonable choices for ϵ, all new methods allow the user to trade off run time for solution quality. Besides that, the formal framework also serves for a comparison in terms of other algorithmic properties of interest, e.g. in terms of a necessary condition for state expansion.

The results of experiments targeting the minimization of *Binary Decision Diagrams* (BDDs) demonstrate large reductions in run time when compared to the best known exact approach for BDD minimization and to previous relaxation methods. Moreover, the quality of the obtained solutions is often much better than the quality guaranteed by the theory.

1 Introduction

In many real-world problems, dominating effort is spent on search, often involving huge state spaces. Therefore in the past many researchers have proposed heuristic and exact search algorithms. The drawbacks of blind methods are overcome by *heuristic search* methods to guide the search on a state space: with every state q a quantity $h(q)$ is associated which allows to search in the direction of the goal states. A prominent guided search algorithm is the well-known A^*-*algorithm* [6]. A^* can be devised to find the minimum cost path in a graph describing the possible transitions from one state to another, i.e. in a state space graph. In its original form, A^* guarantees to find an optimal solution and it is used in many fields of application, including diverse areas such as robotics [7] and logic synthesis [5]. Hereby, two components of information

---

* Current address: German Aerospace Center, Member of the Helmholtz Association, Institute of Transport Research, Rutherfordstrasse 2, 12489 Berlin, Germany, Email: ruediger.ebendt@dlr.de

C. Freksa, M. Kohlhase, and K. Schill (Eds.): KI 2006, LNAI 4314, pp. 331–345, 2007.

are used with every state q: one is $g(q)$, which is the information about the cost of the path already covered. The other is the heuristic function $h(q)$, an estimate of the least cost of the remaining part. The first information, $g(q)$, adds a breadth-first component to the search while the second, $h(q)$, can devise the search to delve deeper into certain paths when they seem promising, i.e. it adds a depth-first component to the search. A^* searches the state space by systematically expanding the most promising state (i.e., a best-first search is performed) and generating states until a match to a goal condition is found. For this purpose a prioritized list OPEN orders the search within the states that are eligible for expansion, and closed states are maintained on a list CLOSED. If certain requirements to the heuristic function guiding the search are met, A^* will find a minimum cost path to a goal state [6].

A serious drawback of A^* is that, in the worst case, the run time as well as the amount of memory required to store OPEN and CLOSED is exponential in the depth of the search. This has led to several extensions of A^*, some of which are memory bounded, e.g. [15], while others mainly target to reduce the run time by allowing for bounded sub-optimality, i.e. they provide A^*-based quasi-exact approximation methods.

E.g., the idea of *Dynamic Weighting* (A^*_{DW}) [10] is to start with a high weighting of the depth first component at the beginning of the search (as this may help to find a promising direction more quickly) and then dynamically weigh the depth-first component less heavily as the search goes deeper (as this may help to prevent too early, i.e. premature termination). In contrast to that, in [9], the *Traveling Salesman Problem* (TSP) has been tackled by an extension of A^* called A^*_ϵ which relaxes the *selection condition* of A^*. This condition triggers the choice of the next state for expansion (i.e. for generating all its successors). More recently, a conceptually much simpler idea has been used in [16]: here, the depth-first component is constantly inflated by a certain factor $1 + \epsilon, \epsilon > 0$. It has also been embedded in a so-called *Anytime Repairing* variant of A^* (ARA*) in [7]. This idea is referred to as $A^*_\uparrow$ throughout the paper.

The contributions of the paper are twofold: On the one hand, a formal analysis aims at deeper theoretical insight and, based on the formal results, new improved methods are devised. On the other, the new methods are compared with each other and with previous methods during an experimental evaluation. First, a formal framework provides a unifying view that describes all of the above mentioned approaches. With the help of this framework, all methods can be identified as special instances of one generic relaxation algorithm. This is of particular interest since the ideas of the methods may seem very different at first glance. As a next step, several interesting properties that are shared among all considered methods can be directly deduced. These are the so-called ϵ-admissibility and a necessary condition for state expansion: the first guarantees that the deviation of the solution must be bounded by $1 + \epsilon$, the second one is important for efficiency considerations.

When searching in large state spaces, a potential source of performance loss is the repeated consideration of the same states, i.e. so-called reopenings. The following question is of interest when considering the efficiency of $A^*_\uparrow$ or A^*_{DW}: how do the methods behave in the case that a so-called *monotone* heuristic function guides the search? And: in this case, can there exist states that are *reopened* and expanded (again and again)? This question is answered in the paper, extending the scope of results given in [4]: here, the effect of relaxing the selection condition of A^* on potential reopenings of states has been considered.

This is of particular interest since the original A^*-algorithm is known to expand every distinct state *at most once* in the case of a monotone heuristic function [6]. If the methods for relaxed best-first search cannot guarantee the same, performance can be degraded. In a worst-case scenario, the overhead as caused by reopened states could even exceed the savings provided by the relaxation.

In this paper it is shown by examples that both discussed relaxation methods in fact can show the above (unwanted) behavior. As a remedy, new revised versions of the methods are suggested that expand each state at most once. The property of ϵ-admissibility must be reconsidered for the new approaches. Using the formal framework, first a general upper bound for the deviation of the solution from the optimum is derived. This upper bound is exponential in the depth of the search. Second, again by use of the framework, the bound can be tightened to an only *linear* bound in the case of a revised version of $A_\uparrow^*$. This result confirms a previous result for this special case, stated in a formal analysis of ARA^* [7]. As a benefit from the provided formal framework, our proof is kept considerably shorter and more concise, further strengthening the framework. Experimental results give a comparison of all methods with each other and with previous methods, showing the efficiency of the suggested approaches.

2   Search by A^*

A g- and an h-component associated with every state q are combined to the so-called evaluation function $\varphi(q) = g(q) + h(q)$. The minimal cost of a path from s to q is denoted $g^*(q)$. The minimal cost of a path from q to a goal state is denoted $h^*(q)$. To guarantee a minimal solution cost, it must be $h(q) \leq h^*(q)$ [6]. In this case, h is called *admissible*. A^* maintains a prioritized queue OPEN which is ordered with respect to increasing values $\varphi(q)$. In the beginning, this queue only contains the initial state s. At each step, a state q with a *minimal* φ-value is expanded, dequeued and put on a list called CLOSED. During expansion, the successor states of q are generated and inserted into the queue OPEN according to their φ-values. For this, the values g and h of the successor states are computed dynamically. The component g accumulates transition costs as the sum of the cost $c(r, r')$ of all transitions $r \longrightarrow r'$ occurring on the cheapest known path to q. If a path between q and q' is optimal, its cost is denoted by $k(q, q')$.

A successor state q' might be generated a second time if q' has more than one predecessor state. If a cheaper path from s to q' is found in this case, $g(q')$ is updated. If q was on the list CLOSED, q is reopened, i.e. it is put on OPEN again. By that, states get a second chance during the search for the minimum cost path when new information about them is available. The algorithm terminates if the next state to expand is a goal state t. The estimate $h(t) = h^*(t)$ must be zero. In this case, the path found up to t is of minimal cost, denoted C^*, and it is reported as solution.

A heuristic function h is said to be *monotone*, if $h(q) \leq k(q, q') + h(q')$ for all descendants q' of q. In [6] it is shown that, in the case of a monotone heuristic function h, A^* finds optimal paths to all expanded nodes. This ensures that every state is expanded at most once.

3   Previous Work

To keep the paper self-contained, next a brief review of previous quasi-exact approaches based on A^* follows. All approaches relax some of the conditions used by A^* to derive a faster algorithm with a provable upper bound on sub-optimality. The ideas, how this is done, vary significantly and in the following three methods are distinguished.

3.1   Dynamic Weighting

The idea of *Dynamic Weighting* (A^*_{DW}) [10] is to relax the fixed weighting of the breadth- and the depth-first component (i.e., of g and h) used by the additive evaluation function $\varphi(q) = g(q) + h(q)$ of A^*. Algorithm A^*_{DW} starts with a high weighting of the depth first component at the beginning of the search (as this may help to find a promising direction more quickly) and then dynamically weighs the depth-first component less heavily as the search goes deeper, preventing premature termination. For $\epsilon > 0$, the evaluation function used by A^*_{DW} is

$$\varphi^{DW}(q) = g(q) + h(q) + \epsilon \cdot \left[1 - \frac{d(q)}{N}\right] \cdot h(q)$$

where $d(q)$ denotes the depth of the node representing state q in the search graph, and N denotes the depth of a goal node, respectively. Often, all paths to a node in this graph are of equal length, and thus this depth is the number of edges on such a path. If N is not known in advance, an upper bound or an estimate can be used instead.

It can be shown that A^*_{DW} is ϵ-*admissible*, i.e. it always finds a solution whose cost does not exceed the optimal cost by more than a factor of $1 + \epsilon$.

3.2   Constant Inflation

More recently, a much simpler idea has been considered in [16] and also within the framework of a so-called *Anytime Repairing A^*-algorithm* (ARA*) [7]: the *constant inflation* of the depth-first component by a fixed factor $1 + \epsilon$ ($\epsilon > 0$). That is, the evaluation function

$$\varphi^\uparrow(q) = g(q) + (1 + \epsilon) \cdot h(q)$$

is used instead of the original evaluation function φ of A^*.

In comparison to A^*_{DW}, no other precautions against premature termination are taken here. However, it can be shown that bounding the inflation of h by the factor $1+\epsilon$ already suffices to guarantee the same bounded sub-optimality as with A^*_{DW}. This method is referred to as $A^*_\uparrow$ and it is further analyzed in Section 4.

3.3   Search Effort Estimates

Experiments have shown the following: during execution of an A^*-algorithm, a large amount of time is spent discriminating among many paths whose cost do not vary significantly from each other. To assure optimality of the final solution, A^* spends a disproportionately long time to select the best of roughly equal candidate states as next state to expand. This behavior raises the idea of equipping A^* with the capability of terminating earlier with a sub-optimal but otherwise perfectly acceptable solution path.

In [9] an extension of A^* called A_ϵ^* has been proposed, that addresses the above problem by adding a second queue FOCAL which maintains a subset of the states on OPEN. This subset is the set of those states whose cost does not deviate from the minimum cost of a state on OPEN by a factor greater than $1 + \epsilon$. Formally,

$$\text{FOCAL} = \{q \mid \varphi(q) \leq (1 + \epsilon) \cdot \min_{r \in \text{OPEN}} \varphi(r)\}). \tag{1}$$

The operation of A_ϵ^* is identical to that of A^* except that A_ϵ^* selects a state q from FOCAL with minimal value $h_F(q)$. The function h_F is a second heuristic estimating the computational effort required to complete the search. By this the nature of h_F differs significantly from that of h since h estimates the *solution cost* of the remaining path whereas h_F estimates the remaining *time* needed to find this solution. The choice of h_F puts a high degree of freedom to the approach which will be subject to further investigation in Section 4. In [9], it has been suggested to use

a) $h_F = h$ or
b) to integrate properties of the subgraph emanating from a given state q.

The motivation behind a) is that minimizing the h-component for the states in the set FOCAL means preferring the states with the highest g-component. Such states are least estimative and a fast completion of the best known path to such a state, i.e., a fast termination can be expected. As a concrete suggestion for b), $h_F(q) = N - d(q)$ will be used later in the experimental evaluation (see Section 7): to minimize $N - d(q)$ means to prefer the deeper states in the search graph. This is done with the motivation that the subgraphs emanating from them tend to be comparatively small and thus the same can be expected for the remaining run time. Also A_ϵ^* is ϵ-admissible, i.e. we have the upper bound $1 + \epsilon$ on sub-optimality.

In [4], an example was given that shows that Algorithm A_ϵ^* has a serious drawback: even when guided by a monotone heuristic h, A_ϵ^* can be doomed to *reopen* many states. This is a source of degradation in run time and contrasts to the behavior of classical A^* which expands every state at most once [6].

The following remedy has been suggested in [4]: instead of maintaining closed states on a list CLOSED, states are simply marked as closed after expansion and removal from OPEN. If the method finds a better path for a state q marked as closed, *this better path is ignored*, i.e. $g(q)$ is *not* updated. Otherwise, method A^{pprox} follows the usual operation of A_ϵ^*. Although ϵ-admissibility can not be guaranteed for A^{pprox} in general, still the following result holds [4].

Theorem 1. *Let N be the maximal length of a solution path. When driven by a monotone heuristic, algorithm A^{pprox} always finds a solution not exceeding the optimal cost by a factor greater than $(1 + \epsilon)^{\lfloor \frac{N}{2} \rfloor}$.*

For smaller values of ϵ, this bound still is useful. Note that during practical operation, A^{pprox} usually is far off this worst-case, i.e. it may yield much better results.

4   Unifying View

In this section, a framework provides a unifying view of the three approaches of the previous section. They are characterized as special instances of one generic relaxation algorithm. As the first step, the next result states a condition that guarantees the conformity of an evaluation function with the strategy described in Section 3.2.

Theorem 2. *Let us consider a state space together with an evaluation function $\varphi = g + h$ and let FOCAL be defined as before in Equation (1). For all states q of the state space, let $\varphi^{\Uparrow}(q) = (1 + \epsilon) \cdot \varphi(q)$ and let $\varphi(q) \leq \varphi'(q) \leq \varphi^{\Uparrow}(q)$. Let*

$$\hat{q} = \arg \min_{q \in \text{OPEN}} \varphi'(q).^{1}$$

Then it must be that $\hat{q} \in$ FOCAL.

Proof. See the Appendix.

In Section 3.3 it has been mentioned that the choice of the heuristic function h_F which estimates the remaining search effort leaves a considerable degree of freedom to the method. Next we go one step further, clarifying that the discussed relaxation methods can be *rediscovered* simply by respective choices for h_F. In detail, Theorem 2 allows to characterize A^*_{DW} and $A^*_{\uparrow}$ as two instantiations of the generic method given in Section 3.3. In this, Pearl and Kims' proposal proves to be more than just another relaxation algorithm: the next result shows that it also serves as a framework for the relaxation of best-first search in general.

Theorem 3. *Let us consider the graph representation of a state space and let g, h be the breadth-first and the depth-first component of a relaxed best-first search algorithm, respectively. For all states q of the state space, let $\varphi^{\uparrow}(q) = g(q) + (1 + \epsilon) \cdot h(q)$, let $d(q)$ denote the depth of the node representing q, let N denote the depth of a goal node, and let $\varphi^{\text{DW}} = g(q) + h(q) + \epsilon \cdot \left[1 - \frac{d(q)}{N}\right] \cdot h(q)$. Further, assume that identical tie-breaking rules are used in the algorithms. Then we have:*

- *The operation of Algorithm A^*_{DW} is identical to that of A^*_{ϵ} with search estimate $h_F = \varphi^{\text{DW}}$, and*
- *The operation of Algorithm $A^*_{\uparrow}$ is identical to that of A^*_{ϵ} with search estimate $h_F = \varphi^{\uparrow}$.*

Proof. See the Appendix.

In brief, the result states that the choice of the next node to expand as performed by A^*_{DW} and $A^*_{\uparrow}$ conforms to the relaxation strategy of A^*_{ϵ} as stated in Equation (1). Notice that, despite the fact that A^*_{DW} and $A^*_{\uparrow}$ are formulated by use of evaluation functions that are different from that of A^* or A^*_{ϵ} (i.e., different from $\varphi = g + h$), they provably

---

[1] Throughout the paper, the following notation is used: $\arg \min_{x \in S} f(x)$ returns one $x \in S$ that minimizes the function f.

act as if $\varphi = g + h$ would be used. This holds since they are also guided by the second heuristic h_F. It is precisely this function h_F that then must be replaced by the respective alternative evaluation function.

From now on, it will be distinguished between the new *framework* provided by A_ϵ^* with this result and *instantiations* as one particular algorithm, e.g. as the algorithm proposed in [9]. Hence, it is denoted A_ϵ^*, $h_F = \dots$ whenever a particular algorithm is addressed, the introduced framework itself however is referred to as A_ϵ^*, i.e. without giving a particular second heuristic function h_F.

The result of Theorem 3 allows to transfer any provable result for A_ϵ^* directly to A_{DW}^* and $A_\uparrow^*$, as the two methods are special instances of A_ϵ^*. This results in the following theorem which considers under what conditions states are eligible for state expansion. It generalizes a known result in [9] and is helpful for efficiency considerations and comparisons.

Theorem 4. *Consider a state space with cost function g. Let us assume an admissible heuristic function h, and consider an optimal path $s, \dots, q'$ where q' is the first state that currently also appears on* OPEN *during a (relaxed) best-first search algorithm A. Further, let $\varphi = g + h$.*

- *$A = A^*$: $\varphi(q) \le C^*$ for all states expanded.*
- *$A = A_\epsilon^*$: $\varphi(q) \le (1 + \epsilon) \cdot C^*$ for all states expanded.*
- *$A = A_\uparrow^*$: for all states expanded, either $\varphi(q) \le C^*$ holds or we have $\varphi(q) > C^*$ and $\varphi(q) \le$ UB where for $h(q) \in [0, h(q')[$, UB ranges from $(1 + \epsilon) \cdot C^*$ down to, but not including C^*.*
- *$A = A_{\mathrm{DW}}^*$: for all states expanded, either $\varphi(q) \le C^*$ holds or we have $\varphi(q) > C^*$ and $\varphi(q) \le$ UB where for $h(q) \in \left[0, \frac{N - d(q')}{N - d(q)} \cdot h(q')\right[$, UB ranges from $(1 + \epsilon) \cdot C^*$ down to, but not including C^*.*

Proof. See the Appendix.

This result indicates the following: Both for $A_\uparrow^*$ and for A_{DW}^*, states q with $C^* < \varphi(q) \le (1 + \epsilon) \cdot C^*$ can only be eligible to expansion if their φ-value also stays below the stated upper bound UB. To achieve the value $(1 + \epsilon) \cdot C^*$ for UB, the h-value of the eligible state must be much less than $h(q')$ and/or the eligible state must reside at a significantly deeper level in the search graph.

This contrasts to the situation in A_ϵ^* where no such additional restriction holds for the eligibility for expansion. Moreover, in many AI problem domains typically a lot more of breadth-fist search than depth-first search will be performed by an exact or quasi-exact best-first search method (this also holds for BDD minimization). As a consequence, during a typical algorithm run, often states with equal or similar h-values and/or depth are expanded in a series of consecutive expansions. Thus, eligible states that really are far enough "below" q' (in terms of the h-value and/or depth) to be chosen for expansion, are rare. Hence, for an eligible state q, $\varphi(q) \le C^*$ often is much more typical.

Consequently it can be expected that the total number of states expanded during run of $A_\uparrow^*$ and A_{DW}^* typically is at least no more than that for A^* (and often much less). This

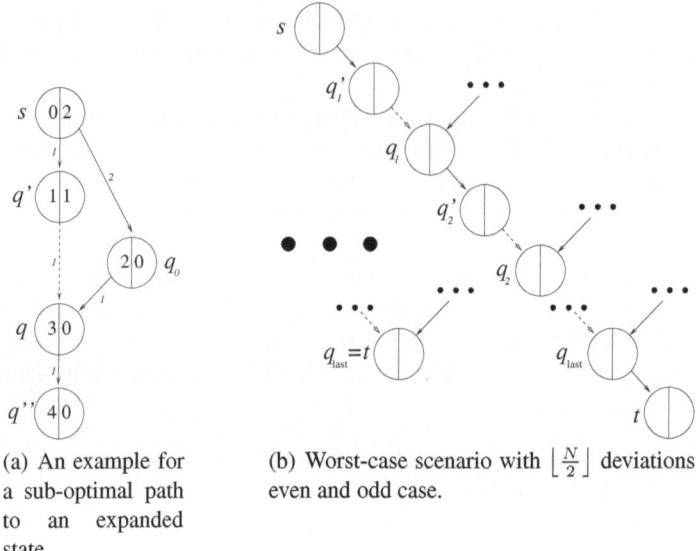

(a) An example for a sub-optimal path to an expanded state.

(b) Worst-case scenario with $\lfloor \frac{N}{2} \rfloor$ deviations, even and odd case.

Fig. 1. Examples for the behavior with and without reopenings

number is expected to still remain this low in the situations where A_ϵ^* using $h_F = h$ or $h_F = N - d(q)$ runs into problems.

5 Monotonicity

In Section 1 the following question has been raised: provided that Algorithm $A = A_\uparrow^*, A_{DW}^*$ is guided by a monotone heuristic h, can states be *reopened*?

Next we give an example which shows that such states may exist for both choices of A. In Fig. 1(a), the left datum annotated at a node is the g-value, the right one is the h-value. Edges depict state transitions and the cost of the transition is annotated at each edge. The heuristic function h is monotone since the series of φ-values is monotonic non-decreasing along every path in the state space graph. In the case $A = A_\uparrow^*$, let $\epsilon = \frac{3}{2}$. In the case $A = A_{DW}^*$, let the anticipated depth of a goal state $N = 3$ and $\epsilon = \frac{9}{4}$. It is easily verified that q is reopened for these choices.

To further analyze the operation of $A_\uparrow^*$, the following new result states an upper bound for the deviation of g from g^* for an expanded state.

Lemma 1. *Let $\epsilon > 0$. The paths to expanded states found by an $A_\uparrow^*$-algorithm that is guided by a monotone heuristic may be sub-optimal. However, this deviation is bounded, in detail:*

$$\forall q \in \text{CLOSED} : g(q) - g^*(q) \leq \epsilon \cdot k(q', q) \qquad (2)$$

where q' is the first state on OPEN on an optimal path $s, \ldots, q', \ldots, q$ at the time of expansion of q.

Proof. See the Appendix.

6 Preventing to Reopen States

The problem discussed in Section 5 can be addressed by the same strategy as for A^{pprox} (see Section 3). The respective approaches that do not reopen any state will be called $A^{pprox}_\uparrow$ and A^{pprox}_{DW}, respectively. As a consequence of Theorem 3, the exponential upper bound of Theorem 1 for A^{pprox} also holds for $A^{pprox}_\uparrow$ and A^{pprox}_{DW}. However, in the case of $A^{pprox}_\uparrow$, the upper bound can be strongly tightened:

Theorem 5. *When driven by a monotone heuristic, algorithm $A^{pprox}_\uparrow$ always finds a solution not exceeding the optimal cost by a factor greater than $1 + \epsilon$.*

Proof. See the Appendix.

Using the introduced framework, the proof is considerably more concise than that of a similar previous result in [7]. Basically, the proof follows a similar flow of arguments as the proof for ϵ-admissibility of A^*_ϵ [9], except that, due to the modified behavior of the algorithm, we have to account for the following consequence. In $A^{pprox}_\uparrow$, the g-value of states on an optimum path may irrecoverably be affected by deviations from the optimum g^*: by Lemma 1, states might be expanded while the best known path to them is still sub-optimal and, due to the modified behavior of $A^{pprox}_\uparrow$, no reopening/improvement can take place later. This effect increases the maximum deviation on an optimal path. To what extent, is determined in the worst-case scenario: let N be the maximal length of a path. Since always *two* nodes must be involved for a deviation of a g-value to occur (see the proof of Lemma 1), the deviation of a g-value from g^* increases at most $\lfloor \frac{N}{2} \rfloor$ times, see Fig. 1(b): dashed transition are "late" transitions, i.e. the state they lead to has already been opened along a sub-optimal path different from p. State q_{last} is the last state that has been prematurely opened along such a sideway and thus is affected by a deviation of the g-value. We have $q_{last} = q_{\lfloor \frac{N}{2} \rfloor}$, regardless whether n is odd or even (see Fig. 1(b)). The proof then is an induction on $i = 1, \ldots, \lfloor \frac{N}{2} \rfloor$.

7 Experimental Results

To evaluate the algorithms described by our framework, respective methods targeting the quasi-exact minimization of reduced ordered *Binary Decision Diagrams* (BDDs) have been implemented. BDDs were introduced in [2] and are well known from hardware verification and logic synthesis. They are *Directed Acyclic Graphs* (DAGs) representing Boolean functions where a Shannon decomposition in a Boolean variable is carried out with each node. Reduced diagrams are considered, derived by removing redundant nodes and merging isomorphic subgraphs. For more details see [2].

Heuristic BDD minimization is done by thumb rules to reorder the Boolean variables, e.g. [11]. The results are often far away from the optimum. For some applications, this is a significant drawback. Especially in applications like logic synthesis targeting multiplexor design styles, e.g. [8, 14], it is important to determine a good ordering, since a reduction in the number of BDD nodes directly transfers to a smaller chip area. For this reason, there is a high demand for faster exact or approximate methods with bounded sub-optimality.

It has been shown that it is NP-complete to decide whether the number of nodes of a given BDD can be improved by variable reordering [1]. Moreover, the existence of a polynomial algorithm to approximate the optimal variable ordering of BDDs implies $P = NP$ [12]. For this reason, as with exact methods, the run time of an approximate method to improve the variable ordering is expected to be much higher than that of heuristics.

All experimental results have been carried out on a machine with a Dual Xeon processor running at 3.2 GHz, with a main memory of 12 GByte and a run time limit of 3,600 CPU seconds. The memory requirement of all evaluated methods never exceeds 500 MBytes, hence no memory limit had to be applied. Three previous methods have been implemented: the first is called A^{pprox} as described in [4], the second is called Dynamic Weighting (A_{DW}^*) [10]. An idea of [16] and of the so-called ARA* algorithm [7], namely the constant inflation of the heuristic function as described in Section 3.2, has been implemented as the approach $A_\uparrow^*$. Moreover, revised versions of the mentioned methods have been implemented as the corresponding methods A_{DW}^{pprox}, a revised version of A_{DW}^*, and as the method $A_\uparrow^{pprox}$, a revised version of $A_\uparrow^*$. To put up a testing environment, all algorithms have been integrated into the CUDD package [13]. By this it is guaranteed that they run in the same system environment. In the experiments, the methods have been applied to BDDs built from a set of standard benchmark circuits of LGSynth93 [3]. The implementation of all algorithms is based on the implementation of the A^*-based approach to exact BDD minimization of [5].

In a first series of experiments A^* and A^{pprox} have been compared to $A_\uparrow^{pprox}$. The results are depicted in Fig. 2. In contrast to the behavior of A^{pprox}, the run time of $A_\uparrow^{pprox}$ is monotonically decreasing. This confirms the result of Theorem 4 and shows that also the revised version of $A_\uparrow^*$, i.e. $A_\uparrow^{pprox}$, behaves according to the upper bounds stated in Theorem 4. For $A_\uparrow^{pprox}$, the degradation of solution quality first increases slowly (e.g., for $\epsilon \in [0, 0.5]$) and later ascends more steeply with increasing ϵ. When comparing the run time of $A_\uparrow^{pprox}$ to that of A^*, Fig. 3 illustrates how the gain in run time grows monotonically with the degree of relaxation (the curve in the space spanned by the percentual gain and the degree ϵ is a convex hyperbola). At the higher relaxation degree of $\epsilon = 3$ the reduction in run time is already more than 90% on average. Taking into account that $A_\uparrow^{pprox}$ also has much more convenient theoretical properties than A^{pprox} (in particular, $A_\uparrow^{pprox}$ guarantees a much tighter upper bound for the deviation of the solution from the optimum), $A_\uparrow^{pprox}$ proves to be clearly superior to A^{pprox} both from a theoretical and a practical standpoint. As Fig. 4 shows, high speed-ups can be obtained at an only small degradation of solution quality. In fact the average degradation is considerably much less than the worst-case degradation by a factor of $1 + \epsilon$ as guaranteed by the theory. Operating at 40% of relaxation, on average the results are only 0.5% larger than the optimum BDD size. When using a degree of relaxation of 100%, i.e. when theoretically allowing for solutions that are twice the minimum size, the average degradation still is only 4.3%. Motivated by these positive results, also very high relaxations have been examined: Fig. 4 shows that the average degradation stays below 20% for a wide range of high relaxation degrees, it first reaches 20.5% for $\epsilon = 20$. Moreover, the resulting plot forms a convex hyperbola where the steepness decreases with ascending degree of relaxation. In a second series of experiments, $A_\uparrow^{pprox}$ has been compared to A_{DW}^{pprox} in terms of quality and run time. Due to space limitation, the results of these experiments have not been included. Summarized, A_{DW}^{pprox} has significantly higher run times than $A_{\uparrow}^{pprox}$ (20-30%) while at the same time slightly better results can be obtained, i.e. there is a significant penalty for small improvements in quality (below 3%) provided by A_{DW}^{pprox}.

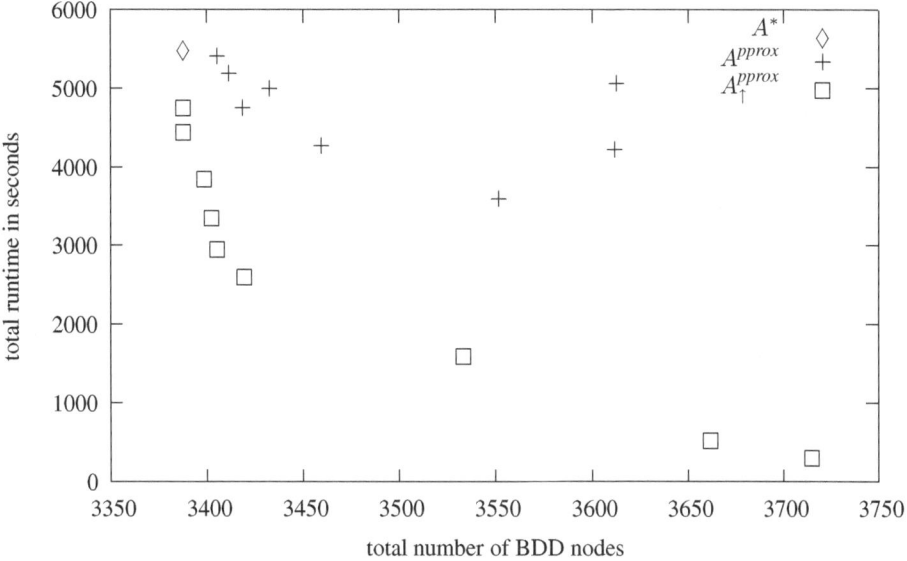

Fig. 2. Trading off run time for solution quality with $\boldsymbol{A}^{pprox}$ and $\boldsymbol{A}_{\uparrow}^{pprox}$

8   Conclusions

A new framework for previous and new extensions of the A^*-algorithm has been presented. It describes a class of generic algorithms that tolerate provably bounded worst-case increases in solution cost. This is achieved by different ideas to relax some of the conditions of A^*, and happens in favour of smaller search efforts required to complete the algorithm. The user has full control of the degree of relaxation and can trade off run time for quality of the solution.

Besides the formal contributions of the paper, two new methods are derived from the framework. They guarantee to expand every state at most once if provided with a so-called monotone heuristic. This can largely reduce the run time and also strengthens the robustness of the approaches.

Experimental results are reported that clearly demonstrate the efficiency of the presented new approaches. A comparison to the best known exact BDD minimization algorithm (which is based on the generic A^* algorithm) and to a previous relaxed method shows reductions in run time by up to one order of magnitude. This is obtained while the degradation of solution quality is provably bounded and stays below a few percents on average.

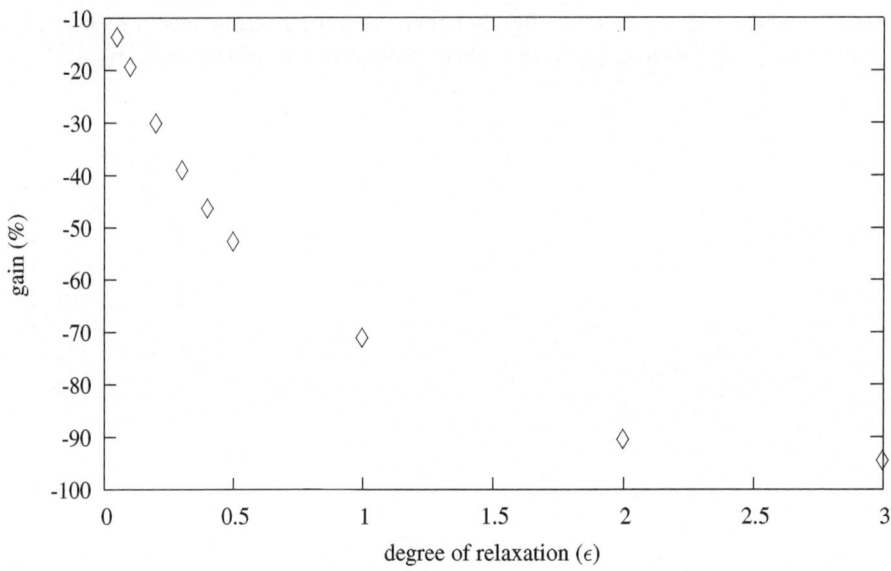

Fig. 3. Degree of relaxation vs. gain in run time of $A_\uparrow^{pprox}$

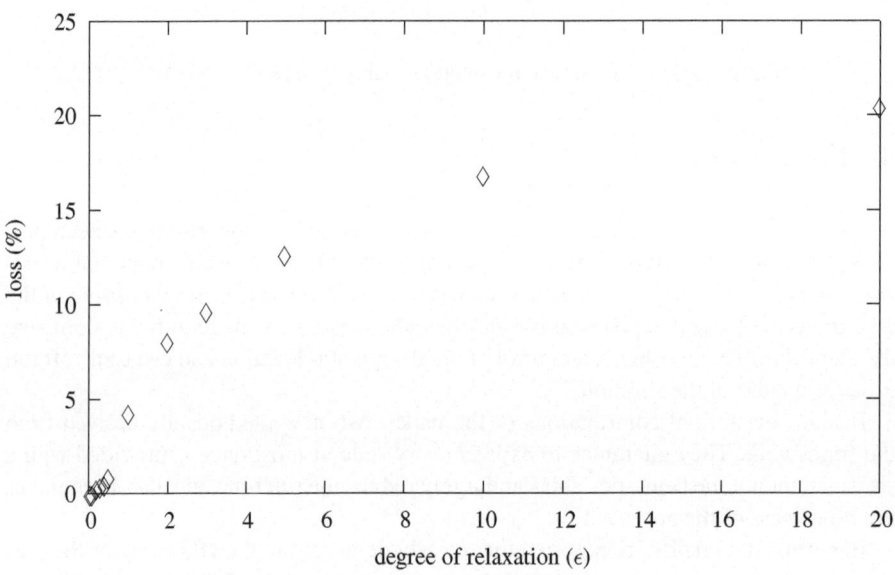

Fig. 4. Degree of relaxation vs. loss in quality of $A_\uparrow^{pprox}$

References

1. B. Bollig and I. Wegener. Improving the variable ordering of OBDDs in NP-complete. *IEEE Trans. on Comp.*, 45(9):993–1002, 1996.
2. R. E. Bryant. Graph-based algorithms for Boolean function manipulation. *IEEE Trans. on Comp.*, 35(8):677–691, 1986.
3. Collaborative Benchmarking Laboratory. *1993 LGSynth Benchmarks*. North Carolina State University, Department of Computer Science, 1993.
4. R. Ebendt and R. Drechsler. Quasi-exact BDD minimization using relaxed best-first search. In *IEEE Annual Symp. on VLSI*, pages 59–64, 2005.
5. R. Ebendt, W. Günther, and R. Drechsler. Combining ordered-best first search with branch and bound for exact BDD minimization. *IEEE Trans. on CAD*, 24(10):1515–1529, 2005.
6. P. Hart, N. Nilsson, and B. Raphael. A formal basis for the heuristic determination of minimum cost paths. *IEEE Trans. Syst. Sci. Cybern.*, 2:100–107, 1968.
7. M. Likhachev, G. Gordon, and S. Thrun. ARA*: Formal analysis. Technical report of the Carnegie Mellon University, 2003.
8. L. Macchiarulo, L. Benini, and E. Macii. On-the-fly layout generation for PTL macrocells. In *Design, Automation and Test in Europe*, pages 546–551, 2001.
9. J. Pearl and J. Kim. Studies in semi-admissible heuristics. *IEEE Trans. on Pattern Analysis and Machine Intelligence*, PAMI-4(4):392–399, 1982.
10. I. Pohl. The avoidance of (relative) catastrophe, heuristic competence, genuine dynamic weighting and computational issues in heuristic problem solving. In *Proc. 3rd Int. Joint Conf. on Artificial Intelligence.*, pages 12–17, 1973.
11. R. Rudell. Dynamic variable ordering for ordered binary decision diagrams. In *Int'l Conf. on CAD*, pages 42–47, 1993.
12. D. Sieling. Nonapproximability of OBDD minimization. *Information and Computation*, 172(2):103–138, 2002.
13. F. Somenzi. *CU Decision Diagram Package Release 2.4.0*. University of Colorado at Boulder, 2004.
14. C. Yang and M. Ciesielski. BDS: a BDD-based logic optimization system. *IEEE Trans. on CAD*, 21(7):866–876, 2002.
15. R. Zhou and E. Hansen. Memory-bounded A^* graph search. In *15th Int. Florida Artificial Intelligence Research Soc. Conf.*, pages 203–209, 2002.
16. R. Zhou and E. Hansen. Multiple sequence alignment using A^*. In *Proc. of the National Conference on Artificial Intelligence, Student Abstract*, 2002.

Appendix

Proof of Theorem 2. Let $q_0 = \arg\min_{q \in \text{OPEN}} \varphi(q)$. We have

$$\varphi(\hat{q}) \leq \varphi'(\hat{q}) \tag{3}$$
$$\leq \varphi'(q_0) \tag{4}$$
$$\leq \varphi^{\Uparrow}(q_0) \tag{5}$$
$$= (1 + \epsilon) \cdot \varphi(q_0)$$
$$= (1 + \epsilon) \cdot \min_{q \in \text{OPEN}} \varphi(q). \tag{6}$$

Equation (3) holds by the definition of φ' in the assumption. Next, Equation (4) holds by definition of $\hat{q}$. Then, Equation (5) holds again with the definition of φ'. By Equation (6), $\hat{q} \in$ FOCAL already follows.   □

Proof of Theorem 3. First it is easily verified that $\varphi(q) \leq \varphi^{DW}(q) \leq \varphi^{\uparrow}(q) \leq \varphi^{\Uparrow}(q)$ for all states q of the considered state space. By Theorem 2, the respective next state expanded by $A_{\uparrow}^{*}$ and A_{DW}^{*} must be contained in FOCAL. Second, A_{ϵ}^{*} chooses a state q_F from FOCAL with $q_F = \arg\min_{q \in \text{FOCAL}} h_F(q)$. As h_F is assigned to the respective evaluation function, and since the same respective tie-breaking rule is used, A_{ϵ}^{*} must act exactly as A_{DW}^{*} and $A_{\uparrow}^{*}$, respectively.   □

Proof of Theorem 4. The results for the cases $A = A^{*}, A_{\epsilon}^{*}$ are already well-known [6,9]. They are merely opposed here to the new results. Because q is expanded before q',

$$\varphi^{\uparrow}(q) = \varphi(q) + \epsilon \cdot h(q) \leq \varphi^{\uparrow}(q') \tag{7}$$

in the case $A = A^{\uparrow}$, and

$$\varphi^{DW}(q) = \varphi(q) + \epsilon \cdot \left[1 - \frac{d(q)}{N}\right] \cdot h(q) \leq \varphi^{DW}(q') \tag{8}$$

in the case of $A = A^{DW}$. To derive the stated upper bounds for $A = A^{\uparrow}, A^{DW}$, it now suffices to separate $\varphi(q)$ on the left side of the two equations (7) and (8), respectively. The upper bounds range within the stated intervals since

- the term $h(q')$ can be bounded by $h^{*}(q')$ because of the admissibility of h, and
- since an optimal path is considered, we have $g(q') = g^{*}(q')$ and finally $\varphi^{*}(q') \leq C^{*}$.   □

Proof of Lemma 1. Consider an optimal path p from s to q. Let q' be the first state on $p = s, \ldots, q', \ldots, q$ which also appears on OPEN[2]. Assume that $q \neq q'$ and assume that q is selected for expansion. Since q is expanded before q', we have $\varphi^{\uparrow}(q) \leq \varphi^{\uparrow}(q')$, and, with the optimality of p and the monotonicity of h

$$\begin{aligned}
g(q) + (1 + \epsilon) \cdot h(q) &\leq g(q') + (1 + \epsilon) \cdot h(q') \\
&\leq g(q') + (1 + \epsilon) \cdot (k(q', q) + h(q)) \\
&= g^{*}(q') + k(q', q) + \epsilon \cdot k(q', q) + (1 + \epsilon) \cdot h(q) \\
&= g^{*}(q) + \epsilon \cdot k(q', q) + (1 + \epsilon) \cdot h(q).
\end{aligned}$$

Hence, $g(q) \leq g^{*}(q) + \epsilon \cdot k(q', q)$ and Equation (2) follows.   □

Proof of Theorem 5. Let N be the maximal length of a search path. The proof uses

$$g(q_i) \leq g(q_i') + (1 + \epsilon) \cdot k(q_i', q_i) \text{ for } 1 \leq i \leq \left\lfloor \frac{N}{2} \right\rfloor. \tag{9}$$

---

[2] Note that it is straightforward to prove that, during operation of the algorithm, at least one state on p must be an open state. The proof is an induction on the length of p which is started by s, the very first state occuring both on OPEN and p.

This follows from the admissibility of h and the fact that q_i is expanded before q'_i: the latter implies $\varphi^\uparrow(q_i) \leq \varphi^\uparrow(q'_i)$, thus $g(q_i) + (1 + \epsilon) \cdot h(q_i) \leq g(q'_i) + (1 + \epsilon) \cdot h(q'_i) \leq g(q'_i) + (1 + \epsilon) \cdot [k(q'_i, q_i) + h(q_i)]$ and Equation (9) follows. We show

$$g(q_i) - g^*(q_i) \leq \epsilon \cdot \sum_{l=1}^{i} k(q'_l, q_l) \text{ for } 1 \leq i \leq \left\lfloor \frac{N}{2} \right\rfloor \tag{10}$$

by induction on i. Then, by the optimality of the considered path, the deviation of $g(q_{\text{last}}) = g(q_{\lfloor \frac{N}{2} \rfloor})$ must be less or equal than $\epsilon \cdot C^*$ and finally also the deviation of the computed solution from the optimum can not be greater than $1 + \epsilon$.

To start the induction, let $i = 1$: the claim of Equation (10) holds by Lemma 1. The lemma can be applied here since the operation of $A_\uparrow^*$ and $A_\uparrow^{pprox}$ is identical before the first state has been reopened. Now let us assume that Equation (10) holds for i. For the step $i \longrightarrow i + 1$ we derive

$$g(q_{i+1}) - g^*(q_{i+1}) \leq g(q'_{i+1}) + (1 + \epsilon) \cdot k(q'_{i+1}, q_{i+1}) - g^*(q_{i+1}) \tag{11}$$

$$= k(q_i, q'_{i+1}) + (1 + \epsilon) \cdot k(q'_{i+1}, q_{i+1}) + g(q_i) - g^*(q_{i+1}) \tag{12}$$

$$= \epsilon \cdot k(q'_{i+1}, q_{i+1}) + g(q_i) - g^*(q_i) \tag{13}$$

$$= \epsilon \cdot k(q'_{i+1}, q_{i+1}) + \epsilon \cdot \sum_{l=1}^{i} k(q'_l, q_l) \tag{14}$$

$$= \epsilon \sum_{l=1}^{i+1} k(q'_l, q_l).$$

Equation (11) holds by Equation (9). Since q'_{i+1} is traversed via q_i on an optimal path, $g(q'_{i+1}) = k(q_i, q'_{i+1}) + g(q_i)$. Thus, Equation (12) follows. We have $g^*(q_{i+1}) = g^*(q_i) + k(q_i, q'_{i+1}) + k(q'_{i+1}, q_{i+1})$ since an optimal path is considered. Thus, Equation (13) follows. Using the induction hypothesis, i.e. Equation (10), next Equation (14) is obtained, completing the proof.   □

Gray Box Robustness Testing of Rule Systems

Joachim Baumeister, Jürgen Bregenzer, and Frank Puppe

Department of Computer Science
University of Würzburg, 97074 Würzburg, Germany
Phone: +49 931 888-6740; Fax: +49 931 888-6732
{baumeister,bregenzer,puppe}@informatik.uni-wuerzburg.de

Abstract. Due to their simple and intuitive manner rules are often used for the implementation of intelligent systems. Besides general methods for the verification and validation of rule systems there exists only little research on the evaluation of their robustness with respect to faulty user inputs or partially incorrect rules. This paper introduces a gray box approach for testing the robustness of rule systems, thus including a preceding analysis of the utilized inputs and the application of background knowledge. The practicability of the approach is demonstrated by a case study.

1   Introduction

Validation and verification are two key issues for evaluating the real world practicability of intelligent systems. In the past, an extensive amount of research was undertaken for the evaluation of rule-based representations of such systems, e.g., [1,2,3]. Alternative knowledge representations like Bayesian networks are also suitable and often more precise for the formalization of domain knowledge. However, using rules is still very popular due to their compact and intuitive manner, e.g., currently the ontology layer of the Semantic Web stack is extended by an appropriate rule representation [4].

Whereas the correct behavior of rule bases (validation) and the correct implementation of rules bases (verification) have been investigated in much detail, there is only little research available when such systems are applied in noisy environments. According to Groot et al. [5] we call such an evaluation task *robustness testing*. In the context of intelligence enriched services on the web, e.g. Semantic Web applications, the robustness is very important, since these services are usually provided for general users.

Robustness testing evaluates the correct behavior of the system with respect to either incorrectly entered input or partially occurring errors in the rule base. Incorrect inputs are the result of user mistakes that can be explained by uncertainty or ignorance of the user when providing the answer to a particular question. A partially incorrect rule base is often the result of a biased or incomplete knowledge acquisition. With robustness testing such effects can be simulated by so called *torture tests*, that gradually decrease the quality of the inputs or the quality of the available knowledge. In general, torture tests are an extension of the well-known empirical testing method. With empirical testing a collection of previously solved and correct test cases is given to the knowledge system as input. Then, for each test case the solutions derived by the knowledge system are compared with the solutions already stored in the cases. Typically, measures like

C. Freksa, M. Kohlhase, and K. Schill (Eds.): KI 2006, LNAI 4314, pp. 346–360, 2007.

the precision, the recall or the E-measure are used for a quantitative comparison of the two solution sets. Torture tests run a series of empirical tests by degrading the quality of settings, e.g. by gradually degrading the input quality or by gradually worsening the quality of the rule base.

In this paper, we present torture tests as a gray box testing technique thus extending the basic work by Groot et al. [5]. We motivate that a sound implementation of robustness testing should be preceded by a thorough analysis of the applied case base and rule base, respectively. With the results of such an analysis additional background knowledge can be applied yielding a reliable simulation of typical system usages.

The rest of the paper is organized as follows: In Section 2 we introduce the basic notions for defining a rule-based system and sketch general measures for evaluating and comparing the robustness of intelligent systems. In Section 3 we introduce the phases of a degradation study, i.e., the implementation of a robustness testing. Furthermore, we define measures for the analysis of the case base and rule base, and we discuss implications that can be drawn from such an analysis. A case study with two rule-based consultation systems is presented in Section 4. The paper is concluded with a summary and an outlook in Section 5.

2   Measuring the Quality of Rule-Based Systems

We consider a rule-based system as a *consultation system*, i.e., a system deriving suitable solutions for a stated problem description. More formally, the input and output of such a system is defined as follows.

Definition 1 (Input and Output). Let Ω_{obs} be the (universal) set of observable *input values*. A tuple $f \in \Omega_{obs}$ with $f = a : v$ is often called a finding, where $a \in \Omega_a$ is an input (attribute) and $v \in dom(a)$ is an assignable value.

Let Ω_{sol} be the universe set of (boolean) *output values*, i.e. solutions derivable by the knowledge system.

A problem solving session is represented by a case containing the specified input and (derived) output values for a given problem.

Definition 2 (Case). A case c is defined as a tuple $c = (\ OBS_c, SOL_c\ )$, where $OBS_c \subseteq \Omega_{obs}$ is the *problem description* of the case, i.e., the observed input values of the case c; $OBS_c = \{f_{1,c}, \ldots, f_{n,c}\}$. The set $SOL_c \subseteq \Omega_{sol}$ contains the (derived) outputs of case c.

A *test suite* is a collection of (test) cases that is used for robustness testing. In general, we define a quality function for a given knowledge system and a collection of test cases as follows.

Definition 3 (Rule Base). A *rule r* is defined as $r : c_r \rightarrow a_r$, where the rule condition c_r is a combination of conjunctions/disjunctions of findings $f \in \Omega_{obs}$, and a_r is the rule action (consequent) of r. In the following, we see that a_r will be used for deriving outputs or for implementing a dialog strategy. A rule base $\mathcal{R}$ is defined as a collection of rules defined in the input and output space.

Examples of rule types are given in Section 3.1 for rule base properties.

Definition 4 (Quality Function). Let C be the universe of test cases, i.e., containing all possible and reasonable combinations of input values, and let $\mathcal{R}$ be the rule base. Then, $q : C \times \mathcal{R} \to [0, 1]$ is a *quality function* comparing the expected solution documented in a case c and the solution of c derived by $\mathcal{R}$.

Examples for a quality function are the *precision*, the *recall* or applications of the *E-measure*, e.g., the F-measure. In the following, we discuss some properties of knowledge systems with respect to the quality of its output. When the input quality is degraded, then the system should show a monotonically degrading output quality, i.e., no fluctuating output quality. In consequence, the output quality of the system is more predictable. More formally we define the monotonic derivation quality of a system as follows.

Definition 5 (Monotonic Derivation Quality). For a rule base $\mathcal{R}$ let $C = (c_1, \ldots, c_n)$ be a sequence of cases sorted according to their input quality in ascending order; further let q be a quality function. Then the system shows a *monotonic derivation quality*, if $q(c_i, \mathcal{R}) \leq q(c_{i+1}, \mathcal{R})$ for all $1 \leq i \leq n$.

This criterion is a necessary requirement for any knowledge system, that should be considered to be *robust*. Groot et al. [5] additionally introduce measures for comparing two knowledge systems discussing their robustness: the quality value and the rate of quality change. We briefly describe them in the following.

Definition 6 (Quality Value). For two rule bases $\mathcal{R}_1$ and $\mathcal{R}_2$ we say that $\mathcal{R}_1$ is more robust than $\mathcal{R}_2$ for a given quality function q, if for any test case c the output quality of $\mathcal{R}_1$ is higher than the output quality of $\mathcal{R}_2$, i.e., $q(c, \mathcal{R}_1) > q(c, \mathcal{R}_2)$.

Whilst the *quality value* considers the isolated behavior of the system for each test case, the *rate of quality change* emphasizes the comparison of dynamic behavior of the systems.

Definition 7 (Rate of Quality Change). For two rule bases $\mathcal{R}_1$ and $\mathcal{R}_2$ and a quality function q, we say that $\mathcal{R}_1$ is more robust than $\mathcal{R}_2$, if for any qualitatively ordered sequence of test cases the average quality of the output of $\mathcal{R}_1$ decreases more slowly than for $\mathcal{R}_2$. The average output quality is computed by a series of degradation sequences (see Section 3).

The measures presented so far are useful for comparing different aspects of the robustness of two knowledge bases. They can be intuitively applied for an analysis of a degradation study that is introduced in the next section.

3   Phases of a Gray Box Degradation Study

The robustness of a knowledge system is investigated by a *degradation study*. Such a study considers one particular aspect of the robustness, e.g., its behavior with respect to noise in the input values or its behavior for biased knowledge. A degradation study consists of a collection of *degradation sequences*. Each degradation sequence is executed with the same settings and the result of the degradation study is determined by the averaged results of the degradation sequences. Every degradation sequence consists of a sequence of torture tests. In such a sequence the settings of the torture tests are subsequently changed, e.g., the input quality is gradually decreased. In every torture test all (possibly modified) cases are passed to the (possibly modified) system, and the accuracy of the system is measured. The results of the single torture tests are then used for computing the mean accuracy of the system in the particular degradation study. The elements of a degradation study are depicted in Figure 1.

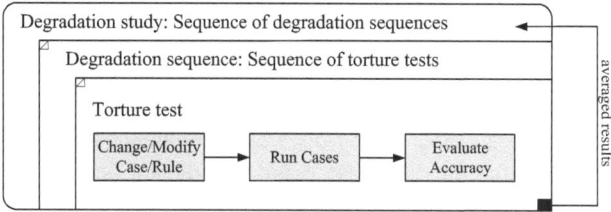

Fig. 1. Overview of the elements of a degradation study

Before discussing the various types of torture tests we first consider the pre-analysis of the used data, i.e., the case base and the rule base.

3.1 The Pre-analysis: Case Base and Rules

The analysis of the properties of the case base and the rule base is essential for a sound implementation of a degradation study. At best, all dimensions of the input-output space are uniformly distributed when investigating the used case base. A biased occurrence of parts of the input-output space would result in unbalanced results of the torture tests; then, some possible input-output combinations are missing and are consequently not investigated.

When concerning the rule base the perfect characteristics are not easy to define: in the best case the possible input space is not restricted by previously answered questions, i.e., the values of some inputs are only asked if values of previous inputs were assigned by the user. A typical example for such a restricted knowledge base is the implementation of a decision tree: depending on given input values other inputs are presented to the user to be answered. With the answers of the user the decision tree is traversed through a path until a leaf is reached which usually contains a solution. It is easy to see that such an interactive structure yields bad characteristics with respect to the robustness of the system, i.e., a falsely answered input will turn the dialog to another path of the tree and consequently will derive another or mostly no solution.

In the following we define measures for formally characterizing the properties of a case base and a rule base, respectively.

Case Base Properties. When investigating the case base we consider the number of cases for each output, i.e., possible solution, and the number of findings typically contained in a case. A sound degradation study would work with a case base, that on the one hand contains equally distributed outputs, and on the other hand consists of equally sized cases (concerning the finding set).

Average Number of Cases (NOC). In a first step, the used cases can be characterized by the *average number of cases* for each output $o \in \Omega_{sol}$, i.e., the mean value with standard deviation of cases c with $o \in SOL_c$. A low number of cases makes it difficult to generalize the obtained results of the torture tests since only a small spectrum of the real world is possibly covered. A high deviation may indicate that some outputs only contain very few cases or a high number of cases. Consequently, the robustness can be very low or high for some outputs.

Average Number of Findings (NOF). A second analysis should consider the *average number of findings* contained in the cases, i.e., the mean value with standard deviation of the number of findings OBS_c for the cases c is computed. A low expectation value of findings can imply that the case base is not suitable for an extensive degradation study, because then large percentages of noise are required in order to actually modify the case; e.g., a noise level above 20% is needed to change at least one input value for a case base with a mean of 5 inputs per case.

Rule Base Properties. A convincing degradation study should be also preceded by a careful investigation of the properties of the rule base. The number of rules for every output, the complexity of the rules, and the usage of the different types of rules are important indicators for the applicability of a degradation study.

Average Number of Rules (NOR). For degradation studies concerning the modification of the rule base the *average number of rules* for each output, i.e., the mean value with standard deviation of derivation rules for each output, is an interesting measure. Derivation rules for an output $o \in \Omega_{sol}$ are rules $r \in \mathcal{R}$ having output o in the rule consequent a_r. A low number of average rules or high deviation values can indicate that the results of the robustness studies may not be representative for all outputs contained in the rule base. The coverage of test cases has been investigated more thoroughly e.g. by Barr [6]. The analysis of the test cases is important for evaluating the sound execution of the degradation studies.

Types of Rules. In classical systems the rule base only contained rules directly deriving an output for a given condition. However, for real-world applications the rule base can contain more refined types of rules. We distinguish three basic types of rules (cf. [7] for a more formal description):

1. *Derivation rules:* For a given condition the rule $r \in \mathcal{R}$ derives a specified output $o \in \Omega_{sol}$, i.e., having o in the rule consequent a_r. For example, with $a \in \Omega_a$, $v \in dom(a)$ and $o \in \Omega_{sol}$, and the rule

$$a : v \rightarrow derive(o) \, ,$$

a solution o is derived for a given finding $a:v$. In detail, we distinguish derivation rules that derive an output categorically (as exemplified above) and derivation rules deriving an output using evidential categories, e.g., weighting scores or probabilities.

2. *Abstraction rules*: For a given condition such rules $r \in \mathcal{R}$ derive the value $v \in dom(a)$ for an intermediate abstraction $a \in \Omega_a$, that in turn can be also used in further rule conditions. For example, with $a_j \in \Omega_a$ and $v_j \in dom(a_j)$, and the rule

$$a_1 : v_1 \wedge a_2 : v_2 \rightarrow set\ value(a_3 : v_3)\,,$$

the finding $a_3 : v_3$ is derived, if the two findings $a_1 : v_1$ and $a_2 : v_2$ are contained in the problem description. Abstraction rules are suitable for improving the reuse of existing knowledge or to enhance the design/understandability of a knowledge base.

3. *Indication rules*: These rules are used to implement an adaptive and efficient dialog of the system with a user. For a given rule condition the rule indicates new questions/inputs $a \in \Omega_a$ to be presented to the user, e.g., for $a_j \in \Omega_a$ and $v_j \in dom(a_j)$, and the rule

$$a_1 : v_1 \wedge a_2 : v_2 \rightarrow indicate(a_3)\,,$$

the input a_3 is presented as a question to the user, when $a_1 : v_1$ and $a_2 : v_2$ holds. Such rules are commonly used for the implementation of a decision tree structure.

We see that a rule base containing not only derivation rules but also abstraction and indication rules is much more difficult to test for robustness. Thus, eliminating a specific question can prevent the system to ask the original questions of the given case, and thus entirely changes the semantics of the case. The availability of abstraction rules introduces rule chains in the knowledge base and therefore the elimination or modification of a specific input value can also change large parts of the original case. For this reason, no accurate evaluation may be possible.

Complexity of Rules. For evaluating the results of degradation studies the complexity of the included rules is an interesting measure. In general, the complexity of a rule is calculated by the number of simple conditions (i.e., evaluating the value of a single input) included in the rule condition. This measure can be integrated in the previously described measure *Average Number of Rules (NOR)* by weighting the single rules with their complexity. Several of complexity measures focusing on the complexity of scoring rules were presented in [8].

3.2 Types of Torture Tests

As described earlier a degradation study considers the application of torture tests within a degradation sequence. We distinguish four different types of torture tests. It is important to notice that for one degradation study only one type of torture test is used. In the following we sketch the idea of the particular torture test types and motivate their applicability.

Torture by Case Input Deletion. For every sequence in a degradation study a number of torture tests is executed: for each subsequent torture test a higher number of randomly selected inputs is not given as an input to the (unchanged) rule system. The test investigates the behavior of the system concerning an incomplete data acquisition, i.e., missing values.

Torture by Case Input Modification. This test is similar to the case input deletion test above, but does not omit an increasing number of inputs from the cases. In contrast an increasing number of input values is modified before passing them to the knowledge system. With this type of torture test the robustness of the system with respect to an incorrect data acquisition can be evaluated, i.e., noise in the input space.

Torture by Rule Deletion. The torture tests executed in degradation sequence are running the (unmodified) test cases with an decreased rule base quality: here, every subsequent torture test omits an increasing number of randomly selected rules during the problem solving session. The test can be useful to evaluate the robustness of the knowledge base with respect to an incomplete knowledge acquisition.

Torture by Rule Modification. In contrast to omitting an increasing number of rules, as for the torture by rule deletion test, an increasing number of rules is *modified*. With this test the robustness of the knowledge system with respect to a biased or a faulty knowledge acquisition can be evaluated. For example, how dramatically do some incorrect rules worsen the accuracy of the system? In the context of our work we changed the weighting of the rule actions, i.e., the weighting of outputs, in order to simulate a biased domain specialist.

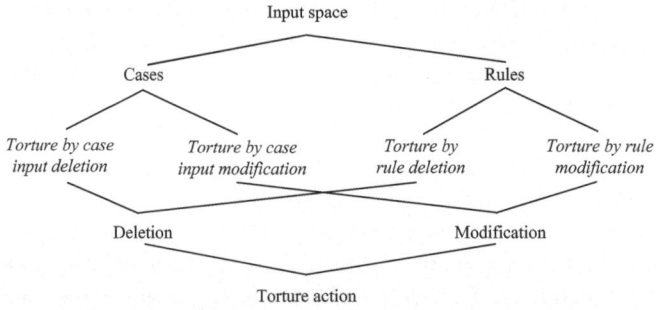

Fig. 2. A two-fold categorization of torture tests.

The four types of torture tests can be categorized in two ways: first, the tests can be classified according to the type of input that is target of the torture, i.e., tests concerning a worsening of the case base and the knowledge base, respectively.

Second, we can classify the tests according to the actual torture action that is performed during the study, i.e., there are tests fully deleting parts of the case base or rule base, and there are tests modifying parts of the case base or knowledge base. The two possible categorizations are depicted in Figure 2; the particular torture tests are written in italics.

3.3   Gray Box Testing with Background Knowledge

In a previous section we have motivated that the analysis of the case base and knowledge base is an essential precondition for robustness testing. The identified characteristics of the knowledge base are useful for predicting general robustness properties of the knowledge. Then, a rule base containing many indication and abstraction rules is more likely to be less robust than a knowledge base without a detailed dialog structure. Additionally, the complexity and type of rule conditions obviously plays a major role for the robustness properties: derivation rules with mostly single conditions should be indicators for a robust inference layer [1].

Furthermore, in a degradation study artificial noise is generated by randomly selecting the elements, i.e., input values or rule parts, for their deletion or modification. However, it is easy to see that in a real world setting not all input values have an equal "probability" to be falsely answered by the user. For example, inputs only relevant for the dialog structure are commonly answered correctly, since deciding about answers for such questions is mostly very easy. In the following we present two disjoint types of background knowledge in order to conduct more realistic degradation studies even for knowledge bases with a dialog structure.

Since the actual implementation of the knowledge base is considered and background knowledge is used for guiding the torture tests we call such a degradation study a *gray box test*.

Preserving Important Inputs. As discussed above some input values are very unlikely to be answered incorrectly by the user but can be very important for the dialog structure. In a gray box degradation study we can provide a set of such important inputs.

Then, the *important inputs* $\mathcal{IV}_a \subset \Omega_a$ are categorically excluded from the elimination and modification procedures of the torture tests. The set of important inputs is disjoint with the set of ambivalent inputs that is introduced in the following.

Ambivalence of Inputs. In contrast to the definition of important inputs the procedure of a degradation study can be also directed by the definition of *ambivalent inputs*. In a real world setting there often exists a set of inputs that are more likely to be mixed up by a user than other inputs. Such inputs commonly have a large range of possible values or the actual answer is difficult to give. For example, in the medical domain some inputs like the age or height of a patient are very simple to obtain, whereas other inputs like the findings of a liver examination are more difficult to acquire.

A set of ambivalent inputs $\mathcal{AI}_a \subset \Omega_a$ can be defined for a degradation study, where $\mathcal{AI}_a \cap \mathcal{IV}_a = \emptyset$. In consequence, these inputs are selected with a higher probability during the modification/deletion procedures of the torture tests. A typical setting for a degradation study would be that inputs $i \in \mathcal{AI}_a$ are selected with a doubled probability for torturing than inputs $i' \in \Omega_a \setminus (\mathcal{AI}_a \cup \mathcal{IV}_a)$.

The presented types of background knowledge only cover the tests modifying or deleting the inputs from cases. Here, for a case c the number of questions to be modified or deleted is not randomly selected from the entire set OBS_c but from a restricted input space, e.g., $OBS_c \setminus \mathcal{IV}_a$ for a degradation study preserving important inputs. The

---

[1] However, it is clear that the formalization of domain knowledge often requires the conjunction of many single conditions.

second type of background knowledge even concerns the randomness of the selection. Up to now we have not considered background knowledge that adapts torture tests concerning the deletion or modification of the rule base.

4   Case Study

We demonstrate the presented work by two case studies, one testing the robustness of a biological consultation system, and one testing the robustness of a medical consultation system. The two corresponding rule bases show different characteristics with respect to their implementation structure: Whereas, the biological rule base mainly consists of simple rules and no complex dialog control, the medical rule base contains many indication rules for the implementation of a sophisticated dialog structure. Furthermore, this rule base has a large portion of complex derivation rules.

4.1   Degradation of a Plant Rule Base

The first case study was conducted in the biological domain with the rule base of a plant consultation system [9]. This system identifies the most common flowering plants vegetating in Franconia, Germany. For a classification of a given plant the user enters findings concerning the flowers, leaves and stalks of the particular plant. Since the system was planned to be used by non-expert users the *diagnostic scores* [10] pattern was applied. This knowledge formalization pattern proposes to implement only derivation rules with single conditions, i.e., conditions over one finding, in order to increase the robustness of the system, e.g., concerning possibly erroneous data acquisition.

The rule base contains 5623 derivation rules, and the used case base consists of 96 cases (with a mean of 40 different inputs and in total 39 distinct outputs). All cases are correctly solved with the original version of the rule base.

For each degradation study 5 iterations of degradation sequences were performed with 20 degrading torture tests in each sequence (plus one initial torture tests with 0% modification, i.e. 100% quality). Since all cases contained only single outputs as solution for a case we used the measures precision and recall for the evaluation of the accuracy. The precision calculates the degree of inferred outputs that are actually correct, and can be defined as follows:

$$
precision(SOL_c, SOL_{c'}) = \begin{cases} \frac{|SOL_c \cap SOL_{c'}|}{|SOL_{c'}|} & \text{if } SOL_{c'} \neq \{\}, \\ 1 & \text{if } SOL_{c'} = \{\} \text{ and } SOL_c = \{\}, \\ 0 & \text{otherwise.} \end{cases}
$$

where $SOL_c, SOL_{c'}$ are the output of the stored case c and the derived output of the problem solving case c', respectively. The *recall* measures the degree of expected outputs that are actually derived, and is defined by

$$
recall(SOL_c, SOL_{c'}) = \begin{cases} \frac{|SOL_c \cap SOL_{c'}|}{|SOL_c|} & \text{if } SOL_c \neq \emptyset \\ 1 & \text{otherwise} \end{cases},
$$

where $SOL_c, SOL_{c'}$ are the output of the stored case c and the derived output of the problem solving case c', respectively.

Case Torture Tests. The first degradation study used the torture test type *case input deletion* in order to see how the system behaves with respect to missing values. The second degradation study used the torture test type *case input modification* simulating incorrect user inputs. In Figure 3 the precision values of the two degradation studies

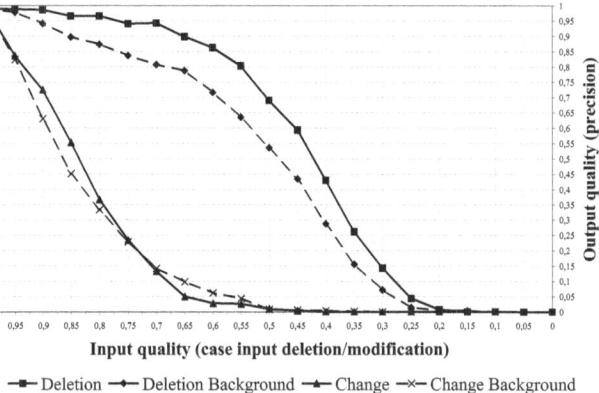

Input quality (case input deletion/modification)

—■— Deletion —◆— Deletion Background —▲— Change —×— Change Background

Fig. 3. Precision values for the plant system without and with ambivalence knowledge (dotted lines): precision values of the case input deletion tests are displayed as lines with squares and rhombus, respectively. Precision values of the case input change test are displayed as triangled and crossed lines, respectively. Degraded input quality is displayed on the x-axis.

are displayed: on the x-axis the degrading input quality (less user inputs vs. less correct user inputs) are displayed; on the y-axis the averaged precisions of the particular torture tests are depicted. Precision values of the case input deletion tests are displayed as lines with squares and rhombus, respectively. Precision values of the case input change test are displayed as triangled and crossed lines, respectively. The tests uncovered that for an incomplete data acquisition the system has an acceptable precision for an input quality between 70% and and 100%, i.e., the system is able to take about 30% missing data. In contrast, the system lacks an acceptable precision for faulty data input, since the precision falls below 85% for an input quality less than 95%. In Figure 4 the recall values of the degradation studies are shown. Recall values of the case input deletion tests are displayed as lines with squares and rhombus, respectively. Recall values of the case input change test are displayed as triangled and crossed lines, respectively. Degraded input quality is displayed on the x-axis. Here, we see a similar behavior as described for the precision of the system: The expected solutions of the cases were also derived by the system for a completeness level greater than 65%. However, the system is very sensitive for changed input values and is only able to derive the expected solutions for a completeness level greater than 95%. Concerning the application of background knowledge we see that the increased omission of ambivalent inputs yield a worse accuracy both for the precision and recall of the plant system. Surprisingly, the (slight) modification of input values of ambivalent inputs had no significant effect on the reasoning accuracy of the system. This behavior can be explained by the fact that there exist similar rules for ambivalent inputs in the rule base. A change of such values therefore resulted in

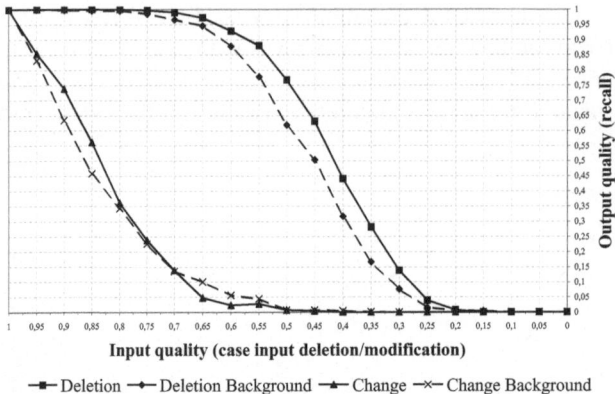

Input quality (case input deletion/modification)

━■━ Deletion ━◆━ Deletion Background ━▲━ Change ━×━ Change Background

Fig. 4. Recall values for the plant system without and with ambivalence knowledge (dotted lines): recall values of the case input deletion tests are displayed as lines with squares and rhombus, respectively. Recall values of the case input change test are displayed as triangled and crossed lines, respectively. Degraded input quality is displayed on the x-axis.

the activation of rules with similar or equivalent consequents. In general, the degradation studies with respect to the case input showed a monotonic derivation quality (see Definition 5).

Rule Torture Tests. Figure 5 displays the results of the torture tests deleting and modifying parts of the rule base. Here, the precision as well as the recall values of the tests are shown. In the upper part of the figure we see the precision and recall values for the torture test *rule modification*. The rule base consists of rules $a : v \rightarrow derive(o)$, where $derive(o)$ either adds a positive or negative weight to the given output o. In this test the weights were randomly reduced or increased by one weight category when selected for the torture test. Interestingly, the modification of the rule weights did not affect the precision and the recall significantly. Moreover we see that the tests uncovered a non-monotonic derivation quality, i.e., a worse input quality can result in a better derivation quality. The lower part of the figure depicts the precision and recall values for the torture test *rule deletion*. Here we see that an increased elimination of rules result in an almost monotonic decrease of the output quality. The improvements of the derivation quality can be explained by the stochastic variance of the tests: for each torture test new rules are randomly selected and removed. Here one collection of selected rules had more impact on the reasoning behavior than another collection of rules.

4.2 Degradation of the SonoConsult Rule Base

The second case study considered the degradation of the medical knowledge system SonoConsult [11], a consultation system for sonographic examinations. Here, 427 inputs and 221 outputs are structured in a taxonomy, where only a small portion of "leaf" inputs and outputs are actually visible to the user. The rule base consists of 4234 derivation rules, 2298 abstraction rules, and 2151 indication rules implementing a complex dialog structure. Due to its different characteristics the system was not supposed to be as

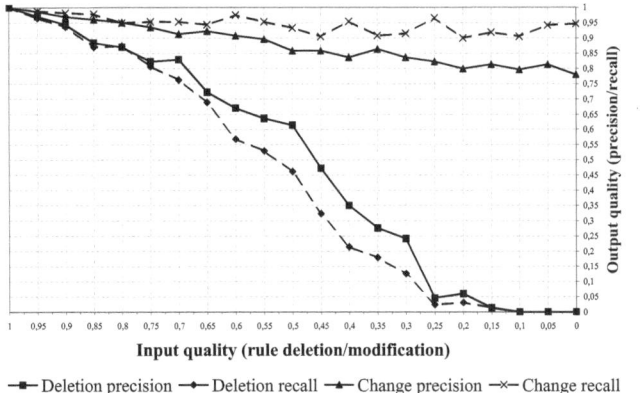

Input quality (rule deletion/modification)

─■─ Deletion precision   ─◆─ Deletion recall   ─▲─ Change precision   ─✕─ Change recall

Fig. 5. Precision and recall values for the plant system while degrading the quality of the rule base: precision values of the rule deletion tests are displayed as lines with squares; the recall for the deletion test is depicted with a dotted line and a rhombus. Precision values of the rule change test are displayed as triangled lines; the recall of this test is displayed with crossed dotted lines. Degraded input quality is displayed on the x-axis.

robust as the previously described plant system. The torture tests were performed using 250 cases, where each case contains about 60 findings as the problem description and a mean of 6 outputs as its solution. Due to the multiple faults (outputs) characteristics of the case base the application of the measures precision and recall seemed to be not appropriate; the E-measure combining both, precision and recall, was used instead. The general E-measure is defined as follows:

$$E(c, c') = \frac{(\beta^2 + 1) \cdot precision(SOL_c, SOL_{c'}) \cdot recall(SOL_c, SOL_{c'})}{\beta^2 \cdot precision(SOL_c, SOL_{c'}) + recall(SOL_c, SOL_{c'})},$$

where $SOL_c, SOL_{c'}$ are the output of the stored case c and the derived output of the problem solving case c', respectively. In the case study we used $\beta = 1$ which is balancing the precision and the recall, i.e., the F-measure.

Case Torture Tests. Figure 6 shows the F-measure values of the degradation studies for the torture test type *case input deletion* and the type *case input change*. In the case study 5 degradation sequences were performed with a quality stepwidth of 10% resulting in 10 torture tests for each degradation sequence (plus one torture test with 100% input quality). On the x-axis the degrading input quality (less user inputs vs. less correct user inputs) are displayed; on the y-axis the averaged F-measure values of the particular torture tests are depicted. F-measure values of the case input deletion tests are displayed as lines with squares and rhombus, respectively. F-measure values of the case input change test are displayed as triangled and crossed lines, respectively. The results confirm the expectations that the rule base only provides a limited robustness. The inclusion of background knowledge weakens the trend only for lower quality values, which can be explained that even more important inputs should be included as background knowledge; up to now only 35 inputs are marked as important and some more

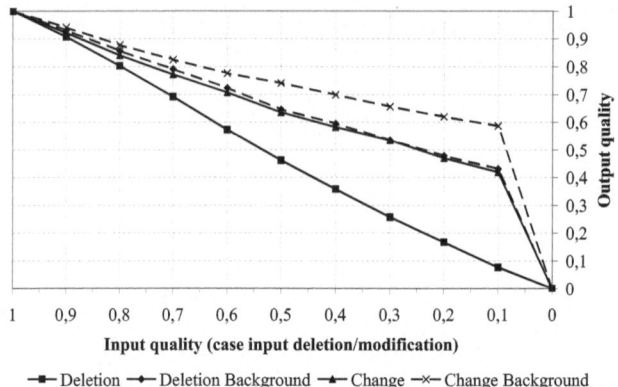

Fig. 6. F-measure values without and with important inputs knowledge (dotted lines): F-measure values of the case input deletion tests are displayed as lines with squares and rhombus, respectively. F-measure values of the case input change test are displayed as triangled and crossed lines, respectively. Degraded input quality is displayed on the x-axis.

indication questions are possibly contained in the rule base. The similar behavior for the delete inputs tests and the change input value tests suggests, that even slightly changing an input value has the same worsening behavior like completely omitting the input.

Rule Torture Tests. Figure 7 shows the F-measure values of the degradation studies of the torture test types *rule deletion* and *rule change*. In the case study 5 degradation sequences were performed with a quality stepwidth of 5% resulting in 20 torture tests for each degradation sequence (plus one torture test with 100% input quality). The results for the torture test *rule change* were quite surprising since slight changes of

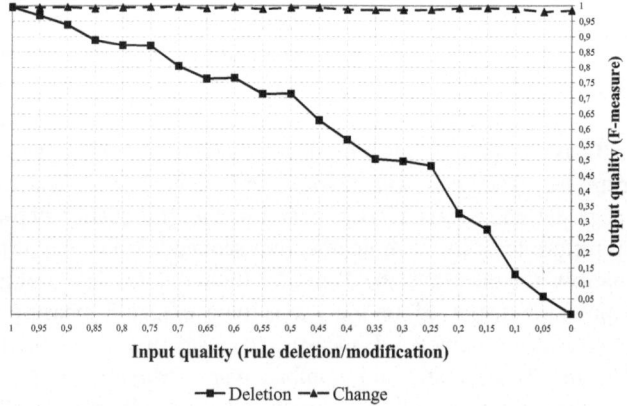

Fig. 7. F-measure values of the rule deletion test are displayed as lines with squares; F-measure values of the rule modification test are displayed as triangled and dotted lines. Degraded input quality is displayed on the x-axis.

the rule weights did not yield a significant decrease of the derivation quality. As for the plant system the rule base also contained redundant knowledge that compensates possible changes of the rule weights. In contrast, the deletion of rules yielded an almost monotonic decrease of the output quality. The slight improvements of the derivation quality for input qualities 0.6 and 0.5 can be explained by the stochastic variance of the tests as already described in the case study with the plant system. We expect that for a larger number of degradation sequences this variation will be removed and the results will show a strict monotonic behavior.

5   Conclusion and Outlook

Rules are an intuitive and simple representation for the formalization of domain knowledge. In the past, there has been much research on general validation and verification methods for rule bases. However, for the practical applicability of intelligent systems also their robustness is of importance. In this paper, we presented an approach for testing the robustness of rule bases extending previous approaches (cf. [5]) by background knowledge and an intensive pre-analysis, thus defining a gray box test. It is worth noticing that on the one hand the pre-analysis of the knowledge base and the case base is necessary for a sound interpretation of the study results. On the other hand, background knowledge is used for a more concise definition of the robustness tests. The work was demonstrated by a case study reporting robustness tests on two larger rule bases taken from the biological and the medical domain. As a general contribution, degradation studies can help to determine the self-confidence level of a system: for example, if a given number of omitted inputs is exceeded, then the system can retain from giving a solution, thus avoiding to give a wrong solution with a high probability.

In the future, we are planning to improve the applicability of background knowledge: first, currently no background knowledge for the adaptation of rule torture tests is considered. For example, when using rules deriving an output with a certain probability we can expect that rules with a medial probability are more likely to be biased or noisy than rules with probabilities near to 1 or 0. Second, the definition of background knowledge can be simplified by automatically generating proposals for important inputs and ambivalent inputs. Important inputs typically are inputs contained in indication rules but not in derivation or abstraction rules. An input can be considered as an ambivalent input if the corresponding input values are occurring very frequently in rules for different outputs.

References

1. Ayel, M., Laurent, J.P.: Validation, Verification and Test of Knowledge-Based Systems. Wiley (1991)
2. Preece, A., Shinghal, R.: Foundation and Application of Knowledge Base Verification. International Journal of Intelligent Systems **9** (1994) 683–702
3. Knauf, R.: Validating Rule-Based Systems: A Complete Methodology. Shaker, Aachen, Germany (2000)
4. Horrocks, I., Patel-Schneider, P.F., Bechhofer, S., Tsarkov, D.: OWL Rules: A Proposal and Prototype Implementation. Journal of Web Semantics **3**(1) (2005) 23–40

5. Groot, P., van Harmelen, F., ten Teije, A.: Torture Tests: A Quantitative Analysis for the Robustness of Knowledge-Based Systems. In: Knowledge Acquisition, Modeling and Management. LNAI 1319, Berlin, Springer (2000) 403–418
6. Barr, V.: Applications of Rule-Base Coverage Measures to Expert System Evaluation. Knowledge-Based Systems **12** (1999) 27–35
7. Baumeister, J.: Agile Development of Diagnostic Knowledge Systems. AKA Verlag, DISKI 284 (2004)
8. Atzmueller, M., Baumeister, J., Puppe, F.: Semi-Automatic Learning of Simple Diagnostic Scores utilizing Complexity Measures. AI in Medicine **37**(1) (2006) 19–30
9. Ernst, R.: Untersuchung verschiedener Problemlösungsmethoden in einem Experten- und Tutorsystem zur makroskopischen Bestimmung krautiger Blütenpflanzen [Analysis of various problem solving methods with an expert and tutoring system for the macroscopic classification of flowers]. Master's thesis, University Würzburg, Biology department (1996)
10. Puppe, F.: Knowledge Formalization Patterns. In: Proceedings of PKAW 2000, Sydney, Australia (2000)
11. Hüttig, M., Buscher, G., Menzel, T., Scheppach, W., Puppe, F., Buscher, H.P.: A Diagnostic Expert System for Structured Reports, Quality Assessment, and Training of Residents in Sonography. Medizinische Klinik **3** (2004) 117–22

A Unifying Framework for Hybrid Planning and Scheduling

Bernd Schattenberg and Susanne Biundo

Dept. of Artificial Intelligence
University of Ulm, Germany
firstname.lastname@uni-ulm.de

Abstract. Many real-world application domains that demand planning and scheduling support do not allow for a clear separation of these capabilities. Typically, an adequate mixture of both methodologies is required, since some aspects of the underlying planning problem imply consequences on the scheduling part and vice versa. Several integration efforts have been undertaken to couple planning and scheduling methods, most of them using separate planning and scheduling components which iteratively exchange partial solutions until both agree on a result.

This paper presents a framework that provides a uniform integration of hybrid planning –the combination of operator based partial order planning and abstraction based hierarchical task network planning– and a hierarchical scheduling approach. It is based on a proper formal account of refinement planning, which allows for the formal definition of hybrid planning, scheduling, and search strategies. In a first step, the scheduling functionality is used to produce plans that comply with time restrictions and resource bounds. We show how the resulting framework is thereby able to perform novel kinds of search strategies that opportunistically interleave what used to be separate planning and scheduling processes.

1 Introduction

Hybrid planning – the combination of hierarchical task network (HTN) planning with partial order causal link (POCL) techniques – turned out to be most appropriate for complex real-world planning applications [1] like crisis management support [2,3], etc. Here, the solution of planning problems often requires the integration of planning from first principles with the utilization of predefined plans to perform certain complex tasks. Beyond the challenge to produce complex courses of action, planning support in these domains has to consider all kinds of resources, ranging from limited time and material to power and supplies, which all define success and efficiency of the mission.

These two aspects of the planning task used to be regarded as two different kinds of problems, called *planning* and *scheduling*, one performed after the other. Newer approaches take into account, that the two problem solving phases interact to a huge extent, and neither can be reasonably carried out without

C. Freksa, M. Kohlhase, and K. Schill (Eds.): KI 2006, LNAI 4314, pp. 361–373, 2007.
© Springer-Verlag Berlin Heidelberg 2007

knowledge and feedback about the progress of the other. This motivates to get more information out of resource analysis into the planning process, e.g. via resource profiles [4] or constraint-based techniques on shared memory [5]. It allows for some corrective steps on the "planner's side", respectively offers alternative plans to pursue. Another frequently used technique is to keep the plan and schedule generating processes completely separate, using the result of one as the input for the other. In this fashion, [6] proposes to perform classic plan generation on a relaxed problem without resource information and to give the result to a scheduler. [7] places scheduling in a pre-planning phase in order to determine necessary overlapping actions and minimal resource capacities. In all these approaches, the plan generation process is not guided by resource demands and vice versa. The IxTeT temporal planning system [8] integrates scheduling in an POCL planner by using temporally qualified expressions throughout the representation formalism, which represent state transitions and state persistences of the planning domain. The authors share our view of opportunistic scheduling as additional plan modification steps which can be interleaved with other planning steps: closing open or unachieved preconditions, resolving (resource) conflicts, and adding constraints to evade bottlenecks. An important feature is the dynamic construction of a resource hierarchy (not to be confused with hierarchical resources as introduced by [9]) based on condition analysis in the current partial plan [10]. The hierarchy represents a partial order on the "importance" of the resources for plan causality, and with that the order in which the different resources should be addressed by the reasoning process. This technique can be considered to be included as an additional strategic advice to our proposed approach.

Our aim is to provide a framework in which planning and scheduling functionality is uniformly integrated. Integration should not be limited to a planner delivering plans to be judged by a scheduler but it should be possible to generate and abandon plans schedule-driven and vice versa. We will show in a first step, what a system configuration under these requirements looks like for generating plans that are resource and time compliant.

To this end, we make use of the hybrid planning approach presented in [11]. It provides a formal framework, in which the plan generation process is functionally decomposed into well-defined flaw detecting and plan modification generating functions. We are going to show, how scheduling capabilities can be integrated and exploited in the plan generation process. *Flexible* planning and scheduling strategies operate opportunistically instead of following a fixed plan generation schema, thereby completely interleaving what used to be separate planning and scheduling processes.

The rest of the paper is organized as follows. In Section 2 we present the refinement-based planning framework on which our integrated hybrid approach relies; we define the necessary refinement operators and flaws for our purposes. Section 3 introduces the resulting system components and Section 4 describes strategies that are capable of guiding search in the integrated planner and scheduler. We conclude in Section 5 with a glimpse on future developments and some final remarks.

2   A Refinement-Based Framework

We employ a hybrid planning formalism that relies on a sorted logical language $\mathcal{L} = \{\mathcal{Z}, \mathcal{P}, \mathcal{C}, \mathcal{V}, \mathcal{O}, \mathcal{T}\}$. $\mathcal{Z}$ describes a hierarchy of sort symbols, $\mathcal{P}$ is a $\mathcal{Z}^*$-indexed set of predicate symbols, and $\mathcal{C}$ and $\mathcal{V}$ are non-empty finite $\mathcal{Z}$-indexed sets of constant symbols and variables. $\mathcal{O}$ denotes a non-empty finite set of *operator* symbols, $\mathcal{T}$ a finite set of *task* symbols; both symbol sets are $\mathcal{Z}^*$-indexed. All sets are assumed to be disjoint. $\mathcal{Z}$ contains two designated sort symbols: Resource and its sub-sort Symbolic for modeling symbolic resources. By specifying sub-sorts of Symbolic, we employ the notion of hierarchical resources in our approach. For presentation purposes we will focus in the following on (shareable) symbolic resources, e.g. Vehicle with sub-sorts Truck and Jeep. The difference between symbolic resources and other objects is subtle: the identity of a resource entity is explicitly not of interest. This allows for efficient reasoning mechanisms that analyze allocation profiles and identify bottlenecks, potential and necessary over-allocations, etc., rather than dealing with equations and un-equations in constraint sets. For a detailed discussion on hierarchical resource representations, also beyond subsumption (including numeric resources), see [9].

An *operator schema* $o(\bar{\tau}) = (\mathrm{prec}(o(\bar{\tau})), \mathrm{add}(o(\bar{\tau})), \mathrm{del}(o(\bar{\tau})),\ d_{o(\bar{\tau})}^{min}, d_{o(\bar{\tau})}^{max})$ specifies the *preconditions* and the *positive* and *negative effects* of that operator ($o \in \mathcal{O}$, $\bar{\tau} = \tau_1, \ldots, \tau_n$ with $\tau_i \in \{\mathcal{C} \cup \mathcal{V}\}$ for $1 \leq i \leq n$ and n being the arity of o). Preconditions and effects are sets of literals and atoms over $\mathcal{P} \cup \mathcal{C} \cup \mathcal{V}$, respectively. Symbolic resources do not have to be allocated explicitly by specific predicates but are implicitly by their use in the condition atoms. $d_{o(\bar{\tau})}^{min}$ is the minimal duration of the operator, $d_{o(\bar{\tau})}^{max}$ the maximal.

A ground instance of an operator schema is called an *operation*. A *state* is a finite set of ground atoms and an operation $o(\bar{c})$ is *applicable* in a state s iff for the positive literals in the precondition of o: $\mathrm{prec}^{\oplus}(o(\bar{c})) \subseteq s$ and for the negative literals: $\mathrm{prec}^{\ominus}(o(\bar{c})) \cap s = \emptyset$. The result of applying $o(\bar{c})$ in state s is a state $s' = (s \cup \mathrm{add}(o(\bar{c}))) \setminus \mathrm{del}(o(\bar{c}))$. Operators are also called *primitive tasks*. Executability of sequences of operations is defined inductively.

Abstract actions are represented by *complex tasks*, which in hybrid planning show preconditions and effects like primitive ones. They are defined by *task schemata* $t(\bar{\tau}) = (\mathrm{prec}(t(\bar{\tau})),\ \mathrm{add}(t(\bar{\tau})),\ \mathrm{del}(t(\bar{\tau})), d_{t(\bar{\tau})}^{min}, d_{t(\bar{\tau})}^{max})$ for $t \in \mathcal{T}$. A *decomposition method* $m = (t(\bar{\tau}), d)$ relates a complex task $t(\bar{\tau})$ to a *task network* d. This task network can be seen as a pre-defined *implementation* of a complex task and therefore the duration intervals of abstract actions have to be valid lower, respectively upper bounds for all their implementing networks. While HTN planning approaches like [12] deduce bounds for resources in abstract tasks from information about resource allocation by more primitive tasks, our notion of hybrid planning has to take into account, that implementing networks might not be complete and that the user's specification might therefore add some estimated extra time overhead, for example. Consequently, as it will be shown in the definition of the task expansion plan modification, the temporal artifacts of abstract task time bounds persist in the respective constraint sets of a plan.

For this presentation we omit axiomatic state refinements for modeling different abstraction levels on preconditions and effects for abstract task [2]. This restricts task networks occuring in methods to those in which all preconditions and effects of the abstract task have to occur explicitly.

A task network –or *partial plan*– over a language $\mathcal{L}$, a set of primitive and complex task schemata T, and a set of decomposition methods M, is a tuple $(TE, \prec, VC, CL, TC)$ with TE being a set of plan steps *task expressions* $te = l : t(\tau_1, \ldots, \tau_m)$, where l represents a unique label in TE and $t(\tau_1, \ldots, \tau_m)$ an instance of a task schema in T, using variables unique in TE. $\prec$ is a set of *ordering constraints* imposing a partial order on the steps in TE. VC denotes a set of *variable constraints*, which *codesignate* and *non-codesignate* variables that occur in TE with each other or constants. It also contains sort restrictions: Constraints of the form $v \dot{\in} Z$ and $v \dot{\notin} Z$, $Z \in \mathcal{Z}$, include or exclude variables from being assigned to terms of the specified sort. CL is a set of *causal links* $te_i \xrightarrow{\phi} te_j$ with $te_i, te_j \in TE$ and ϕ being a literal with $\sigma(\phi) \in \sigma(\text{prec}(te_j))$ and $\sigma(\phi) \in \sigma(\text{add}(te_i))$, if ϕ is positive, and $\sigma(|\phi|) \in \sigma(\text{del}(te_i))$, if ϕ is negative. σ is a VC-compatible variable substitution, i.e. a substitution that is consistent with the variable constraints. A causal link indicates that a task $(l : t_i(\bar{\tau}))$ *establishes* a precondition of a task $(l' : t_j(\bar{\tau}'))$ and is used in the usual sense as a book keeping entity.

Finally, TC represents the temporal information as a *simple temporal problem* [13]. TC is a constraint system (Z, D, C) with Z being a set of variables that represent time points and $D : Z \to R^+$ a function for assigning sets of real numbers (including the symbol ∞ for representing an infinite amount of time) to each variable in Z. The set of real numbers D_{x_i} that is assigned by D to a variable $x_i \in Z$ is called the *domain* of that variable. C is a set of unary and binary constraints. A binary constraint represents the temporal distance between two time point variables x_i and x_j by an interval $[d_{min}, d_{max}]$, which stands for the equation $d_{min} \leq x_j - x_i \leq d_{max}$. A unary temporal constraint specifies a time point x by an interval $[early, late]$, which means that $early \leq x \leq late$. The temporal network specifies for each task expression $te \in TE$ two time points that denote that beginning and the end of the action: $start_{te}$ in $[0, \infty)$ and end_{te} in $[start_{te} + d_{te}^{min}, start_{te} + d_{te}^{max}]$. For any two task expressions $te_i, te_j \in TE$ with $te_i \prec^* te_j$ (the transitive closure of $\prec$), the temporal relation $end_{te_i}^{max} \leq start_{te_j}$ holds, i.e. their temporal distance is given by the interval $[0, \infty)$. Conversely, for every two tasks with $end_{te_i}^{max} \leq start_{te_j}$, the transitive closure of the ordering relation $\prec^*$ contains $te_i \prec te_j$. A causal link $te_i \xrightarrow{\phi} te_j \in CL$ is reflected by a temporal relation $start_{te_i} \leq end_{te_j}^{max}$.

An *integrated planning and scheduling problem* $(d, T, M, s_{init}, s_{goal})$ consists of an initial task network d, a set of task and operator schemata T, and a set M of decomposition methods for implementing the complex tasks in T. The state s_{init} represents the initial world state, including resource capacities, and s_{goal} is a specification of the desired goal state and the overall makespan limit. Like it is common in partial order planning, we encode the initial and (optional) goal state as artificial task expressions te_{init} and te_{goal} in $TE(d)$ with respective effects and preconditions, artifacts in the temporal constraint system, and the obligation to order any other task between them.

Given a problem specification $(d, \mathsf{T}, \mathsf{M}, s_{init}, s_{goal})$, a hybrid planning and scheduling system transforms the initial task network d into a task network that is considered a *solution* to the problem. A partial plan $P = (TE, \prec, VC, CL, TC)$ is a solution to the problem, if and only if P is obtained from d by the application of plan modification steps, if TE includes only primitive task expressions, and if P is executable in s_{init} and generates s_{end}. P is called *executable* in a state s and generates a state s', if all ground linearizations of P, that means all linearizations of all ground instances of the task expressions in TE that are compatible with $\prec$ and VC, are executable in s and generate a state $s'' \supseteq s'$. No linearization may exceed the capacities specified in s_{init} in any intermediate state by its accumulated allocations, nor it may exceed the time limit specified in s_{goal} by its makespan in TC.

The presented framework makes violations of the solution criteria explicit by introducing *flaws*, data structures that literally "point" to deficiencies in the plan and allow for the problems' classification. This will allow us to guide the search process in particular to address specific problems at a specific time.

Definition 1 (Flaws). *For a given planning and scheduling problem specification and a plan P that is no solution to the problem, a flaw $\mathtt{f}$ is a pair $(flaw, E)$ with "flaw" denoting the flaw class and E being the set of components in P to which the flaw refers.*

The set of flaws is denoted by $\mathcal{F}$ with subsets $\mathcal{F}_{flaw}$ for given labels $flaw$. The set $\mathcal{F}$ of flaw classes in a partial plan $P = (TE, \prec, VC, CL, TC)$ for a given problem $(d, \mathsf{T}, \mathsf{M}, s_{init}, s_{goal})$ addresses the solution criteria by the following sub-sets (some of which being classical plan generation flaws, some related to a scheduling view):

1. $(\mathtt{AbstractTask}, \{te\})$ with $te = l : t(\bar{\tau}) \in TE, t \in \mathcal{T}$ being an abstract task expression. P is not yet primitive and this flaw class is typically associated with hybrid planning.
2. $(\mathtt{OrdInconsistency}, \{te_1, \ldots, te_k\})$ with $te_i \in TE, te_i \prec^* te_i, 1 \leq i \leq k$, i.e. if $\prec^*$ defines a cyclic partial order. There exists no defined linearization of P, which makes this flaw related to planning and scheduling likewise.
3. $(\mathtt{VarInconsistency}, \{v\})$ with $v \in \mathcal{V}$ being a variable for which $VC \models v \neq v$ holds. Since VC is inconsistent, no VC-compatible ground substitutions can be deduced to gain grounded operations from TE. This criterion is needed in both paradigms.
4. $(\mathtt{OpenVarBinding}, \{v\})$ with $v \in \mathcal{V}$ being a variable occurring in TE and there exists a constant $c \in \mathcal{C}$ with $VC \not\models v = c$ and $VC \not\models v \neq c$. The solution criterion requests all ground linearizations to be executable, therefore it has to be decided whether an operation is compatible with VC or not, for plan as well as for schedule generation.
5. $(\mathtt{OpenPrecondition}, \{te, \phi\})$ with $\phi \in \mathrm{prec}(te), te \in TE$, denotes a not fully supported task, i.e., for the subset of te-supporting causal links $\{te_i \xrightarrow{\phi_i} te | 1 \leq i \leq k\} \subseteq CL$ we find $\bigcup_{1 \leq i \leq k} \phi_i \subset \mathrm{prec}(te)$. If not all necessary

precondition establishers have yet been identified in P, some, if not all ground linearizations might not be executable. This is a typical planning-only flaw.

6. $(\texttt{OrdInconsistency}, \{te_1, te_2\})$ with $te_1, te_2 \in TE$ being causally linked task expressions, say $te_1 \xrightarrow{\phi} te_2 \in CL$, for which $te_2 \prec^* te_1$ does hold. If an identified establisher for a given precondition is ordered after the respective consumer, no linearization will be executable. Like the previous flaw, this one belongs to the area of classical planning.

7. $(\texttt{Threat}, \{te_i \xrightarrow{\phi} te_j, te_k\})$ with $te_k \not\prec^* te_i$ or $te_j \not\prec^* te_k$ and there exists a VC-compatible substitution σ such that $\sigma(\phi) \in \sigma(\text{del}(te_k))$ for positive literals ϕ and $\sigma(|\phi|) \in \sigma(\text{add}(te_k))$ for negative literals. A classical planning flaw, because te_k will corrupt executability of at least some ground linearizations.

8. Temporal constraint inconsistencies belong to scheduling and occur if intervals for time variables collapse or temporal distances become negative. The flaw structure is either $(\texttt{TempInconsistency}, \{te\})$ with $te \in TE$ being the task expression for which $start_{te}$ or end_{te} time point intervals have collapsed, or it is $(\texttt{TempInconsistency}, \{te_1, te_2\})$ with $te_1, te_2 \in TE$ being two task expressions for which the temporal distance between associated variables became negative. An overrun of the specified maximum makespan is covered by the interval for $start_{te_{goal}}$.

9. $(\texttt{SymbolicOverAllocation}, \{v_1, \ldots, v_n\})$ with $v_1, \ldots, v_n \in \mathcal{V}$ being variables for which $VC \not\models v_i = v_j$, $1 \leq i < j \leq n$ and n exceeding the capacity of any common (sub-) sort of the $v_1, \ldots, v_n$ (cf. *potential* allocations and resource profiles discussed in [9]). It is a classical scheduling aspect of the problem, that too many objects of one kind might be required at one point in time.

It can be shown that the above flaw definitions are complete in the sense, that for any given planning problem $(d, \mathsf{T}, \mathsf{M}, s_{init}, s_{goal})$ and plan P that is not flawed, P is a solution to the problem.

We now define the refinement operators for the integrated system, some of them origin in hybrid planning, some in scheduling.

Definition 2 (Plan Modifications). *For a given partial plan P over a language $\mathcal{L}$, a set of primitive and complex task schemata T, and a set of decomposition methods M, a plan modification m is a pair $(mod, E^{\oplus} \cup E_{\ominus})$. "mod" denotes the modification class, $E^{\oplus}$ and $E_{\ominus}$ are sets of elementary additions and deletions of plan components over $\mathcal{L}$, P, T, and M. These two sets are assumed to be disjoint and $E^{\oplus} \cup E_{\ominus} \neq \emptyset$.*

The set of all plan modifications is denoted by $\mathcal{M}$ and grouped into subsets $\mathcal{M}_{mod}$ for given classes mod. The application of a plan modification is characterized by the generic plan transformation function $app : \mathcal{M} \times \mathsf{P} \to \mathsf{P}$, which takes a plan modification $\mathsf{m} = (mod, E^{\oplus} \cup E_{\ominus})$ and a plan P, and returns a plan P' in which all elements of $E^{\oplus}$ have been added to and that of $E_{\ominus}$ have been removed from P.

In the integrated hybrid planning and scheduling approach, the following classes of correct plan modifications are defined for manipulating a given partial plan $P = (TE, \prec, VC, CL, TC)$ over a given language $\mathcal{L}$ and sets of task schemata T and expansion methods M:

1. $(\texttt{InsertTask}, \{\oplus te, \oplus (te \xrightarrow{\phi} te'), \oplus (v_1 = \tau_1), \ldots, \oplus (v_k = \tau_k)\})$ with $te = l : t(\bar{\tau}) \notin TE, t \in \texttt{T}$, being a new task expression to be added, $te' \in TE$, and $\sigma'(\phi) \in \sigma'(\mathrm{add}(te))$ for positive literals ϕ, $\sigma'(|\phi|) \in \sigma'(\mathrm{del}(te))$ for negative literals, and $\sigma'(\phi) \in \sigma'(\mathrm{prec}(te'))$ with σ' being a VC'-compatible substitution for $VC' = VC \cup \{v_i = \tau_i | 1 \le i \le k\}$.

   We focus on symbolic resources, which cannot be "produced". Adding a new task is therefore only done for planning purposes.

2. $(\texttt{AddOrdConstraint}, \{\oplus (te_i \prec te_j)\})$ for $te_i, te_j \in TE$. This lies in the domain of planning as well as scheduling.

3. $(\texttt{AddVarConstraint}, \{\oplus (v = \tau)\})$ for codesignating variables $v \in \mathcal{V}$ with terms $\tau \in \mathcal{V} \cup \mathcal{C}$ and $(\texttt{AddVarConstraint}, \{\oplus (v \neq \tau)\})$ for a corresponding non-codesignation. Cotyping and non-cotyping constraints are added by $(\texttt{AddVarConstraint}, \{\oplus (v \dot{\in} Z)\})$ and $(\texttt{AddVarConstraint}, \{\oplus (v \dot{\notin} Z)\})$ for $Z \in \mathcal{Z}$. Variable constraints are in the focus of plan generation as well as scheduling methods.

4. $(\texttt{AddCausalLink}, \{\oplus (te_i \xrightarrow{\phi} te_j), \oplus (v_1 = \tau_1), \ldots, \oplus (v_k = \tau_k)\})$, with $te_i, te_j \in TE$. The codesignations represent necessary variable substitutions, such that after the modification execution, they induce a VC'-compatible substitution σ' for $VC' = VC \cup \{(v_1 = \tau_1), \ldots, (v_k = \tau_k)\}$ for which $\sigma'(\phi) \in \sigma'(\mathrm{add}(te_i))$ for positive literals ϕ, $\sigma'(|\phi|) \in \sigma'(\mathrm{del}(te_i))$ for negative literals, and $\sigma'(\phi) \in \sigma'(\mathrm{prec}(te_j))$. This treatment of causal links is originated in classical partial order planning.

5. Given an abstract task expression $te = l : t(\bar{\tau})$ in TE, $t \in \mathcal{T}$ and an expansion method $m = (t, (TE_m, \prec_m, VE_m, CL_m, TC_m))$ in M, the expansion of te is defined as:

   $(\texttt{Expansion}, \{\ominus te\} \cup \{\oplus te_m | te_m \in TE_m\} \cup$
   $\quad \{\oplus (te_{m1} \prec te_{m2}) | (te_{m1} \prec_m te_{m2})\} \cup$
   $\quad \{\oplus (te_{m1} \xrightarrow{\phi} te_{m2}) | (te_{m1} \xrightarrow{\phi} te_{m2}) \in CL_m\}$
   $\quad \{\oplus (v = \tau) | (v = \tau) \in VC_m\} \cup \{\oplus (v \neq \tau) | (v \neq \tau) \in VC_m\} \cup$
   $\quad \{\ominus (te' \prec te), \oplus (te' \prec te_m) | (te' \prec te), te_m \in TE_m\} \cup$
   $\quad \{\ominus (te \prec te'), \oplus (te_m \prec te') | (te \prec te'), te_m \in TE_m\} \cup$
   $\quad \{\ominus (te \xrightarrow{\phi} te'), \oplus (te_m \xrightarrow{\phi} te') | (te \xrightarrow{\phi} te') \in CL,$
   $\qquad te_m \in TE_m, |\phi| \in \mathrm{add}(te_m) \cup \mathrm{del}(te_m)\} \cup$
   $\quad \{\ominus (te' \xrightarrow{\phi} te), \oplus (te' \xrightarrow{\phi} te_m) | (te' \xrightarrow{\phi} te) \in CL,$
   $\qquad te_m \in TE_m, \phi \in \mathrm{prec}(te_m)\} \cup$
   $\quad \{\oplus (d_{min} \le x_j - x_i \le d_{max}) | (d_{min} \le x_j - x_i \le d_{max}) \in TC_m\}$

   During an expansion the abstract task is replaced by the decomposition network with all its sub-tasks being ordered between the predecessors and successors of the abstract task and with all the causalities re-distributed among the appropriate sub-tasks. If the causal links cannot be re-distributed unambiguously, that means if there is more than one task in the expansion network that carries the respective precondition, one expansion modification has to be generated for each such permutation. This modification is clearly associated with (hierarchical) planning.

6. $(\texttt{AddTempConstraint}, \{\oplus(d_{min} \leq x_j - x_i \leq d_{max})\}$ adds the specified binary distance constraint $[d_{min}, d_{max}]$ between two temporal variables x_i and x_j in TC. The handling of temporal constraints is in the focus of scheduling. It is a refinement that is used to narrow the temporal distance interval between two time point variables x_i and x_j in TC. The variant $(\texttt{AddTempConstraint}, \{\oplus(d'_{min} \leq x \leq d'_{max})\}$ defines, respectively contracts, the interval for a time point x.

These plan modifications are the canonical plan transformation generators in a refinement-based planner: starting from an initial task network, the current plan can be checked against the solution criterion, and while that is not met, all refinements are applied. If no applicable modification exists, backtracking is performed. In order to make the search more systematic and efficient, the algorithm should focus on those modification steps which are appropriate to overcome the deficiencies in the current plan. Based on the formal notions of plan modifications and flaws, a generic algorithm and planning strategies can be defined. A strategy specifies *how* and *which* flaws in a partial plan are eliminated through appropriate plan modification steps. We therefore need to define the conditions under which a plan modification can *in principle* eliminate a given flaw.

Definition 3 (Appropriate Modifications). *A class of plan modifications* $\mathcal{M}_m \subseteq \mathcal{M}$ *is* appropriate *for a class of flaws* $\mathcal{F}_f \subseteq \mathcal{F}$ *iff there exist partial plans* P, *which contain flaws* $\texttt{f} \in \mathcal{F}_f$, *and modifications* $\texttt{m} \in \mathcal{M}_m$ *such that the refined plans* $P' = app(\texttt{m}, P)$ *do not contain* $\texttt{f}$.

The defined plan modifications perform a strict refinement, i.e., a subsequent application of them does never result in the same plan twice; the plan development is inherently a-cyclic. Given that, the same flaw cannot be re-introduced once it has been eliminated. This qualifies the appropriateness relation as a valid strategic advice for plan and schedule generation and motivates its use as the following trigger function for plan modifications:

Definition 4 (Modification Triggering Function). *Flaws in a partial plan can be removed by triggering the application of suitable plan modification steps according to the following function:* $\alpha(\mathcal{F}_x) =$

$$
\begin{array}{ll}
\mathcal{M}_{\texttt{Expansion}} & \textit{if } \mathcal{F}_x = \mathcal{F}_{\texttt{AbstractTask}} \\
\mathcal{M}_{\texttt{AddVarConstraint}} & \textit{if } \mathcal{F}_x = \mathcal{F}_{\texttt{OpenVarBinding}} \\
\mathcal{M}_{\texttt{AddCausalLink}} \cup \mathcal{M}_{\texttt{InsertTask}} \cup \mathcal{M}_{\texttt{Expansion}} & \textit{if } \mathcal{F}_x = \mathcal{F}_{\texttt{OpenPrecondition}} \\
\mathcal{M}_{\texttt{Expansion}} \cup \mathcal{M}_{\texttt{AddOrdConstraint}} \cup \mathcal{M}_{\texttt{AddVarConstraint}} & \textit{if } \mathcal{F}_x = \mathcal{F}_{\texttt{Threat}} \\
\mathcal{M}_{\texttt{Expansion}} \cup \mathcal{M}_{\texttt{AddVarConstraint}} & \textit{if } \mathcal{F}_x = \mathcal{F}_{\texttt{SymbolicOverAllocation}} \\
\emptyset & \textit{else}
\end{array}
$$

Modification class 7 is missing intentionally in this line-up: the manipulation of temporal constraints, $\mathcal{M}_{\texttt{AddTempConstraint}}$, is not used "actively" or as a "point of choice" for this presentation, since temporal reasoning is only performed in order to check for schedule executability. Their utilization for additional plan inferences will be shown later.

We have to omit the appropriateness proofs for the above function due to a lack of space, we would however like to sketch the arguments for two particular relationship expressed in α which displays the outstanding flexibility of the unified hybrid planning and scheduling: a) Threats of causal links can not only be addressed as usual by relocating the threatening task outside the scope of the causal link or by decoupling variable constraints. If an abstract task is involved in the threat situation, hybrid planning can alternatively make use of task expansion for producing "overlapping" task networks that may offer less strict promotion or demotion opportunities, since the causalities in the expansion network are typically linked from and to several of the introduced sub-tasks. As a side effect, the variable constraints of such a network may also rule out the threat. b) The flaws and associated modifications reflect the interplay between planning and scheduling aspects. E.g., in a plan with an abstract task, there are n different symbolic resources allocated of a given sort $Z \in \mathcal{Z}$. This implies a potential need of n such objects for every sub-sort of Z – which are not available. An expansion now concretizes a specific resource demand by co-typing constraints in its implementing task network, thereby assigning some of the allocations to one of the sub-sorts of Z, which lowers the need in other sub-sorts.

Ordering cycles $\mathcal{F}_{\texttt{OrdInconsistency}}$ and variable $\mathcal{F}_{\texttt{VarInconsistency}}$ and temporal inconsistencies $\mathcal{F}_{\texttt{TempInconsistency}}$ obviously cannot be resolved by our modifications and do therefore not trigger any modification.

3   Integrated Hybrid Planning and Scheduling

It is an important property of this approach, that the trigger function allows to completely separate the computation of flaws from that of modifications, and in turn both computations to be independent from search issues. The system architecture relies on this separation and exploits it in two ways: module invocation and interplay are specified through α while reasoning about search can be performed on the basis of flaws and modifications without taking their actual computation (or even their origin in the planning or scheduling field) into account. The issued flaws can *only* be addressed by the assigned modification generators; if none can solve the flaw, the system has to backtrack. Hence, we can map flaw and modification classes directly onto groups of modules which are responsible for their computation.

Definition 5 (Detection Modules). *A detection module x is a function that, given a partial plan P, returns all flaws of type x in P:*

$$f_x^{det} : \mathrm{P} \rightarrow 2^{\mathcal{F}_x}$$

It may rank the flaws according to local priorities. E.g., $f_{\texttt{OpenPrecondition}}^{det}$ prioritizes its detections according to the number of literals in the tasks' preconditions.

Definition 6 (Modification Modules). *A modification module y is a function which computes all plan modifications of type y that are appropriate for given flaws:*

$$f_y^{mod} : \mathrm{P} \times 2^{\mathcal{F}_x} \to 2^{\mathcal{M}_y} \text{ for } \mathcal{M}_y \subseteq \alpha(\mathcal{F}_x)$$

These modules also prioritize their answers by local preferences. Priorities for modifications in $\mathcal{M}_{\mathtt{InsertTask}}$ correlate with the number of available task schemata and implied variable constraints, for example. Scheduling related modules can quantify their expected gain in plan quality or simply use their local cost estimates, e.g. preferring variable co-typing modifications in the least allocated sort.

Definition 7 (Strategy Modules). *A strategy module z is a function that selects plan modifications for their application to the current plan, possibly taking into account the detected flaws. It is defined by the projection*

$$f_z^{strat} : \mathrm{P} \times 2^{\mathcal{F}} \times 2^{\mathcal{M}} \to \mathcal{M} \cup \epsilon$$

Strategies discard a current plan P if any flaw remains un-addressed by the associated modification modules, i.e., if for any f_x^{det} and $f_{y_1}^{mod}, \ldots, f_{y_n}^{mod}$ with $\mathcal{M}_{y_1} \cup \ldots \cup \mathcal{M}_{y_n} = \alpha(\mathcal{F}_x)$:

$$\bigcup_{1 \leq i \leq n} f_{y_i}^{mod}(P, f_x^{det}(P)) = \emptyset$$

A very important consequence of the last definition is, that planning and scheduling flaws can force a backtracking at any time, in contrast to approaches where a plan has to be fully developed before it can be checked by the scheduler, etc.

In order to keep the design as flexible as possible, it is necessary to provide additional inference capabilities to the system, which may be shared by the participating modules. For all inference tasks on the plan which are not subject to choice, we define inference rules in the following way:

Definition 8 (Inference Modules). *An inference module ρ is a function that computes plan modifications of type ρ which represent necessary changes on the plan to uncover implicit information:*

$$f_\rho^{inf} : \mathrm{P} \to 2^{\mathcal{M}_\rho}$$

These inferences are used in hybrid planning to add ordering constraints between causally linked primitive tasks (cf. modification classes InsertTask and AddCausalLink, which both do not add ordering constraints in order to maintain flexibility in the case of a later overlapping of abstract task expansions). In the presented integrated hybrid planning and scheduling system, an AC-3 based constraint engine keeps the TCSP arc-B-consistent (cf. [14]), includes the implicit constraints respectively narrows down intervals, and synchronizes temporal and ordering constraints. This is done by the two specific inference modules $f_{\mathtt{AddTempConstraint}}^{inf}$ and $f_{\mathtt{AddOrdConstraint}}^{inf}$.

The following generic algorithm implements a stepwise refinement of partial plans by applying plan modifications according to detected deficiencies, i.e., flaws. It is used as the core component of any integrated planning and scheduling system that is to be implemented within our architecture:

$\mathbf{plan}(P, \mathtt{T}, \mathtt{M})$:
$F \leftarrow \emptyset$
$\mathbf{for\ all}\ f_x^{det}\ \mathbf{do}$
$\quad F \leftarrow F \cup f_x^{det}(P)$
$\mathbf{if}\ F = \emptyset\ \mathbf{then}$
$\quad \mathbf{return}\ P$
$M \leftarrow \emptyset$
$\mathbf{for\ all}\ F_x = F \cap \mathcal{F}_x\ \text{with}\ F_x \neq \emptyset\ \mathbf{do}$
$\quad \text{answered} \leftarrow \mathbf{false}$
$\quad \mathbf{for\ all}\ f_y^{mod}\ \text{with}\ \mathcal{M}_y \subseteq \alpha(\mathcal{F}_x)\ \mathbf{do}$
$\quad\quad M' \leftarrow f_y^{mod}(P, F_x)$
$\quad\quad \mathbf{if}\ M' \neq \emptyset\ \mathbf{then}$
$\quad\quad\quad M \leftarrow M \cup M'$
$\quad\quad\quad \text{answered} \leftarrow \mathbf{true}$
$\quad \mathbf{if}\ \text{answered} = \mathbf{false}\ \mathbf{then}$
$\quad\quad \mathbf{return\ fail}$
$\mathbf{return}\ \text{plan}(\text{infer}(\text{apply}(P, f_z^{strat}(P, F, M))), \mathtt{T}, \mathtt{M})$

The procedure `infer` recursively calls all provided inference modules and applies their modifications on the plan, until no further inferences are issued.

Please note, that the algorithm is formulated independently from the deployed modules, since the options to address existing flaws by appropriate plan modifications is defined via α. The call of the strategy function z is of course the backtracking point of the system.

4   Search Strategies

The translation of existing search strategies for hybrid planning revealed that practically all of them are fixed in the sense, that they represent a preference schema on the flaw type they want to get eliminated primarily and then select appropriate modification methods. For example, it is very common to care for the plan to become primitive first and then to deal with causal interactions. A similar situation can be observed in integrated planning and scheduling systems, where the typical process is first to generate a plan and then to verify whether it can be scheduled or not. We propose the use of *flexible* strategies [11], which are capable of operating on a more general level by exploiting flaw/modification information: they are neither *flaw-dependent* as they do not primarily rely on a flaw type preference schema, nor *modification-dependent* as their do not have to be biased in favor of specific modification types. An example is the following strategy in the least commitment fashion that has proven to be very effective in the context of hybrid planning alone.

Definition 9 (Least Committing First). *Let* $f \in \mathcal{F}$ *be a flaw and* $m_1, \ldots, m_n \in \mathcal{M}$ *a set of modifications that has been issued for the current plan. The* committing level l_c *of such a flaw is defined as follows:*

$$l_c(f, \{m_1, \ldots, m_n\}) = \begin{cases} 0 & \text{for } n = 0 \\ 1 + l_c(f, \{m_1, \ldots, m_{n-1}\}) & \text{for } m_n \text{ answering } f \\ l_c(f, \{m_1, \ldots, m_{n-1}\}) & \text{otherwise} \end{cases}$$

The Least Committing First *strategy selects from the set of modifications those, which deal with flaws that have a minimal* l_c *value.*

$$f_{LCF}^{strat}(P, F, M) = m \in \{m_f | f \in \min(l_c(f, M))\}$$

It can easily be seen, that this is a *flexible* strategy, since it does not depend on the actual types of issued flaws and modifications: it just compares answer set sizes in order to keep the branching in the search space low.

More importantly, it is also opportunistic with respect to planning and scheduling, since it selects whatever modification has the lowest commitment level; planning and scheduling flaws and modifications will be addressed, respectively solved, in an interleaving manner. In this way, a planning process guides the scheduling and vice versa. And if one of the two formerly separate processes finds a reason to discard the current plan, the system performs backtracking immediately.

5   Conclusions and Future Developments

We have presented a novel unifying framework and architecture for integrated planning and scheduling systems. It relies on a formal account of hybrid planning and scheduling, which allows to decouple flaw detection, modification computation, and search control. Problem solving capabilities – in this case HTN, POCL, and scheduling – can easily be combined by orchestrating respective elementary modules via an appropriate strategy module. In particular it can be configured as a classical partial order planner, an HTN planner, a resource scheduler, or any hybrid of these methods. The implemented system can be employed as a platform to implement and evaluate various planning methods and strategies. It can be easily extended to additional functionality, e.g. probabilistic reasoning [15,16], without implying changes to the deployed modules – in particular flexible strategy modules – and without jeopardizing system consistency through interfering activity.

Future work includes experimental evaluation of search strategies, e.g., the flexible HotSpot technology [11] as well as providing modification modules with local optimization techniques.

References

1. Estlin, T.A., Chien, S.A., Wang, X.: An argument for a hybrid HTN/operator-based approach to planning. In Steel, S., Alami, R., eds.: Proceedings of the 4th European Conference on Planning. Volume 1348 of LNAI., Springer (1997) 182–194

2. Biundo, S., Schattenberg, B.: From abstract crisis to concrete relief – A preliminary report on combining state abstraction and HTN planning. In Cesta, A., Borrajo, D., eds.: Proceedings of the 6th European Conference on Planning. LNCS, Springer (2001) 157–168

3. Castillo, L., Fdez-Olivares, J., González, A.: On the adequacy of hierarchical planning characteristics for real-world problem solving. In: Proceedings of the 6th European Conference on Planning. (2001)

4. Drabble, B., Tate, A.: The use of optimistic and pessimistic resource profiles to inform search in an activity based planner. In Hammond, K., ed.: Proceedings of the 2nd International Conference on Artificial Intelligence Planning Systems, AAAI (1994) 243–248

5. Garrido, A., Salido, M.A., Barber, F.: Scheduling in a planning environment. In Sauer, J., Köhler, J., eds.: Proceedings of the 14th European Conference on Artificial Intelligence Workshop on New Results in Planning, Scheduling and Design. (2000) 36–43

6. Srivastava, B., Kambhampati, S.: Scaling up planning by teasing out resource scheduling. In Biundo, S., Fox, M., eds.: Proceedings of the 5th European Conference on Planning. Volume 1809 of LNCS., Springer (2000) 172–186

7. El-Kholy, A., Richards, B.: Temporal and resource reasoning in planning: The parcPLAN approach. In Wahlster, W., ed.: Proceedings of the 12th European Conference on Artificial Intelligence, John Wiley & Sons (1996) 614–618

8. Laborie, P., Ghallab, M.: Planning with sharable resource constraints. In Mellish, C.S., ed.: Proceedings of the 14th International Joint Conference on Artificial Intelligence, Morgan Kaufmann (1995) 1643–1651

9. Schattenberg, B., Biundo, S.: On the identification and use of hierarchical resources in planning and scheduling. In Ghallab, M., Hertzberg, J., Traverso, P., eds.: Proceedings of the 6th International Conference on Artificial Intelligence Planning Systems, AAAI (2002) 263–272

10. Garcia, F., Laborie, P.: Hierarchisation of the seach space in temporal planning. In Ghallab, M., Milani, A., eds.: New Directions in AI Planning, Proceedings of the 3rd European Workshop on AI Planning. Volume 31 of Frontiers in Artificial Intelligence., IOS Press (1996) 217–232

11. Schattenberg, B., Weigl, A., Biundo, S.: Hybrid planning using flexible strategies. In Furbach, U., ed.: Proceedings of the 28th German Conference on Artificial Intelligence. Volume 3698 of LNAI., Springer (2005) 258–272

12. Clement, B.J., Barrett, A.C., Rabideau, G.R., Durfee, E.H.: Using abstraction in planning and scheduling. In Cesta, A., Borrajo, D., eds.: Proceedings of the 6th European Conference on Planning. LNCS, Springer (2001) 145–156

13. Dechter, R., Meiri, I., Pearl, J.: Temporal constraint networks. Artificial Intelligence 49 (1991) 61–91

14. Lhomme, O.: Consistency techniques for numeric CSPs. In Bajcsy, R., ed.: Proceedings of the 13th International Joint Conference on Artificial Intelligence, Morgan Kaufmann (1993) 232–238

15. Biundo, S., Holzer, R., Schattenberg, B.: Dealing with continuous resources in AI planning. In: Proceedings of the 4th International Workshop on Planning and Scheduling for Space (IWPSS'04), European Space Agency Publications Division (2004) 213–218

16. Biundo, S., Holzer, R., Schattenberg, B.: Project planning under temporal uncertainty. In Castillo, L., Borrajo, D., Salido, M.A., Oddi, A., eds.: Planning, Scheduling, and Constraint Satisfaction: From Theory to Practice. Volume 117 of Frontiers in Artificial Intelligence and Applications. IOS Press (2005) 189–198

A Hybrid Time Management Approach to Agent-Based Simulation

Dirk Pawlaszczyk[1] and Ingo J. Timm[2]

[1] Technische Universität Ilmenau, Institute for Information Systems,
Postfach 10 05 65, 98684 Ilmenau, Germany
Dirk.Pawlaszczyk@tu-ilmenau.de
[2] Center of Computing Technologies (TZI), University of Bremen
Postfach 33 04 40, 28334 Bremen, Germany
i.timm@tzi.de

Abstract. In this paper we describe a time management approach to distributed agent-based simulation. We propose a new time management policy by joining optimistic synchronization techniques and domain-specific knowledge based on agent communication protocols. With respect to our experimental results, we assume that our approach helps to prevent too optimistic event execution. Consequently, the probability of time consuming rollbacks is reduced in comparison to a pure time warp based solutions. The approach has been implemented as a synchronization service for the JADE agent platform *SimJade*. The paper concludes by the discussion of our experimental results and future improvements.

1 Introduction

Research on systems of autonomous agents, called multiagent systems (MAS), has received much interest in the domain of (distributed) artificial intelligence, in recent years. MAS are most suitable for the development of distributed applications, with uncertain and a dynamically changing environment. For validation of those systems agent based simulation seems to be well-suited [1].

Simulation is the imitation of a system's behaviour and structure in an experimental model to reach findings, which are transferable to reality. In multiagent-based simulation (MABS) real world systems are modelled using multiple agents. The system emerges by interaction of the individual agents as well as their collective behaviour. Agents send messages with respect to some communication protocol and are disposed at some discrete point in time (see Definition 1). In this context, a software agent is defined as, a program that acts autonomously, communicates with other agents, is goal-oriented (pro-active) and uses explicit knowledge [2]. Beyond this, MABS is influenced and grounded on existing simulation techniques such as object oriented simulation and distributed simulation [3]. Agent-based simulation has reached a growing attention from both science and industry in recent years. Agent-based modeling and therefore agent-based simulation seems to be the right tool for domains characterized by discrete decisions and distributed local decision makers [4],[5]. MABS is an appropriate method if we wish to understand the evolution of a distributed system, minted with non-linear dynamics. For instance, a rising number of contributions are dealing with agent-based simulation models in the context of supply chain management since this domain provides distributed entities with autonomous

C. Freksa, M. Kohlhase, and K. Schill (Eds.): KI 2006, LNAI 4314, pp. 374–388, 2007.
© Springer-Verlag Berlin Heidelberg 2007

behaviour [6]. Furthermore agents based simulation can be applied to evaluate decentralized decision policies for example within agile or holonic manufacturing systems [4]. As a third point agent-based simulation can help to analyze the behavior of complex self-organizing and emergent systems [8]. Last but not least MABS can be used to validate multiagent systems within the simulation model, before they are deployed to real environments, as part of the software development process [1],[9].

Definition 1. (Simulation Model):
A Simulation model S is defined as a tuple $<A^*, M^*, P^*, T^*>$ where
  $A^* = \{a_0, a_1, \dots, a_n\}$ *is a set of agents,*
  $M^* = \{m_0, m_1, \dots, m_n\}$ *is a set of messages,*
  $P^* = \{p_0, p_1, \dots, p_n\}$ *is a set of agent communication protocols, and*
  $T^* = \{t_0, t_1, \dots, t_n\}$ *is a set of time stamps within simulation time ($n \in N$).*

Let $M(p) \subseteq M^*$ define a set of messages that belongs to some protocol p ($p \in P^*$).

Let $T(a) \subseteq T^*$ denote a set of time stamps that some agent a ($a \in A^*$) processes during a simulation run.

Obviously, MABS has strong requirements with respect to its inherent computational complexity. Nevertheless, in many domains, when distributed systems or high complex world models are in question, MABS seems to be the most adequate form of distributing the simulation model resp. to run agents on multiple computational nodes in a parallel manner. The distribution of the simulation model, however, leads to an additional challenge: Correctness of experimental results generated by a distributed simulation run mainly depends on the accuracy of the underlying synchronization mechanism. Event-based simulation is the 'gold standard' for simulation. In distributed simulation, the distributed simulation processes could compute progress in the simulation in an asynchronous way, i.e., simulation events in the distributed computational node are happening at different time points with respect to the time in simulation as well as the current time in the real world. Events, which were scheduled to happen after each other, could happen in a varying order if the computational nodes would not be synchronized. However, complete synchronization would create an almost sequential simulation behaviour, which is not desirable and would prevent the system to scale with respect to speed-up. In Consequence, synchronization is the key challenge in distributed simulation [10],[11],[12]. In this paper, we propose a hybrid event driven time management approach based on local knowledge to provide efficient distributed simulation. Therefore, we first give a short introduction on time management protocols and why they have relevance in distributed simulation. In section 3 we propose a new synchronization approach to MABS using constrained optimistic behaviour. Furthermore, we introduce *SimJade* as a prototypical implementation of the resulting synchronization service with our approach to time management policy. Finally, we present experiential results, discuss related work, and conclude with a discussion and a brief outline of future work.

2   Synchronization – Technical Background

The main component of simulation is the simulation model. Commonly, it is represented by a set of variables and specific behaviour for their value changes, i.e., induced by some input, each variable is modified with respect to the input and its value over time. Another key component of simulation is the time model. There are different notions of time within simulation [10]. The *physical* time denotes the time of the physical system, whereas the *simulation* time is used to represent the physical time within the simulation model. Beyond this, we have to discern the *wallclock* time, which refers to the processing or lead-time of the simulation program.

Depending on their timely fashion, simulation approaches are distinguished as *continuous* and *discrete* simulation techniques. Within continuous simulation, changes to the system state occur continuously in time, whereas in discrete models states are modified at discrete time points only. Continuous simulation is normally performed by using a system of differential equations.

Discrete simulation can further be slit into *time-stepped* models, where simulation advances in equidistant time steps, and *event-driven* models, where the system state is changed only if an event is fired [10]. In a time-stepped model, simulation time can only advance systematically. In fact, not every state variable is really changed in each time step, but must be synchronized anyway. Actions occurring at the same time step are normally considered to be simultaneous and therefore independent of each other. For many problem domains, such an assumption seems to be rather insufficient. By choosing a high granularity of the time steps such situation may be avoided. However, this again must be paid with a high synchronization overhead. Nevertheless, many agent-based simulation test beds rely on this synchronization approach, since it is easy to implement [5],[9],[13].

Within event-driven simulation, a time stamp is assigned to each event, to indicate the point in simulation time, when the event occurs. Discrete event simulation can easily be realized on a single machine using one global event list that manages the events of all *Logical Processes* (LP), i.e. the agent process. All agents in the simulation model are synchronized by a centralized data structure. Hence, it is ensured that events are processed in the correct order of time stamps (*sequential* simulation approach). But this solution is clearly contradictory to the underlying idea of MABS, i.e., the inherent parallelism is hardly used. The dominant approach in parallel event-driven simulation (PEDS) on the other hand is based on the approach, that each LP maintains its own local simulation time (*Local Virtual Time* (LVT)).

Definition 2. (local virtual time): *LVT(a, t) denotes the current local virtual time, i.e., how far an agent has progressed in simulation time, with $a \in A^*$, $t \in T(a)$.*

During simulation, the progress of the agents varies from agent to agent. Additionally, it is assumed, that there is no global event list but each LP contains an individual event list, locally. As soon as simulation is processed on different processors, there is the necessity of synchronizing event execution to preserve the correct event order with respect to their time stamps. Since every agent manages its event list independently, the correct order of events with respect to their time stamps is not ensured. This is commonly known as the *causality problem* within distributed simulation [10]. *Figure1* gives an example for violation of causality. Corresponding to this figure, we can define a straggler message (see definition 3a).

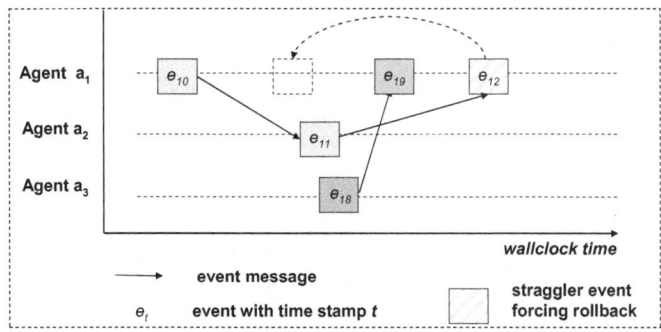

Fig. 1. Scenario in distributed simulation causing violation of event ordering (causality error). When event e_{12} arrives, e_{19} has already been processed by agent a_1 *(straggler event)*.

Definition 3a. (straggler message): *If a message m is received by agent a with time stamp t_2 which is less than current local time of the agent LVT(a,t_1), this "late" message is referred to as straggler message:*

$$\forall a, t_1, t_2, m: (LVT(a, t_1) \wedge received(a, m, t_2) \wedge (t_2 < t_1) \rightarrow isStraggler(m))$$
$$a \in A^*, m \in M^*, t_1, t_2 \in T(a)$$

The ternary relation received(a, m, t) denotes that agent a has received message m on time stamp t.

see definition (1),(2)

There are two main approaches in place to ensure correct time stamp order: *conservative* and *optimistic* synchronization. Conservative synchronization algorithms introduce constraints on events and therefore avoid violation and ensure local causality Conservative algorithms prevent the event ordering from being violated. Accordingly, a situation as illustrated in Figure 1 should never occur when using a conservative synchronization scheme. Therefore conservative protocols cannot fully take advantage of parallelism within the application resp. the parallel infrastructure of multiple computational nodes, as in the conservative approach, the guidance is addressed to the worst-case scenario only, i.e. the incoming of a straggler message, which may rarely actually occur in practice. *Optimistic* algorithms like Jefferson's *Time Warp* mechanism on the other hand explicitly allow causality errors, and provide suitable techniques to recover from an incorrect system state. In the example introduced above, agent a_1 has to go back in time and rollback its execution state from *LVT 19* to *LVT 12*. Occasionally, an incorrect message that has already been processed by an agent can result in the generation of additional incorrect messages that have been processed by other agents, which results in still other incorrect messages. Undoing all effects of incorrect computations, the agent has to unsent all previously sent (possibly) incorrect messages using so called *anti-messages*[10]. Each agent that receives such an anti-message has to rollback its state as well. Therefore a rollback within optimistic simulation can be defined as:

Definition 3b. (rollback): *Let m be an anti-message resp. a straggler message re-ceived in t_2, the rollback is defined as the behavior initiated by the receiving agent a. Doing so, the agent reactivates its state to a former point in time t_3 (new LVT) that is less then the current local virtual time t_1, and less or equal t_2 (cf. definition (1),(2),(3a)):*

$$\forall a, t_1, t_2, t_3, m: (LVT(a, t_1) \wedge (isAntiMessage(m) \vee isStraggler(m)) \wedge$$
$$received(a, m, t_2) \rightarrow rollback(a, t_3) \wedge newLVT(a, t_3))$$

$$a \in A^*, m \in M^*, t_1, t_2, t_3 \in T(a), with\ t_3 \leq t_2 < t_1$$

Conservative as well as optimistic time management protocols have gained remark-able speedups in the recent years. Various approaches have been proposed to control the optimism of the simulation model. Conservative synchronization scales well, if the lookahead, i.e. the ability of an LP to make predictions about its own future, is high. But, in dynamic environments with frequently changing relationships between model entities optimistic synchronization insures a better scalability, since it does not rely on the lookahead of a certain simulation model. Additionally, the model devel-oper has not to be familiar with details of synchronization, as within conservative simulation. Since we are looking for a suitable time management approach to agent-based simulation that provides enough performance for a wide range of models opti-mistic synchronization seems to be a good choice.

3   The Hybrid Approach

In conservative synchronization, algorithms prevent the event ordering from being violated. Within pure Time Warp implementations, no constraints exist on the dis-tance in time an agent process resp. the advance ahead of others into future. Conse-quently, the probability of incorrect computations increases. As shown above, each straggler message causes one or even more rollbacks. If the time to perform a rollback is high, i.e. many states have to be rolled back, the performance of the simulation decreases significantly. This commonly known performance hazard within optimistic synchronization is caused by too optimistic event execution [10]. A good time man-agement protocol should avoid such situations. At the same time, it has to be ensured, that parallelism is not fully lost within the simulation model, as it is caused by a too restrictive time management policy for example. The question that arises is how to prevent these potential performance hazards? Communication is one of the key fea-tures in agent technology. Messages are sent out from a sender to a receiver resp. a list of receivers. Messages are encoded in an *Agent Communication Language (ACL)*, an external language that defines intended meaning of a message by using performa-tives. A series of messages produces a dialog. A dialog normally follows a predefined structure – the *Interaction Protocol* (IP). Commonly, communication of agents is based on such protocols. The FIPA specification, as an internationally agreed agent standard, defines an agent communication language as well as a set of interaction protocols, most commonly used [14]. The FIPA *Request Interaction Protocol* for example allows an agent to request another agent to perform some action. The par-ticipant needs to decide whether to accept or refuse the request. In any case, the mes-sage receiver has to respond with a reply to a request message. Even if the receiver does not have any clue how to deal with a message, the specification prescribes to send a least a *not-understood* message. With this in mind we define policies for time management (see def. 4.) Furthermore we have to consider some special cases, where the agent exceptionally is allowed to do something, even if it is waiting for a reply-message (see def.5a,5b).

Definition 4. (wait for rule): *Given agent a_1 which has sent a message m_1 to agent a_2 and assuming that there is at least one valid reply m_2 for m_1. The rule "wait for" is defined as follows: If the expected reply message was not received yet, the agent should wait for this particular message, before going on with the next message:*

$$\forall m_1, m_2, a_1, a_2, p: (sent(a_1, m_1, a_2) \wedge validReply(m_2, m_1) \notin \emptyset \wedge$$
$$\neg received(a_1, m_2, t) \rightarrow wait\text{-}for(a_1, m_2))$$
$$a_1, a_2 \in A^*, a_1 \quad a_2, m_1, m_2 \in M(p), m_1 \quad m_2, p \in P^*, t \in T(a)$$

Whereby the ternary relation $sent(a_1, m_1, a_2)$ denotes that agent a_1 has sent message m_1 to the receiver a_2.

see definition (1)

Definition 5a. (execution condition): *Given some message m_2 which has been received while agent a is waiting for message m_1 of the same sender, and message m_2 matches to m_1 (message performative and conversation ID of m_2 are the same as in m_1) the execution condition is defined as the following behavior of the agent: agent a does not need to rollback, process the message and remove wait-for condition:*

$$\forall a, m_1, m_2, p, t: (wait\text{-}for(a, m_1) \wedge received(a, m_2, t) \wedge matched(m_1, m_2) \rightarrow$$
$$\neg rollback(a) \wedge process(m_2) \wedge \neg wait\text{-}for(a, m_1))$$

$$a \in A^*, m_1, m_2 \in M(p), p \in P^*, t \in T^*$$
see definition (1),(3b),(4)

Definition 5b. (delayed execution condition): *A delayed execution condition is defined by some message m_2 which is received while agent a is waiting for message m_1 from a sender different to the sender of m_2, whereby m_2 is not part of the current interaction protocol p. In this case the message is buffered.*

$$\forall a, m_1, m_2, p, t: (wait\text{-}for(a, m_1) \wedge received(a, m_2, t) \wedge sender(m_1) \quad sender(m_2) \wedge$$
$$m_1 \in M(p) \wedge m_2 \notin M(p) \rightarrow bufferMessage(a, m_2))$$

$$a \in A^*, m_1, m_2 \in M^*, p \in P^*, t \in T^*$$
see definition (1),(3b),(4)

We have to avoid *deadlock* situations, where agent a_1 waits for agent a_2, and to the same time agent a_2 waits for agent a_1, since both independently have sent a message to each other. Both agents are blocked; each is waiting for a message event which will never occur. Hence, the agent must also process a message from its opponent even if it is not belonging to the protocol it currently processes:

Definition 6. (deadlock avoidance condition): *If some message m_2 is received while agent a is waiting for message m_1 of the same sender, but m_2 does not belong to the current interaction protocol ($m_2 \notin M(p)$), and agent a doesn't need to rollback, then the deadlock avoidance condition is defined as the behavior of processing the new m_2 despite of any wait-for condition:*

$$\forall a, m_1, m_2, p, t: (wait\text{-}for(a, m_1) \wedge received(a, m_2, t) \wedge$$
$$(sender(m_1) = sender(m_2)) \wedge m_1 \in M(p) \wedge m_2 \notin M(p) \wedge \neg rollback(a)$$
$$\rightarrow process(a, m_2))$$

$$a \in A^*, m_1, m_2 \in M^*, p \in P^*, t \in T(a)$$
See definition (1),(3b),(4)

Definition 7. (consideration of cyclic dependencies): *If some message m_2 is received while agent a is waiting for message m_1 with a sender different to the sender of m_2, and the new message m_2 does belong to the current interaction protocol ($m_2 \in M(p)$), and agent a doesn't need to rollback, then the condition consideration of cyclic dependencies is defined as the behavior of processing m_2 in despite of any wait-for condition, since the protocol could not proceed otherwise:*

$$\forall a, m_1, m_2, p, t: (wait\text{-}for(a, m_1) \wedge received(a, m_2, t) \wedge sender(m_1) \neq sender(m_2) \wedge$$
$$(m_1, m_2 \in M(p)) \wedge \neg rollback(a) \rightarrow process(a, m_2))$$

$$a \in A^*, m_1, m_2 \in M^*, p \in P^*, t \in T(a)$$
see definition (1),(3b),(4)

Finally, cyclic dependencies have to be handled adequately. Although situations as described in the following are not very probable, there may occur situations when for example in multi-staged-protocols an agent receives a request within the same conversation, from a new communication partner different from its original opponent. In such a case, this message of course must be processed before going to wait state again (see def. 7).

Accordingly, on each message send agent a_i executes:

```
send(message msg) to a_n{...
    if (msg.requiresReply()∉∅) then{
        waitForReply:=true;          //[def.4]
        MsgTemplate = msg.createReplyTemplate();
    }...
}
```

When a message is received by agent a_i it exectues:

```
receive(message msg){...
        if (LVT(a_i)>msg.Time)∨ msg.isANTIEVENT = true)
            then rollback();          //[def.3]
        else
```

```
if(waitForReply = true) then
    if (MessageTemplate.match(msg)=true) then
    {
        waitForReply:=false;
        process(msg);                //[def.5a]
    }
    else
    if(msg.Sender = MsgTemplate.Sender)
        then process(msg);       //[def.6]
    else
    if(msg.ProtocolID = MsgTemplate.ProtocolID)
        then process(msg);       //[def.7]
    else
        bufferMessage(msg);      //[def.5b]
}
else process(msg); //no wait condition was set
}
```

Now remembering the example from *section 2*, where a straggler message caused the agent to rollback and recover to an earlier point in time. With the new policy in place such situations could easily be avoided. Now the agent waits until it gets a reply message, and not simply processes with the next message (see *figure 2*). As shown above, delayed execution helps to preserve event order and therefore avoids wrong computation. Of course, such a policy cannot fully prevent rollback situations. In fact, there are conceivable cases, where this policy may fail. But it seems to be at least fairly better than a pure TW solution, without fully loss of parallelism, and is considered within the evaluation of this approach. Furthermore the implementation effort of this solution is considerable low. The agent has to be provided with information about the structure of the used protocols at initialization stage only. Depending on the protocol length, the policy is applied more frequently. Particular long interaction

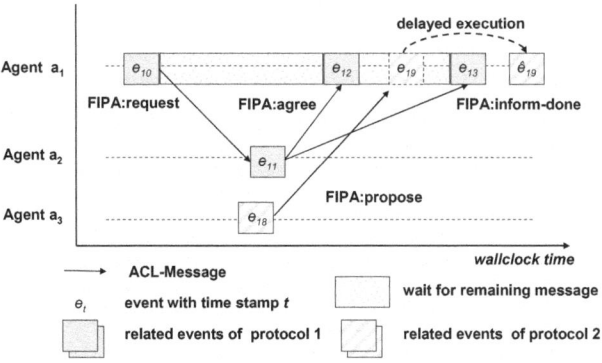

Fig. 2. Delayed event execution based on protocol information. Agent a_1 receives a proposal from Agent a_3, while he is waiting for an *inform-done* message of Agent a_2. Instead of immediately processing the incoming message, the execution is delayed. Thus, event order is preserved and still valid.

Table 1. Communication complexity for some standard FIPA Agent Interaction Protocols (m – number of participants) [14]

Protocol name	Min. message number	Max. message number
Propose	2	2
Request	2	3
Query	2	3
Contract Net	2m	5m

protocols, like the *fipa-contract-net* are most eligible (see *Table 1*). Since this approach is joining optimistic techniques with constrained optimism, we actually pursue a *hybrid* time management approach.

4 The *SimJade* Synchronization Service

A prototype implementation of the proposed synchronization service was realized by using the *Java Agent Development Environment* (JADE) [15]. This framework offers an appropriate middleware to simplify the implementation of multiagent systems. Beyond this, it is widely used in academia. As one part of this agent toolkit, there are ready-to-use behaviour objects for standard interaction protocols such as *fipa-request* and *fipa-query-ref*. By supporting generic interaction protocols, application developers just need to implement domain specific actions, while the framework will carry out all application independent protocol logic. Since JADE is FIPA compliant, a high degree of interoperability is guaranteed. To test our approach, we have implemented an extension named *SimJade* to support a local synchronization scheme. This service implements optimistic Time Warp based synchronization algorithm, first introduced by *Jefferson* and discussed in [10]. Each agent is equipped with a local control mechanism for event scheduling. Furthermore, a dedicated synchronization service, which is integrated into the agent platform, is provided. This service offers the functionality of computing state copies of an agent as well as recovering a former agent state. A specialized simulation manager agent implements global control mechanisms, like memory reclaiming, starting and stopping simulation of experiments, as well as detecting termination of simulation runs. For distributed computation of the global virtual time, a procedure based on a snapshot algorithm first proposed by *Mattern* has been implemented [16]. Unused memory is reclaimed by using fossil collection [10]. Most of the described functionality is transparent to the agent developer. This is realized by encapsulating all time management functionality within a single agent superclass. Using the new service only requires that the domain agents are inherited from this new agent class instead of the default agent class.

Since *SimJade* synchronization service is based on a widespread agent toolkit it enables the testing of multiagent systems developed within JADE before they are deployed in the real world. Moreover using an optimistic synchronization scheme relieves the agent developer from most technical issues associated with time management. Thus, the programmers can concentrate their efforts on implementing the domain specific application logic.

5 Evaluation

To evaluate the proposed time management policy, we performed a number of experiments using the *SimJade* service together with a self-defined agent model. Our test environment consists of a mini-cluster with five P4 2.8 GHz workstations (256 MB RAM, *Suse Linux 8*) which are connected by a 100 Mbit switched Ethernet.

The test model comprises of 30 agents. In a single run those agents together exchange over 2000 event messages. To emulate a realistic workload, each agent in our evaluation scenario implements standard reasoning capabilities using *JESS*-engine and the respective behaviour for integration in JADE [17]. The *JESS* contribution can easily be combined with the generic interaction protocols provided by JADE [15]. If an agent process receives a message, the JESS inference engine, to create a suitable reply-message, first interprets this message. A combination of different standard FIPA protocols (see *Table 1*) with mixed communication lengths where used within each run. We tracked the wall clock time required to finish as simulation run, as well as the total number of rollbacks to measure the efficiency of our approach. *Figure 3* shows the obtained lead times. For orientation, the dotted line shows the cycle time using sequential synchronization scheme. As can be seen, our time management policy clearly outperforms the pure time warp implementation in this scenario. Even more, a remarkable speedup could be gained, compared to sequential synchronization approach. The last fact is not self-evident, since results are crucially determined by the parallelism of the application and even more by the synchronization overhead caused by distribution. The main reason for these good performance results probably grounds on the small number of rollbacks compared to none-constrained optimistic synchronization policy (*Figure 4*). Within the pure Time Warp solution agents tend to be too far ahead of each other in simulation time, and therefore more frequently causing rollbacks. By deploying message delays we could minimize the probability of time consuming rollbacks, at least for the test scenario. To summarize, there is a clear connection of rollback frequency and obtained speed up, whereby our time management policy seems to have an advantage over the pure time warp based solution.

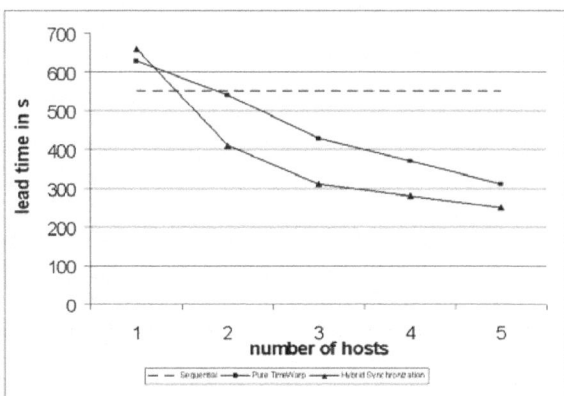

Fig. 3. The figure shows the arising average execution times with and without an advanced time management policy in place. The dotted line indicates the lead time reached by using a sequential (none distributed) synchronization policy.

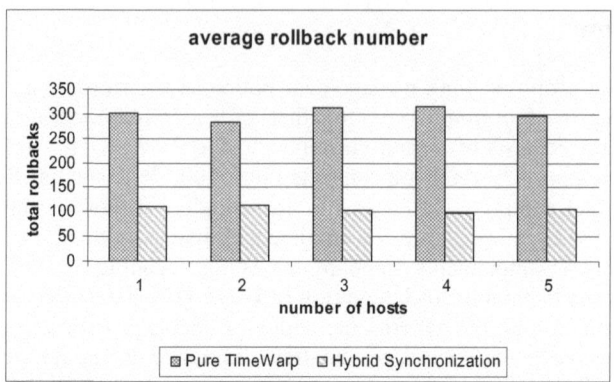

Fig. 4. The figure depicts the total number of rollbacks within a single simulation run comparing pure time warp solution with our advanced synchronization approach

6   Related Work

Many different agent-based simulation toolkits have been proposed in recent years. So far, the majority neither offer a support to distributed simulation nor provide any advanced time management schemes.

One of the first general purpose frameworks for agent-based simulation is the GENSIM system by Anderson and Evans [18], where agents are given perceptions at fixed intervals, and with a fixed amount of time to react to each perception. Within the distributed version, DGENSIM a time-stepped synchronization mode is used [19]. A support for swarm-like simulation is realized within SWAGES simulation environment [20]. This framework gives support to parallel execution of agent-based models and dynamic load balancing. Therefore, simulated agents are referred to a spatial model. Agents are allowed to act asynchronously within their event horizon, but have to consider whenever a non-local entity could potentially influence them, and potentially affects the sphere of influence of an agent.

The SYNCER framework is another remarkable contribution, that enables running a distributed simulation with the well known SWARM agent toolkit based on a time stepped time management approach, whereas remote communication is realized via proxy objects located on each computational node [21]. The simulators used in the RoboCup Simulation Leagues like rescue and 2D soccer simulator use a fixed step discrete time model [22]. In [5], a sequential synchronization service for linking different agent simulation test beds is presented. Popov et al [223] again describe a parallel sequential simulation approach to simulation of 10^6 agents to capture the behaviour of web users. This vast number of simulated agents is reached by keeping the agent implementation unchanged. Furthermore they are using a relative weak notion of agency, and do not consider deliberative agent structures.

A conservative synchronization approach is discussed as part of the MPADES framework, a middleware for building (distributed) simulation environments [24]. Furthermore there are some contributions in place concerning the distributed simulation through federation of agent-based simulation environments using the

High Level Architecture (HLA), a generic, language-independent specification, that allows the integration of different sequential simulators, and was originally initiated by the *Department of Defence* and meanwhile committed as an ISO standard [25],[26]. With HLA simulators, referred to as federates, can be integrated into a global simulation context known as federation. A strict hierarchical tree-oriented model is used to structure respective federations. Communication between certain simulators is enabled by predefined gateways. Thus HLA can be considered as a centralized coordination approach to distributed simulation resources. HLA_AGENT for example introduces support for distributed simulation to the SIMAGENT toolkit [26]. HLA_REPAST is another distributed simulation environment that uses HLA to parallelize simulation of the artificial life-toolkit REPAST [25]. HLA clearly focuses on the interoperability between different sequential simulation toolkits and is not designed to gain speedups.

To enable the simulation3 of large-scale agent system the MACE3J system by Gasser and Kakugawa provides services, for registration, scheduling and messaging of so-called *ActivationGroups* [27]. These groups again consist of *Active Objects*, i.e. the agents within the system. Scaling up is on main design criteria of the system. The time management relies on a conservative synchronization regime. MACE3J has been run with up to 5000 agents on a shared memory system too prove scalability; admittedly, agents have not changed any messages in this test. In [28] scalable multi agent simulation using the grid approach is discussed. This contribution clearly aims on providing an infrastructure for distributed simulation, without treating synchronization issues in detail.

The first simulation framework that gives support to optimistic time management is the well-known JAMES system [29]. Actually, there is no report about using any adaptive optimism schemes. Furthermore, JAMES does not support a general simulation model, since it comes with predefined agent architecture. Another contribution from Logan et al proposes a metric to compute the degree of optimism based on the shared state of the agent system. This metric is used to define a *moving time window* for constraining optimism. To test their approach they are using an external library together with the SIM_AGENT toolkit [11]. This approach is comparable to our simulation service, since it relies on the idea of constraining the optimism of the simulation model.

To enable testing to a wide range of real world agent applications Helleboogh et al propose a semantic duration model to capture timing requirements that reflect the semantic of agent activities in an explicit way [30]. This time management approach is primary meant to ensure causality within a simulation run. Beyond this, it is not supporting efficient simulation or distributed simulation by default. At last several time management policies have been introduced within the context of conventional parallel discrete event simulation [10]. However, these approaches of course do not consider the deviations of agent-based simulation approach effectively, but could be easily combined with our policy.

Although MABS has received a lot of attention in recent years there are only a few contributions that deal with the problem of efficient time management. Moreover, only a small set of agent-based simulation toolkits does support distribution of simulation over multiple hosts. In fact, most currently existing simulation environments support a simple time-stepped model, which is inappropriate to simulate multitude real world system in reliable manner (see section 3). Namely, the JAMES simulation toolkit [29] and contributions by Lees and Logan [31] explicitly make use of an optimistic synchronization scheme within the context of multi-agent-based simulation. Beside this, nearly all existing general-purpose simulation frameworks are lack to be compatible with the FIPA standard. Mostly, they refer to a particular type of agent model respectively application field, like artificial life or social science. Since they oblige to some particular agent architecture, there is a low degree of freedom to the application developer left.

7 Conclusion

Simulation is one of the key features for testing and evaluating distributed systems [1]. However, the simulation of multiagent systems or the simulation using agent-based models is still under research. It is commonly assumed, that the inherent distribution of multiagent systems could also be used for scale-up resp. speed-up simulation. However, in practice this does not has to be true. The objective of our approach outlined in this paper was to introduce a new time management approach to agent-based simulation. The approach integrates time warp synchronization with constraints. Therefore a hybrid framework for synchronization in simulation has been introduced consisting of policies to avoid excessive rollbacks as well as too high deviations in simulation time of the various agent processes. As we've pointed out, using interaction protocols within agent-based simulation can offer an appropriate way to constrain optimism of the underlying simulation model. First experimental results show a significant benefit of this hybrid approach compared to pure time warp based solution.

The evaluation of the proposed algorithm is still going on; concurrently with the integration of new simulation models from manufacturing and logistics domain. Additional performance improvements, as well as the introduction of advanced load balancing schemes are planed, to support scalable simulation for a wide range of agent based models.

References

1. Timm, I.J.; Scholz, T.; Fürstenau, H.: From Testing to Theorem Proving. Chapter IV.8 in Kirn, S. et al. (Hrsg.): Multiagent Engineering - Theory and Application in Enterprises. Springer-Verlag (Handbuch): Berlin, (2006), pp. 531-554.
2. Weiss, G.: Multiagent Systems – A Modern Approach to Distributed Artificial Intelligence. The MIT Press: Cambridge, Massachusetts, (1999)
3. Davidsson, P.: Multiagent Based Simulation: Beyond Social Simulation. In: Moss, S., Davidsson, P. (eds.): 3. Workshop on Multiagent Based Simulation (MABS) 2000, LNAI Vol.1997. Springer-Verlag Berlin Heidelberg New York, (2000) 97-107
4. Gentile, M., Paolucci, M., Sacile, R.: Agent-Based Simulation. In: Paolucci, M., Sacile, R.: Agent-Based Manufacturing and Control Systems: new agile manufacturing solution for achieving peak performance. CRC Press LCC, (2005)
5. Braubach, L., Pokahr, A. et al: A Generic Simulation Service for Distributed Multi-Agent Systems. In: Trappl, R. (eds.): Cybernetics and Systems 2004 (Vol. 2), Vienna, Austria, (2004) 576-581

6. Dangelmaier, W., Franke, H., et al: Agent-based Simulation of Transportation Nets. In: Coelho, H., Espinasse, B.(eds.): 5 th Workshop on Agent-based Simulation. Lisboa, Portugal, (2004) 174-179
7. Kádár, B., Pfeifer, A., Monostori, L.: Building Agent-Based Systems in a Discrete-Event Simulation Environment. In: Pechoucek, M., Petta, P., Varga L.Z. (eds.): CEEMAS 2005, LNAI Vol. 3690, Springer-Verlag Berlin Heidelberg New York, (2005) 595-599
8. Serugendo, G.D.: Engineering Emergent Behaviour: A Vision. In: D. Hales et al (eds.): Multiagent Based Simulation (MABS) 2003, LNAI 2927, (2003) 1-7
9. Klügl, F.; Herrler, R.; Oechslein, C.: From Simulated to Real Environments: How to use SeSAm for software development In: M. Schillo et al. (eds) Multiagent System Technologies - 1st German Conferences MATES, (LNAI 2831), (2003) 13-24
10. Fujimoto, R.M.: Parallel and Distributed Simulation Systems. John Wiley & Sons Inc. (2000)
11. Lees, M., Logan, B. et al: Time Windows in Multi-Agent Distributed Simulation. (2004)
12. Wang, F., Turner, S. J., Wang L.: Agent Communication in Distributed Simulations. In: Davidsson, P. et al (eds.): Workshop on Multiagent Based Simulation (MABS) 2004, LNAI Vol.3415. Springer-Verlag Berlin Heidelberg New York, (2005) 11-24
13. Timm, I.J.: Dynamisches Konfliktmanagement als Verhaltenssteuerung Intelligenter Agenten. DISKI 283 – Dissertationen Künstliche Intelligenz, infix-AKA Verlagsgruppe, (2004)
14. Foundation For Intelligent Physical Agents (FIPA): Interaction Protocol Specification Document no. SC00026H-SC00036H, http://www.fipa.org/specs/, (2002)
15. JADE Framework, http://sharon.cselt.it/projects/jade/
16. Mattern, F.: Efficient algorithms for distributed snapshots and global virtual time approximation. In: Journal of Parallel and Distributed Computing 18(4), (1993) 423-434
17. Friedman-Hill, E. J.: Jess, The Rule Engine for the Java Platform. Distributed Computing Systems, Sandia National Laboratories, Livermore, CA21 http://herzberg.ca.sandia.gov/jess
18. Anderson, J., Evans, M.: A Generic Simulation System for Intelligent Agent Designs. In: Applied Artificial Intelligence 9:5, (1995) 527-562
19. Anderson, J.: A Generic Distributed Simulation System For Intelligent Agent Design And Evaluation. http://www.citeseer.ist.psu.edu/399301.html, (2000)
20. Scheutz, M., Schermerhorn, P.: Adaptive Algorithms for the Dynamic Distribution and Parallel Execution of Agent-Based Models, (2005) http://www.nd.edu/%7Eairolab/publications/scheutzschermerhorn06pardist.pdf
21. Goic, J., Sauter, J. A., Toth-Fejel, T.: Syncer: Distributed simulations using swarm. In: SwarmFest 2001, Santa Fe, NM, (2001)
22. Homepage ot the RoboCup-Rescue Simulation Project: http://www.rescuesystem.org/robocuprescue/simulation.html
23. Popov, K. et al: Parallel Agent-Based Simulation on a Cluster of Workstation. In: Kosch, H., Böszörményi, H. Hellwagner, H. (eds.): Euro-Par 2003, Springer-Verlag Berlin Heidelberg New York, (2003) 470-480
24. Riley, P.: MPADES: Middleware for Parallel Agent Discrete Event Simulation. In: Kaminka, G.A., Lima, P.U. and Rojas, R. (eds.): RoboCup 2002, LNAI Vol. 2752, Springer-Verlag Berlin Heidelberg New York, (2003) 162-178
25. Minson, R., Theordoropoulos, G.: Distributing repast agent based simulations with HLA. In: Proceedings of the 2004 European Simulation Interoperability Workshop, Edinburgh, UK, (2004)

26. Lees, M., Logan, B.: Simulating Agent-Based Systems with HLA: The Case of SIM_AGENT - Part II (03E-SIW-076), (2003)
27. Gasser, L., Kakugawa, K. et al: Smooth Scaling Ahead: Progressive MAS Simulation from Single PCs to Grids. In: Proceedings of the Joint Workshop on Multi-Agent & Multi-Agent-Based Simulation, Autonomous Agents & Multiagent Systems (AAMAS) New York, USA, (2004) 1-10
28. Ingo J. Timm, I.J., Pawlaszczyk, D.: Large scale multiagent simulation on the grid. In: IEEE International Symposium on Cluster Computing and the Grid <5, 2005, Cardiff>, NJ: IEEE Operations Center, (2005) 334-341
29. Uhrmacher, A.M., Schattenberg, B.: Agents in Discrete Event Simulation. In: Bargiela, A., Kerckhoffs, E. (eds.): Proceedings of the 10TH ESS'98, SCS Publications Ghent, (1998) 129-136
30. Helleboogh, Holvoet, T. Weyns, D.: Extending Time Mangement Support for Multiagent Systems. In: Davidsson et al. (eds): Multi Agent Based Simulation (MABS) 2004, LNAI 3415, (2005) 37-48
31. Lees, M., Logan, B., Theodoropoulos, G.: Adaptive optimistic synchronisation for multi-agent distributed simulation. In: Proceedings of the 5th EUROSIM Congress on Modelling and Simulation (EuroSim'04), (2004)

Adaptive Multi-agent Programming in GTGolog

Alberto Finzi[1,2] and Thomas Lukasiewicz[2,1]

[1] Institut für Informationssysteme, Technische Universität Wien
Favoritenstraße 9-11, A-1040 Vienna, Austria
[2] Dipartimento di Informatica e Sistemistica, Università di Roma "La Sapienza"
Via Salaria 113, I-00198 Rome, Italy
{finzi,lukasiewicz}@dis.uniroma1.it

Abstract. We present a novel approach to adaptive multi-agent programming, which is based on an integration of the agent programming language GTGolog with adaptive dynamic programming techniques. GTGolog combines explicit agent programming in Golog with multi-agent planning in stochastic games. A drawback of this framework, however, is that the transition probabilities and reward values of the domain must be known in advance and then cannot change anymore. But such data is often not available in advance and may also change over the time. The adaptive generalization of GTGolog in this paper is directed towards letting the agents themselves explore and adapt these data, which is more useful for realistic applications. We use high-level programs for generating both abstract states and optimal policies, which benefits from the deep integration between action theory and high-level programs in the Golog framework.

1 Introduction

In the recent years, the development of controllers for autonomous agents has become increasingly important in AI. One way of designing such controllers is the programming approach, where a control program is specified through a language based on high-level actions as primitives. Another way is the planning approach, where goals or reward functions are specified and the agent is given a planning ability to achieve a goal or to maximize a reward function. An integration of both approaches for multi-agent systems has recently been proposed through the language GTGolog [7] (generalizing DTGolog [3]), which integrates explicit agent programming in Golog [20] with game-theoretic multi-agent planning in stochastic games [16]. It allows for partially specifying a high-level control program (for a system of two competing agents or two competing teams of agents) in a high-level language as well as for optimally filling in missing details through game-theoretic multi-agent planning.

However, a drawback of GTGolog (and also of DTGolog) is that the transition probabilities and reward values of the domain must be known in advance and then cannot change anymore. But such data often cannot be provided in advance in the model. It would thus be more useful for realistic applications to make the agents themselves capable of estimating, exploring, and adapting these data.

This is the main motivating idea behind this paper. We present a novel approach to adaptive multi-agent programming, which is an integration of GTGolog with reinforcement learning as in [13]. We use high-level programs for generating both abstract states

C. Freksa, M. Kohlhase, and K. Schill (Eds.): KI 2006, LNAI 4314, pp. 389–403, 2007.

and policies over these abstract states. The generation of abstract states exploits the structured encoding of the domain in a basic action theory, along with the high-level control knowledge in a Golog program. A learning process then incrementally adapts the model to the executive context and instantiates the partially specified behavior.

To our knowledge, this is the first adaptive approach to Golog interpreting. Differently from classical Golog, here the interpreter generates not only complex sequences of actions, but also an abstract state space for each machine state. Similarly to [2,11], we rely on the situation calculus machinery for state abstraction, but in our system the state generation is driven by the program structure. Here, we can take advantage from the deep integration between the action theory and programs provided by Golog: deploying the Golog semantics and the domain theory, we can produce a tailored state abstraction for each program state. In this way, we can extend the scope of programmable learning techniques [4,17,5,1,14] to a logic-based agent [12,20,21] and multi-agent [6] programming framework: the choice points of partially specified programs are associated with a set of state formulas and are instantiated through reinforcement learning and dynamic programming constrained by the program structure.

2 Preliminaries

In this section, we recall the basics of the situation calculus and Golog, matrix games, stochastic games, and reinforcement learning.

2.1 The Situation Calculus and Golog

The situation calculus [15,20] is a first-order language for representing dynamically changing worlds. Its main ingredients are *actions*, *situations*, and *fluents*. An *action* is a first-order term of the form $a(u_1, \ldots, u_n)$, where the function symbol a is its *name* and the u_i's are its *arguments*. All changes to the world are the result of actions. For example, the action $moveTo(r, x, y)$ may stand for moving the agent r to the position (x, y). A *situation* is a first-order term encoding a sequence of actions. It is either a constant symbol or of the form $do(a, s)$, where do is a distinguished binary function symbol, a is an action, and s is a situation. The constant symbol S_0 is the *initial situation* and represents the empty sequence, while $do(a, s)$ encodes the sequence obtained from executing a after the sequence of s. For example, the situation $do(moveTo(r, 1, 2), do(moveTo(r, 3, 4), S_0))$ stands for executing $moveTo(r, 1, 2)$ after executing $moveTo(r, 3, 4)$ in the initial situation S_0. We write $Poss(a, s)$, where $Poss$ is a distinguished binary predicate symbol, to denote that the action a is possible to execute in the situation s. A *(relational) fluent* represents a world or agent property that may change when executing an action. It is a predicate symbol whose most right argument is a situation. For example, $at(r, x, y, s)$ may express that the agent r is at the position (x, y) in the situation s. In the situation calculus, a dynamic domain is represented by a *basic action theory* $AT = (\Sigma, \mathcal{D}_{una}, \mathcal{D}_{S_0}, \mathcal{D}_{ssa}, \mathcal{D}_{ap})$, where:

- Σ is the set of (domain-independent) foundational axioms for situations [20].
- $\mathcal{D}_{una}$ is the set of unique names axioms for actions, which express that different actions are interpreted in a different way.

- $\mathcal{D}_{S_0}$ is a set of first-order formulas describing the initial state of the domain (represented by S_0). For example, $at(r, 1, 2, S_0) \wedge at(r', 3, 4, S_0)$ may express that the agents r and r' are initially at the positions $(1, 2)$ and $(3, 4)$, respectively.
- $\mathcal{D}_{ssa}$ is the set of *successor state axioms* [20]. For each fluent $F(\boldsymbol{x}, s)$, it contains an axiom of the form $F(\boldsymbol{x}, do(a, s)) \equiv \Phi_F(\boldsymbol{x}, a, s)$, where $\Phi_F(\boldsymbol{x}, a, s)$ is a formula with free variables among $\boldsymbol{x}, a, s$. These axioms specify the truth of the fluent F in the next situation $do(a, s)$ in terms of the current situation s, and are a solution to the frame problem (for deterministic actions). For example, the axiom $at(r, x, y, do(a, s)) \equiv a = moveTo(r, x, y) \vee (at(r, x, y, s) \wedge \neg \exists x', y' (a = moveTo(r, x', y')))$ may express that the agent r is at the position (x, y) in the situation $do(a, s)$ iff either r moves to (x, y) in the situation s, or r is already at the position (x, y) and does not move away in s.
- $\mathcal{D}_{ap}$ is the set of *action precondition axioms*. For each action a, it contains an axiom of the form $Poss(a(\boldsymbol{x}), s) \equiv \Pi(\boldsymbol{x}, s)$, which characterizes the preconditions of the action a. For example, $Poss(moveTo(r, x, y), s) \equiv \neg \exists r' \, at(r', x, y, s)$ may express that it is possible to move the agent r to the position (x, y) in the situation s iff no other agent r' is at (x, y) in s.

We use the concurrent version of the situation calculus [20,18], which is an extension of the standard situation calculus by concurrent actions. A *concurrent action* c is a set of standard actions, which are concurrently executed when c is executed.

The *regression* of a formula ϕ through an action a, denoted $Regr(\phi)$, is a formula ϕ' that holds before executing a, given that ϕ holds after executing a. The regression of ϕ whose situations are all of the form $do(a, s)$ is defined inductively using the successor state axioms $F(\boldsymbol{x}, do(a, s)) \equiv \Phi_F(\boldsymbol{x}, a, s)$ as follows:

$$Regr(F(\boldsymbol{x}, do(a, s))) = \Phi_F(\boldsymbol{x}, a, s), \quad Regr(\neg \phi) = \neg Regr(\phi),$$
$$Regr(\phi_1 \wedge \phi_2) = Regr(\phi_1) \wedge Regr(\phi_2), \quad \text{and} \quad Regr(\exists x \, \phi) = \exists x \, (Regr(\phi)).$$

Golog [12,20] is an agent programming language that is based on the situation calculus. It allows for constructing complex actions from the primitive actions defined in a basic action theory AT, where standard (and not so standard) Algol-like control constructs can be used, in particular, (i) sequence: $p_1; p_2$; (ii) test action: $\phi?$; (iii) nondeterministic choice of two programs: $(p_1 \mid p_2)$; (iv) nondeterministic choice of an argument: $\pi x \, (p(x))$; and (v) conditional, while-loop, and procedure. For example, the Golog program **while** $\neg at(r, 1, 2)$ **do** $\pi x, y \, (moveTo(r, x, y))$ repeats moving the agent r to a nondeterministically chosen position (x, y) while r is not at $(1, 2)$. The semantics of a Golog program p is specified by a situation-calculus formula $Do(p, s, s')$, which encodes that s' is a situation which can be reached from s by executing p. That is, Do represents a macro expansion to a situation calculus formula. For example, the action sequence is defined through $Do(p_1; p_2, s, s') = \exists s''(Do(p_1, s, s'') \wedge Do(p_2, s'', s'))$. For more details on the core situation calculus and Golog, we refer the reader to [20].

2.2 Matrix Games

Matrix games from classical game theory [23] describe the possible actions of two agents and the rewards that they receive when they simultaneously execute one action each. Formally, a *matrix game* $G = (A, O, R_a, R_o)$ consists of two nonempty finite sets of *actions* A and O for two agents a and o, respectively, and two *reward functions* $R_a, R_o: A \times O \to \mathbf{R}$ for a and o. The matrix game G is *zero-sum* iff $R_a = -R_o$; we then often omit R_o and abbreviate R_a by R.

A *pure* (resp., *mixed*) *strategy* specifies which action an agent should execute (resp., which actions an agent should execute with which probability). If the agents a and o play the pure strategies $a \in A$ and $o \in O$, respectively, then they receive the *rewards* $R_a(a, o)$ and $R_o(a, o)$, respectively. If the agents a and o play the mixed strategies $\pi_a \in PD(A)$ and $\pi_o \in PD(O)$, respectively, then the *expected reward* to agent $k \in \{a, o\}$ is $R_k(\pi_a, \pi_o) = \mathbf{E}[R_k(a, o)|\pi_a, \pi_o] = \sum_{a \in A, o \in O} \pi_a(a) \cdot \pi_o(o) \cdot R_k(a, o)$.

We are especially interested in pairs of mixed strategies (π_a, π_o), called *Nash equilibria*, where no agent has the incentive to deviate from its half of the pair, once the other agent plays the other half: (π_a, π_o) is a *Nash equilibrium* (or *Nash pair*) for G iff (i) $R_a(\pi_a', \pi_o) \leq R_a(\pi_a, \pi_o)$ for any mixed π_a', and (ii) $R_o(\pi_a, \pi_o') \leq R_o(\pi_a, \pi_o)$ for any mixed π_o'. Every two-player matrix game G has at least one Nash pair among its mixed (but not necessarily pure) strategy pairs, and many have multiple Nash pairs.

2.3   Stochastic Games

Stochastic games [16], or also called Markov games [22,13], generalize both matrix games [23] and (fully observable) Markov decision processes (MDPs) [19].

They consist of a set of states S, a matrix game for every state $s \in S$, and a transition function that associates with every state $s \in S$ and combination of actions of the agents a probability distribution on future states $s' \in S$. Formally, a *(two-player) stochastic game* $G = (S, A, O, P, R_a, R_o)$ consists of a finite nonempty set of states S, two finite nonempty sets of actions A and O for two agents a and o, respectively, a transition function $P: S \times A \times O \to PD(S)$, and two *reward functions* $R_a, R_o: S \times A \times O \to \mathbf{R}$ for a and o. The stochastic game G is *zero-sum* iff $R_a = -R_o$; we then often omit R_o.

Assuming a finite horizon $H \geq 0$, a *pure* (resp., *mixed*) time-dependent *policy* associates with every state $s \in S$ and number of steps to go $h \in \{0, \dots, H\}$ a pure (resp., mixed) matrix-game strategy. The H-*step reward* to agent $k \in \{a, o\}$ under a start state $s \in S$ and the pure policies α and ω, denoted $G_k(H, s, \alpha, \omega)$, is $G_k(0, s, \alpha, \omega) = R_k(s, \alpha(s, 0), \omega(s, 0))$ and $G_k(H, s, \alpha, \omega) = R_k(s, \alpha(s, H), \omega(s, H)) + \sum_{s' \in S} P(s'|s, \alpha(s, H), \omega(s, H)) \cdot G_k(H-1, s', \alpha, \omega)$ for $H > 0$. The notions of an *expected H-step reward* for mixed policies and of a *Nash pair* can then be defined in a standard way. Every two-player stochastic game G has at least one Nash pair among its mixed (but not necessarily pure) policy pairs, and it may have exponentially many Nash pairs.

2.4   Learning Optimal Policies

Q-learning [24] is a reinforcement learning technique, which allows to solve an MDP without a model (that is, transition and reward functions) and can be used on-line. The value $Q(s, a)$ is the expected discounted sum of future payoffs obtained by executing a from the state s and following an optimal policy. After being initialized to arbitrary numbers, the Q-values are estimated through the agent experience. For each execution of an action a leading from the state s to the state s', the agent receives a reward r, and the Q-value update is $Q(s, a) := (1 - \alpha) \cdot Q(s, a) + \alpha \cdot (r + \gamma \cdot \max_{a' \in A} Q(s', a'))$, where γ (resp., α) is the discount factor (resp., the learning rate). This algorithm converges to the correct Q-values with probability 1 assuming that every action is executed in every state infinitely many times and α is decayed appropriately.

Littman [13] extends Q-learning to learning an optimal mixed policy in a zero-sum two-player stochastic game. Here, the Q-value update is $Q(s, a, o) := (1 - \alpha) \cdot Q(s, a, o) + \alpha \cdot (r + \gamma \cdot \max_{\pi \in PD(A)} \min_{o' \in O} \sum_{a' \in A} Q(s', a', o') \cdot \pi(a'))$, where the "maxmin"-term gives the expected reward of a Nash pair for a zero-sum matrix game.

3   Adaptive GTGolog (AGTGolog)

In this section, we first define the domain theory behind Adaptive GTGolog (AGT-Golog) and then the syntax of AGTGolog.

3.1   Domain Theory of AGTGolog

A *domain theory* $DT = (AT, ST, OT)$ of AGTGolog consists of a basic action theory AT, a *stochastic theory* ST, and an *optimization theory* OT, as defined below.

We first give some preliminaries. We assume two zero-sum competing agents a and o (called *agent* and *opponent*, respectively, where the former is under our control, while the latter is not). The set of primitive actions is partitioned into the sets of primitive actions A and O of agents a and o, respectively. A *two-player action* is any concurrent action c over $A \cup O$ such that $|c \cap A| \leq 1$ and $|c \cap O| \leq 1$. For example, the concurrent actions $\{moveTo(a, 1, 2)\} \subseteq A$, $\{moveTo(o, 2, 3)\} \subseteq O$, and $\{moveTo(a, 1, 2), moveTo(o, 2, 3)\} \subseteq A \cup O$ are all two-player actions. We often write a, o, and $a \| o$ to abbreviate $\{a\} \subseteq A$, $\{o\} \subseteq O$, and $\{a, o\} \subseteq A \cup O$, respectively.

A *state formula* over $\boldsymbol{x}$, s is a formula $\phi(\boldsymbol{x}, s)$ in which all predicate symbols are fluents, and the only free variables are the non-situation variables $\boldsymbol{x}$ and the situation variable s. A *state partition* over $\boldsymbol{x}$, s is a nonempty set of state formulas $P(\boldsymbol{x}, s) = \{\phi_i(\boldsymbol{x}, s) \mid i \in \{1, \ldots, m\}\}$ such that (i) $\forall \boldsymbol{x}, s\,(\phi_i(\boldsymbol{x}, s) \Rightarrow \neg \phi_j(\boldsymbol{x}, s))$ is valid for all $i, j \in \{1, \ldots, m\}$ with $j > i$, (ii) $\forall \boldsymbol{x}, s\,(\bigvee_{i=1}^{m} \phi_i(\boldsymbol{x}, s))$ is valid, and (iii) every $\exists \boldsymbol{x}, s\,(\phi_i(\boldsymbol{x}, s))$ is satisfiable. For state partitions P_1 and P_2, we define their *product* as follows:

$$P_1 \otimes P_2 = \{\psi_1 \wedge \psi_2 \mid \psi_1 \in P_1,\ \psi_2 \in P_2,\ \psi_1 \wedge \psi_2 \neq \bot\}.$$

We often omit the arguments of a state formula when they are clear from the context.

We next define the stochastic theory. As usual [3,10,2], stochastic actions are expressed by a finite set of deterministic actions. When a stochastic action is executed, then "nature" chooses and executes with a certain probability exactly one of its deterministic actions. We use the predicate $stochastic(a, s, n)$ to associate the stochastic action a with the deterministic action n in situation s. We also specify a state partition $P_{pr}^{a,n}(\boldsymbol{x}, s) = \{\phi_j^{a,n}(\boldsymbol{x}, s) \mid j \in \{1, \ldots, m\}\}$ to group together situations s with common p such that "nature" chooses n in s with probability p, denoted $prob(a(\boldsymbol{x}), n(\boldsymbol{x}), s) = p$:

$$\exists p_1, \ldots, p_m\,(\bigwedge_{j=1}^{m}(\phi_j^{a,n}(\boldsymbol{x}, s) \Leftrightarrow prob(a(\boldsymbol{x}), n(\boldsymbol{x}), s) = p_j)).$$

A stochastic action s is indirectly represented by providing a *successor state axiom* for each associated nature choice n. Thus, AT is extended to a probabilistic setting in a minimal way. We assume that the domain is *fully observable*. For this reason, we introduce *observability axioms*, which disambiguate the state of the world after executing a stochastic action. For example, after executing $moveS(a, x, y)$, we test $at(a, x, y, s)$ and $at(a, x, y+1, s)$ to see which of the deterministic actions was executed (that is, $moveTo(a, x, y)$ or $moveTo(a, x, y+1)$). This condition is denoted $condSta(a, n, s)$. For example, $condSta(moveS(a, x, y), moveTo(a, x, y+1), s) \equiv at(a, x, y+1, s)$. Similar axioms are introduced to observe which actions the two agents have chosen.

As for the optimization theory, for every two-player action a, we specify a state partition $P^a_{rw}(\boldsymbol{x}, s) = \{\phi^a_k(\boldsymbol{x}, s) \mid k \in \{1, \ldots, q\}\}$ to group together situations s with common r such that $a(\boldsymbol{x})$ and s assign the reward r to $\boldsymbol{a}$, denoted $reward(a(\boldsymbol{x}), s) = r$:

$$\exists r_1, \ldots, r_q \left(\bigwedge_{k=1}^q (\phi^a_k(\boldsymbol{x}, s) \Leftrightarrow reward(a(\boldsymbol{x}), s) = r_k) \right).$$

Moreover, a utility function associates with every reward v and success probability pr a real-valued utility $utility(v, pr)$. We assume that $utility(v, 1) = v$ and $utility(v, 0) = 0$ for all v. An example of such a function is $utility(v, pr) = v \cdot pr$.

Example 3.1 (Stratagus Domain). Consider the following scenario inspired by [14]. The stratagus field consists of 9×9 positions (see Fig. 1). There are two agents, denoted a and o, which occupy one position each. The stratagus field has designated areas representing two *gold-mines*, one *forest*, and one *base* for each agent (see Fig. 1). The two agents can move one step in one of the directions N, S, E, and W, or remain stationary. Each of the two agents can also pick up one unit of wood (resp., gold) at the forest (resp., gold-mines), and drop these resources at its base. Each action of the two agents can fail, resulting in a stationary move. Any carried object drops when the two agents collide. After each step, the agents $\boldsymbol{a}$ and $\boldsymbol{o}$ receive the (zero-sum) rewards $r_a - r_o$ and $r_o - r_a$, respectively, where r_k for $k \in \{\boldsymbol{a}, \boldsymbol{o}\}$ is 0, 1, and 2 when k brings nothing, one unit of wood, and one unit of gold to its base, respectively.

The domain theory $DT = (AT, ST, OT)$ for the above stratagus domain is defined as follows. As for the basic action theory AT, we assume the deterministic actions $move(\alpha, m)$ (agent α performs m among N, S, E, W, and $stand$), $pickUp(\alpha, o)$ (agent α picks up the object o), and $drop(\alpha, o)$ (agent α drops the object o), as well as the relational fluents $at(q, x, y, s)$ (agent or object q is at the position (x, y) in the situation s), and $holds(\alpha, o, s)$ (agent α holds the object o in the situation s), which are defined through the following successor state axioms:

$$at(q, x, y, do(c, s)) \equiv agent(q) \wedge (at(q, x, y, s) \wedge move(q, stand) \in c \vee$$
$$\exists x', y' (at(q, x', y', s) \wedge \exists m (move(\alpha, m) \in c \wedge \phi(x, y, x', y', m)))) \vee$$
$$object(q) \wedge (at(q, x, y, s) \wedge \neg \exists \alpha (pickUp(\alpha, q) \in c) \vee$$
$$\exists \alpha ((drop(\alpha, q) \in c \vee collision(c, s)) \wedge at(\alpha, x, y, s) \wedge holds(\alpha, q, s)));$$
$$holds(\alpha, o, do(c, s)) \equiv holds(\alpha, o, s) \wedge$$
$$drop(\alpha, o) \notin c \wedge \neg collision(c, s) \vee pickUp(\alpha, o) \in c.$$

Here, $\phi(x, y, x', y', m)$ represents the coordinate change due to m, and $collision(c, s)$ encodes that the concurrent action c causes a collision between the agents $\boldsymbol{a}$ and $\boldsymbol{o}$ in

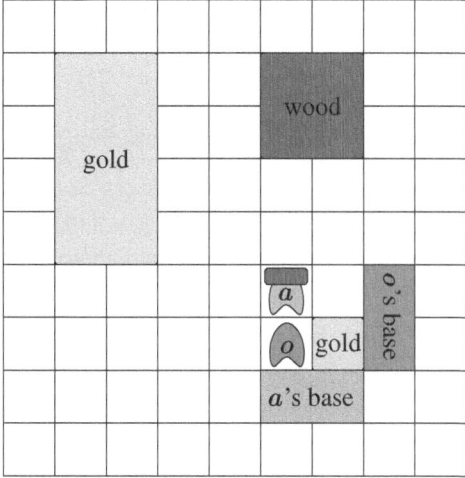

Fig. 1. Stratagus Domain

the situation s. The deterministic actions $move(\alpha, m)$, $drop(\alpha, o)$, and $pickUp(\alpha, o)$ are associated with precondition axioms as follows:

$Poss(move(\alpha, m), s) \equiv \top$;
$Poss(drop(\alpha, o), s) \equiv holds(\alpha, o, s)$;
$Poss(pickUp(\alpha, o), s) \equiv \neg \exists x \, holds(\alpha, x, s)$.

Furthermore, we assume the following additional precondition axiom, which encodes that two agents cannot pick up the same object at the same time (where $\alpha \neq \beta$):

$Poss(\{pickUp(\alpha, o), pickUp(\beta, o)\}, s) \equiv$
$\quad \exists x, y, x', y' (at(\alpha, x, y, s) \wedge at(\beta, x', y', s) \wedge (x \neq x' \vee y \neq y'))$.

As for the stochastic theory ST, we assume the stochastic actions $moveS(\alpha, m)$ (agent α executes m among N, S, E, W, and $stand$), $pickUpS(\alpha, o)$ (agent α picks up the object o), $dropS(\alpha, o)$ (agent α drops the object o), which may succeed or fail. We assume the state partition $P_{pr}^{a,n} = \{\top\}$ for each pair consisting of a stochastic action and one of its deterministic components:

$\exists p \, (prob(pickUpS(\alpha, o), pickUp(\alpha, o), s) = p)$;
$\exists p \, (prob(pickUpS(\alpha, o), move(\alpha, stand), s) = p)$;
$\exists p \, (prob(dropS(\alpha, o), drop(\alpha, o), s) = p)$;
$\exists p \, (prob(dropS(\alpha, o), move(\alpha, stand), s) = p)$;
$\exists p \, (prob(moveS(\alpha, d), move(\alpha, d), s) = p)$;
$\exists p \, (prob(moveS(\alpha, d), move(\alpha, stand), s) = p)$;
$\exists p \, (prob(a\|o, a'\|o', s) = p \equiv \exists p_1, p_2 \, (prob(a, a', s) = p_1 \wedge$
$\quad prob(o, o', s) = p_2 \wedge p = p_1 \cdot p_2))$.

As for the optimization theory OT, we use the product as the utility function $utility$. Furthermore, we define the reward function $reward$ as follows:

$$reward(\alpha, a, s) = r \equiv \exists r_\alpha, r_\beta \, (rewAg(\alpha, a, s) = r_\alpha \wedge$$
$$\exists \beta \, (rewAg(\beta, a, s) = r_\beta) \wedge r = r_\alpha - r_\beta) \, ;$$
$$\exists r_1, \ldots, r_m \, (\textstyle\bigwedge_{j=1}^m (\phi_j^{\alpha,a}(s) \Leftrightarrow rewAg(\alpha, a, s) = r_j)) \, .$$

Here, $\phi_j^{\alpha,a}(x, s)$ belongs to $P_{rw}^{\alpha,a}$, which is defined as follows. If $a = moveS(\alpha, x, y)$, then $P_{rw}^{\alpha,a} = \{\top\}$; if $a = pickUpS(\alpha, o)$, then $P_{rw}^{\alpha,a} = \{\neg h \wedge atg, \neg h \wedge atw, \neg h \wedge ato, h\}$; if $a = dropS(\alpha, o)$, then $P_{rw}^{\alpha,a} = \{hw \wedge atb, hg \wedge atb, \neg atb \wedge h, \neg h\}$, where h, atg, atw, atb, hg, hw, ato are formulas that stand for α holding something, being at the gold-mine, being at the wood, being at the base, holding gold, holding wood, and being close to an object, respectively.

3.2   Syntax of AGTGolog

AGTGolog has the same syntax as standard GTGolog: Given the actions specified by a domain theory DT, a *program* p in AGTGolog has one of the following forms (where α is a two-player action, ϕ is a condition, p, p_1, p_2 are programs, and $a_1, \ldots, a_n$ and $o_1, \ldots, o_m$ are actions of agents a and o, respectively):

1. *Deterministic or stochastic action:* α. Do α.
2. *Nondeterministic action choice of a:* **choice**$(a \colon a_1 | \cdots | a_n)$.
   Do an optimal action (for agent a) among $a_1, \ldots, a_n$.
3. *Nondeterministic action choice of o:* **choice**$(o \colon o_1 | \cdots | o_m)$.
   Do an optimal action (for agent o) among $o_1, \ldots, o_m$.
4. *Nondeterministic joint action choice:*
   choice$(a \colon a_1 | \cdots | a_n) \, \| \, $**choice**$(o \colon o_1 | \cdots | o_m)$.
   Do any action $a_i \| o_j$ with an optimal probability $\pi_{i,j}$.
5. *Test action:* $\phi?$. Test the truth of ϕ in the current situation.
6. *Sequence:* $p_1; p_2$. Do p_1 followed by p_2.
7. *Nondeterministic choice of two programs:* $(p_1 | p_2)$. Do p_1 or p_2.
8. *Nondeterministic choice of an argument:* $\pi x \, (p(x))$. Do any $p(x)$.
9. *Nondeterministic iteration:* $p^\star$. Do p zero or more times.
10. *Conditional:* **if** ϕ **then** p_1 **else** p_2.
11. *While-loop:* **while** ϕ **do** p.
12. *Procedures, including recursion.*

Example 3.2 (Stratagus Domain cont'd). We define some AGTGolog programs relative to the domain theory $DT = (AT, ST, OT)$ of Example 3.1. The following AGT-Golog procedure *carryToBase* describes a partially specified behavior where agent a is trying to move to its base in order to drop down an object:

> **proc** *carryToBase*
> **choice**$(a \colon moveS(a, N) | moveS(a, S) | moveS(a, E) | moveS(a, W))$;
> **if** *atBase* **then** $\pi x \, (dropS(a, x))$
>   **else** *carryToBase*
> **end**.

The subsequent procedure $pickProc(x)$ encodes that if the two agents a and o are at the same location, then they have to compete in order to pick up an object, otherwise agent a can directly use the primitive action $pickUpS(a, x)$:

proc $pickProc(x)$
if $atSameLocation(\boldsymbol{a}, \boldsymbol{o})$ **then** $tryToPickUp(x)$
    else $pickUpS(\boldsymbol{a}, x)$
end.

Here, the joint choices of the two agents $\boldsymbol{a}$ and $\boldsymbol{o}$ when they are at the same location are specified by the following procedure $tryToPickUp(x)$ (which will be instantiated by a *mixed policy*):

proc $tryToPickUp(x)$
choice($\boldsymbol{a}$: $pickUpS(\boldsymbol{a}, x) \mid moveS(\boldsymbol{a}, stand)$) $\|$
    choice($\boldsymbol{o}$: $pickUpS(\boldsymbol{o}, x) \mid moveS(\boldsymbol{o}, stand)$)
end.

4   Learning Optimal Policies

We now define state partitions SF for finite-horizon AGTGolog programs p. We then show how to learn an optimal policy for p. Intuitively, given a horizon $h \geq 0$, an h-step policy π of p relative to a domain theory is obtained from the h-horizon part of p by replacing every single-agent choice by a single action, and every multi-agent choice by a collection of probability distributions, one over the actions of each agent.

4.1   State Partition Generation

Given a GTGolog program p, a *machine state* consists of a subprogram p' of p and a horizon h. A *joint state* (ϕ, p', h) consists of a state formula ϕ and a machine state (p', h). Note that the joint state represents both the state of the environment and the executive state of the agent. Every machine state (p, h) is associated with a state partition, denoted $SF(p, h) = \{\phi_1(\boldsymbol{x}, s), \ldots, \phi_m(\boldsymbol{x}, s)\}$, which is inductively defined relative to the main constructs of AGTGolog (and naturally extended to all the other constructs) by:

1. Null program or zero horizon:

$$SF(nil, h) = SF(p, 0) = \{\top\}.$$

    At the program or horizon end, the state partition is given by $\{\top\}$.
2. Deterministic first program action:

$$SF(a; p', h) = P_{rw}^a(\boldsymbol{x}, s) \otimes \{Regr(\phi(\boldsymbol{x}, do(a, s)) \wedge Poss(a, s) \mid$$
$$\phi(\boldsymbol{x}, s) \in SF(p', h-1)\} \cup \{\neg Poss(a, s)\} \setminus \{\bot\}.$$

    Here, the state partition for $a; p'$ with horizon h is obtained as the product of the reward partition $P_{rw}^a(\boldsymbol{x}, s)$, the state partition $SF(p', h-1)$ of the next machine state $(p', h-1)$, and the executability partition $\{\neg Poss(a, s), Poss(a, s)\}$.
3. Stochastic first program action (nature choice):

$$SF(a; p', h) = \bigotimes_{i=1}^{k}(SF(n_i; p', h) \otimes P_{pr}^{a, n_i}(\boldsymbol{x}, s)),$$

    where $n_1, \ldots, n_k$ are the deterministic components of a. That is, the partition for $a; p'$ in h, where a is stochastic, is the product of the state partitions $SF(n_i; p', h)$ relative to the deterministic components n_i of a combined with the partitions P_{pr}^{a, n_i}.

4. Nondeterministic first program action (choice of agent k):

$$SF(\textbf{choice}(k\colon a_1|\cdots|a_n); p', h) = \bigotimes_{i=1}^{n} SF(a_i; p', h),$$

where $a_1, \ldots, a_n$ are two-player actions. That is, the state partition for a single choice of actions is the product of the state partitions for the possible choices. Note that the state partition for the joint choice of both agents is defined in a similar way.

5. Nondeterministic choice of two programs:

$$SF((p_1 \,|\, p_2); p', h) = SF(p_1; p', h) \otimes SF(p_2; p', h).$$

The state partition for a nondeterministic choice of two programs is obtained as the product of the state partitions associated with the possible programs.

6. Test action:

$$SF(\phi?; p', h) = \{\phi, \neg\phi\} \otimes SF(p', h).$$

The partition for $(\phi?; p', h)$ is obtained by composing the partition $\{\phi, \neg\phi\}$ induced by the test $\phi?$ with the state partition for (p', h).

4.2   Learning Algorithm

The main learning algorithm is *Learn* in Algorithm 1. For each joint state $\sigma = (\phi, p, h)$, where (p, h) is a machine state and $\phi \in SF(p, h)$, it generates an optimal h-step policy of p in ϕ, denoted $\pi(\sigma)$. We use a hierarchical version of Q-learning.

More concretely, the algorithm takes as input a program state (p, h) and generates as output an optimal policy π for each associated joint state (ϕ, p, h). In line 1, we initialize the *learning rate* α to 1; it decays at each learning cycles according to *decay*. In line 2, we also initialize to $\langle 1, 1 \rangle$ the variables $\langle v, pr \rangle$ representing the current *value function* (or *v-function*). At each cycle, the current state $\phi \in SF(p, h)$ is estimated (that is, the agent evaluates which of the state formulas describes the current state of the world). Then, from the joint state $\sigma = (\phi, p, h)$, the procedure $Update(\phi, p, h)$ (see Section 4.3) executes the program p with horizon h, and updates and refines the v-function $\langle v, pr \rangle$ and the policy π. At the end of the execution of *Update*, if the learning rate is greater than a suitable threshold ε, then the current state ϕ is estimated and a new learning cycle starts. At the end of the algorithm *Learn*, for suitable *decay* and ε, each possible execution of (p, h), from each ϕ, is performed often enough to obtain the convergency. That is, the agent executes the program (p, h) several times refining its v-function $\langle v, pr \rangle$ and policy π until an optimal behavior is reached.

4.3   Updating Step

The procedure $Update(\phi, p, h)$ in Algorithms 2 and 3 (parts 1 and 2, respectively) implements the execution and update step of a Q-learning algorithm. Here, each joint state σ of the program is associated with a variable $\langle v, pr \rangle(\sigma)$, which stores the current value of the v-function, and the variable $\pi(\sigma)$, which contains the current optimal policy at σ. Notice here that $\langle v, pr \rangle(\sigma)$ collects the cumulated reward v and probability pr of

---

Algorithm 1. $Learn(p, h)$

Require: AGTGolog program p and finite horizon h.
Ensure: optimal policy $\pi(\phi, p, h)$ for all $\phi \in SF(p, h)$.

1:   $\alpha := 1$;
2:   **for each** joint state σ **do** $\langle v, pr \rangle(\sigma) := \langle 1, 1 \rangle$;
3:   **repeat**
4:      estimate $\phi \in SF(p, h)$;
5:      $Update(\phi, p, h)$;
6:      $\alpha := \alpha \cdot decay$
7:   **until** $\alpha < \varepsilon$;
8:   **return** $(\pi(\phi, p, h))_{\phi \in SF(p,h)}$.

---

successful execution of σ, and $utility(\langle v, pr \rangle)$ is the associated utility. The procedure $Update(\phi, p, h)$ updates these value during an execution of a program p with horizon h, from a state $\phi \in SF(p, h)$. It is recursive, following the structure of the program.

Algorithm 2 describes the first part of the procedure $Update(\phi, p, h)$. Lines 1–4 encode the base of the induction: if the program is empty or the horizon is 0, then we set the v-function to $\langle 0, 1 \rangle$, that is, reward 0 and success probability 1. In lines 5–8, we consider the nonexecutable cases: if a primitive action a is not executable in the current state (here, $\neg Poss(a, \phi)$ abbreviates $DT \cup \phi \models \neg Poss(a, s)$) or a test failed in the current situation (here, $\neg\psi[\phi]$ stands for $DT \cup \phi \models \neg\psi(s)$), then we have the reward 0 and the success probability 0. In lines 9–15, we describe the execution of a primitive action a from $(\phi, a; p, h)$ (here, $Poss(a, \phi)$ is a shortcut for $DT \cup \phi \models Poss(a, s)$): after the execution, the agent receives a $reward$ from the environment. Here, the update of the v-function and of the policy π is postponed to the execution $Update(do(a, \phi), p', h-1)$ of the rest of the program, $(p', h-1)$, from the next state formula $do(a, \phi)$, that is, the state formula $\phi' \in SF(p', h-1)$ such that $Regr(\phi'(do(a, s)))$ equals to $\phi(s)$ relative to DT. Then, the v-function $\langle v, pr \rangle$ is updated as for Q-learning ($v(do(a, \phi), p', h)$ is for a Q-value for a in σ), while the success probability pr is inherited from the next joint state. In lines 16–22, we consider the stochastic action execution: after the execution, we observe a reward and the executed deterministic component n_q, then we update as in the deterministic case. The generated strategy is a conditional plan where each possible execution is considered. Here, ϕ_i are the conditions to discriminate the executed component (represented by the *observability axioms*).

The core of the learning algorithm (lines 24–55) is in Algorithm 3, where we show the second part of the procedure $Update(\phi, p, h)$. This code collects the agent choice constructs and describes how the agent learns an optimal probability distribution over the possible options in the choice points. Here, the algorithm selects one possible choice with the exploration strategy *explore*: with probability α, the agent selects randomly, and with probability $1-\alpha$, the agent selects according to the current policy $\pi(\sigma)$. Upon the execution of the selected action through the procedure $Update$, the v-function $\langle v, pr \rangle(\sigma)$ is updated. In the case of an agent (resp., opponent) choice (see lines 24–30 (resp., 31–38)), the current policy $\pi(\sigma)$ selects the current maximal (resp., minimal) choice; in the case of joint choices (see lines 39–48), following [13], an optimal

---

Algorithm 2. $Update(\phi, p, h)$: Part 1

Require: state formula ϕ, AGTGolog program p, and finite horizon h.
Ensure: updates $\langle v, pr \rangle(\sigma)$ and $\pi(\sigma)$, where $\sigma = (\phi, p, h)$.
 1: **if** $p = nil \lor h = 0$ **then**
 2:      $\langle v, pr \rangle(\sigma) := \langle 0, 1 \rangle$;
 3:      $\pi(\sigma) := stop$
 4: **end if**;
 5: **if** $p = a; p' \land \neg Poss(a, \phi) \lor p = \psi?; p' \land \neg \psi[\phi]$ **then**
 6:      $\langle v, pr \rangle(\sigma) := \langle 0, 0 \rangle$;
 7:      $\pi(\sigma) := stop$
 8: **end if**;
 9: **if** $p = a; p' \land Poss(a, \phi)$ and a is deterministic **then**
10:      execute a and observe $reward$;
11:      $Update(do(a, \phi), p', h-1)$;
12:      $\langle v, pr \rangle(\sigma) := \langle (1 - \alpha) \cdot v(\sigma) + \alpha \cdot (reward +$
13:                              $\gamma \cdot v(do(a, \phi), p', h-1)), pr(do(a, \phi), p', h-1) \rangle$;
14:      $\pi(\sigma) := a; \pi'(do(a, \phi), p', h-1)$
15: **end if**;
16: **if** $p = a; p' \land Poss(a, \phi)$ and a is stochastic **then**
17:      "nature" selects any deterministic action n_q of the action a;
18:      $Update(\phi, n_q; p', h)$;
19:      $\langle v, pr \rangle(\sigma) := \langle v, pr \rangle(\phi, n_q; p', h)$;
20:      $\pi(\sigma) := a;$ **if** ϕ_1 **then** $\pi(\phi, n_1; p', h)$ ...
21:                  **else if** ϕ_k **then** $\pi(\phi, n_k; p', h)$
22: **end if**;
23:      $\triangleright$ The algorithm is continued in Alg. 3, where the agent choices are described.

---

current mixed policy is given by the Nash pair computed by a Nash selection function *selectNash* from the matrix game defined by the possible joint choices. Then, depending on the case, the v-function is updated accordingly. Lines 49–55 encode the agent choice among programs. Finally, lines 55–60 define the successful test execution.

4.4 Example

We now illustrate the learning algorithm in the Stratagus Domain.

Example 4.1 (Stratagus Domain cont'd). Let the AGTGolog program p and the horizon h be given by $p = PickProc(x); carryToBase$ and $h = 3$, respectively. The learning algorithm for this input (that is, $Learn(p, 3)$) then works as follows. The agent runs several times p with horizon 3, playing against the opponent, until the learning ends and the variables $\langle v, pr \rangle$ are stabilized for each joint state (ϕ, p, h) associated with $(p, 3)$ obtaining the relative policies $\pi(\phi, p, h)$.

The state partition of $(p, 3)$ is given by $SF(p, 3) = SF(p_1, 3) \otimes \{asl\} \cup SF(p'_1, 3) \otimes \{\neg asl\}$, where $p_1 = tryToPickUp(x); carryToBase$ and $p'_1 = pickUpS(\boldsymbol{a}, x); carry-ToBase$. In the machine state $(p_1, 3)$, we have the joint choices $c_a^h \| c_o^k$, where

Algorithm 3. $Update(\phi, p, h)$: Part 2

24: **if** $p = \mathbf{choice}(\boldsymbol{a}: a_1 | \cdots | a_n); p'$ **then**

25:      select any $q \in \{1, \ldots, n\}$ with strategy *explore*;

26:      $Update(\phi, \boldsymbol{a}{:}a_q; p', h)$;

27:      $k := \mathrm{argmax}_{i \in \{1,\ldots,n\}} utility(\langle v, pr \rangle(\phi, \boldsymbol{a}{:}a_i; p', h))$;

28:      $\langle v, pr \rangle(\sigma) := \langle v, pr \rangle(\phi, \boldsymbol{a}{:}a_k; p', h)$;

29:      $\pi(\sigma) := \pi(\phi, \boldsymbol{a}{:}a_k; p', h)$

30: **end if**;

31: **if** $p = \mathbf{choice}(\boldsymbol{o}: o_1 | \cdots | o_m); p'$ **then**

32:      select any $q \in \{1, \ldots, m\}$ with strategy *explore*;

33:      $Update(\phi, \boldsymbol{o}{:}o_q; p', h)$;

34:      $k := \mathrm{argmin}_{i \in \{1,\ldots,m\}} utility(\langle v, pr \rangle(\phi, \boldsymbol{o}{:}o_i; p', h))$;

35:      $\langle v, pr \rangle(\sigma) := \langle v, pr \rangle(\phi, \boldsymbol{o}{:}o_k; p', h)$;

36:      $\pi(\sigma) := \mathbf{if}\ \phi_1\ \mathbf{then}\ \pi(\phi, \boldsymbol{o}{:}o_1; p', h)\ \cdots$

37:                  $\mathbf{else\ if}\ \phi_m\ \mathbf{then}\ \pi(\phi, \boldsymbol{o}{:}o_m; p', h)$

38: **end if**;

39: **if** $p = \mathbf{choice}(\boldsymbol{a}: a_1 | \cdots | a_n) \,\|\, \mathbf{choice}(\boldsymbol{o}: o_1 | \cdots | o_m); p'$ **then**

40:      select any $r \in \{1, \ldots, n\}$ and $s \in \{1, \ldots, m\}$ with strategy *explore*;

41:      $Update(\phi, \boldsymbol{a}{:}a_r \,\|\, \boldsymbol{o}{:}o_s; p', h)$;

42:      $(\pi_a, \pi_o) := selectNash(\{r_{i,j} = utility(\langle v, pr \rangle(\phi, \boldsymbol{a}{:}a_i \,\|\, \boldsymbol{o}{:}o_j; p', h)) \,|\, i, j\}$;

43:      $\langle v, pr \rangle(\sigma) := \sum_{i=1}^{n} \sum_{j=1}^{m} \pi_a(a_i) \cdot \pi_o(o_j) \cdot \langle v, pr \rangle(\phi, \boldsymbol{a}{:}a_i \,\|\, \boldsymbol{o}{:}o_j; p', h)$;

44:      $\pi(\sigma) := \pi_a \,\|\, \pi_o; \mathbf{if}\ \phi_1 \wedge \psi_1\ \mathbf{then}\ \pi(\phi, \boldsymbol{a}{:}a_1 \,\|\, \boldsymbol{o}{:}o_1; p', h)\ \cdots$

45:                  $\mathbf{else\ if}\ \phi_n \wedge \psi_m\ \mathbf{then}\ \pi(\phi, \boldsymbol{a}{:}a_n \,\|\, \boldsymbol{o}{:}o_m; p', h)$

46: **end if**;

47: **if** $p = (p_1 \,|\, p_2); p'$ **then**

48:      select any $i \in \{1, 2\}$ with strategy *explore*;

49:      $Update(\phi, p_i; p', h)$;

50:      $k := \mathrm{argmax}_{i \in \{1,2\}} utility(\langle v, pr \rangle(\phi, p_i; p', h))$;

51:      $\langle v, pr \rangle(\sigma) := \langle v, pr \rangle(\phi, p_k; p', h)$;

52:      $\pi(\sigma) := \pi(\phi, p_k; p', h)$

53: **end if**;

54: **if** $p = \psi?; p' \wedge \psi[\phi]$ **then**

55:      $Update(\phi, p', h)$;

56:      $\langle v, pr \rangle(\sigma) := \langle v, pr \rangle(\phi, p', h)$;

57:      $\pi(\sigma) := \pi(\phi, p', h)$

58: **end if**.

$c_\alpha^{pk} = pickUpS(\alpha, x)$ and $c_\alpha^{pk} = moveS(\alpha, stand)$, and the agent is to learn the probability distributions of the relative mixed policies. The choices $c_a^h \,\|\, c_o^k$ are associated with the programs $p_{h,k} = c_a^h \,\|\, c_o^k; p_2$, where $p_2 = carryToBase$. In the machine state $(p_2, 2)$, we have another choice point over the possible moves $m(q) = moveS(\boldsymbol{a}, q)$ towards the base. Each choice $m(q)$ is associated with the program $p_q = m(q); p_3$, where $p_3 =$ **if** atb **then** $dropS(a, x)$ **else** p_2 (atb abbreviates $atBase$). Here, the partition is $SF(p_q, 2)$ $= \{atb^q \wedge hw, atb^q \wedge hg, \neg atb^q, \neg h\}$, where atb^q represents atb after $m(q)$ obtained form $Regr(ab, m(q))$. If $\neg atb$ is represented by atb^0, and $atb^{q_1} \wedge \cdots \wedge atb^{q_4}$ is represented by $atb^{q_1,\dots,q_4}$, then $SF(p_2, 2) = \otimes_q SF(p_q, 2) = \{\neg h, atb^{0,\dots,0}\} \cup \{atb^{k_1,\dots,k_4} \wedge hg \mid \exists i : k_i \neq 0\} \cup \{atb^{h_1,\dots,h_4} \wedge hw \mid \exists i : k_i \neq 0\}$. For each state formula $\phi \in SF(p_2, 2)$, the algorithm $Learn(p, 3)$ continuously refines, through the $Update$ step, the probability distribution over the policies $\pi(\phi, p_2, 2)$. For example, training the agent against a random opponent, in the state $\phi = atb^{S,0,0,0} \wedge hg$, the algorithm produces a policy $\pi(\phi, p_2, 2)$ assigning probability 1 for the component $q = S$, and probability 0 for $q \neq S$. Analogously, in the choice point $(p_1, 3)$, $Learn(p_1, 3)$ defines a mixed policy $\pi(\phi, p_1, 3)$ for each state $\phi \in SF(p_1, 3) = \otimes_{h,k} SF(p_{h,k}, 3)$. For example, given $\phi_1 \in SF(p_1, 3)$ equal to $\neg asl \wedge ag \wedge atb^{k_1,\dots,k_4} \wedge \neg h_a \wedge h_o$, we get probability 1 for the choice $c_{pk,s}$ in $\pi(\phi_1, p_1, 3)$, instead, for $\phi_2 \in SF(p_1, 3)$ equal to $asl \wedge ag \wedge atb^{k_1,\dots,k_4} \wedge \neg h_a \wedge \neg h_o$, we get probability 0 for $c_{pk,s}$, $c_{s,pk}$, $c_{pk,pk}$, and probability 1 for $c_{s,s}$.

5   Summary and Outlook

We have presented a framework for adaptive multi-agent programming, which integrates high-level programming in GTGolog with adaptive dynamic programming. It allows the agent to on-line instantiate a partially specified behavior playing against an adversary. Differently from the classical Golog approach, here the interpreter generates not only complex sequences of actions (the policy), but also the state abstraction induced by the program at the different executive stages (machine states). In this way, we show how the Golog integration between action theory and programs allows to naturally combine the advantages of symbolic techniques [2,11] with the strength of hierarchical reinforcement learning [17,5,1,14]. This work aims at bridging the gap between programmable learning and logic-based programming approaches. To our knowledge, this is the first work exploring this very promising direction.

An interesting topic of future research is to explore whether the presented approach can be extended to the partially observable case.

Acknowledgments. This work was supported by the Austrian Science Fund Project P18146-N04 and by a Heisenberg Professorship of the German Research Foundation (DFG). We thank the reviewers for their comments, which helped to improve this work.

References

1. D. Andre and S. J. Russell. State abstraction for programmable reinforcement learning agents. In *Proceedings AAAI-2002*, pp. 119–125.
2. C. Boutilier, R. Reiter, and B. Price. Symbolic dynamic programming for first-order MDPs. In *Proceedings IJCAI-2001*, pp. 690–700.
3. C. Boutilier, R. Reiter, M. Soutchanski, and S. Thrun. Decision-theoretic, high-level agent programming in the situation calculus. In *Proceedings AAAI-2000*, pp. 355–362.
4. P. Dayan and G. E. Hinton. Feudal reinforcement learning. In *Proc. NIPS-1993*, pp. 271–278.
5. T. G. Dietterich. The MAXQ method for hierarchical reinforcement learning. In *Proceedings ML-1998*, pp. 118–126.

6. A. Ferrein, C. Fritz, and G. Lakemeyer. Using Golog for deliberation and team coordination in robotic soccer. *Künstliche Intelligenz*, 1:24–43, 2005.
7. A. Finzi and T. Lukasiewicz. Game-theoretic agent programming in Golog. In *Proceedings ECAI-2004*, pp. 23–27.
8. A. Finzi and T. Lukasiewicz. Relational Markov games. In *Proceedings JELIA-2004*, Vol. 3229 of *LNCS/LNAI*, pp. 320–333.
9. A. Finzi and T. Lukasiewicz. Game-theoretic Golog under partial observability. In *Proceedings AAMAS-2005*, pp. 1301–1302.
10. A. Finzi and F. Pirri. Combining probabilities, failures and safety in robot control. In *Proceedings IJCAI-2001*, pp. 1331–1336.
11. C. Gretton and S. Thiebaux. Exploiting first-order regression in inductive policy selection. In *Proceedings UAI-2004*, pp. 217–225.
12. H. J. Levesque, R. Reiter, Y. Lespérance, F. Lin, and R. Scherl. GOLOG: A logic programming language for dynamic domains. *J. Logic Program.*, 31(1–3):59–84, 1997.
13. M. L. Littman. Markov games as a framework for multi-agent reinforcement learning. In *Proceedings ICML-1994*, pp. 157–163.
14. B. Marthi, S. J. Russell, D. Latham, and C. Guestrin. Concurrent hierarchical reinforcement learning. In *Proceedings IJCAI-2005*, pp. 779–785.
15. J. McCarthy and P. J. Hayes. Some philosophical problems from the standpoint of Artificial Intelligence. In *Machine Intelligence*, Vol. 4, pp. 463–502. Edinburgh University Press, 1969.
16. G. Owen. *Game Theory: Second Edition*. Academic Press, 1982.
17. R. Parr and S. J. Russell. Reinforcement learning with hierarchies of machines. In *Proceedings NIPS-1997*, Vol. 10, pp. 1043–1049.
18. J. Pinto. Integrating discrete and continuous change in a logical framework. *Computational Intelligence*, 14(1):39–88, 1998.
19. M. L. Puterman. *Markov Decision Processes: Discrete Stochastic Dynamic Programming*. Wiley, 1994.
20. R. Reiter. *Knowledge in Action: Logical Foundations for Specifying and Implementing Dynamical Systems*. MIT Press, 2001.
21. Michael Thielscher. Programming of reasoning and planning agents with FLUX. In *Proceedings KR-2002*, pp. 435–446.
22. J. van der Wal. *Stochastic Dynamic Programming*, Vol. 139 of *Mathematical Centre Tracts*. Morgan Kaufmann, 1981.
23. J. von Neumann and O. Morgenstern. *The Theory of Games and Economic Behavior*. Princeton University Press, 1947.
24. C. Watkins. *Learning from Delayed Rewards*. PhD thesis, King's College, Cambridge, UK, 1989.

Agent Logics as Program Logics: Grounding KARO

Koen V. Hindriks[1] and John-Jules Ch. Meyer[2]

[1] Nijmegen Institute for Cognition and Information, Radboud University Nijmegen,
The Netherlands
k.hindriks@nici.ru.nl

[2] Utrecht University, Department of Information and Computing Sciences,
The Netherlands
jj@cs.uu.nl

Abstract. Several options are available to relate agent logics to computational agent systems. Among others, one can try to find useful executable fragments of an agent logic or use a model checking approach. In this paper, an alternative approach is explored based on the view that *an agent logic is a program logic*. Using the same starting point, one of the established agent logics, we ask instead if it is possible to construct a programming language for that agent logic. We show that the programming language and the agent logic are formally related by constructing a denotational semantics. As a result, the agent logic can be used as as a design tool to specify and verify the corresponding agent programs.

In particular, we construct an agent programming language that is formally related to the KARO agent logic. The KARO logic is an agent logic that builds on top of dynamic logic. The approach is based on mapping worlds in the modal semantics for KARO onto a state-based semantics. The state-based semantics can be used to define an operational semantics for KARO programs. In this way, we obtain a computationally grounded semantics for a significant part of the KARO logic, including the operators for knowledge or beliefs, motivational attitudes and belief revision actions of a rational KARO agent.

1   Introduction

Various agent logics have been proposed as models of so-called intelligent or rational agents. Some of the more influential ones have been those of [1,2,3]. These agent logics have played a guiding role in the research on the so-called strong notion of intelligent agent.

At the same time, the use of modal agent logics for engineering rational agents has been questioned (see for an extensive discussion also [4]). Agent logics have not been as useful as was hoped for in the specification and verification of agent systems. Consequently, the development of agent architectures and progamming languages has been inspired by operational systems such as the PRS system, which motivated the definition of AgentSpeak(L) [5].

The issue concerning the relation between agent logics and systems goes both ways: (i) can a suitable logical framework for reasoning about existing agent

C. Freksa, M. Kohlhase, and K. Schill (Eds.): KI 2006, LNAI 4314, pp. 404–418, 2007.

systems be constructed? and (ii) can a suitable operational framework for engineering agent systems be constructed that is related to an existing agent logic "in the right way"? These questions have been referred to as the gap between theory and practice. In this paper, we will not pursue the first question (see e.g. [6,7]) but only consider the second. There are good reasons to continue the search for a solution to bridge the gap. Agent systems are inherently complex due to the many different components and mechanisms such systems consist of. To be able to build such complex systems, at least two conditions should be met. First, it should be unambiguously clear how these components and mechanisms operate. One of the proven approaches to provide such an unambiguous interpretation is to provide a mathematical semantics that specifies the operation of an agent system. Second, tools for the design, specification and verification of agent systems should be available. This additionally requires the development of a design method including proof techniques for agent systems. Both conditions require that a precise relation between agent logics and systems is established.

As discussed in [4], various approaches to demonstrate the applicability of agent logics are available. Each of these approaches proposes techniques for relating agent logics to computational agent systems. For example, one method is to apply techniques for *directly executing* the logic. The goal of such methods is to find fragments that can be executed *efficiently*. Relevant work in this area is, for example, that of [8,9]. A drawback, however, is that these techniques are applicable only to relatively small fragments of agent logics.

In this paper, an alternative approach is proposed. Using the same starting point, one of the established agent logics, we ask instead if it is possible to construct an agent programming framework for that agent logic. The idea is to construct an agent programming framework for engineering agent systems that is related to an agent logic in a way such that the logic can be used to prove properties of the agent system. The idea promoted here is that *agent logics are program logics*. Program logics are used as a design tool to specify and verify agent programs instead as directly executable frameworks. It is shown that this approach can be successfully applied to close the gap between theory and practice by constructing a programming framework for the KARO logic [3].

2   Grounding Agent Logics

Agent logics typically are *modal* logics. Since the associated possible world semantics is abstract, in [4], the issue has been framed as the question whether it is possible to *ground* the semantics of agent logics. As the authors explain, "there is usually no precise relationship between the abstract accessibility relations that are used to characterize an agent's state, and any concrete computational model." Such a precise relationship, in a mathematical sense, is exactly what we will be looking for in this paper. To achieve this objective, we briefly clarify what we mean by (i) an agent logic, (ii) a concrete computational system, and (iii) the relation between the two.

What is an Agent Logic? Any logic that is explicitly constructed as a tool for modeling rational agents and is rich enough to model agents that derive their choice of action from their beliefs and motivations is an agent logic. In this paper, agent logics are single agent modal logics. Some of the better known agent logics such as [1,2,3] fit this definition and are reference examples. Agent logics based on *dynamic logic* (e.g. [3]) can be distinguished from agent logics based on *temporal logic* (e.g. [2]). Since in dynamic logic programs are explictly represented, agent logics based on dynamic logic provide a good starting point for our purposes.

What is a Concrete Computational System? To be able to construct a concrete computational system that relates to an agent logic in the right way, we start with providing a rather abstract definition and then move on to provide a concrete instance of this definition.

The main assumption that we introduce here is that *computational systems are state-based*. We take this to mean that the possible behavior of such a system can be uniquely predicted given its current state. Moreover, states are *extensional* and do not have an intensional flavour. That is, computational systems behave identically whenever they are in the same state at different times.

In fact, we will be more concrete and take a computational system to be a system that is programmed using a particular *programming language*. A programming language is a set of programming constructs to perform operations on specific data structures. Here, we are particularly interested in the set of data structures of agent systems, i.e. agent states, and associated agent programming frameworks for dynamically changing such states.

Programming frameworks typically have features for inspecting states of a computational system. For example, the language AgentSpeak(L) includes tests on the beliefs of an agent. The power that such tests have, however, may vary considerably. We distinguish so-called *poor tests* from *rich tests* (cf. [10]). Poor tests only allow inspection of the *current* state of a system whereas rich tests allow inspection of potential *future* states as well. Rich tests thus presuppose capabilities, called *look ahead facilities*, to perform tests on future states. In general, it is not clear how to provide a computational interpretation for rich tests. The second assumption we introduce is that *computational systems do not have look ahead facilities*. It will turn out, in fact, that the latter assumption will require the most effort in constructing a suitable agent programming framework. Computations thus are *local* in the sense that actions and tests are performed on the current state and do not require additional resources. The approach to define a state-based semantics is inspired by [11], but differs in its aim to derive a programming language from an agent logic. Any computational framework that introduces programming constructs to build computational systems and satisfies the two assumptions discussed is called a *programming framework*. As far as we know, all agent programming languages in the literature are programming frameworks in this sense (e.g. [5,12,13]).

How are Agent Logics and Computational Systems Related? The view promoted here is that agent logics are program logics for the specification and verification of agent programs. Agent logics provide declarative specifications of *what* an agent program should compute, whereas agent programming languages provide operational specifications *how* to execute an agent program. The semantics of the first is provided by *possible worlds semantics*. *Structural operational semantics* is used here to define computation steps that provide an interpretation of the operations of an agent program [14]. The precise relation between the two is established by proving that both semantics are equivalent. Formally, a *denotational* semantics for programs is derived from the logical semantics of an agent logic and is shown to be equivalent to the operational semantics. This is a standard technique in programming theory to show that a logic can be used to verify (partial correctness) properties of programs. The approach differs from directly executing an agent logic since the logic itself is only used to *verify properties of an executable agent program*. It differs from a model checking approach in that the logic is not used to check whether execution traces of a program satisfy a specification, but instead is used to *axiomatically verify program properties based on the program text*.

3   Grounding KARO

To demonstrate the approach discussed in the previous section, an agent programming language for KARO is presented. The exposition of KARO is based on [3]. For additional explanation about the logic and some of the choices made in modeling rational agents in KARO the reader is referred to this paper. For an example application of KARO to specify agents see e.g. [9]. KARO is an integrated logical framework for modeling rational agents that offers a logical theory of how actions, information and motivation of agents are related. All of these notions are formalised in a modal logic that is a blend of dynamic and epistemic logic extended with operators that model several motivational attitudes of agents.

It is shown that a substantial fragment of KARO is the corresponding program logic for a particular agent programming language. KARO agent programs ground the KARO logic, and, consequently, the KARO logic can be applied as a specification and verification tool for KARO agent systems that are built using that programming language.

KARO is a very expressive logic in which several concepts are defined using non-local constraints. Such constraints refer to potential future states of a system and do not naturally fit into a state-based approach. In particular, KARO introduces three non-local constraints:

- In the definition of the concept of *ability*, a constraint is included to verify the ability of an agent in potential future states. Complex abilities of an agent are defined in terms of a dynamic operator $\langle do(\pi) \rangle$ that models the opportunities and results achieved by actions. Because of the dynamic operators in the definition of abilities that involve tests on potential future states, the ability operator $A\pi$ has been excluded from the fragment that is discussed here. It is not clear by inspection of the logical semantics *how* the abilities of an agent change and a computational interpretation is not obtained by providing a state-based semantics for KARO.

- In the definition of the concept of a *goal*, quantification over actions is used to verify that a goal can be achieved via the execution of some plan by the agent. This verification involves reference to potential future states and some mechanism to test the possibility of performing an action in those states. It is not clear *how* to implement such look ahead facilities in a state-based approach. Instead of this non-local definition of goals in the KARO framework, a slightly weaker notion is defined that is implied by the KARO definition of goals, but not vice versa.
- In the definition of *commitments*, a test whether a plan can be executed is included. The actions commit_to to make a commitment and uncommit to remove a commitment similarly are defined in terms of the possibility to execute a plan. These definitions presuppose that KARO agents have look ahead facilities which do not straightforwardly translate into a computational semantics. Therefore, the commitment operator $\mathsf{Com}\pi$ is excluded here.

The computational complexity of KARO has been located primarily in the non-local constraints that are made use of in the definition of some operators. The logical semantics does not provide a clue on how to operationalize such constraints. It is an interesting question whether these constraints can somehow be operationalized in a state-based approach.

In the remainder, it is shown that the KARO fragment excluding operators that are defined by non-local constraints can be grounded. This fragment includes the modal operators $[do(\pi)]$ to represent the actions, $\mathsf{B}^k, \mathsf{B}^o$ to represent innate and observational knowledge, W to represent the wishes or desires, and C to represent the choices or goals of an agent. It also includes the informational actions expand φ and contract φ to add or remove a proposition φ from the observational knowledge base, and the action select φ to select φ as goal. Additionally, program constructs for tests, written as confirm φ, conditional composition if _ then _ else _ and repetition while _ do _ are part of the KARO language. Due to space restrictions, the latter is not discussed. The label KARO will be used below to refer to this fragment.

A distinguishing feature of KARO is the distinction of various belief clusters. Due to space restrictions, we only discuss two of the four clusters. First, built-in knowledge represents the fixed, objective options an agent considers possible. Second, observational knowledge is based on perceptual sources and may change through time. The built-in knowledge is restricted to objective propositional formulae. This seems reasonable since built-in knowledge is supposed to pertain to the external world and not to states of the agent itself. Moreover, this restriction will allow us to model the relation between built-in knowledge and observational knowledge in the state-based semantics. In this context, we do not allow an agent to have wishes about the fixed built-in knowledge, and allow an agent only to have wishes to obtain knowledge through its perceptual apparatus.

Definition 1. (*KARO* Propositions)
Let $\mathcal{L}_0$ be a classical propositional language, built from an infinite set At of

propositional atoms, *the connectives* $\neg, \wedge$ *and let* $\mathcal{L}_{obs}$ *be the standard extension of* $\mathcal{L}_0$ *to an epistemic modal language with epistemic operator* B^o. *Let Act be a set of atomic actions. Then:*

The KARO language $\mathcal{L}$ *is defined by:*

$-$ $At \subseteq \mathcal{L}$,
$-$ *if* $\varphi, \psi \in \mathcal{L}, \chi \in \mathcal{L}_0$, *then* $\neg\varphi, \varphi \wedge \psi, \mathsf{B}^k\chi, \mathsf{B}^o\varphi \in \mathcal{L}$,
$-$ *if* $\varphi \in \mathcal{L}$ *and* $\pi \in \Pi$, *then* $[do(\pi)]\varphi \in \mathcal{L}$,
$-$ *if* $\varphi \in \mathcal{L}_{obs}$, *then* $\mathsf{W}\varphi$ *in*$\mathcal{L}$, $\mathsf{C}\varphi \in \mathcal{L}$.

The set of KARO programs Π *is defined by:*

$-$ $Act \subseteq \Pi$,
$-$ *if* $\varphi \in \mathcal{L}_0$, *then* $\texttt{expand}\ \varphi, \texttt{contract}\ \varphi \in \Pi$,
$-$ *if* $\varphi \in \mathcal{L}_{obs}$, *then* $\texttt{select}\ \varphi \in \Pi$,
$-$ *if* $\varphi \in \mathcal{L}, \pi_1, \pi_2 \in \Pi$, *then* $\texttt{confirm}\ \varphi$,
   $\texttt{if}\ \varphi\ \texttt{then}\ \pi_1\ \texttt{else}\ \pi_2, \pi_1; \pi_2 \in \Pi$.

The semantics of KARO is defined as usual by Kripke structures.

Definition 2. (KARO Structure)
A KARO structure M *is a tuple* $\langle W, R, B^k, B^o, D, C, V \rangle$ *with:*

$-$ W *a non-empty set of worlds, typically denoted by* w,
$-$ R *a partial function such that for each* $a \in Act$, $R_a : W \rightharpoonup W$,
$-$ $B^k, B^o \subseteq W \times W$ *equivalence relations, such that* $B^o \subseteq B^k$,
$-$ $D \subseteq W \times W$,
$-$ $C : W \rightarrow \wp(\mathcal{L}_{obs})$, *a mapping of worlds to subsets of* $\mathcal{L}_{obs}$, *and*
$-$ V *a truth function such that (s.t.)* $V(p, w) \in \{1, 0\}$ *for* $p \in At$.

The knowledge operators are modal S5 operators as usual in agent logics. Knowledge obtained by perception always extends the agent's built-in knowledge since $B^o \subseteq B^k$. Observe that wishes modeled by the relation D may be inconsistent.

Definition 3. (KARO Semantics)
Let $M = \langle W, R, B^k, B^o, D, C, V \rangle$, $w \in W$ *and* $x \in \{k, o\}$.
The truth conditions for KARO propositions are defined by:

$-$ $M, w \models p$ *iff* $V(p, w) = 1$,
$-$ $M, w \models \neg\varphi$ *iff* $M, w \not\models \varphi$,
$-$ $M, w \models \varphi \wedge \psi$ *iff* $M, w \models \varphi$ *and* $M, w \models \psi$,
$-$ $M, w \models \mathsf{B}^x\varphi$ *iff* $M, w' \models \varphi$, $\forall w'$ *s.t.* wB^xw',
$-$ $M, w \models [do(\pi)]\varphi$ *iff* $M, w' \models \varphi$, $\forall w'$ *s.t.* $wR_\pi w'$,
$-$ $M, w \models \mathsf{W}\varphi$ *iff* $M, w' \models \varphi$, $\forall w'$ *s.t.* wDw',
$-$ $M, w \models \mathsf{C}\varphi$ *iff* $\varphi \in C(w)$.

The meaning of KARO programs is defined by:

R_a,
$R_{\texttt{confirm}\ \varphi}$ $= \{(w, w) \mid M, w \models \varphi\}$,
$R_{\texttt{if}\ \varphi\ \texttt{then}\ \pi_1\ \texttt{else}\ \pi_2}$ $= (R_{\pi_1} \cap (\llbracket\varphi\rrbracket \times W)) \cup$
                                 $(R_{\pi_2} \cap (\llbracket\neg\varphi\rrbracket \times W))$,
$R_{\pi_1; \pi_2}$ $= R_{\pi_1} \circ R_{\pi_2}$.

where $[\![\varphi]\!]_M = \{w \,|\, M, w \models \varphi\}$, *written* $[\![\varphi]\!]$ *when the structure* M *is clear from the context, and* $R \circ S = \{(a, c) \,|\, \exists b(aRb \wedge bSc)\}$.

Note that all KARO programs are deterministic since the transition relation R for actions is a function. In KARO, complex motivational attitudes such as goals are defined in terms of the basic motivational operators W and C. KARO defines goals as those wishes that are (a) *selected* by the agent, (b) *not (yet) fulfilled* and (c) *implementable*. Condition (c) is defined by quantifying over actions and involves tests on potential future states. It does not satisfy the constraints on state-based systems. A weaker version, by dropping (c), can be defined, however. The goal operator G is then defined as: $G\varphi \equiv W\varphi \wedge \neg\varphi \wedge C\varphi$. Note that tautologies cannot be goals and neither are goals closed under implication.

Changing One's Mind. One interesting feature of KARO is that it incorporates specific instances of atomic actions to change the knowledge or motivations of an agent. KARO thus not only formalizes the logic of propositional attitudes but also provides a theory about how an agent can change its knowledge or motivations by performing mental "actions".

Since mental actions do not change the (external) world, a natural way in modal semantics to model these actions is to change the knowledge or motivational components B^k, B^o, D, C in a KARO structure instead of a world w (see for extensive discussion [3]). This type of action semantics extends the component R in a structure to apply to pairs (M, w) as well as worlds w. Actions thus are interpreted as structure transformers. Some notation is introduced to facilitate the semantic definition of the KARO actions. Given an equivalence relation S, the equivalence class of w is defined as: $[w]_S = \{w' \,|\, wSw'\}$. Due to space restrictions, the semantics of contract φ is not discussed.

Definition 4. (Semantics of KARO Actions)
Let $M = \langle W, R, B^k, B^o, D, C, V \rangle$ *be a structure and* $w \in W$. R *is extended to a structure transforming semantics as follows:*
$$R_a(M, w) \qquad\quad = (M, R_a(w)),$$
$$R_{\texttt{expand}\,\varphi}(M, w) =$$
$$\begin{cases} (\langle W, R, B^k, B^{o'}, D, C, V \rangle, w) \text{ such that:} \\ \quad [w']_{B^{o'}} = [w']_{B^o} \cap [\![\varphi]\!], \forall w' \in [w]_{B^k} \cap [\![\varphi]\!], \\ \quad [w']_{B^{o'}} = [w']_{B^o} \cap [\![\neg\varphi]\!], \forall w' \in [w]_{B^k} \cap [\![\neg\varphi]\!], \text{ and} \\ \quad [w']_{B^{o'}} = [w']_{B^o}, \forall w' \notin [w]_{B^k}, \qquad \text{if } M, w \models \varphi, \\ \emptyset \qquad\qquad\qquad\qquad\qquad\qquad\qquad\qquad\quad \text{otherwise.} \end{cases}$$
$$R_{\texttt{select}\,\varphi}(M, w) =$$
$$\begin{cases} (\langle W, R, K, D, C', V \rangle, w) \text{ such that:} \\ \quad C'(w) = C(w) \cup \{\varphi\}, \\ \quad C'(w') = C(w'), \forall w' \neq w, \qquad \text{if } M, w \models W\varphi, \\ \emptyset \qquad\qquad\qquad\qquad\qquad\qquad\quad \text{otherwise.} \end{cases}$$
The extension for compound constructs can be defined analogously.

The semantics for expand φ ensures that an agent after (successfully) performing it knows φ, i.e. $B^o\varphi$, still knows what it knew before, and doesn't change anything when the agent already knew φ. This semantics validates the propositions $\varphi \rightarrow [do(\texttt{expand } \varphi)]\mathsf{B}^o\varphi$, $\mathsf{B}^o\psi \rightarrow [do(\texttt{expand } \varphi)]\mathsf{B}^o\psi$, and $\mathsf{B}^o\varphi \rightarrow (\mathsf{B}^o\psi \leftrightarrow [do(\texttt{expand } \varphi)]\mathsf{B}^o\psi)$, for $\varphi, \psi \in \mathcal{L}_0$, which are expected properties of adding true information to ones knowledge.

A State-Based Semantics for KARO. A programming language with KARO as its associated program logic must account for each of the components in a KARO structure. Except for the relation R all components will be represented in the state of a KARO agent program. These states, additionally, must be extensional, i.e. they can be uniquely identified by their syntactic content.

To avoid overly complex states, an additional axiom on top of the KARO axioms is introduced to relate the knowledge and wishes of an agent. As for knowledge, we assume that an agent is also able to introspect its motivational attitudes. If an agent has a wish, consequently, it knows it has a wish. Formally, we introduce the axioms $\mathsf{B}^o\mathsf{W}\varphi \leftrightarrow \mathsf{W}\varphi$ and $\mathsf{B}^o\neg\mathsf{W}\varphi \leftrightarrow \neg\mathsf{W}\varphi$. From now on, we will assume that these axioms are part of the KARO logic.

The assignment function V and possible worlds W in the modal semantics are replaced by so-called *world states* $v \subseteq At$ representing the external world in the state-based semantics. A corresponding world state can be defined for each $w \in W$ by $v = \{p \mid V(p, w) = 1\}$. The informational components B^k and B^o are replaced by *knowledge bases*. Corresponding knowledge and observational bases can be defined for each $w \in W$ respectively by $k = \{\varphi \in \mathcal{L}_0 \mid M, w \models \mathsf{B}^k\varphi\}$ and $o = \{\varphi \in \mathcal{L}_0 \mid M, w \models \mathsf{B}^o\varphi\}$. The motivational components D and C are replaced by a set of wishes $d \subseteq \mathcal{L}_{obs}$ closed under logical consequence and a set of choices $c \subseteq \mathcal{L}_{obs}$. Finally, the function R that provides the meaning of actions is quite straightforwardly replaced by a similar function R^c defined on KARO states instead. A KARO state can be viewed as the agent's internal, mental state.

Definition 5. (State-Based KARO Structure)
A state-based KARO structure M^c is a tuple $\langle W^c, R^c \rangle$ with:

- *W^c a set of states of the form (v, k, o, d, c), with v a world state, k, o knowledge bases such that $\models_v k$, $\models_v o$ and $o \models k$, $d, c \subseteq \mathcal{L}_{obs}$ a set of wishes and choices respectively, and such that W^c satisfies the following closure conditions: if $(v, k, o, d, c) \in W^c$, then:*
  - *$(v', k, o, d, c) \in W^c$ for all v' such that $\models_{v'} k$,*
  - *$(v', k, o, d, c) \in W^c$ for all v' such that $\models_{v'} o$,*
  - *$(v', k, o', d, c) \in W^c$ for all v', o' such that $\models_{v',o'} d$.*
- *R^c a partial function such that for each $a \in Act$ or $a \in \{\texttt{expand } \varphi, \texttt{select } \varphi\}$, $R_a^c : W^c \rightharpoonup W^c$.*

The definition of states clarifies the nature of the states in the state-based semantics. States are tuples of various databases which are the data structures that a KARO program operates on. To ensure a proper relation with the modal semantics, these components need to be related in the right way, which explains the various constraints on states. These constraints correspond to e.g. the relations between the accessibility relations in the modal semantics.

The truth conditions using state-based structures are defined next. The semantic clauses for atomic actions, tests and compound actions are the same as in definition 3 and are not repeated. Typically, states (v, k, o, d, c) are denoted by s, s' and we write $s[v'/v]$, $s[k'/k]$, ... to denote the state that results from replacing v by v', k by k', etc. Note the subscript c to distinguish $\models_c$ from the standard relation $\models$.

Definition 6. (State-Based Semantics for KARO)
Let $M^c = \langle W^c, R^c \rangle$ and $s = (v, k, o, d, c) \in W^c$. Then:
The truth conditions for KARO propositions φ are defined by:

- $M^c, s \models_c p$ *iff* $p \in v$,
- $M^c, s \models_c \neg\varphi$ *iff* $M^c, s \not\models_c \varphi$,
- $M^c, s \models_c \varphi \wedge \psi$ *iff* $M^c, s \models_c \varphi$ and $M^c, s \models_c \psi$,
- $M^c, s \models_c [do(\pi)]\varphi$ *iff* $M^c, s' \models_c \varphi$, $\forall s'$ s.t. $sR^c_\pi s'$,
- $M^c, s \models_c \mathsf{B}^k\varphi$ *iff* $M^c, s[v'/v] \models_c \varphi$, $\forall v'$ s.t. $\models_{v'} k$,
- $M^c, s \models_c \mathsf{B}^o\varphi$ *iff* $M^c, s[v'/v] \models_c \varphi$, $\forall v'$ s.t. $\models_{v'} o$,
- $M^c, s \models_c \mathsf{W}\varphi$ *iff* $M^c, s[v'/v, o'/o] \models_c \varphi$,
  $\forall v', o'$ s.t. $\models_{v',o'} d$,
- $M^c, s \models_c \mathsf{C}\varphi$ *iff* $\varphi \in c$.

where $\models_v \varphi$ is $v \models \varphi$ and $\models_{v,o} \varphi$ is defined by the first three clauses above and the clause for B^o.
The semantics of KARO programs π is defined by:
$$R^c_{\text{expand}\,\varphi} = \{(s, s[exp(o, \varphi)/o']) \mid M^c, s \models \varphi\},$$
$$R^c_{\text{select}\,\varphi} = \{(s, s[c \cup \{\varphi\}/c]) \mid M^c, s \models \mathsf{W}\varphi\}$$
$$R^c_{\text{confirm}\varphi} = \{(s, s) \mid M^c, s \models \varphi\}.$$
with $exp(o, \varphi)$ defined as $\{\psi \mid \forall v'(\models_{v'} o \wedge \varphi \Rightarrow \models_{v'} \psi)\}$.

The modal and state-based semantics are equivalent, i.e. the expressive power is not reduced by introducing a state-based semantics.

Theorem 7. (Equivalence of $\models$ and $\models_c$ for KARO)
The standard and the state-based semantics for KARO are equivalent. That is, assuming the set of propositional atoms At is infinite, for any $\varphi \in \mathcal{L}$:

$$\models \varphi \quad \text{iff} \quad \models_c \varphi$$

Proof. We give a sketch of the proof, the full proof is available in the full paper. First, observe that the structure-transforming action semantics in the standard modal semantics can be replaced with a standard Kripke semantics, by defining a super structure with worlds w_M for each pair (M, w). The right to left implication then is proved by a straightforward mapping from state-based structures to standard structures. For the left to right implication, use the finite model property for the (super structure) Kripke semantics to show the equivalence with the state-based semantics. We need to prove that if $M^c, s \not\models \varphi$, then also $M, w \not\models \varphi$. Since *the truth of φ can only depend on a finite number of propositional atoms, an infinite number of atoms remains that can be used as names for possible worlds in the standard structure to keep track of these worlds in a state-based structure. Using this observation, then set up a correspondence between possible worlds and states to define a state-based structure and show their equivalence.*   □

Note that no restrictions are imposed on the KARO language in theorem 7. The state-based semantics is defined for arbitrary formulae of the KARO language. To provide an operational interpretation, as discussed, however, we need to restrict the tests that are allowed in programs. The KARO language is a rich test version of a dynamic agent logic. To avoid the introduction of undecidable look ahead facilities into the programming framework, a poor test variant of KARO is introduced. The tests that can be allowed are those that can be evaluated in the current state. Consequently, propositions without occurrences of dynamic operators $[do(\pi)]$, called *intentional propositions* (since they refer to intentional or mental states), can be used as tests since KARO states contain the information needed to evaluate such propositions. We use $\mathcal{L}_i$ to denote the set of intentional propositions; note that $\mathcal{L}_{obs} \subset \mathcal{L}_i$. The fragment of KARO with restricted tests confirm φ such that φ is an intentional proposition is called *poor test* KARO and denoted by $\mathcal{L}^p$. $\mathcal{L}^p$ is strictly less expressive as $\mathcal{L}$.

The computational interpretation for intentional propositions, and, consequently, poor tests, is provided by the state-based semantic clauses for the non-dynamic operators. The definition below provides a computational interpretation since it identifies concrete data structures on which operations can be performed by a computer. Moreover, the computational interpretation is state-based and thus fits our definition of a computational system.

Definition 8. (Computational Interpretation of Poor Tests)
Let $s = (v, k, o, d, c)$ be a state such that $\models_v k$, $\models_v o$, and $o \models k$. Then the truth conditions for intentional propositions are defined by:

- $\models_s p$ *iff* $p \in v$,
- $\models_s \neg\varphi$ *iff* $\not\models_s \varphi$,
- $\models_s \varphi \wedge \psi$ *iff* $\models_s \varphi$ *and* $\models_s \psi$,
- $\models_s B^k\varphi$ *iff* $\models_{s[v'/v]} \varphi$, $\forall v'$ *s.t.* $\models_{v'} k$,
- $\models_s B^o\varphi$ *iff* $\models_{s[v'/v]} \varphi$, $\forall v'$ *s.t.* $\models_{v'} o$,
- $\models_s W\varphi$ *iff* $\models_{s[v'/v,,o'/o]} \varphi$, $\forall v', o'$ *s.t.* $\models_{v',o'} d$,
- $\models_s C\varphi$ *iff* $\varphi \in c$.

In case an intentional proposition is of the form $X\varphi$ with X some non-dynamic operator and $\varphi \in \mathcal{L}_0$ a state proposition, we have $\models_{v,k,o,d,c} \varphi$ iff $x \models \varphi$ for $x = k, o, d, c$ respectively. For example, it is easy to show that $\models_{v,k,o,d,c} W\varphi$ iff $d \models \varphi$.

Lemma 9. *Let $\varphi \in \mathcal{L}_0$ be a state proposition. Then we have:*

$$\models_{v,k,o,d,c} B^o\varphi \quad \text{iff} \quad o \models \varphi,$$
$$\models_{v,k,o,d,c} W\varphi \quad \text{iff} \quad d \models \varphi,$$
$$\models_{v,k,o,d,c} C\varphi \quad \text{iff} \quad \varphi \in c.$$

Proof. *The last item follows immediately from the semantic definition of* C. *The proof of the other three statements is similar. We prove the case for* $W\varphi$. *By definition, we have that* $\models_{v,k,o,d,c} W\varphi$ *iff for all* v', k', o' *such that* $\models_{v',k',o'} d$ *we have* $\models_{v',k',o',d,c} \varphi$. *Since* $\varphi \in \mathcal{L}_0$ *is a state proposition, its truth evaluation depends only on the world state component* v'. *Now suppose that* $d \not\models \varphi$. *Then there is a model of* d *in which* d *is true, but* φ *is not true. I.e., there are* v', k', o' *such that* $\models_{v',k',o'} d$ *and* $\not\models_{v',k',o'} \varphi$, *contrary to the assumption that* $\models_{v,k,o,d,c} W\varphi$. $\square$

In the remainder of this section, we use the computational interpretation of intentional propositions to provide a transition semantics for KARO programs. A transition semantics, defined in terms of a computation step relation $\longrightarrow$, provides a computational semantics for KARO programs.

Interestingly, in the transition semantics for KARO programs transition rules are required for the specific actions expand φ and select φ. Agents that explicitly represent their knowledge and motivational attitudes in their mental states need capabilities to modify these mental structures. In this respect, KARO contributes to an understanding of such capabilities, for both informational as well as motivational attitudes. Of course, agent logics allow for the *specification* of additional actions and the fact that KARO can be used as a program logic for KARO programs, proven below, shows that such specifications can be usefully applied to build agent programs.

In the transition semantics a transition function $\mathcal{T}(a, s)$ maps actions and a state to their successor state. This function must respect the constraints on states from definition 5. This translates into the following condition on transition functions $\mathcal{T}$: if $\mathcal{T}(a, v, k, o, d, c) = (v', k', o', d', c')$, then we have $\models_{v'} k', \models_{v'} o'$ and $o' \models k'$. KARO states in state-based structures are tuples (v, k, o, d, c) such that $\models_v k, \models_v o$ and $o \models k$ to ensure that beliefs of an agent are always true in the current world state and the belief clusters k and o are properly related. In the transition semantics, this relation between world states and knowledge bases also needs to be enforced.

If a KARO program starts in an initial state that satisfies this condition, then the constraint on transition relations ensures that states during the execution of the program invariantly satisfy this condition. Next, the transition semantics is defined. The rules for sequential composition and the if_then_else_ are standard and not included here (cf. [14]).

Definition 10. (Transition Semantics for KARO Programs)
Let s, s' *be states,* $\mathcal{T}$ *a transition function,* $\varphi \in \mathcal{L}_i$, $\psi \in \mathcal{L}_0$ *and* $\chi \in \mathcal{L}_{obs}$. *We use the symbol* E *to denote successful program termination. Then the transition semantics for KARO programs is defined by:*

$$\frac{\mathcal{T}(a, s) = s'}{\langle a, s \rangle \longrightarrow \langle E, s' \rangle} \qquad \frac{\models_s \varphi}{\langle \mathtt{confirm}\,\varphi, s \rangle \longrightarrow \langle E, s \rangle}$$

$$\frac{\models_s \psi}{\langle \texttt{expand } \psi, s \rangle \longrightarrow \langle E, s[exp(o, \psi)/o] \rangle}$$

$$\frac{\models_s \mathsf{W}\chi}{\langle \texttt{select } \chi, s \rangle \longrightarrow \langle E, s[c \cup \{\chi\}/c] \rangle}$$

The operational semantics for KARO programs defines the input-output relation on KARO states for arbitrary KARO programs in terms of the transitive closure $\longrightarrow^*$ of the transition relation $\longrightarrow$.

Definition 11. (Operational Semantics for KARO Programs)
The operational semantics for KARO programs is defined by:

$$\mathcal{O}(\pi)(s) = s' \ \textit{ for } s' \ \textit{ such that } \langle \pi, s \rangle \longrightarrow^* \langle E, s' \rangle$$

Note that the definition of the operational semantics is well-defined since KARO programs are deterministic.

The denotational semantics for KARO programs is derived from the logical semantics for KARO. In the definition of the denotational semantics the state-based semantics for KARO is used, which is justified by theorem 7. To provide a definition of the denotational semantics, we need to fix an interpretation of atomic actions. It will be convenient to ensure that this interpretation is equivalent to the one that is fixed in the transition semantics by the transition function $\mathcal{T}$ since we need this later on to prove the equivalence of both types of semantics.

Definition 12. ($\mathcal{T}$-compatible)
Let $M^c = \langle W^c, R^c \rangle$ be a state-based KARO structure. We say that M^c and the accessibility relation R^c are $\mathcal{T}$-compatible if the following condition is satisfied: $sR_a^c s'$ iff $\mathcal{T}(a, s) = s'$ for all KARO actions.

The denotational semantics can now be defined using the concept of a $\mathcal{T}$-compatible KARO structure; that is, a unique, well-defined and compositional semantic function can be defined for KARO programs.

Definition 13. (Denotational Semantics for KARO Programs)
Let $M^c = \langle W^c, R^c \rangle$ be a state-based, $\mathcal{T}$-compatible KARO structure. Then the denotational semantics for KARO programs is defined by:

$$[\![a]\!](s) \ = \begin{cases} s' & , \ if \ sR_a^c s', \\ undefined & , \ otherwise, \end{cases}$$

$$[\![\texttt{expand } \varphi]\!](s) \ = \begin{cases} s[exp(o, \varphi)/o'] & , \ if \ M^c, s \models_c \varphi, \\ undefined & , \ otherwise \end{cases}$$

$$[\![\texttt{select } \varphi]\!](s) \ = \begin{cases} s[c \cup \{\varphi\}/c] & , \ if \ M^c, s \models_c \mathsf{W}\varphi, \\ undefined & , \ otherwise \end{cases}$$

$$[\![\texttt{confirm } \varphi]\!](s) \ = \begin{cases} s & , \ if \ M^c, s \models_c \varphi \\ undefined & , \ otherwise \end{cases}$$

$$[\![\texttt{if } \varphi \texttt{ then } \pi_1 \texttt{ else } \pi_2]\!](s) = \begin{cases} [\![\pi_1]\!](s) & , \ if \ M^c, s \models_c \varphi, \\ [\![\pi_2]\!](s) & , \ otherwise \end{cases}$$

$$[\![\pi_1; \pi_2]\!](s) \qquad = [\![\pi_2]\!]([\![\pi_1]\!](s)),$$

It is easy to show that $[\![\_]\!]$ is well-defined. The next step is to show that the denotational and operational semantics are equivalent. This provides the precise relationship of the computation step relation $\longrightarrow$ and the logical semantics of KARO that we were looking for and shows that the logic can be used as a program logic to verify properties of KARO programs.

Theorem 14. (Denotational Equivalent to Operational Semantics)
The denotational and operational semantics of KARO programs are equivalent, i.e. $[\![\pi]\!](s) = \mathcal{O}(\pi)(s)$.

Proof. Use induction on the structure of programs.   □

The equivalence of the denotational and operational semantics shows that KARO has an application in the verification of KARO programs, in particular, to prove *partial correctness properties* of KARO programs. This fact is expressed mathematically in the following corollary:

Corollary 15. (Proving Partial Correctness Properties)
Let π *be a poor test KARO program,* $\varphi, \psi \in \mathcal{L}_i$, *and* $\mathcal{M}_{\mathcal{T}}$ *be the set of all* $\mathcal{T}$-*compatible KARO structures. Then we have:*

$$\forall s, s' : \ if \models_s \varphi \ and \ \langle \pi, s \rangle \longrightarrow^* \langle E, s' \rangle \ , \ then \models_{s'} \psi$$
$$iff$$
$$\models^{\mathcal{M}_{\mathcal{T}}} \varphi \to [\pi]\psi$$

Proof. Immediate from theorem 7 and 14.   □

By using the techniques explored in the previous section, we have been able to define a state-based programming framework that corresponds to a substantial fragment of the original KARO logic. In other words, the KARO logic is a program logic for the KARO programming framework introduced in this section.

The KARO programming framework introduced can be compared to existing agent programming languages in the literature. It is instructive, for example, to compare the transition semantics of the KARO programming framework with similar approaches. For example, this style of semantics has been proposed for the closely related programming languages AgentSpeak(L), ConGolog, and 3APL (cf. e.g. [15]). Even though the KARO programming framework introduced here does not include all aspects of these languages (notably concurrency is absent), it also includes features that are not present in one of the mentioned programming languages. The most important distinguishing features are the presence of *declarative* motivational attitudes and the definition of specific actions to change the agent's mind. For example, in AgentSpeak(L) the structures called intentions or plans are similar to the KARO programs and do not have a declarative interpretation. Instead, the motivational components in KARO program states are declarative. Another way to illustrate the same point is the observation that the corresponding program logic for 3APL proposed in [7] includes an operator for beliefs but not for motivational operators as the KARO logic does.

4   Conclusion

In this paper, we have explored formal techniques for relating agent logics to agent programming frameworks. We showed that this is a viable approach which clarifies the use of agent logics in the practice of agent engineering as specification and verification tools. The approach has been illustrated by constructing a programming framework for KARO, an agent logic that extends dynamic logic.

One of the benefits of our approach is that it explores the space of programming languages from a logical point of view. Taking an agent logic as our starting point, we showed what a programming language related to that logic looks like. This clarifies at least partly which agent logics are related to which agent programming languages.

A precise relationship between agent logics and agent programming frameworks will clarify what an agent programming language should be like from a logical agent perspective. The precise analysis of states of agents and the associated operations, moreover, facilitates a comparison between various agent logics as well as between various agent programming frameworks.

In the paper, it is shown that the approach based on the view that agent logics are useful as program logics resolves at least part of the gap between theory and practice. As discussed in [4], various approaches to show the usefulness of agent logics are available. These approaches are not each as practical as the other. Directly executing agent logics, for example, has to face the high computational complexity of agent logics. Model checking approaches suffer from other problems, as highlighted in [4]. By taking another view and viewing agent logics not as executable frameworks but as program logics we were able to circumvent some of these problems.

There have been other attempts to provide a computational grounding of KARO. In [9] a reduction approach is presented, based on translating KARO to first order logic. Alternatively, a translation of a fragment of the KARO logic to a combination of branching time logic CTL and a modal S5 logic, has been proposed (cf. [9]). In this approach, the *core* - as it is called - of the KARO framework is first translated into another logical formalism, to obtain an executable fragment. However, the fragment that can be translated into executable form is smaller and does not include multiple belief clusters, the wishes and choices of an agent, nor the specific KARO actions that are included here.

References

1. Cohen, P.R., Levesque, H.J.: Intention is choice with commitment. Artificial Intelligence **42** (1990) 213–261
2. Rao, A., Georgeff, M.: Decision Procedures for BDI Logics. Journal of Logic and Computation **8**(3) (1998) 293–343
3. van der Hoek, W., van Linder, B., Meyer, J.-J.Ch.: An Integrated Modal Approach to Rational Agents. In Wooldridge, M., ed.: Foundations of Rational Agency. Kluwer, Dordrecht (1999) 133–168

4. van der Hoek, W., Wooldridge, M.: Towards a Logic of Rational Agency. Logic Journal of the IGPL **11**(2) (2003) 133–157
5. Rao, A.S.: AgentSpeak(L): BDI Agents Speak Out in a Logical Computable Language. In van der Velde, W., Perram, J., eds.: Agents Breaking Away. Number 1038 in LNAI, Springer (1996) 42–55
6. de Boer, F., Hindriks, K., van der Hoek, W., Meyer, J.-J.Ch.: A Verification Framework for Agent Programming with Declarative Goals. Accepted for the Journal of Applied Logic (2006)
7. Hindriks, K., de Boer, F., van der Hoek, W., Meyer, J.-J.Ch.: A Programming Logic for part of the Agent Language 3APL. In Rash, J., ed.: Proceedings of the First Goddard Workshop on Formal Approaches to Agent-Based Systems. Number 1871 in LNCS, Springer (2001) 78–89
8. Fisher, M.: A Survey of Concurrent MetateM. In: Proceedings of the First International Conference on Temporal Logic (ICTL). Number 827 in LNCS, Springer (1994) 480–505
9. Hustadt, U., Dixon, C., Schmidt, R., Fisher, M., Meyer, J.-J.Ch., van der Hoek, W.: Reasoning about Agents in the KARO Framework. In Bettini, C., Montanari, A., eds.: Proc. of the Eighth Int. Symposium on Temporal Representation and Reasoning. (2001) 206–213
10. Harel, D., Kozen, D., Tiuryn, J.: Dynamic Logic. MIT Press (2000)
11. Fagin, R., Halpern, J.Y., Moses, Y., Vardi, M.Y.: Knowledge-based programs. Distributed Computing **10**(4) (1997) 199–225
12. Bordini, R., Dastani, M., Dix, J., Seghrouchni, A.E.F., eds.: Multi-Agent Programming: Languages, Platforms and Applications. Springer (2005)
13. de Giacomo, G., Lespérance, Y., Levesque, H.J.: ConGolog, a Concurrent Programming Language Based on the Situation Calculus. Artificial Intelligence **121**(1-2) (2000) 109–169
14. Plotkin, G.D.: A Structural Approach to Operational Semantics. Technical Report DAIMI FN-19, Computer Science Department, Aarhus University (1981)
15. Hindriks, K.V., de Boer, F.S., van der Hoek, W., Meyer, J.-J.Ch.: A Formal Embedding of AgentSpeak(L) in 3APL. In Antoniou, G., Slaney, J., eds.: Advanced Topics in Artificial Intelligence. Number 1502 in LNAI. Springer (1998) 155–166

On the Relationship Between Playing Rationally and Knowing How to Play: A Logical Account

Wojciech Jamroga

Department of Informatics,
Clausthal University of Technology, Germany
wjamroga@in.tu-clausthal.de

Abstract. Modal logics of strategic ability usually focus on capturing what it means for an agent to have a feasible strategy that brings about some property. While there is a general agreement on abilities in scenarios where agents have perfect information, the right semantics for ability under incomplete information is still debated upon. Epistemic Temporal Strategic Logic, an offspring of this debate, can be treated as a logic that captures properties of agents' rational play.

In this paper, we provide a semantics of ETSL that is more compact and comprehensible than the one presented in the original paper by van Otterloo and Jonker. Second, we use ETSL to show that a rational player knows that he will succeed if, and only if, he knows how to play to succeed– while the same is not true for rational coalitions of players.

Keywords: multi-agent systems, theories of agency, game-theoretical foundations, modal logic.

1   Introduction

Modal logics of strategic ability usually focus on capturing what it means for an agent to have a feasible strategy that brings about some property. While there is a general agreement on abilities in scenarios where agents have perfect information, the right semantics for ability under incomplete information is still debated upon. Epistemic Temporal Strategic Logic, proposed by van Otterloo and Jonker [13], is an offspring of this debate, but one that leads in an orthogonal direction to the mainstream solutions. The central operator of ETSL can be read as: "if A play *rationally* to achieve φ (meaning: they never play a dominated strategy), they will achieve φ". Thus, one may treat ETSL as a logic that captures properties of agents' rational play in a sense.

This paper contains two main messages. First, we provide a semantics of ETSL that is more compact and comprehensible than the one presented in [13]. ETSL is underpinned by several exciting concepts. Unfortunately, its semantics is also quite hard to read due to a couple non-standard solutions and a plethora of auxiliary functions, which is probably why the logic never received the attention it deserves. Second, and perhaps more importantly, we use ETSL to show that a rational player knows that he will succeed if, and only if, he knows how to play to succeed – while the same is not true for rational coalitions of players.

C. Freksa, M. Kohlhase, and K. Schill (Eds.): KI 2006, LNAI 4314, pp. 419–433, 2007.

2   Reasoning About Abilities of Agents

Modal logics of strategic ability [1,2] form one of the fields where logic and game theory can successfully meet. The logics have clear possible worlds semantics, are axiomatizable, and have some interesting computational properties. Moreover, they are underpinned by intuitively appealing conceptual machinery for modeling and reasoning about systems that involve multiple autonomous agents.

2.1   ATL: Ability in Perfect Information Games

Alternating-time Temporal Logic (ATL) [1, 2] can be seen as a logic for systems involving multiple agents, that allows one to reason about what agents can achieve in game-like scenarios. Since ATL does not include incomplete information in its scope, it can be seen as a logic for reasoning about agents who always have perfect information about the current state of affairs. Formula $\langle\!\langle A \rangle\!\rangle \varphi$, where A is a coalition of agents, expresses that A have a collective strategy to enforce φ. ATL formulae include temporal operators: "$\bigcirc$" ("in the next state"), $\square$ ("always from now on") and $\mathcal{U}$ ("until"). Operator $\Diamond$ ("now or sometime in the future") can be defined as $\Diamond\varphi \equiv \top \mathcal{U} \varphi$. Like in CTL, every occurrence of a temporal operator is preceded by exactly one cooperation modality $\langle\!\langle A \rangle\!\rangle$.[1] Formally, the recursive definition of ATL formulae is:

$$\varphi ::= p \mid \neg\varphi \mid \varphi \wedge \varphi \mid \langle\!\langle A \rangle\!\rangle \bigcirc \varphi \mid \langle\!\langle A \rangle\!\rangle \square \varphi \mid \langle\!\langle A \rangle\!\rangle \varphi \, \mathcal{U} \, \varphi$$

A number of semantics have been defined for ATL, most of them equivalent [3]. In this paper, we use a variant of *concurrent game structures*,

$$M = \langle \mathrm{Agt}, St, \Pi, \pi, Act, d, o \rangle,$$

which includes a nonempty finite set of all agents $\mathrm{Agt} = \{1, ..., k\}$, a nonempty set of states St, a set of atomic propositions Π, a valuation of propositions $\pi :$ $\Pi \rightarrow \mathcal{P}(St)$, and a nonempty set of (atomic) actions Act. Function $d : \mathrm{Agt} \times St \rightarrow$ $\mathcal{P}(Act)$ defines actions available to an agent in a state, and o is a deterministic transition function that assigns an outcome state $q' = o(q, \alpha_1, \ldots, \alpha_k)$ to state q, and a tuple of actions $\langle \alpha_1, \ldots, \alpha_k \rangle$ that can be executed by Agt in q. A *strategy* of agent a is a conditional plan that specifies what a is going to do for every possible situation ($s_a : St \rightarrow Act$ such that $s_a(q) \in d(a, q)$). A *collective strategy* (called also a *strategy profile*) S_A for a group of agents A is a tuple of strategies S_a, one per agent $a \in A$. A *path* Λ in M is an infinite sequence of states that can be effected by subsequent transitions, and refers to a possible course of action (or a possible computation) that may occur in the system; by $\Lambda[i]$, we denote the ith position on path Λ. Function $out(q, S_A)$ returns the set of all paths that may result from agents A executing strategy S_A from state q onward:

---

[1] The logic to which such a syntactic restriction applies is sometimes called *"vanilla"* ATL (resp. "vanilla" CTL etc.).

$out(q, S_A) = \{\lambda = q_0 q_1 q_2 ... \mid q_0 = q$ and for every $i = 1, 2, ...$ there exists a tuple of actions $\langle \alpha_1^{i-1}, ..., \alpha_k^{i-1} \rangle$ such that $\alpha_a^{i-1} = S_a(q_{i-1})$ for each $a \in A$, $\alpha_a^{i-1} \in d(a, q_{i-1})$ for each $a \notin A$, and $o(q_{i-1}, \alpha_1^{i-1}, ..., \alpha_k^{i-1}) = q_i\}$.

Now, the semantics of ATL formulae can be given via the following clauses:

$M, q \models p$ iff $q \in \pi(p)$ (where $p \in \Pi$);

$M, q \models \neg\varphi$ iff $M, q \not\models \varphi$;

$M, q \models \varphi \wedge \psi$ iff $M, q \models \varphi$ and $M, q \models \psi$;

$M, q \models \langle\!\langle A \rangle\!\rangle \bigcirc \varphi$ iff there is a collective strategy S_A such that, for every $\Lambda \in out(q, S_A)$, we have $M, \Lambda[1] \models \varphi$;

$M, q \models \langle\!\langle A \rangle\!\rangle \square \varphi$ iff there exists S_A such that, for every $\Lambda \in out(q, S_A)$, we have $M, \Lambda[i]$ for every $i \geq 0$;

$M, q \models \langle\!\langle A \rangle\!\rangle \varphi \mathcal{U} \psi$ iff there is S_A st. for every $\Lambda \in out(q, S_A)$ there is $i \geq 0$, for which $M, \Lambda[i] \models \psi$, and $M, \Lambda[j] \models \varphi$ for every $0 \leq j < i$.

2.2 Strategic Ability and Incomplete Information

ATL is unrealistic in a sense: real-life agents seldom possess complete information about the current state of the world. *Alternating-time Temporal Epistemic Logic* (ATEL) [12] enriches the picture with an epistemic component, adding to ATL operators for representing agents' knowledge: $K_a\varphi$ reads as "agent a knows that φ". Additional operators $E_A\varphi$, $C_A\varphi$, and $D_A\varphi$ refer to *mutual knowledge* ("everybody knows"), *common knowledge*, and *distributed knowledge* among the agents from A. Models for ATEL extend concurrent game structures with epistemic accessibility relations $\sim_1, ..., \sim_k \subseteq Q \times Q$ (one per agent) for modeling agents' uncertainty; the relations are assumed to be equivalences. We will call such models *concurrent epistemic game structures* (CEGS) in the rest of the paper. Agent a's epistemic relation is meant to encode a's inability to distinguish between the (global) system states: $q \sim_a q'$ means that, while the system is in state q, agent a cannot determine whether it is not in q'. Then:

$M, q \models K_a\varphi$ iff φ holds for every q' such that $q \sim_a q'$.

Relations $\sim_A^E$, $\sim_A^C$ and $\sim_A^D$, used to model group epistemics, are derived from the individual relations of agents from A. First, $\sim_A^E$ is the union of relations $\sim_a$, $a \in A$. Next, $\sim_A^C$ is defined as the transitive closure of $\sim_A^E$. Finally, $\sim_A^D$ is the intersection of all the $\sim_a$, $a \in A$. The semantics of group knowledge can be defined as below (for $\mathcal{K} = C, E, D$):

$M, q \models \mathcal{K}_A\varphi$ iff φ holds for every q' such that $q \sim_A^{\mathcal{K}} q'$.

Example 1. (**Gambling Robots**)   Two robots (a and b) play a simple card game. The deck consists of Ace, King and Queen (A, K, Q); it is assumed that A beats K, K beats Q, but Q beats A. First, the "environment" agent env deals a random card to both robots (face down), so that each player can see his own hand, but he does not know the card of the other player. Then robot a can

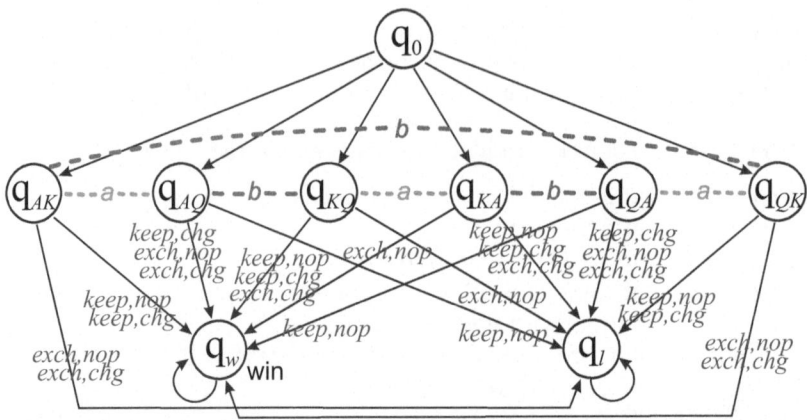

Fig. 1. Gambling Robots game. Arrows represent possible transitions of the system (labeled with tuples of agents' actions); dashed lines connect states that are indiscernible for particular agents.

exchange his card for the one remaining in the deck (action *exch*), or he can keep the current one (*keep*). At the same time, robot b can change the priorities of the cards, so that A becomes better than Q (action *chg*) or he can do nothing (*nop*). If a has a better card than b after that, then a win is scored, otherwise the game ends in a "losing" state. A CEGS for the game is shown in Figure 1; we will refer to the model as M_0 throughout the rest of the paper. Note that $M_0, q_0 \models \langle\langle a \rangle\rangle \Diamond$win (and even $M_0, q_0 \models K_a \langle\langle a \rangle\rangle \Diamond$win), although, intuitively, a has no feasible way of ensuring a win. This is a fundamental problem with ATEL, which we discuss briefly below.

It was pointed out in several places that the meaning of ATEL formulae is somewhat counterintuitive [5, 6, 10]. Most importantly, one would expect that an agent's ability to achieve property φ should imply that the agent has enough control and knowledge to *identify* and *execute* a strategy that enforces φ (cf. also [11]). This problem is closely related to the well known distinction between knowledge *de re* and knowledge *de dicto*.

A number of frameworks were proposed to overcome this problem [5, 6, 11, 10, 13, 4], yet none of them seems the ultimate definitive solution. Most of the solutions agree that only *uniform* strategies (i.e., strategies that specify the same choices in indistinguishable states) are really executable. However, in order to identify a successful strategy, the agents must consider not only the courses of action, starting from the current state of the system, but also from states that are indistinguishable from the current one. There are many cases here, especially when group epistemics is concerned: the agents may have common, ordinary or distributed knowledge about a strategy being successful, or they may be hinted the right strategy by a distinguished member (the "boss"), a subgroup ("headquarters committee") or even another group of agents ("consulting company"). Most existing solutions [11, 13, 4] treat only some of the cases (albeit rather in an elegant way), while others [6, 10] offer a more general treatment of the problem at the expense of an overblown logical language (which is by no means elegant).

Recently, a new, non-standard semantics for ability under incomplete information has been proposed in [8,9], which we believe to be both intuitive, general and elegant. We summarize the proposal in the next section, as we will use it further to capture strategic abilities of agents.

2.3   An Intuitive Semantics for Ability and Knowledge

In [8,9], a non-standard semantics for the logic of strategic ability and incomplete information has been proposed, which we believe to be finally satisfying. In the semantics, formulae are interpreted over *sets of states* rather than single states. Moreover, we introduce "constructive knowledge" operators $\mathbb{K}_a$, one for each agent a, that yield the set of states, indistinguishable from the current state from a's perspective. Constructive common, mutual, and distributed knowledge is formalized via operators $\mathbb{C}_A, \mathbb{E}_A$, and $\mathbb{D}_A$. The language, which we tentatively call Constructive Strategic Logic (CSL) here, is defined as follows:

$$\varphi ::= p \mid \neg\varphi \mid \sim\varphi \mid \varphi \wedge \varphi \mid \langle\!\langle A \rangle\!\rangle \bigcirc \varphi \mid \langle\!\langle A \rangle\!\rangle \Box \varphi \mid \langle\!\langle A \rangle\!\rangle \varphi \mathcal{U} \varphi \mid C_A\varphi \mid E_A\varphi \mid D_A\varphi \mid$$
$$\mathbb{C}_A\varphi \mid \mathbb{E}_A\varphi \mid \mathbb{D}_A\varphi.$$

Individual knowledge operators can be derived as: $K_a\varphi \equiv E_{\{a\}}\varphi$ and $\mathbb{K}_a\varphi \equiv \mathbb{E}_{\{a\}}\varphi$. Moreover, we define $\varphi_1 \vee \varphi_2 \equiv \neg(\neg\varphi_1 \wedge \neg\varphi_2)$, and $\varphi_1 \rightarrow \varphi_2 \equiv \neg\varphi_1 \vee \varphi_2$.

The models are concurrent epistemic game structures again, and we consider only memoryless uniform strategies. Let $\text{img}(q, \mathcal{R})$ be the image of state q with respect to relation $\mathcal{R}$, i.e. the set of all states q' such that $q\mathcal{R}q'$. Moreover, we use $out(Q, S_A)$ as a shorthand for $\cup_{q \in Q} out(q, S_A)$, and $\text{img}(Q, \mathcal{R})$ as a shorthand for $\cup_{q \in Q}\text{img}(q, \mathcal{R})$. The notion of a formula φ being satisfied by a set of states $Q \subseteq St$ in a model M is given through the following clauses.

$M, Q \models p$ iff $q \in \pi(p)$ for every $q \in Q$;

$M, Q \models \neg\varphi$ iff $M, Q \not\models \varphi$;

$M, Q \models \sim\varphi$ iff $M, q \not\models \varphi$ for every $q \in Q$;

$M, Q \models \varphi \wedge \psi$ iff $M, Q \models \varphi$ and $M, Q \models \psi$;

$M, Q \models \langle\!\langle A \rangle\!\rangle \bigcirc \varphi$ iff there exists S_A such that, for every $\Lambda \in out(Q, S_A)$, we have that $M, \{\Lambda[1]\} \models \varphi$;

$M, Q \models \langle\!\langle A \rangle\!\rangle \Box \varphi$ iff there exists S_A such that, for every $\Lambda \in out(Q, S_A)$ and $i \geq 0$, we have $M, \{\Lambda[i]\} \models \varphi$;

$M, Q \models \langle\!\langle A \rangle\!\rangle \varphi \mathcal{U} \psi$ iff there exists S_A such that, for every $\Lambda \in out(Q, S_A)$, there is $i \geq 0$ for which $M, \{\Lambda[i]\} \models \psi$ and $M, \{\Lambda[j]\} \models \varphi$ for every $0 \leq j < i$;

$M, Q \models \mathcal{K}_A\varphi$ iff $M, q \models \varphi$ for every $q \in \text{img}(Q, \sim^{\mathcal{K}}_A)$ (where $\mathcal{K} = C, E, D$);

$M, Q \models \hat{\mathcal{K}}_A\varphi$ iff $M, \text{img}(Q, \sim^{\mathcal{K}}_A) \models \varphi$ (where $\hat{\mathcal{K}} = \mathbb{C}, \mathbb{E}, \mathbb{D}$ and $\mathcal{K} = C, E, D$, respectively).

We will also write $M, q \models \varphi$ as a shorthand for $M, \{q\} \models \varphi$, and this is the notion of satisfaction (in single states) that we are ultimately interested in – but that notion is defined in terms of the satisfaction in sets of states.

Now, $\mathbb{K}_a \langle\!\langle a \rangle\!\rangle \varphi$ expresses the fact that a has a single strategy that enforces φ from *all* states indiscernible from the current state, instead of stating that φ can be achieved from *every* such state *separately* (what $K_a \langle\!\langle a \rangle\!\rangle \varphi$ says, which is very much in the spirit of standard epistemic logic). More generally, the first kind of formulae refer to *having a strategy "de re"* (i.e. having a successful strategy and knowing the strategy), while the latter refer to *having a strategy "de dicto"* (i.e. only knowing that *some* successful strategy is available; cf. [6]). Note also that the property of having a winning strategy in the current state (but not necessarily even knowing *about* it) is simply expressed with $\langle\!\langle a \rangle\!\rangle \varphi$. Capturing different ability levels of coalitions is analogous, with various "epistemic modes" of collective recognizing the right strategy.

Example 2. Robot a has no winning strategy in the starting state of the game: $M_0, q_0 \models \neg \langle\!\langle a \rangle\!\rangle \Diamond \mathsf{win}$, which implies that it has neither a strategy "de re" nor "de dicto" $(M_0, q_0 \models \neg \mathbb{K}_a \langle\!\langle a \rangle\!\rangle \Diamond \mathsf{win} \land \neg K_a \langle\!\langle a \rangle\!\rangle \Diamond \mathsf{win})$. On the other hand, he has a successful strategy in q_{AK} (just play *keep*) and he knows he has one (because another action, *exch*, is bound to win in q_{AQ}); still, the knowledge is not constructive, since a does not know which strategy is the right one in the current situation: $M_0, q_{AK} \models \langle\!\langle a \rangle\!\rangle \bigcirc \mathsf{win} \land K_a \langle\!\langle a \rangle\!\rangle \bigcirc \mathsf{win} \land \neg \mathbb{K}_a \langle\!\langle a \rangle\!\rangle \bigcirc \mathsf{win}$. Also, b's playing *chg* enforces a transition to q_w for both q_{AQ}, q_{KQ}, so $M_0, q_{AQ} \models \mathbb{K}_b \langle\!\langle b \rangle\!\rangle \bigcirc \mathsf{win}$ (robot b has a strategy "de re" to enforce a win from q_{AQ}).

Finally, $q_{QK} \models \langle\!\langle a, b \rangle\!\rangle \Diamond \mathsf{win} \land E_{\{a,b\}} \langle\!\langle a, b \rangle\!\rangle \Diamond \mathsf{win} \land C_{\{a,b\}} \langle\!\langle a, b \rangle\!\rangle \Diamond \mathsf{win} \land \neg \mathbb{E}_{\{a,b\}} \langle\!\langle a, b \rangle\!\rangle \Diamond \mathsf{win} \land \mathbb{D}_{\{a,b\}} \langle\!\langle a, b \rangle\!\rangle \Diamond \mathsf{win}$: in q_{QK}, the robots have a collective strategy to enforce a win, and they all know it (they even have common knowledge about it); on the other hand, they cannot identify the right strategy as a team – they can only see one if they share knowledge at the beginning (i.e., in q_{QK}).

3   Epistemic Temporal Strategic Logic

A very interesting variation on the theme of combining strategic, epistemic and temporal aspects of a multi-agent system was proposed in [13]. Epistemic Temporal Strategic Logic (ETSL) digs deeper in the repository of game theory, and focuses on the concept of *undominated strategies*. Thus, its variant of cooperation modalities has a different flavor than the ones from ATL, ATEL, CSL etc. In a way, formula $\langle\!\langle A \rangle\!\rangle \varphi$ in ETSL can be summarized as:

"If A play *rationally* to achieve φ (meaning: they never play a dominated strategy), they will achieve φ".

ETSL can be treated as a logic that describes the outcome of *rational play* under incomplete information,[2] in the same way as CSL can be seen as a logic that

---

[2] We emphasize that this is a specific notion of rationality (i.e., agents are assumed to *play only undominated strategies*). Game theory proposes several other rationality criteria as well, based e.g. on Nash equilibrium, dominant strategies, or Pareto efficiency. In fact, it is easy to imagine ETSL-like logics based on these notions instead.

captures agents' strategic abilities (regardless of whether the agents play rationally or not). The main claim we propose in this paper is that a rational player knows that he will succeed if, and only if, he has a strategy "de re" to succeed – while the same is not true for rational coalitions of players. However, before we present and discuss the claim formally in Section 4, we must re-write the semantics of ETSL in several respects.

First, the original semantics of ETSL is defined only for finite turn-based acyclic game models with epistemic accessibility relations, and we will generalize the semantics to concurrent epistemic game structures. Next, the semantics comes with a plethora of auxiliary functions and definitions (and a couple of omissions), which makes it rather hard to read. In fact, this is probably the reason why the logic never received the attention it deserves, and it is definitely worth trying to make the semantics more compact. Finally, the authors of [13] propose that a model should include also a "grand strategy profile" S_{Agt}, defining the actual strategies of all agents (or at least constraining them in some way, since non-deterministic strategies are allowed in ETSL). While the idea seems interesting in itself (a similar idea was later exploited e.g. in [7] to allow for explicit analysis of strategies and reasoning about strategy revision), we will show that it does not introduce a finer-grained analysis of "vanilla" ETSL formulas: if a formula holds in M, q for one strategy profile, it holds in M, q for all the other strategy profiles, too. Moreover, it can be proved that the semantics of cooperation modalities $\langle\langle A \rangle\rangle$ is the same regardless of whether we consider non-deterministic strategies or not. In consequence, we will be able to show a "vanilla" ETSL semantics expressed entirely in terms of concurrent epistemic game structures and their states.

3.1   The Semantics Made Easier to Read

Formulae of ETSL come with no restriction wrt grouping of temporal operators:

$$\varphi := p \mid \neg\varphi \mid \varphi \wedge \psi \mid \langle\langle A \rangle\rangle\varphi \mid \bigcirc\varphi \mid \Box\varphi \mid \varphi\,\mathcal{U}\,\psi \mid K_a\varphi.$$

After some re-writing (and having it generalized to general game structures, not only turn-based trees), the semantics can be given as follows. Strategies are allowed to be non-deterministic, i.e. $S_a : St \to \mathcal{P}(Act)$.[3] We require strategies to be uniform, although [13] does not do it explicitly (we take it as a simple omission, because otherwise many claims in that paper seem to be false). A collective strategy (strategy profile) S_A is a tuple of strategies, one per agent from A. S_a^0 is the "neutral strategy" with no restriction on a's actions ($S_a^0(q) = Act$ for each $q \in St$), and strategy profile S_A^0 assigns neutral strategies to agents from A. Moreover, we generalize function $out(q, S_A)$ to handle nondeterministic strategies too; in $out'(q, S_A)$, "$\alpha_a^{i-1} = S_a(q_{i-1})$" is replaced with $\alpha_a^{i-1} \in S_a(q_{i-1})$.

Now, the semantics can be given through the following clauses (the semantics for p, $\neg\varphi$ and $\varphi \wedge \psi$ is analogous to the one presented in Section 2.1):

---

[3]   To preserve seriality ("time flows forever"), we assume that $S_a(q) \neq \varnothing$ for all $q \in St$.

$M, S_{\text{Agt}}, q \models \langle\!\langle A \rangle\!\rangle \varphi$ iff for all strategies T_A, undominated wrt q, φ, we have $M, (T_A, S^0_{\text{Agt}\backslash A}), q \models \varphi$;

$M, S_{\text{Agt}}, q \models \bigcirc \varphi$ iff for every $\Lambda \in out'(q, S_{\text{Agt}})$ we have $M, S_{\text{Agt}}, \Lambda[1] \models \varphi$;

$M, S_{\text{Agt}}, q \models \Box \varphi$ iff for every $\Lambda \in out'(q, S_{\text{Agt}})$ and $i \geq 0$ we have $M, S_{\text{Agt}}, \Lambda[i] \models \varphi$;

$M, S_{\text{Agt}}, q \models \varphi \mathcal{U} \psi$ iff for every $\Lambda \in out'(q, S_{\text{Agt}})$ there is $i \geq 0$ such that $M, S_{\text{Agt}}, \Lambda[i] \models \psi$ and for all j such that $0 \leq j < i$ we have $M, S_{\text{Agt}}, \Lambda[j] \models \varphi$;

$M, S_{\text{Agt}}, q \models K_a \varphi$ iff for all $q \sim_a q'$ we have $M, (S_{\text{Agt}}(a), S^0_{\text{Agt}\backslash\{a\}}), q' \models \varphi$.

Definition 1. *Strategy S_A dominates T_A with respect to formula φ, model M, and state q, if S_A achieves φ better then T_A, i.e. iff:*

1. *for every q' such that $q \sim_A q'$: if $M, (T_A, S^0_{\text{Agt}\backslash A}), q' \models \varphi$ then also $M, (S_A, S^0_{\text{Agt}\backslash A}), q' \models \varphi$, and*

2. *there exists q' such that $q \sim_A q'$, and $M, (S_A, S^0_{\text{Agt}\backslash A}), q' \models \varphi$, and $M, (T_A, S^0_{\text{Agt}\backslash A}), q \not\models \varphi$.*

Remark 1. Definition 1 uses epistemic relation $\sim_A$. However, epistemic accessibility relations are defined only for individual agents in [13], which is perhaps another omission. In this study, we take the liberty to fix $\sim_A$ as $\sim_A^E$.

We also point out that ETSL can be extended with collective epistemic operators E_A, C_A, D_A in a straightforward manner.

Example 3. Consider the gambling robots again. Robot a has two undominated strategies wrt $\bigcirc$win, M, q_{AK}: namely, to play *exch* in both q_{AK}, q_{AQ}, or to play *keep* in both (other choices do not matter). Since playing *exch* fails in q_{AK}, so: $M_0, q_{AK} \not\models \langle\!\langle a \rangle\!\rangle \bigcirc$win. Furthermore, playing *keep* is the only undominated strategy in q_{KQ} and q_{KA} (and it succeeds only in q_{KQ}). Thus, $M_0, q_{KQ} \models \langle\!\langle a \rangle\!\rangle \bigcirc$win, and $M_0, q_{KA} \not\models \langle\!\langle a \rangle\!\rangle \bigcirc$win. Hence, $M_0, q_{KQ} \not\models K_a \langle\!\langle a \rangle\!\rangle \bigcirc$win.

3.2   A Few Properties

In this section, we present several properties of ETSL formulae that will allow us to give an even simpler semantic definition of "vanilla" ETSL.

Proposition 1. *For every "vanilla" ETSL formula φ, concurrent epistemic game structure M, and state q in M: $M, S_{\text{Agt}}, q \models \varphi$ iff $M, S'_{\text{Agt}}, q \models \varphi$ for any pair of "grand" strategy profiles $S_{\text{Agt}}, S'_{\text{Agt}}$.*

Proof. By induction on the structure of φ. Note that it is sufficient to prove the implication one way, as the choice of $S_{\text{Agt}}, S'_{\text{Agt}}$ is completely arbitrary.

Case $\varphi \equiv p$: $M, S_{\text{Agt}}, q \models p$, so $q \in \pi(q)$, so $M, S'_{\text{Agt}}, q \models p$.
Case $\varphi \equiv \neg\psi$: $M, S_{\text{Agt}}, q \models \neg\psi$, so $M, S_{\text{Agt}}, q \not\models \psi$, so (by induction hypothesis) $M, S'_{\text{Agt}}, q \not\models \psi$, so $M, S'_{\text{Agt}}, q \models \neg\psi$. (As the choice of $S_{\text{Agt}}, S'_{\text{Agt}}$ was completely arbitrary, the implication holds the other way too.)

Case $\varphi \equiv \psi_1 \wedge \psi_2$: analogous.

Case $\varphi \equiv \langle\!\langle A \rangle\!\rangle \bigcirc \psi$: $M, S_{\mathrm{Agt}}, q \models \langle\!\langle A \rangle\!\rangle \bigcirc \psi$ iff $M, (T_A, S^0_{\mathrm{Agt} \setminus A}), \Lambda[1] \models \varphi$ for all undominated T_A and $\Lambda \in out'(q, (T_A, S^0_{\mathrm{Agt} \setminus A}))$. Note that the latter condition does not refer to S_{Agt}, so $M, S'_{\mathrm{Agt}}, q \models \langle\!\langle A \rangle\!\rangle \bigcirc \psi$ too.

Cases $\varphi \equiv \langle\!\langle A \rangle\!\rangle \Box \psi$ **and** $\varphi \equiv \langle\!\langle A \rangle\!\rangle \psi_1 \mathcal{U} \psi_2$: analogous.

Case $\varphi \equiv K_a \psi$: $M, S_{\mathrm{Agt}}, q \models K_a \psi$, so $M, (S_{\mathrm{Agt}}(a), S^0_{\mathrm{Agt} \setminus \{a\}}), q' \models \psi$ for all $q \sim_a q'$. By induction hypothesis, also $M, (S'_{\mathrm{Agt}}(a), S^0_{\mathrm{Agt} \setminus \{a\}}), q' \models \psi$ for all $q \sim_a q'$, so $M, S'_{\mathrm{Agt}}, q \models K_a \psi$.

Remark 2. We point out that restricting the scope of Proposition 1 to "vanilla" ETSL formulae is important. In particular, the epistemic opertor K_a has a non-standard interpretation when the full language of ETSL is considered.

Proposition 2. *Let $\Phi \equiv \bigcirc \psi, \Box \psi,$ or $\psi_1 \mathcal{U} \psi_2$ where ψ, ψ_1, ψ_2 are "vanilla" ETSL formulae. Moreover, let $|\Phi|$ denote the set of paths for which Φ holds; formally,*
$$|\bigcirc \psi| \ = \ \{\Lambda \mid M, \Lambda[1] \models \psi\}, \quad |\Box \psi| \ = \ \{\Lambda \mid \forall_i M, \Lambda[i] \models \psi\}, \quad \text{and}$$
$$|\psi_1 \mathcal{U} \psi_2| = \{\Lambda \mid \exists_i (M, \Lambda[i] \models \psi_2 \wedge \forall_{0 \le j < i} M, \Lambda[j] \models \psi_1\}.$$

Then, S_A dominates T_A wrt $\Phi, M,$ and q iff:

1. *for every q', $q \sim_A^E q'$: if $out(q', T_A) \subseteq |\Phi|$ then also $out(q', S_A) \subseteq |\Phi|$, and*
2. *there exists q', $q \sim_A^E q'$, such that $out(q', S_A) \subseteq |\Phi|$ and $out(q', T_A) \not\subseteq |\Phi|$.*

Proof. Straightforward from the definition.

Remark 3. Note that dominance can be characterized in an even more compact way. Let $succ_{q, \Phi}(S_A) = \{q \in img(q, \sim_A^E) \mid out(q, S_A) \subseteq |\Phi|\}$ be the set of states from $img(q, \sim_A^E)$, for which s_a succeeds to enforce Φ. Now, S_A dominates T_A wrt Φ, M, q iff $succ_{q, \Phi}(T_A) \subsetneq succ_{q, \Phi}(S_A)$.

Proposition 3. *Let $\Phi \equiv \bigcirc \psi, \Box \psi,$ or $\psi_1 \mathcal{U} \psi_2$ where ψ, ψ_1, ψ_2 are "vanilla" ETSL formulae. Strategy T_A is dominated wrt Φ, M, q by a strategy S_A iff it is dominated wrt Φ, M, q by a deterministic strategy S'_A.*

Proof. $\Rightarrow$: Let T_A be dominated by S_A (wrt φ, M, q). We construct the deterministic strategy S'_A by fixing arbitrary (uniform) choices out of S_A. Formally, for every agent $a \in A$ and abstraction class $img(q', \sim_a) \subseteq St$ such that $S_a(q') = \{\alpha, \alpha', ...\}$, we fix $S'_a(q'') = \alpha$ for all $q'' \in img(q', \sim_a)$. (By uniformity of S_A, we have $\alpha \in S_a(q'')$ for all $q'' \in img(q', \sim_a)$, so S'_A is a valid strategy.) First, this enforces uniformity of S'_A. Second, $out(\bar{q}, S'_A) \subseteq out(\bar{q}, S_A)$ for all $\bar{q} \in St$ (by definition of out). Thus, we can use Proposition 2 to show that S'_A dominates T_A, which concludes the proof.

$\Leftarrow$: Straightforward.

Proposition 4. *Let Φ be as above. Then, $M, S_{\mathrm{Agt}}, q \models \langle\!\langle A \rangle\!\rangle \Phi$ iff for all deterministic strategies T_A, undominated wrt Φ, we have $M, (T_A, S^0_{\mathrm{Agt} \setminus A}), q \models \Phi$.*

Proof. ⇒:   Straightforward.

⇐:   Assume that $M, (T_A, S^0_{\text{Agt} \setminus A}), q \models \Phi$ for all deterministic strategies T_A, un-dominated wrt Φ, and suppose that there is a nondeterministic undominated S_A such that $M, (S_A, S^0_{\text{Agt} \setminus A}), q \not\models \Phi$. Let us fix a deterministic uniform strategy S'_A out of S_A in a similar way as in Proposition 3. Now, $out(\bar{q}, S'_A) \subseteq out(\bar{q}, S_A)$ for all $\bar{q} \in St$, so $out(q', S_A) \subseteq |\Phi|$ implies $out(q', S'_A) \subseteq |\Phi|$ (S'_A is never worse than S_A wrt Φ). Moreover, $out(q, S'_A) \subseteq |\Phi|$ and $out(q, S_A) \not\subseteq |\Phi|$. By Proposition 2, S'_A dominates S_A, so S_A is dominated – a contradiction.

3.3   ETSL in Terms of Concurrent Epistemic Game Structures

We have shown that, for "vanilla" ETSL, strategies do not have to be referred explicitly in the interpretation of formulae (Propositions 1 and 2). Moreover, we can restrict the set of considered strategies to deterministic strategies (Propositions 3 and 4). In consequence, we can express the semantics of "vanilla" ETSL equivalently in ATL-like fashion:

$M, q \models \langle\!\langle A \rangle\!\rangle \bigcirc \varphi$ iff for every strategy S_A, undominated wrt $q, \bigcirc \varphi$, and every $\Lambda \in out(q, S_A)$, we have that $M, \Lambda[1] \models \varphi$;

$M, q \models \langle\!\langle A \rangle\!\rangle \square \varphi$ iff for every strategy S_A, undominated wrt $q, \square \varphi$, and every $\Lambda \in out(q, S_A)$ and $i \geq 0$ we have $M, \Lambda[i] \models \varphi$;

$M, q \models \langle\!\langle A \rangle\!\rangle \varphi \mathcal{U} \psi$ iff for every strategy S_A, undominated wrt $q, \varphi \mathcal{U} \psi$, and every $\Lambda \in out(q, S_A)$, there is $i \geq 0$ such that $M, \Lambda[i] \models \psi$ and for all j such that $0 \leq j < i$ we have $M, \Lambda[j] \models \varphi$.

Only uniform deterministic strategies are taken into account. The semantics of p, $\neg \varphi$, $\varphi \wedge \psi$, and the epistemic operators is the same as for ATL and ATEL.

4   Playing Rationally vs. Knowing How to Play

We can finally present the main result of this paper, namely, that a rational player knows that he will succeed if, and only if, he has a strategy "de re" to succeed. The result holds under the assumption that the model is finite,[4] or more generally, that it includes at least one undominated strategy.

Moreover, we show that having common knowledge how to succeed is, in general, a stronger property than knowing that one will succeed for rational coalitions of players. That is, if rational agents have common knowledge about a winning strategy, then they have common knowledge that they will succeed – but the converse is not true any more. Surprisingly enough, it turns out that the relationship is strictly reverse for distributed knowledge: if a rational coalition has distributed knowledge that it will succeed, then it has distributed knowledge about a winning strategy – but not necessarily the other way around. For mutual knowledge, the relationship holds neither way.

In what follows, we use $\models_{\text{ETSL}}$ and $\models_{\text{CSL}}$ to denote the ETSL and CSL satis-faction relation, respectively.

---

[4] We use the term "finite model" to denote a CEGS with a *finite set of states St*.

4.1   Rational Play of Individual Agents

We begin with two important lemmas.

Lemma 1. *Given a finite model M, state q in M, formula Φ and agent a, there is a strategy s_a which is undominated wrt M, q, Φ.*

Proof. First, we consider the simpler case when the set of actions Act is finite. In such a case, the set of strategies is also finite, and the dominance relation is transitive and antireflexive. Suppose that every strategy is dominated; then, there must be a strategy which is dominated by itself – a contradiction.

We sketch the proof for infinite Act as follows. We partition the infinite set of strategies into equivalence classes, such that strategies in the same class have the same outcome paths for every state q (i.e., $s_a \approx t_a$ iff $\forall_q out(q, s_a) = out(q, t_a)$). Obviously, if s_a dominates t_a, then all strategies $s'_a \approx s_a$ dominate t_a too. Now, at every state q (and therefore at every point on a path from $out(q', s_a)$) there is a finite number of possible sets of successor states (the actual set being determined by the choice $s_a(q)$). Moreover, the same choice must be taken at every further occurrence of the same state q on a path, since s_a is a memoryless strategy. In consequence, there is only a finite number of different sets of outcome paths, and hence a finite number of the equivalence classes. Again, dominance is transitive and antireflexive, so an undominated strategy must exist.

Remark 4. Note that the result in Lemma 1 does not extend to CEGS with infinite state spaces. Consider the game of "Fuzzy Blackjack" (called so all the more because our robots play it usually after having consumed too much machine oil). Only a single player is necessary, and we use positive real numbers as states and actions (i.e., $St = Act = \mathbb{R}_+$). When the player chooses a number in state q, the number is added to the state: $o(q, \alpha) = q + \alpha$. The values below 1 are the winning ones, i.e. $\pi(\mathsf{win}) = (0, 1)$ (it should be 21, but this would make the game too complicated for a drunken robot). Moreover, the robot cannot distinguish between the states below 1: $q \sim_a q'$ for all $q, q' \in (0, 1)$. Now, there is no undominated strategy wrt $0.5, \bigcirc \mathsf{win}$.

To prove this, suppose that a strategy s_a is undominated. The strategy is uniform, so $s_a(q) = \alpha$ for some $\alpha \in \mathbb{R}_+$ and all $q \in (0, 1)$. Obviously, $\alpha \in (0, 1)$, because else s_a never succeeds. Now, the set of states in which s_a is successful is: $succ_{0.5, \bigcirc \mathsf{win}}(s_a) = (0, 1 - \alpha)$. Let $t_a(q) = q + \alpha/2$. Now, $succ_{0.5, \bigcirc \mathsf{win}}(t_a) = (0, 1 - \alpha/2) \not\subseteq succ_{0.5, \Phi}(s_a)$ – a contradiction. Note also that:

- If we replace $\mathbb{R}_+$ with the set of positive rational numbers, the result is the same. So, there may be no undominated strategies even when we restrict St and Act to countable sets.
- In order to show the same for countable St and *finite* Act, it is sufficient to modify the example so that $Act = \{0, 1, call\}$, and the initial state and every subsequent action $\alpha = 0, 1$ are simply stored in the resulting state. Now $o(q, call)$ takes the initial state q_0 and the string of 0s and 1s $\alpha_1, ..., \alpha_n$ stored in q, and returns $q' = q_0 + (0.\alpha_1...\alpha_n 1)_2$. For such a game, there is no undominated strategy wrt $0.5, \Diamond \mathsf{win}$.

Lemma 2. *Given M, q, Φ, a, if there is an undominated strategy wrt M, q, Φ, then there is also an undominated strategy wrt M, q', Φ for every $q' \in \mathrm{img}(q, \sim_a)$.*

Proof. Take any s_a undominated wrt M, q, Φ (*). Suppose now that s_a is dominated by some strategy t_a wrt another state $q' \in \mathrm{img}(q, \sim_a)$ (**).

1. By (*) and Prop. 2: $\forall_{q'' \in \mathrm{img}(q, \sim_a)} (out(q'', t_a) \subseteq |\Phi| \Rightarrow out(q'', s_a) \subseteq |\Phi|)$.
2. By (**) and Prop. 2: $\exists_{q'' \in \mathrm{img}(q', \sim_a)} (out(q'', t_a) \subseteq |\Phi| \wedge out(q'', s_a) \not\subseteq |\Phi|)$.

Moreover, $\mathrm{img}(q, \sim_a) = \mathrm{img}(q', \sim_a)$ because is $\sim_a$ is an equivalence relation – which gives a contradiction between (1) and (2). ∎

Remark 5. We note that Lemma 2 may hold even for indistinguishability relations that are not equivalences. In fact, it is sufficient to require that $\sim_a$ is transitive. In that case, $q' \in \mathrm{img}(q, \sim_a)$ and $q'' \in \mathrm{img}(q', \sim_a)$ implies that $q'' \in \mathrm{img}(q, \sim_a)$, and we also get the contradiction.

We are ready to prove the main claim of this paper now.

Theorem 1. *Let us consider only finite models, and formulae $\Phi \equiv \bigcirc\psi, \Box\psi$, or $\psi_1 \,\mathcal{U}\, \psi_2$ where ψ, ψ_1, ψ_2 are "vanilla" ETSL formulae. An agent has a strategy "de re" to enforce Φ if, and only if, he knows that his rational play will bring about Φ. Formally, for every finite M and state q in M:*

$$M, q \models_{\mathrm{ETSL}} K_a \langle\!\langle a \rangle\!\rangle \Phi \quad \text{iff} \quad M, q \models_{\mathrm{CSL}} \mathbb{K}_a \langle\!\langle a \rangle\!\rangle \Phi.$$

Proof. Induction on the structure of Φ. We prove the theorem for the case $\Phi \equiv \Box\psi$. Other cases are analogous.

$\Rightarrow$: Let $M, q \models_{\mathrm{ETSL}} K_a \langle\!\langle a \rangle\!\rangle \Box\psi$. Then, $\forall_{q' \in \mathrm{img}(q, \sim_a)} M, q' \models_{\mathrm{ETSL}} \langle\!\langle a \rangle\!\rangle \Box\psi$, and hence $M, q \models_{\mathrm{ETSL}} \langle\!\langle a \rangle\!\rangle \Box\psi$ in particular. By Lemmas 1 and 2, there is a strategy s_a, undominated wrt $M, q', \Box\psi$ for every $q' \in \mathrm{img}(q, \sim_a)$.

Then: $\forall_{q' \in \mathrm{img}(q, \sim_a)} \forall_{\Lambda \in out(q', s_a)} \forall_i M, \Lambda[i] \models_{\mathrm{ETSL}} \Box\psi$. By the induction hypothesis, also $\forall_{q' \in \mathrm{img}(q, \sim_a)} \forall_{\Lambda \in out(q', s_a)} \forall_i M, \Lambda[i] \models_{\mathrm{CSL}} \psi$. Thus, $\forall_{\Lambda \in out(\mathrm{img}(q, \sim_a), s_a)} \forall_i M, \Lambda[i] \models_{\mathrm{CSL}} \psi$ and so $M, \mathrm{img}(q, \sim_a) \models_{\mathrm{CSL}} \langle\!\langle a \rangle\!\rangle \Box\psi$, and finally $M, q \models_{\mathrm{CSL}} \mathbb{K}_a \langle\!\langle a \rangle\!\rangle \Box\psi$.

$\Leftarrow$: Let $M, q \models_{\mathrm{CSL}} \mathbb{K}_a \langle\!\langle a \rangle\!\rangle \Box\psi$, i.e. $M, \mathrm{img}(q, \sim_a) \models_{\mathrm{CSL}} \langle\!\langle a \rangle\!\rangle \Box\psi$. Consider $q' \in \mathrm{img}(q, \sim_a)$. By transitivity of $\sim_a$, we have $\mathrm{img}(q', \sim_a) \subseteq \mathrm{img}(q, \sim_a)$, so also $\forall_{q' \in \mathrm{img}(q, \sim_a)} M, \mathrm{img}(q', \sim_a) \models_{\mathrm{CSL}} \langle\!\langle a \rangle\!\rangle \Box\psi$. Then, for every $q' \in \mathrm{img}(q, \sim_a)$, there must be s_a such that $\forall_{q'' \in \mathrm{img}(q', \sim_a)} \forall_{\Lambda \in out(q'', s_a)} \forall_i M, \Lambda[i] \models_{\mathrm{CSL}} \psi$, and hence (by induction) $\forall_{q'' \in \mathrm{img}(q', \sim_a)} \forall_{\Lambda \in out(q'', s_a)} \forall_i M, \Lambda[i] \models_{\mathrm{ETSL}} \psi$. So, $succ_{q', \Box\psi}(s_a) = \mathrm{img}(q', \sim_a)$, and therefore $succ_{q', \Box\psi}(t_a) = \mathrm{img}(q', \sim_a)$ for every other undominated strategy t_a (otherwise t_a would be dominated by s_a). Thus, $M, q' \models_{\mathrm{ETSL}} \langle\!\langle a \rangle\!\rangle \Box\psi$ for every $q' \in \mathrm{img}(q, \sim_a)$, and finally $M, q \models_{\mathrm{ETSL}} K_a \langle\!\langle a \rangle\!\rangle \Box\psi$. ∎

Theorem 2. *More generally, for every Φ as above, and M, q such that there exists an undominated strategy wrt M, q, Φ: $M, q \models_{\mathrm{ETSL}} K_a \langle\!\langle a \rangle\!\rangle \Phi$ iff $M, q \models_{\mathrm{CSL}} \mathbb{K}_a \langle\!\langle a \rangle\!\rangle \Phi$.*

4.2   Rational Coalitions Are at Disadvantage

Beside some philosophical insight into the nature of knowledge and rational play, Theorems 1 and 2 provide us with an alternative way of decomposing strategic abilities under incomplete information into a strategic and epistemic part. The definition of the strategic dimension is more sophisticated and less straightforward than usually; on the other hand, we do not pay the price of a non-standard satisfaction relation. Unfortunately, such decomposition is not valid any more when abilities of collective agents are concerned. Now, the relationship is much more limited: if a coalition has *common* knowledge how to play, then it has also common knowledge that rational play will be successful; the same does *not* hold for other types of collective knowledge. Moreover, the converse relationship is guaranteed for distributed knowledge, but *not* for common nor mutual knowledge.

Theorem 3. *Let* $\Phi \equiv \bigcirc\psi, \square\psi,$ *or* $\psi_1 \mathcal{U} \psi_2$ *where* ψ, ψ_1, ψ_2 *are "vanilla"* ETSL *formulae. Then, if a coalition has common knowledge how to play, then it has common knowledge that rational play will be successful:*

$$\text{if}\quad M, q \models_{\text{CSL}} \mathbb{C}_A \langle\!\langle A \rangle\!\rangle \Phi \quad \text{then}\quad M, q \models_{\text{ETSL}} C_A \langle\!\langle A \rangle\!\rangle \Phi.$$

The same holds for neither mutual nor distributed knowledge.

Proof. **Common knowledge:** Let $M, q \models_{\text{CSL}} \mathbb{K}_A \langle\!\langle A \rangle\!\rangle \square\psi$, i.e. $M, \text{img}(q, \sim_A^C) \models_{\text{CSL}} \langle\!\langle A \rangle\!\rangle \square\psi$. Consider $q' \in \text{img}(q, \sim_A^C)$. We have $\text{img}(q', \sim_A^E) \subseteq \text{img}(q', \sim_A^C) \subseteq \text{img}(q, \sim_A^C)$, so also $\forall_{q' \in \text{img}(q, \sim_A^C)} M, \text{img}(q', \sim_A^E) \models_{\text{CSL}} \langle\!\langle A \rangle\!\rangle \square\psi$. Then, for every $q' \in \text{img}(q, \sim_A^C)$, there must be S_A such that $\forall_{q'' \in \text{img}(q', \sim_A^E)} \forall_{\Lambda \in out(q'', S_A)} \forall_i M, \Lambda[i] \models_{\text{CSL}} \psi$, and hence (by induction) $\forall_{q'' \in \text{img}(q', \sim_A^E)} \forall_{\Lambda \in out(q'', S_A)} \forall_i M, \Lambda[i] \models_{\text{ETSL}} \psi$. So, $succ_{q', \square\psi}(S_A) = \text{img}(q', \sim_A^E)$, and therefore $succ_{q', \square\psi}(T_A) = \text{img}(q', \sim_A^E)$ for every other undominated strategy T_A (otherwise T_A would be dominated by S_A). Thus, $M, q' \models_{\text{ETSL}} \langle\!\langle A \rangle\!\rangle \square\psi$ for every $q' \in \text{img}(q, \sim_A^C)$, and finally $M, q \models_{\text{ETSL}} C_A \langle\!\langle A \rangle\!\rangle \square\psi$.

Mutual knowledge: for a counterexample, consider a modification of the game from Figure 1, in which a third robot c is introduced. The robot can only execute *nop*, and its epistemic relation $\sim_c = \{(q, q) \mid q \in St\} \cup \{(q_{KQ}, q_{KA}), (q_{KA}, q_{KQ})\}$, i.e. c can distinguish all states except q_{KQ}, q_{KA}. Moreover, the transition function is slightly changed: now, $o(q_{KA}, keep, nop) = q_w$. For the resulting system M_1, we have that $M_1, q_{AQ} \models_{\text{CSL}} \mathbb{E}_{\{b,c\}} \langle\!\langle b, c \rangle\!\rangle \bigcirc \text{win}$, but at the same time $M_1, q_{AQ} \not\models_{\text{ETSL}} E_{\{a,c\}} \langle\!\langle a, c \rangle\!\rangle \bigcirc \text{win}$ because $M_1, q_{KQ} \not\models_{\text{ETSL}} \langle\!\langle a, c \rangle\!\rangle \bigcirc \text{win}$.

Distributed knowledge: analogously, $M_1, q_{KQ} \models_{\text{CSL}} \mathbb{D}_{\{b,c\}} \langle\!\langle b, c \rangle\!\rangle \bigcirc \text{win}$, yet $M_1, q_{KQ} \not\models_{\text{ETSL}} D_{\{a,c\}} \langle\!\langle a, c \rangle\!\rangle \bigcirc \text{win}$ because $M_1, q_{KQ} \not\models_{\text{ETSL}} \langle\!\langle a, c \rangle\!\rangle \bigcirc \text{win}$.

Theorem 4. *Let* $\Phi \equiv \bigcirc\psi, \square\psi,$ *or* $\psi_1 \mathcal{U} \psi_2$ *where* ψ, ψ_1, ψ_2 *are "vanilla"* ETSL *formulae, and let M be a finite* CEGS.[5] *Then, if A have distributed knowledge*

---

[5] Alternatively, we can request that A have at least one undominated strategy for every relevant state.

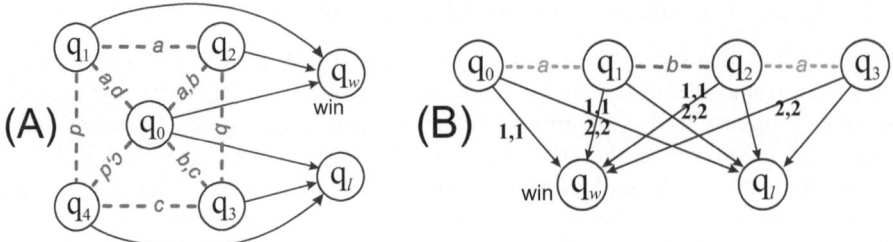

Fig. 2. (A) Model M_2: four agents a, b, c, d, epistemic relations shown with the dashed lines, $Act = \{1, 2, 3, 4\}$. Transitions: $o(q_i, j, j, j, j) = q_w$ for $j \neq i$, otherwise the system proceeds to the "losing" state q_l;   **(B)** Model M_3: two agents a, b, two actions $1, 2$. The tuples of actions that are absent in the graph lead to q_l.

that rational play will bring about Φ, then they have distributed knowledge how to play to bring about Φ. Formally:

$$\text{if}\quad M, q \models_{\text{ETSL}} D_A \langle\!\langle A \rangle\!\rangle \Phi \quad \text{then}\quad M, q \models_{\text{CSL}} D_A \langle\!\langle A \rangle\!\rangle \Phi.$$

The same holds for neither mutual nor common knowledge.

Proof. (sketch) **Distributed knowledge:** the proof is analogous to the proofs of Lemma 2 and Theorem 1 (part $\Rightarrow$), as we can exploit the fact that $\sim_A^D$ is transitive, and $\text{img}(q, \sim_A^D) \subseteq \text{img}(q, \sim_A^E)$.

Mutual knowledge: for a counterexample, consider model M_2 from Figure 2A. Let $\overline{q}$ denote the state "opposite" to q, i.e. $\overline{q_1} = q_3$, $\overline{q_2} = q_4$ etc. Furthermore, let S_{Agt}^i denote the strategy of playing $\langle i, i, i, i \rangle$ in all states. Now, S_{Agt}^i is the only undominated strategy wrt $\overline{q_i}$, $\bigcirc$win for $i = 1, ..., 4$, and $S_{\text{Agt}}^1, ..., S_{\text{Agt}}^4$ are exactly the strategies undominated wrt q_0, $\bigcirc$win. So, $M_2, q_i \models_{\text{ETSL}} \langle\!\langle \text{Agt} \rangle\!\rangle \bigcirc$win for every $i = 0, 1, ..., 4$, and therefore $M_2, q_0 \models_{\text{ETSL}} E_{\text{Agt}} \langle\!\langle \text{Agt} \rangle\!\rangle \bigcirc$win. On the other hand, there is no single strategy that succeeds for all $q_0, q_1, ..., q_4$.

Common knowledge: consider model M_3 from Figure 2B. Let $S_{\{a,b\}}$ be the strategy "play $\langle 1, 1 \rangle$ everywhere", and $T_{\{a,b\}}$ be "play $\langle 2, 2 \rangle$ everywhere". Note that $S_{\{a,b\}}$ is the only undominated strategy wrt q, $\bigcirc$win for $q = q_0, q_1$, and $T_{\{a,b\}}$ is the only undominated strategy wrt q, $\bigcirc$win for $q = q_2, q_3$. Thus, for every $q = q_0, ..., q_3$: $M_3, q \models_{\text{ETSL}} \langle\!\langle a, b \rangle\!\rangle \bigcirc$win, and hence $M_3, q_1 \models_{\text{ETSL}} C_{\{a,b\}} \langle\!\langle a, b \rangle\!\rangle \bigcirc$win. On the other hand, $M_3, q_1 \not\models_{\text{CSL}} \mathbb{C}_{\{a,b\}} \langle\!\langle a, b \rangle\!\rangle \bigcirc$win.

5   Conclusions

In this paper, the relationship between rational play and knowing how to play is investigated in a formal way. To this end, we dust off Epistemic Temporal Strategic Logic by van Otterloo and Jonker [13], and propose a simpler semantics expressed entirely in terms of concurrent epistemic game structures and their states; we prove that the new semantics is equivalent to the original one for "vanilla" ETSL formulae. ETSL serves as a device for talking about the outcome of rational play (in the sense that agents are assumed to play only undominated strategies). To capture properties of the other kind ("knowing how to play"), we use the recent proposal of Constructive Strategic Logic [8, 9].

The main result of this paper states that, for finite models, *a rational player knows that he will succeed if, and only if, he knows how to succeed*. We also show that the relationship is much more limited for rational coalitions. That is, if rational agents have common knowledge about a winning strategy, then they have common knowledge that they will succeed – but the converse is not guaranteed any more. Moreover, it turns out that the relationship is *strictly reverse* for distributed knowledge: if a rational coalition has distributed knowledge that it will succeed, then it has distributed knowledge about a winning strategy – but not necessarily the other way around. Finally, for mutual knowledge, the relationship does not hold either way in general. This is a curious result, and one that may lead to interesting philosophical conclusions.

References

1. R. Alur, T. A. Henzinger, and O. Kupferman. Alternating-time Temporal Logic. In *Proceedings of FOCS*, pages 100–109. IEEE Computer Society Press, 1997.
2. R. Alur, T. A. Henzinger, and O. Kupferman. Alternating-time Temporal Logic. *Journal of the ACM*, 49:672–713, 2002.
3. V. Goranko and W. Jamroga. Comparing semantics of logics for multi-agent systems. *Synthese*, 139(2):241–280, 2004.
4. A. Herzig and N. Troquard. Knowing how to play: Uniform choices in logics of agency. In *Proceedings of AAMAS'06*, 2006. To appear.
5. W. Jamroga. Some remarks on alternating temporal epistemic logic. In *Proceedings of FAMAS 2003*, pages 133–140, 2003.
6. W. Jamroga and W. van der Hoek. Agents that know how to play. *Fundamenta Informaticae*, 63(2–3):185–219, 2004.
7. W. Jamroga, W. van der Hoek, and M. Wooldridge. Intentions and strategies in game-like scenarios. In *Proceedings of EPIA 2005*, volume 3808 of *Lecture Notes in Artificial Intelligence*, pages 512–523. Springer Verlag, 2005.
8. W. Jamroga and Thomas Ågotnes. Constructive knowledge: What agents can achieve under incomplete information. Technical Report IfI-05-10, Clausthal University of Technology, 2005.
9. Wojciech Jamroga and Thomas Ågotnes. What agents can achieve under incomplete information. In *Proceedings of AAMAS'06*, 2006.
10. G. Jonker. Feasible strategies in Alternating-time Temporal Epistemic Logic. Master thesis, University of Utrecht, 2003.
11. P. Y. Schobbens. Alternating-time logic with imperfect recall. *Electronic Notes in Theoretical Computer Science*, 85(2), 2004.
12. W. van der Hoek and M. Wooldridge. Tractable multiagent planning for epistemic goals. In *Proceedings of AAMAS-02*, pages 1167–1174, 2002.
13. S. van Otterloo and G. Jonker. On Epistemic Temporal Strategic Logic. *ENTCS*, 126:77–92, 2005. Proceedings of LCMAS'04.

Special Event

50 Years Artificial Intelligence

1956-1966 How Did It All Begin? - Issues Then and Now

Marvin Minsky

MIT Media Lab and MIT AI Lab
Professor Emeritus, Media Arts and Sciences Professor of E.E. and C.S., M.I.T
minsky@media.mit.edu

Extended Abstract

Many computer programs today show skills that appear to rival those of outstanding human consultants. However, while each such program does certain things well, it is helpless at doing anything else. Why do our present-day programs lack the versatility and resourcefulness that a typical person shows? Clearly, those programs are deficient in both commonsense knowledge and commonsense reasoning. I'll argue that this has happened because the field of AI has evolved in a backwards direction, as compared with how a typical person develops-and that this is because our AI programmers have not appreciated the importance of making their system able to use more 'reflective' ways to think.

We can see this backwards trend in the earliest years. Consider the following list of AI accomplishments.

1957 Arthur Samuel: A machine that plays master-level Checkers.
1957 Newell, Shaw and Simon. Proving theorems in Propositional Logic.
1960 Herbert Gelernter: Proving theorems in Euclidean Geometry
1960 James Slagle: Symbolic Integral Calculus
1963 Lawrence G. Roberts: 3-D Visual Perception
1964 Thomas G. Evans: Solving Geometry Analogy problems.
1965 Daniel Bobrow: Solving word problems in Algebra
1969 C. Engelman, W. A. Martin and J. Moses: the MACSYMA project.
1969 Minsky, Papert, et al: A robot builds structures with wooden blocks.
1970 Patrick Winston: A robot that learns to recognize such structures.
1970 Terry Winograd: A program that understands many sentences.
1972 Gerald Sussman: A program that recognizes some bugs in programs.
1974 Eugene Charniak: A program that understands a few simple stories

Although there are many exceptions to this, one can discern a trend in which the early programs made progress at 'expert' tasks-whereas the later programs attempted to do things that typical four-year-olds do. In retrospect, we can clearly see that those 'more advanced' skills were easier because they required less commonsense knowledge and reasoning. The situation is still the same today: no visual program can recognize the objects in a typical room, or answer simple-seeming questions about the stories in a typical first-grade storybook.

In subsequent years most AI researchers aimed to discover some single problem-solving technique could keep extending itself. Consequently the field of AI divided

C. Freksa, M. Kohlhase, and K. Schill (Eds.): KI 2006, LNAI 4314, pp. 437–438, 2007.

itself into such specialties as Rule-Based Systems, Artificial Neural Networks, Statistical Inference Systems, Formal Mathematical Logic, Genetic Programs, and so on. Each of these were effective in certain types of situations, but never became very competent in other kinds of realms of domains.

It seems to me that what went wrong was that too many researchers became advocates and too few attempted to discover the limitations of their favorite method. Instead, most such researchers publish only instances in which their favorite scheme solves some particular problem, but the literature shows little discussion or classification of the realms in which each such system fails.

For example, in 1969 Seymour Papert and I published a book that showed some serious limitations of certain three-layer non-reentrant neural networks-but most researchers in that field have wrongly assumed that those limitations would not hold for such networks with more layers. However, so far as I can see, almost all of our theorems still apply, in the sense that the size of the networks and their coefficients still grow exponentially with the scale of the problem. The basic problem is that such networks compute only non-recursive functions. In other words, they simply cannot do any 'reflective thinking' about why their recent activities failed or succeeded-and this limited the effectiveness of what they could learn from experience. An attempt to describe a more reflective system appeared in a 1960 paper by Newell, Shaw, and Simon, but I have seen no later references to that proposed approach.

My lecture will describe some ideas about how we could build more resourceful machines, by designing system that can learn when and how to switch between many different ways to think. More details about these ideas are discussed on my website at http://web.media.mit.edu/~minsky/E8/eb8.html. If there is time, I will also discuss the early years of working with Warren McCulloch and other pioneers.

Fundamental Questions

Aaron Sloman

School of Computer Science
The University of Birmingham
Birmingham, B15 2TT England, UK
A.Sloman@cs.bham.ac.uk

Extended Abstract

I first heard about AI in 1969 from Max Clowes, then an AI vision researcher, when I was a philosophy lecturer at Sussex University (with a background in mathematics and physics). Gradually I came to realise that the best way to make progress in most areas of philosophy (e.g. philosophy of mind, epistemology, philosophy of language, philosophy of science, philosophy of mathematics, and probably even aesthetics) was to do AI.

Attempting (and usually failing) to design and implement working fragments of minds with human-like capabilities is a much more rapid route to understanding the real problems than the typical arm-chair analysis and smoke-filled seminar discussions of philosophers (in those days). That was partly because apriori philosophical analyses are usually based on ignorance of requirements and constraints that must be met by working systems and also ignorance of the full range of possible mechanisms, architectures, forms of representation, virtual machine types, etc.

For example, philosophical discussions about free will are often based on simplistic assumptions about the nature of human decision making and the kinds of mechanisms that might support such processes. This leads to spurious oppositions between determinism and freedom. By exploring a wide variety of information processing architectures, whether produced by evolution or by engineers and philosopher-designers, we can show that there are more varied and complex cases than philosophers had previously considered, and explain why desirable forms of freedom and responsibility depend on deterministic mechanisms rather than being incompatible with them.

Likewise by investigating architectures involving multiple concurrent sub-architectures, including some that monitor and modulate others, we can begin to understand more varieties of consciousness and self consciousness than philosophers were able to dream up in their arm chairs.

During those early years it became clear that whereas much of AI research in the past had been focused on algorithms and representations, it was also necessary to start thinking about how to put all the pieces together in an *architecture* combining multiple kinds of functionality, in concurrently active components, especially if we are to explain or model the kind of autonomy and creativity found in humans and other animals. (This was the topic of Chapter 6 of The Computer Revolution in Philosophy (1978) now online http://www.cs.bham.ac.uk/research/cogaff/crp/)

Two other philosophers whose interest in AI grew in that period were Dan Dennett, whose book Brainstorms (1978) also attempted to build bridges between the two

C. Freksa, M. Kohlhase, and K. Schill (Eds.): KI 2006, LNAI 4314, pp. 439–441, 2007.

disciplines, and Margaret Boden whose two books (Purposive Explanation in Psychology (1972) and Artificial Intelligence and Natural Man (1978)) helped to spread the word to wider audiences. [OUP will shortly publish her new 2 volume History of Cognitive Science which will help to illuminate the early years of AI.] Other philosophers also became interested.

Apart from the impact of AI on philosophy there was also a need for AI researchers to develop philosophical expertise in order to help them in their work. One reason was that they were often insensitive to the crudeness of the questions they asked (e.g. how can we model emotions? learning? creativity? consciousness?) because they did not know how to analyse complex concepts, and tended (and still tend) to assume over-simple analyses, as a result of which they often make inflated claims (e.g. to have modelled learning, or emotions, or scientific discovery, when all they have modelled are very simple and shallow special cases).

One example was the strong tendency among many AI researchers in the first decade to think that all reasoning or problem solving had to make use of essentially logical or sentential information structures -- an assumption I challenged in my first AI paper in 1971, claiming that Fregean and analogical modes of representation and reasoning are both important. Many others have made the same point, but I think it is fair to say that whereas logicist AI has many important achievements there has been little success in modelling visual/spatial/diagrammatic reasoning: mainly because most of the problems of vision are still unsolved in AI, even though there has been a lot of work on sub-problems, such as recognition, tracking and route-finding. There is far more to seeing a spanner than recognising it, as you can tell by watching a 3-year old trying to use one.

[http://www.cs.bham.ac.uk/research/projects/cosy/papers/#tr0603]

Another reason for AI researchers to learn philosophy is that old philosophical problems can inspire new AI research. One example is the old philosophical debate between empiricists (e.g. Hume) and apriorists (mainly Kant) which, reformulated in modern terms, leads to investigations of nature-nurture tradeoffs. Unfortunately many AI theorists just assume that any learning system must start off with as little prior knowledge as possible, and must derive all its concepts by abstraction from experienced instances (concept empiricism/symbol grounding) --- as if proposing that the human genome should discard millions of years of learning about the nature of the environment: unlike all the many animals that start off highly competent at birth. The time is ripe to re-open that discussion in collaboration with biologists studying varieties of animal cognition.

One of the philosophical conflicts between Hume and Kant that drove my own research interests concerned the nature of mathematics. Kant criticised Hume for allowing nothing to exist between the empirical knowledge acquired through the senses and trivial tautologies that are true by definition. Kant thought mathematical discoveries were not empirical and truly expand our knowledge. He was right of course.

If we can shift from attempting to model the theorem-proving done by adult mathematicians (which many AI researchers have attempted to do) to attempting to model the processes of learning about numbers, shape, motion, and operations such as counting, grouping, constructing things, and learning to apply such operations to themselves (e.g. counting counting operations), as happens in many human children during the first decade, we may both come to understand better what needs to go on in a robot with human-like intelligence, including mathematical intelligence, and also understand better what goes wrong in much mathematical education in primary schools because it is based on incorrect models of learning and discovery. (Piaget tried, but lacked the conceptual tools.)

There were many technical achievements in AI in the 1970s, many of them concerned with new engineering applications including the early development of expert systems and many tools now taken for granted by researchers (e.g. Matlab, Mathematica). A major robotic achievement, now generally forgotten, was Freddy the Edinburgh robot, which could assemble a toy wooden car in 1973, though it could not see and act at the same time. Minsky's frame-systems paper was very influential, and inspired many formalisms and toolkits. Logic programming started to take off.

AI vision research was also starting to get off the ground, at last moving away from pattern recognition. E.g. pioneering work was done by Barrow and Tennenbaum, published in 1978, and by others working on ways of getting 3-D structure from static or moving image data. However many did not appreciate the importance of the third dimension and merely tried to classify picture regions -- a task that still occupies far too many researchers who could be doing something deeper.

Gibson's ideas were just beginning to be noticed around that time, especially his emphasis on the importance of optical flow and texture gradients. Some people were already trying to resurrect neural nets. Many worked on new higher level languages and toolkits (though not architecture toolkits). Prolog was an example. There was much work on natural language processing, including European translation projects and the DARPA speech understanding project. My own vision project (POPEYE) based on a multi-level multi-processing visual architecture made some progress then hit a funding wall. I also started trying, without much success, to get people to think about surveying spaces of possibilities and the tradeoffs therein instead of (vainly) competing to find the single best solution to a problem.

During the following decade the field started increasingly to fragment for several different reasons (including rapid growth in numbers), with many bad effects, including killing off some major promising developments (e.g. research on 3-D vision). AI has become far more a collection of narrow specialisms with most researchers barely aware of anything going on outside their own sub-fields. Perhaps we can now start re-integrating AI, both as engineering and as the most general science of mind. At least the hardware support is more powerful than ever before. For a 'Grand Challenge' proposal see http://www.cs.bham.ac.uk/research/cogaff/gc/ And for a suggested means to re-integrate AI see
http://www.cs.bham.ac.uk/research/cogaff/gc/aisb06/sloman-gc5.pdf

Towards the AI Summer

Wolfgang Bibel

FG Intellektik, FB Informatik
Technische Universität Darmstadt
Hochschulstr. 10, D-64283 Darmstadt, Germany
bibel@informatik.tu-darmstadt.de

Extended Abstract

The talk summarizes the beginnings of AI as a discipline in Germany until around 1982. This includes the formation of a community in 1975 by establishing a German AI conference series, a quarterly newsletter, and a representative body for AI within the GI, all of which is still in existence and thriving today. It also includes the embedding of the national activities within the international AI community by organizing the first "ECAI" conference as an AISB/GI conference in Hamburg, the foundation of ECCAI, the European AI umbrella society, and many other initiatives of German AI researchers.

AI as pursued in Germany is rooted mainly in the GOFAI (good old- fashioned AI) vision which is vigorously developed until this day (without neglecting novel methods in addition). A fundamental part thereof is the processing of coded knowledge in artificially intelligent systems. Because it has become apparent that perhaps hundreds of millions of knowledge chunks are needed to enable truly intelligent behavior at a human level of breadth and depth the author takes the position that we are just now beginning to approach the blossoming period of GOFAI. He argues that the time has come for putting AI (converging with other technologies) into the service of enhancing science and the humanities in a fundamental and revolutionary way and of overcoming the numerous problems mankind is facing in our times such as climate change and ecological destruction, energy and resources crisis, public problem solving and governance including social justice, global epidemics, social disintegration, cultural communication, education, and so forth. In other words, AI could and should make the decisive difference in the "promise and perish" dichotomy which is at stake.

As representatives for such a change of paradigm the challenge problems of a competitive door to door public transportation system and of a harmonized legal system out of the currently 26+ different legal systems in Europe are proposed (as companions of the RoboCup challenge).

References

W. Bibel, The Beginnings of AI in Germany (2006).
W. Bibel, Information Technology. Report, European Commission (2005). ftp://ftp.cordis.lu/pub/foresight/docs/kte_informationtech.pdf

C. Freksa, M. Kohlhase, and K. Schill (Eds.): KI 2006, LNAI 4314, pp. 443–444, 2007.
© Springer-Verlag Berlin Heidelberg 2007

W. Bibel et al., Converging Technologies and the Natural, Social and Cultural World. Report, European Commission (2004). http:// europa.eu.int/comm/research/conferences/2004/ntw/ pdf/sig4_en.pdf

W. Bibel, AI and the Conquest of Complexity in Law. Artificial Intelligence and Law Journal 12, 159-180 (2004).

W. Bibel, Lehren vom Leben - Essays Ã¼ber Mensch und Gesellschaft. Deutscher Universitäts-Verlag, Wiesbaden (2003).

History of AI in Germany
and
The Third Industrial Revolution

Jörg Siekmann

DFKI GmbH,
Deduktion und Multiagentensysteme,
Stuhlsatzenhausweg 3,
D-66123 Saarbrücken
Joerg.Siekmann@dfki.de

Extended Abstract

This forthcoming book, jointly written by Corinna Elsenbroich (history and philosophy of science) and Jörg Siekmann (one of the founders of AI in Germany) explores the history of the establishment of AI in Germany with respect to two central themes:

(A)   The pattern of development of a high-tech academic discipline (like AI) with substantial industrial and economic potential, follows an interesting, but different pattern from the well known and much older research subjects.

(B)   The historical development of AI should not be viewed in isolation, but within the context of the so called "Third Industrial Revolution", i.e. the age of information processing and the computer, with all its societal consequences, such as economic globalization as well as the global information on the web.

Due to its loss of academic excellence after 1933, the emigration or loss of life of a whole generation of leading scientists and the aftermath of the Second World War, Germany had a late start in almost all modern scientific subjects.

In particular, computer science started about two decades later than in the Anglo-Saxon countries.

On the other hand, in the span of about thirty years, German research in AI and its industrial exploitation succeeded in closing the gap to the now dominant worldwide scientific standards and became again a well recognized scientific and technological country with not only the largest national AI society, the most substantial AI funding (when calibrated against the size of the population), the worldwide largest AI institute (the DFKI), but also measured against its number of publications, projects or citation index.

Thus under the magnifying glass of less than half a century, we can observe a unique pattern of development which we claim appears to be typical for many modern scientific fields.

The talk will concentrate and outline the general pattern of development which currently forms our hypothesis.

C. Freksa, M. Kohlhase, and K. Schill (Eds.): KI 2006, LNAI 4314, p. 445, 2007.
© Springer-Verlag Berlin Heidelberg 2007

Three Decades of Human Language Technology in Germany

Wolfgang Wahlster

German Research Center for AI,
DFKI GmbH,
Stuhlsatzenhausweg 3, D-66123 Saarbrücken, Germany
wahlster@dfki.de

Extended Abstract

Natural language understanding is one the most challenging goals of artificial intelligence. Since almost everyone speaks and understands a language, the development of natural language systems allows the average person to interact with computer systems anytime and anywhere without special skills or training, using common devices such as a cell phone. Full natural language understanding is AI-complete, in other words it requires solutions to all other core AI problems like knowledge representation, reasoning, vision, learning, and action planning. Nevertheless, after three decades of intensive and successful research, every day millions of users experience human language technology by calling directory assistance, getting train table or account information, dictating an SMS or a patient record, or telling a navigation system their destination. Human language technology has grown from an esoteric research area, 30 years ago, to a multi-billion euro market with a total revenue of more than two billion euro just for spoken dialog systems.

Thirty years ago, in March 1976, the first official workshop of the German Informatics Association's (GI) special interest group on AI brought together 48 participants in the town of Freudenstadt in the black forest. It is typical for the development and strength of AI in Germany, that the first official workshop, which was chaired by Joachim Laubsch, dealt with "Natural Language Dialog and Knowledge Representation", since these are core topics of AI, to which German researchers have made seminal contributions.

The 1976 proceedings include papers about the two pioneering German question-answering systems HAM-RPM (v. Hahn, Hoeppner, Jameson, Wahlster) and PLIDIS. Both systems were implemented in LISP. HAM-RPM included many advanced features like user modelling or meta-communication and could deal with certain aspects of vagueness, spatial reference, presuppositions, and why-questions. This early system used already a precursor to description logics with a conceptual (today: T-Box) and referential semantic (today: A-Box) network based on model-theoretic semantics. PLIDIS was based on a theorem prover coupled with a database that was used to answer typed natural language questions about wastewater control.

Ten years later, in 1986, the first international conference on user modelling was held in the medieval abbey of Maria Laach jointly chaired by Alfred Kobsa and myself, which led to a worldwide breakthrough of this AI subfield. Today, there is an international journal (User Modeling and User-Adapted Interaction), ranked among

C. Freksa, M. Kohlhase, and K. Schill (Eds.): KI 2006, LNAI 4314, pp. 447–448, 2007.
© Springer-Verlag Berlin Heidelberg 2007

the 5% computer science journals, and a biannual conference series (User Modeling), which has been already held ten times. From its beginning, user modeling was mainly motivated by the research on natural language dialog systems as the title of the first comprehensive book "User Models for Dialog Systems (Kobsa/Wahlster 1989) suggests. The largest project on text understanding the world has ever seen, LILOG supported by IBM, was also started in 1986, combining advanced methods of linguistics and logic (Herzog/Rollinger 1991) in a Prolog-based system. LILOG produced seminal results in unification-based parsing, discourse representation theory, and the processing of temporal and spatial expressions.

Another 10 years later, in 1996, the second phase of the large speech-to-speech translation project VERBMOBIL was started. VERBMOBIL was a speaker-independent and bidirectional speech-to-speech translation system for spontaneous dialogs in mobile situations. VERB¬- MOBIL used a multi-engine and multi-blackboard approach, e.g. it used five concurrent trans¬lation engines: statistical translation, case-based translation, substring-based transla¬tion, dialog-act based translation, and semantic transfer. Other distinguishing features were the multilingual prosody module and the generation of dialog summaries. VERBMOBIL has successfully met the project goals with more than 80% of approximately correct translations and a 90% success rate for dialog tasks.

Another 10 years later, in 2006, the largest German project on Human Language Technology is SmartWeb. It is the follow-up project to the SmartKom project, which led to the first fully symmetric multimodal dialog system. In face-to-face situations, human dialogue is not only based on speech but also on nonverbal communication including gesture, gaze, facial expression, and body posture. Multimodal dialogue systems exploit one of the major characteristics of human-human interaction: the coordinated use of different modalities. Although SmartKom worked in multiple domains (e.g. TV program guide, telecommunication assistant, travel guide), it supports only restricted-domain dialogue understanding. The follow-up project SmartWeb goes beyond SmartKom in supporting open-domain question answering using the entire Web as its knowledge base.

One of the main lessons learned from all the research during the past three decades is that the problem of natural language understanding can only be cracked by the combined muscle of deep and shallow processing approaches. This means that corpus-based and probabilistic methods must be integrated with logic-based and linguistically inspired approaches to achieve true progress on this AI-complete problem.

1996-2006 Autonomous Robots

Sebastian Thrun

Stanford AI Lab,
Stanford University,
353 Serra Mall, Stanford, CA 94305, USA
thrun@ai.stanford.edu

Extended Abstract

The "fifth decade of Artificial Intelligence", the period from 1996 to 2006, has been a decade of success for AI. Computers beat the residing world chess champion. AI companies like Google and Yahoo! have been taking over the world. And AI research changed forever the nature of other scientific fields, like biology and cognitive science. So how could this happen?

In my humble opinion, the fifth decade of AI has been the decade of data and statistics. Data has been available for many years, but somehow the amount of available data exploded in the past decade. The advent of the World Wide Web made huge numbers of documents, images, and videos available online. Hundreds of AI researchers refocused their energy on the Web. AI systems were developed for learning people's browsing patterns, for raking online music, for finding and parsing job advertisements, and for making search more effective. Out of this grew a billion-dollar industry, with Google being the most significant example.

Data has also exploded in biology. Early in fifth decade of AI, the human genome was decoded, and now the race is up to actually understand it. AI techniques were developed to find genes in the DNA string, to model gene expression in cells, to link DNA information to diseases, and to find commonalities among the DNA in different species. Today, we have a field of biocomputation that has profoundly changed the field of biology. And AI has played a major role in all this.

Data has also become abundant in robotics and in physical spaces. In the beginning of this decade, there were less than one million robots operating around the world. iRobot alone, an AI company, has sold over two million robots in the past five years. Thanks to much improved sensor technology, it has become easy to equip cars with sensors for driver and environment perception. Five of these cars just drove autonomously across a 131-miles long desert course, fueled by some very advanced AI. And the field of sensor networks, which is still in its infancy, is beginning to create some interesting data sets worthy of AI research.

But data along would not have been sufficient.

The other ingredient is statistics. AI has embraced the field of statistics in great many ways. Early examples include statistical approaches to machine translation, which outperformed then-popular hand-crafted grammars in a stunning series of experiments. Since then, many other fields have gone statistic: computer vision, machine learning, inference, information retrieval, robotics, speech recognition, to name a few. The advent of statistics in AI has led to great new ways to extract

C. Freksa, M. Kohlhase, and K. Schill (Eds.): KI 2006, LNAI 4314, pp. 449–450, 2007.
© Springer-Verlag Berlin Heidelberg 2007

information from data. To do all this, we needed faster computers. Somehow in the mid-1990s, we reached the critical speed that makes it possible to run complex statistical techniques on large amount of data in reasonable time.

In my talk, I intend to focus on some of these advances. I will in large parts focus on robotics, in which statistics, data, and fast computers have had a profound effect in the recent decade. Robots are now programmed probabilistically, and they use extensive numerical simulations of statistical equations to make robust decisions. In doing so, they are now much more robust to noise in the sensor data, and to the uncertainly that naturally exists in physical spaces.

One of my own robots, named Stanley, just won the DARPA Grand Challenge, the 131-miles long autonomous robot desert race. In developing this robot, AI played a significant role, as did data and statistics. Stanley uses probabilistic techniques to reason about sensor data. We, the developers, extensively relied on machine learning to endow the robot with advanced perception and control abilities ahead of the race. During the race, the robot used statistical machine learning to continuously adapt its perceptual routines to the terrain ahead---which gave it a key advantage. For a field like robotics, which has traditionally been dominated by non-AI research (despite the perception of many AI researchers!), this is a treat.

So what's next? It is time for AI to become mainstram in computer science. Why don't we have programming languages that make it possible to handle uncertain data? Why can't software diagnose itself and fix its bug based on data? What don't we routinely train our computers with examples, and why can't our databases adapt to the type unstructured information available in places like the Web?

And what have we learned about human-level AI lately? How can we understand brain activities recoded by fMRI, a technology that has been rapidly improving in the past few years? And where are those universal robots that can do many different task, not just one, and act as if endowed with common sense?

These are all challenging topics. But with all the success in the fifth decade of AI, and all those new technical insights, it's time to try again to solve the big problems in AI.

Projects and Vision in Robotics

Hiroshi Ishiguro[1,2,3]

[1] Department of Adaptive Machine Systems, Osaka University
[2] ATR Intelligent Robotics and Communications Laboratories
[3] JST ERATO ASADA Synergistic Intelligence Project
ishiguro@ams.eng.osaka-u.ac.jp

Extended Abstract

Current Robotics and Research Projects

The history of intelligent robotics started with Shakey developed at SRI in 1965. Shakey provided us several important research issues and we, robotics researchers, focused on the fundamental issues for making it more intelligent, such as Artificial Intelligence, Computer Vision, and Language Recognition. After Shakey, we have spent 40 years and developed humanoids as new Shakeys by using the developed technologies. The humanoids provide us new important research issues as Shakey did. The research issues lies on interdisciplinary areas among Robotics, Cognitive Science, Neuroscience, and Social Science.

Current projects running in the world are categorized as follows. Each category aims at a different new purpose. Note the list is not perfect and there are many other projects. In these projects, we are trying to identify new fundamental issues and establish new research areas by integrating the interdisciplinary areas.

Studies on Hardware Mechanisms of Humanoids

– ASHIMO Project, Honda Co., Ltd.
– HRP Project, National Institute of Advanced Industrial Science and Technology (AIST), Japan
– Humanoid Project, University of Karlsruhe

Studies on Cognitive Mechanism of Humanoids

– Cog Project, MIT
– QRIO Project, SONY Intelligence Dynamics Laboratories, Inc.
– COGNIRON Project, European research institutes including LAAS, EPFL, FhG IPA, GPS, KTH, …

Studies on Interactive Robots

– Robovie Project, ATR Intelligent Robotics and Communications Laboratories
– KISMIT Project, MIT

C. Freksa, M. Kohlhase, and K. Schill (Eds.): KI 2006, LNAI 4314, pp. 451–453, 2007.
© Springer-Verlag Berlin Heidelberg 2007

Studies on Social and Network Robots

- Network Robot Forum, Japanese companies including ATR, Toshiba, Panasonic, NTT,
  MHI, …
- Ubiquitous Robotic Companion Project, Ministry of Information and Communication, Korea

One of the Future Research Directions

Obviously, our robot is getting close to a human as the technologies progress. Why are we attracted in humanoids? The answer is simple. It is because of our tendency to anthropomorphize non-human things. We, humans, always anthropomorphize targets of communication and interaction. Therefore, we expect much with humanoids. In other words, we find a human itself in the humanoid.

One of the future research directions to directly dealing with this fundamental issue is to develop androids. Here, Let me briefly introduce our android projects. The project has been stated with a neglected issue in the human-robot interaction study. That is "appearance vs. behavior problem." The interactive robots that have been developed thus far are non-android types. Evidently, the appearance of the robot influences impressions of the subjects, and it is a very important factor in the evaluation of the interaction. Although there are many technical reports that compare robots with different behaviors, the appearance of the robots has not been focused upon. There are many empirical discussions on very simplified robots such as dolls. However, the design of a robot's appearance, particularly to make it appear a humanoid, has always been a role of industrial designers. This is a serious problem for developing and evaluating interactive robots. The appearance and the behavior are tightly coupled, and further, the results of the evaluation change with the appearance.

Android Science that Bridges Science and Engineering

One of the methods to tackle to the neglected issue is to develop a very humanlike robot, i.e., an android, and use it for studying human-robot interaction. The right figure of Figures 1 shows the developed android by Ishiguro. The android has 42 air actuators for the upper torso, excluding fingers. In the development, we have determined the positions of the actuators by analyzing the movements of a real human by using a precise 3D motion capture system. The actuators can represent the unconscious movements such as chest movements due to breathing in addition to conscious large movements of the head and arms. Furthermore, the android has a function for generating facial expressions that are important for interaction with humans.

The following figure shows a humanoid and android. The left picture shows a humanoid Eveliee P1 based on WAKAMARU that was developed by Mitsubishi Heavy Industry Co. Ltd.; and the right one shows an android Repliee Q2 developed by cooperation with KOKORO (www.kokoro-dreams.co.jp) Co. Ltd.

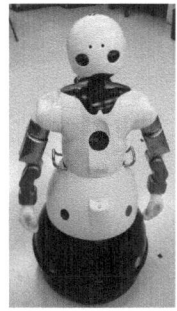

The development of the androids requires contributions from both robotics and cognitive science. In order to realize a more humanlike android, knowledge from human science is necessary. This new framework is called "android science" Thus, android science is an interdisciplinary framework between engineering and cognitive science. Robotics attempts to build very humanlike robots based on the knowledge from cognitive science. Cognitive science employs the robot for verifying hypotheses for understanding humans.

In the past, robotics research used knowledge from cognitive science, while research in cognitive science utilized robots. However, the contribution from robotics to cognitive science has not been adequate. Appearance and behavior could not be handled separately and non-android type robots were not sufficient as tools of cognitive science. We expect that this problem can be solved by using the android that has a very humanlike appearance. On the other hand, robotics research based on the cues from knowledge in cognitive science faces a similar problem, since it is difficult to recognize properly whether the cues pertain solely to robot behaviors, isolated from their appearance, or the combination of its appearance and behavior. In the framework of android science, androids enable us to directly share knowledge between the development of androids in engineering and the understanding of humans in cognitive science.

Let me summarize major research issues in android science here. The issues in Robotics are to develop the very humanlike appearance with silicon, the humanlike movement, and the humanlike perception by integrating with ubiquitous sensor systems. On the other hand, the issue in cognitive science is "conscious and unconscious recognition." The goal of android science is to realize a humanlike robot and find the essential factors of human likeness. How can we define human likeness? Further, how do we perceive human likeness? It is well known that a human has conscious and unconscious recognition. When we observe objects, various modules are activated in our brain. Each of them matches the input sensory data with the human models, and then they affect on reactions. A typical example is that even if we recognize the robot as an android, we react to it as a human. This issue is fundamental both for the engineering and scientific approaches. It will be an evaluation criterion in the development of the android, and it provides us cues for understanding the human brain mechanism of recognition.

What Will Happen in Algorithm Country?

Simon Schmitt

Kippenberg-Gymnasium

During the last few days I learned a little about artificial intelligence and it was fascinating. I read an heard about things like machine learning and the idea of artificial emotions as the basis of artificial intelligence, which I had considered to be deepest science fiction some weeks before. In fact I just touched the surface, but the glimpse behind the curtain, that I got, left me stunned and made me think about the future. I thought how life would be in times, when you have to talk of intelligent artificial beings or even individuals instead of computer programs, because they seem to be living.

One of my next thoughts was whether those artificial intelligent individuals will just exist to fulfil certain tasks or if they might have fun together or with us.

Of course we are used to play chess against computers, but to play the violin together with an android, who changes the speed if I would play to slow or maybe an android member of a jazz band playing his solo does not fit in our picture of toady's real life.

But why should our communication just contain giving commands and getting debugging information or further questions on the artificial intelligent individuals task? Why could we not talk and interact together just because we want to? But could you call that a social relationship or is friendship just possible between humans?

How deep are our feelings towards these intelligent machines? Suppose you have one of those RoboCup dogs (AIBO). Of course it would be very old then, but it is your best friend. Assume his accumulator being very old and while playing with the garbage it switches off, so that it is taken by the garbage collector because it seemed to be very old.

Of course all of us would be sad, but one could say: "Just buy a new one." In fact you might not be satisfied, because it is not the same mixture between the artificial mind and the physical engine you are used to. Shall we treat artificial beings like living ones or even like humans, because of our own feelings? Would we have to redefine our laws in this case? If the hard disc of an several years old android will be destroyed, due to an update or just because it is to old. Is that a crime? Maybe a kind of electronic murder, because we wipe out the mind - or the data, which forms what we consider to be the mind?

Suppose we decide because of the well modelled nature-like mind of artificial beings to avoid deleting anything. As a consequence artificial beings can not die. They continue consuming resources and might even become depressed immortals. Think of Marvin the paranoid android form "The Hitchhiker's Guide to the Galaxy". On the other hand if we keep them running, we do not know how the artificial intelligent being would react, if its third owner dies?

C. Freksa, M. Kohlhase, and K. Schill (Eds.): KI 2006, LNAI 4314, pp. 455–456, 2007.

The other question that keeps bugging me is what could happen with us humans in such a future? Currently machines and the artificial intelligence are used in order to do our work and to solve our problems. But what happens if all our major problems are solved? What will our task be? Will we just have to become engineers keeping the system running? Will we otherwise become philosophers searching for problems we do not have? Or will we just get bored, depressed and longing for happy times in cyberspace or searching for perfect computer games with full body simulators or with direct connections to our brain.

But doesn't that mean, that in our assumed world, which we consider being Utopian, we would voluntarily enter a virtual reality, which makes us dull? All in all the hope that one day, we might be able to hand all problems over to artificial beings seems to be unnatural for humans born to solve problems.

But I am not afraid because it will take very long if it is even possible until we might have the psychological problem of having no real life problems anymore.

Author Index

Ahn, Byung-Ha   317
Arcos, Josep-Lluís   1

Bach, Joscha   7
Bauer, Colin   7
Baumeister, Joachim   346
Benzmüller, Christoph   159
Bibel, Wolfgang   443
Biundo, Susanne   361
Bredeweg, Bert   33
Bregenzer, Jürgen   346

Cardoso Rodrigues de Souza,
    Renata Maria   260

de Boer, Viktor   202
Drechsler, Rolf   331
Dylla, Frank   274

Ebendt, Rüdiger   331
Ekenel, Hazım Kemal   302

Finzi, Alberto   113, 389
Furbach, Ulrich   174

Gottfried, Björn   289
Grachten, Maarten   1
Grinberg, Maurice   76
Güngör, Tunga   91

Haarslev, Volker   188
Herzog, Otthein   289
Hindriks, Koen V.   404
Hofmann, Thomas   244
Holzapfel, Hartwig   302
Horacek, Helmut   159

Ishiguro, Hiroshi   451

Jamroga, Wojciech   419
Jannach, Dietmar   49
Jeon, Moon-Gu   317

Kim, Jong-Hwan   317
Kokinov, Boicho   76

Kruijff-Korbayová, Ivana   159
Kusper, Gábor   128

Lesourd, Henri   159
Liem, Jochem   33
Lisetti, Christine L.   19
López de Mántaras, Ramon   1
Lukasiewicz, Thomas   113, 389

Marpaung, Andreas   19
Mehta, Bhaskar   244
Meyer, John-Jules Ch.   404
Minsky, Marvin   437
Möller, Ralf   188

Naydenov, Tchavdar   76

Obermaier, Claudia   174
Ohlbach, Hans Jürgen   214
Özgür, Arzucan   91

Park, Jong-Hwan   317
Pawlaszczyk, Dirk   374
Petkov, Georgi   76
Pizzato, Daniel F.   260
Puppe, Frank   346

Saad, Emad   143
Schaa, Christoph   302
Schaaf, Thomas   302
Schattenberg, Bernd   361
Schiller, Marvin   159
Schmid, Ute   64
Schmitt, Simon   455
Schuldt, Arne   289
Siekmann, Jörg   445
Sloman, Aaron   439
Stonier, Daniel   317

Tang, Lappoon R.   102
Tenorio de Carvalho, Francisco de
    Assis   260
Thrun, Sebastian   449
Timm, Ingo J.   374

van Someren, Maarten   202
Veloso, Manuela   229

von Hundelshausen, Felix   229
Vuine, Ronnie   7

Wahlster, Wolfgang   447
Waibel, Alex   302

Wallgrün, Jan Oliver   274
Weller, Stephan   64
Wessel, Michael   188
Wielinga, Bob J.   202
Wolska, Magdalena   159

Lecture Notes in Artificial Intelligence (LNAI)

Vol. 4384: T. Washio, K. Satoh, H. Takeda, A. Inokuchi (Eds.), New Frontiers in Artificial Intelligence. IX, 401 pages. 2007.

Vol. 4371: K. Inoue, K. Satoh, F. Toni (Eds.), Computational Logic in Multi-Agent Systems. X, 315 pages. 2007.

Vol. 4369: M. Umeda, A. Wolf, O. Bartenstein, U. Geske, D. Seipel, O. Takata (Eds.), Declarative Programming for Knowledge Management. X, 229 pages. 2006.

Vol. 4342: H. de Swart, E. Orłowska, G. Schmidt, M. Roubens (Eds.), Theory and Applications of Relational Structures as Knowledge Instruments II. X, 373 pages. 2006.

Vol. 4335: S.A. Brueckner, S. Hassas, M. Jelasity, D. Yamins (Eds.), Engineering Self-Organising Systems. XII, 212 pages. 2007.

Vol. 4334: B. Beckert, R. Hähnle, P.H. Schmitt (Eds.), Verification of Object-Oriented Software. XXIX, 658 pages. 2007.

Vol. 4333: U. Reimer, D. Karagiannis (Eds.), Practical Aspects of Knowledge Management. XII, 338 pages. 2006.

Vol. 4327: M. Baldoni, U. Endriss (Eds.), Declarative Agent Languages and Technologies IV. VIII, 257 pages. 2006.

Vol. 4314: C. Freksa, M. Kohlhase, K. Schill (Eds.), KI 2006: Advances in Artificial Intelligence. XII, 458 pages. 2007.

Vol. 4304: A. Sattar, B.-H. Kang (Eds.), AI 2006: Advances in Artificial Intelligence. XXVII, 1303 pages. 2006.

Vol. 4303: A. Hoffmann, B.-H. Kang, D. Richards, S. Tsumoto (Eds.), Advances in Knowledge Acquisition and Management. XI, 259 pages. 2006.

Vol. 4293: A. Gelbukh, C.A. Reyes-Garcia (Eds.), MICAI 2006: Advances in Artificial Intelligence. XXVIII, 1232 pages. 2006.

Vol. 4289: M. Ackermann, B. Berendt, M. Grobelnik, A. Hotho, D. Mladenič, G. Semeraro, M. Spiliopoulou, G. Stumme, V. Svátek, M. van Someren (Eds.), Semantics, Web and Mining. X, 197 pages. 2006.

Vol. 4285: Y. Matsumoto, R.W. Sproat, K.-F. Wong, M. Zhang (Eds.), Computer Processing of Oriental Languages. XVII, 544 pages. 2006.

Vol. 4274: Q. Huo, B. Ma, E.-S. Chng, H. Li (Eds.), Chinese Spoken Language Processing. XXIV, 805 pages. 2006.

Vol. 4265: L. Todorovski, N. Lavrač, K.P. Jantke (Eds.), Discovery Science. XIV, 384 pages. 2006.

Vol. 4264: J.L. Balcázar, P.M. Long, F. Stephan (Eds.), Algorithmic Learning Theory. XIII, 393 pages. 2006.

Vol. 4259: S. Greco, Y. Hata, S. Hirano, M. Inuiguchi, S. Miyamoto, H.S. Nguyen, R. Słowiński (Eds.), Rough Sets and Current Trends in Computing. XXII, 951 pages. 2006.

Vol. 4253: B. Gabrys, R.J. Howlett, L.C. Jain (Eds.), Knowledge-Based Intelligent Information and Engineering Systems, Part III. XXXII, 1301 pages. 2006.

Vol. 4252: B. Gabrys, R.J. Howlett, L.C. Jain (Eds.), Knowledge-Based Intelligent Information and Engineering Systems, Part II. XXXIII, 1335 pages. 2006.

Vol. 4251: B. Gabrys, R.J. Howlett, L.C. Jain (Eds.), Knowledge-Based Intelligent Information and Engineering Systems, Part I. LXVI, 1297 pages. 2006.

Vol. 4248: S. Staab, V. Svátek (Eds.), Managing Knowledge in a World of Networks. XIV, 400 pages. 2006.

Vol. 4246: M. Hermann, A. Voronkov (Eds.), Logic for Programming, Artificial Intelligence, and Reasoning. XIII, 588 pages. 2006.

Vol. 4223: L. Wang, L. Jiao, G. Shi, X. Li, J. Liu (Eds.), Fuzzy Systems and Knowledge Discovery. XXVIII, 1335 pages. 2006.

Vol. 4213: J. Fürnkranz, T. Scheffer, M. Spiliopoulou (Eds.), Knowledge Discovery in Databases: PKDD 2006. XXII, 660 pages. 2006.

Vol. 4212: J. Fürnkranz, T. Scheffer, M. Spiliopoulou (Eds.), Machine Learning: ECML 2006. XXIII, 851 pages. 2006.

Vol. 4211: P. Vogt, Y. Sugita, E. Tuci, C.L. Nehaniv (Eds.), Symbol Grounding and Beyond. VIII, 237 pages. 2006.

Vol. 4203: F. Esposito, Z.W. Raś, D. Malerba, G. Semeraro (Eds.), Foundations of Intelligent Systems. XVIII, 767 pages. 2006.

Vol. 4201: Y. Sakakibara, S. Kobayashi, K. Sato, T. Nishino, E. Tomita (Eds.), Grammatical Inference: Algorithms and Applications. XII, 359 pages. 2006.

Vol. 4200: I.F.C. Smith (Ed.), Intelligent Computing in Engineering and Architecture. XIII, 692 pages. 2006.

Vol. 4198: O. Nasraoui, O. Zaïane, M. Spiliopoulou, B. Mobasher, B. Masand, P.S. Yu (Eds.), Advances in Web Mining and Web Usage Analysis. IX, 177 pages. 2006.

Vol. 4196: K. Fischer, I.J. Timm, E. André, N. Zhong (Eds.), Multiagent System Technologies. X, 185 pages. 2006.

Vol. 4188: P. Sojka, I. Kopeček, K. Pala (Eds.), Text, Speech and Dialogue. XV, 721 pages. 2006.

Vol. 4183: J. Euzenat, J. Domingue (Eds.), Artificial Intelligence: Methodology, Systems, and Applications. XIII, 291 pages. 2006.

Vol. 4180: M. Kohlhase, OMDoc – An Open Markup Format for Mathematical Documents [version 1.2]. XIX, 428 pages. 2006.

Vol. 4177: R. Marín, E. Onaindía, A. Bugarín, J. Santos (Eds.), Current Topics in Artificial Intelligence. XV, 482 pages. 2006.

Vol. 4160: M. Fisher, W. van der Hoek, B. Konev, A. Lisitsa (Eds.), Logics in Artificial Intelligence. XII, 516 pages. 2006.

Vol. 4155: O. Stock, M. Schaerf (Eds.), Reasoning, Action and Interaction in AI Theories and Systems. XVIII, 343 pages. 2006.

Vol. 4149: M. Klusch, M. Rovatsos, T.R. Payne (Eds.), Cooperative Information Agents X. XII, 477 pages. 2006.

Vol. 4140: J.S. Sichman, H. Coelho, S.O. Rezende (Eds.), Advances in Artificial Intelligence - IBERAMIA-SBIA 2006. XXIII, 635 pages. 2006.

Vol. 4139: T. Salakoski, F. Ginter, S. Pyysalo, T. Pahikkala (Eds.), Advances in Natural Language Processing. XVI, 771 pages. 2006.

Vol. 4133: J. Gratch, M. Young, R. Aylett, D. Ballin, P. Olivier (Eds.), Intelligent Virtual Agents. XIV, 472 pages. 2006.

Vol. 4130: U. Furbach, N. Shankar (Eds.), Automated Reasoning. XV, 680 pages. 2006.

Vol. 4120: J. Calmet, T. Ida, D. Wang (Eds.), Artificial Intelligence and Symbolic Computation. XIII, 269 pages. 2006.

Vol. 4118: Z. Despotovic, S. Joseph, C. Sartori (Eds.), Agents and Peer-to-Peer Computing. XIV, 173 pages. 2006.

Vol. 4114: D.-S. Huang, K. Li, G.W. Irwin (Eds.), Computational Intelligence, Part II. XXVII, 1337 pages. 2006.

Vol. 4108: J.M. Borwein, W.M. Farmer (Eds.), Mathematical Knowledge Management. VIII, 295 pages. 2006.

Vol. 4106: T.R. Roth-Berghofer, M.H. Göker, H.A. Güvenir (Eds.), Advances in Case-Based Reasoning. XIV, 566 pages. 2006.

Vol. 4099: Q. Yang, G. Webb (Eds.), PRICAI 2006: Trends in Artificial Intelligence. XXVIII, 1263 pages. 2006.

Vol. 4095: S. Nolfi, G. Baldassarre, R. Calabretta, J.C.T. Hallam, D. Marocco, J.-A. Meyer, O. Miglino, D. Parisi (Eds.), From Animals to Animats 9. XV, 869 pages. 2006.

Vol. 4093: X. Li, O.R. Zaïane, Z. Li (Eds.), Advanced Data Mining and Applications. XXI, 1110 pages. 2006.

Vol. 4092: J. Lang, F. Lin, J. Wang (Eds.), Knowledge Science, Engineering and Management. XV, 664 pages. 2006.

Vol. 4088: Z.-Z. Shi, R. Sadananda (Eds.), Agent Computing and Multi-Agent Systems. XVII, 827 pages. 2006.

Vol. 4087: F. Schwenker, S. Marinai (Eds.), Artificial Neural Networks in Pattern Recognition. IX, 299 pages. 2006.

Vol. 4068: H. Schärfe, P. Hitzler, P. Øhrstrøm (Eds.), Conceptual Structures: Inspiration and Application. XI, 455 pages. 2006.

Vol. 4065: P. Perner (Ed.), Advances in Data Mining. XI, 592 pages. 2006.

Vol. 4062: G.-Y. Wang, J.F. Peters, A. Skowron, Y. Yao (Eds.), Rough Sets and Knowledge Technology. XX, 810 pages. 2006.

Vol. 4049: S. Parsons, N. Maudet, P. Moraitis, I. Rahwan (Eds.), Argumentation in Multi-Agent Systems. XIV, 313 pages. 2006.

Vol. 4048: L. Goble, J.-J.C.. Meyer (Eds.), Deontic Logic and Artificial Normative Systems. X, 273 pages. 2006.

Vol. 4045: D. Barker-Plummer, R. Cox, N. Swoboda (Eds.), Diagrammatic Representation and Inference. XII, 301 pages. 2006.

Vol. 4031: M. Ali, R. Dapoigny (Eds.), Advances in Applied Artificial Intelligence. XXIII, 1353 pages. 2006.

Vol. 4029: L. Rutkowski, R. Tadeusiewicz, L.A. Zadeh, J.M. Zurada (Eds.), Artificial Intelligence and Soft Computing – ICAISC 2006. XXI, 1235 pages. 2006.

Vol. 4027: H.L. Larsen, G. Pasi, D. Ortiz-Arroyo, T. Andreasen, H. Christiansen (Eds.), Flexible Query Answering Systems. XVIII, 714 pages. 2006.

Vol. 4021: E. André, L. Dybkjær, W. Minker, H. Neumann, M. Weber (Eds.), Perception and Interactive Technologies. XI, 217 pages. 2006.

Vol. 4020: A. Bredenfeld, A. Jacoff, I. Noda, Y. Takahashi (Eds.), RoboCup 2005: Robot Soccer World Cup IX. XVII, 727 pages. 2006.

Vol. 4013: L. Lamontagne, M. Marchand (Eds.), Advances in Artificial Intelligence. XIII, 564 pages. 2006.

Vol. 4012: T. Washio, A. Sakurai, K. Nakajima, H. Takeda, S. Tojo, M. Yokoo (Eds.), New Frontiers in Artificial Intelligence. XIII, 484 pages. 2006.

Vol. 4008: J.C. Augusto, C.D. Nugent (Eds.), Designing Smart Homes. XI, 183 pages. 2006.

Vol. 4005: G. Lugosi, H.U. Simon (Eds.), Learning Theory. XI, 656 pages. 2006.

Vol. 4002: A. Yli-Jyrä, L. Karttunen, J. Karhumäki (Eds.), Finite-State Methods and Natural Language Processing. XIV, 312 pages. 2006.

Vol. 3978: B. Hnich, M. Carlsson, F. Fages, F. Rossi (Eds.), Recent Advances in Constraints. VIII, 179 pages. 2006.

Vol. 3963: O. Dikenelli, M.-P. Gleizes, A. Ricci (Eds.), Engineering Societies in the Agents World VI. XII, 303 pages. 2006.

Vol. 3960: R. Vieira, P. Quaresma, M.d.G.V. Nunes, N.J. Mamede, C. Oliveira, M.C. Dias (Eds.), Computational Processing of the Portuguese Language. XII, 274 pages. 2006.

Vol. 3955: G. Antoniou, G. Potamias, C. Spyropoulos, D. Plexousakis (Eds.), Advances in Artificial Intelligence. XVII, 611 pages. 2006.